# Understanding Engineering Systems

## Conservation Principles
### and
## Computer Modeling

*Volume 1–Excludes Thermodynamics*

**Fifth Edition**

Louis J. Everett
*Texas A&M University*
*College Station, Texas*

Lynn Bellamy
*Arizona State University*

The McGraw-Hill Companies, Inc.
Primis Custom Publishing

New York  St. Louis  San Francisco  Auckland  Bogotá
Caracas  Lisbon  London  Madrid  Mexico  Milan  Montreal
New Delhi  Paris  San Juan  Singapore  Sydney  Tokyo  Toronto

**McGraw·Hill**

A Division of The McGraw·Hill Companies

Understanding Engineering Systems
Conservation Principles and computer Modeling
Volume 1–Excludes Thermodynamics

McGraw-Hill's Primis Custom Publishing consists of products that are produced from camera-ready copy. Peer review, class testing, and accuracy are primarily the responsibility of the author(s).

1 2 3 4 5 6 7 8 9 0 BKM BKM 9 0 9 8

ISBN 0-07-230574-6

Editor: Sharon Noble
Cover Design: Maggie Lytle
Printer/Binder: Bookmart Press

# Contents

# PREFACE

## FOR THE STUDENT

Use what you know about accounting and conservation to solve the following problem:

---

**PROBLEM:**

We want to use the motor and pulleys in figure 0.1 to lift the second mass $(m_2)$. The motor generates a tension in the cable proportional to the current $(i)$ flowing in the electrical circuit. The polarity of the motor is such that a positive current in the circuit produces a tension of $F$ in the cable. The fuse in the circuit is rated at $i_f$. This means that the fuse will melt if the current through it exceeds $i_f$. The mechanical motion of the motor generates a back voltage of $V_b$. This voltage is proportional to the rotational speed of the motor. The motion of the first mass $(m_1)$ to the left produces a positive back voltage $(V_b)$ in the motor. Determine the maximum speed that the second mass can achieve if the voltage $(V)$ applied to the circuit is instantaneous and constant.

---

After reading this problem, you probably found it impossible to solve. This is why this course is necessary, even if you have a prior knowledge of the accounting and conservation laws you may not have the skills to organize and apply that knowledge. The purpose of the course is to help you apply what you know about accounting and conservation to solve complex problems.

If you just consider all of the complex systems that you used this morning on your way to school – the water system that produced clean warm water for your shower, the washer and dryer that allowed you to have clean dry clothes, the refrigerator that kept your breakfast fresh, or the car that allowed you to get to school on time, you should begin to see why this type of analysis is important. All of these systems contain multidisciplinary components. Consider the car. Not long ago, only mechanical principles were used in its design. Today, electronics, mechanics, and thermodynamics all play an essential role. Yet, these aspects of design, when considered separately, fall into different disciplines' areas of expertise. So how do engineers, educated in a disciplinary manner, design products like cars? They work in multidisciplinary teams to design these products.

If you consider all the processes that contain multidisciplinary components, you should begin to see why engineers with interdisciplinary abilities are important. The goal of this course is to develop engineers with these abilities. To accomplish this, the course will introduce a generalized approach for solving engineering problems based on the accounting and conservation laws. You will begin by studying this approach. Then, you will explore strategies of applying the accounting and conservation laws to solve problems in electronics, mechanics, and thermodynamics. Finally, you will use

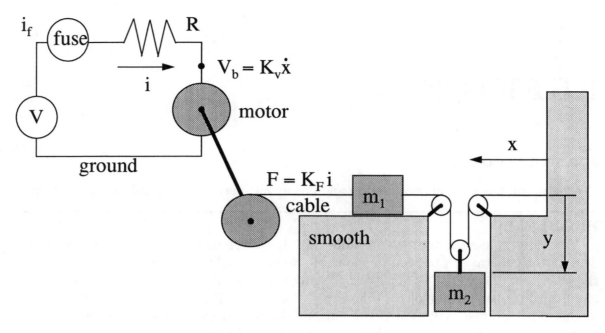

Figure 0.1: An Electric Hoist

your knowledge of these systems to solve multidisciplinary problems.

**THE FUTURE**

It is important to recognize that not long ago none of the complex systems discussed above existed – people cleaned and dried their clothes by hand, they used iceboxes to keep their food fresh, and they walked to school. With time, however, engineering intuition and consumer desire caused the products that perform these tasks to be invented and improved. The evolution of these products illustrates three interrelated points that are important to the nature of this course.

1. Technology advances with time. This means that if we focus only on current technology in our academic pursuits, the skills that we learn will most likely be obsolete within a decade. We must be able to *change* with technology. In other words, we must either continuously learn entirely new concepts or we must change the focus of our education. Since the fundamental laws of nature will not change with time, if we concentrate on the fundamentals, we can *evolve* with technology because our skills will be based on invariant concepts.

2. In design, it is important to avoid focusing exclusively on existing technological configurations. The first cars invented had very poor steering. As a result, early cars were not enormously popular. By delving into the mechanics of motion, however, inventors eventually developed excellent steering. Another example, which better illustrates this point, is flight. In designing the first aircraft, inventors modeled the existing method of flight in their world – birds. As a result, most early aircraft had wings. However, it is important to realize that the winged aircraft is only one configuration. If we concentrate on studying this configuration without contemplating the physics of flight itself, it may be difficult for us to conceive of an aircraft that does not have wings. What about helicopters,

kites, and Frisbees?

3. When we design something, we must have insight into how systems behave so that we can select a configuration that has a chance of working. Once we select a configuration, it is necessary to determine if the process will operate as intended. Most early inventors built prototypes and used a trial and error approach. Unfortunately, processes have become much more complex and expensive. Physical prototypes are often impractical or are very expensive. Also, there is a degree of danger associated with the trial and error approach. Do we really want to test our newly designed aircraft off the edge of a cliff? Engineers, therefore, often must be able to predict the performance of a process before they construct it.

## TERMINOLOGY

Terms sometimes mean different things in different engineering disciplines. This discrepancy in terminology can be confusing to engineers from different disciplines who are trying to work together. A complete understanding of the accounting and conservation laws generally helps to clear up this confusion. If an engineer understands how a different discipline's terminology relates to the accounting and conservation laws, then they should be able to relate to the terminology of that field.

An example of the way that engineers of different disciplines use different terminology is the term degree of freedom. Degree of freedom means one thing in thermodynamics and something entirely different in systems modeling and design. Another example is the term input. Control engineers often define an input as a known variable or as a variable that can be modified at will. For example, a control engineer might consider the volume knob on a television to be an input. To communicate effectively with practicing engineers of the different disciplines, we must often translate the terminology (jargon) of the discipline into terms that everyone can understand.

A firm grasp of the accounting and conservation laws makes it easier to relate to engineers in different disciplines. The following concept illustrates this point.

**Concept 0.1**

### Control Jargon

In controls engineering, it is customary to refer to variables with terms like inputs, outputs, and disturbances. Consider the device in figure 0.2. The pump removes liquid from the tank and the pipes allow liquid to flow through the tank. Julie Smith, a control engineer, is concerned with the height of the fluid in the tank and has the ability to adjust the flow through the pump but cannot adjust the flow in the pipes. She might call the pump flow rate an *input* (even though the fluid is exiting the tank) because its rate can be freely adjusted. If she is computing the height of the fluid in the tank, she might call the height an *output* because the height changes as she adjusts the pump flow rate. Since the flow in the pipes cannot be adjusted, they merely *disturb* the height. Thus, Julie may call them *disturbances*. Notice that the terminology has nothing to do with the flow of the substance into or out of the system. Nevertheless, when she models the tank's height, she uses conservation of mass.

The point to get out of all this is that even if other engineers define things differently, they still use the same basic laws as we do. By learning how their terminology relates to the laws, we can use our knowledge of the laws to understand their terminology.

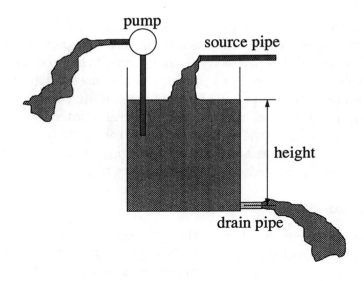

Figure 0.2: A Fluid Filled Tank

**ACKNOWLEDGEMENTS by L. Everett**

The authors acknowledge the help of Dr. Charles Glover – the force behind the conservation concept. Special thanks go to Dr. Alan Barhorst, a colleague and friend, for his technical advise, critical review of the manuscript, and creative contribution in several areas. I also want to thank Leo Hanus one of the course sequence students for his efforts on this text. Among other things, Leo is responsible for the conservation boxes and assumption figures of the text. The intent of these reader aids is to clearly show the conservation basis of the equations used in the text. Thanks are also due to the students of Dr. Lynn Bellamy at Arizona State University and mine at Texas A&M for their feedback and advice.

Additionally, I acknowledge the sacrifices made by Michael and Sandra, and those of my wife June. Most of all, I thank the personal God and Savior who carried me not only through this project, but through my entire life.

# FOREWORD

This text is the third in a series developed for use in sophomore level engineering. Its purpose is to use the concepts that were introduced in the series' previous courses to model more advanced engineering systems. Therefore, we have limited the amount of this text that we devote to reviewing the concepts that were the focus of the series previous courses.

The accounting and conservation principles introduced in *Conservation Principles* [12] are the basis of this textbook. We have focused the chapters of the text on developing strategies of applying these principles to analyze a variety of engineering systems. The difference between this course and the series' previous courses is that we analyze more complicated systems. The primary objective of this text is to develop a general problem solving methodology that students can use to analyze all types of engineering systems. The steps we will use focus more on the details of when and why we apply certain conservation principles rather than on the details of how it is done. We have designed the text with an emphasis on:

1. providing strategies and conceptual tools that enhance the application of the conservation principles in analyzing various types of engineering systems

2. imparting an understanding of systems and their behavior using the conservation principles. By understanding, we mean the ability to qualitatively deduce the fundamental behavior of systems.

**KNOWLEDGE TYPES**

Another way to understand the scope of this course is to categorize the type of knowledge one might possess. Throughout this course, we will introduce three types of knowledge. We will explain what these three knowledge types are and how they are related using the pendulum in figure 0.3.

The first type of knowledge[1] is *content*. Content knowledge comprises the basic principles that govern a system. For example, in the pendulum (as in all real systems), angular momentum, linear momentum, and total energy are conserved. To account for these properties, we have to have an understanding of what momentum and energy are. This understanding must include how factors like air resistance and gravity enter into the accounting process. The content knowledge, in this text, builds on fundamentals established in your previous courses. Our intent is not to introduce vast amounts of content.

The second and most prevalent knowledge type in this text is *procedural*. Procedural knowledge is the process or methodology that we apply to understand systems.

---

[1] First in the sense of the order we choose to discuss it not in its supremacy to other forms.

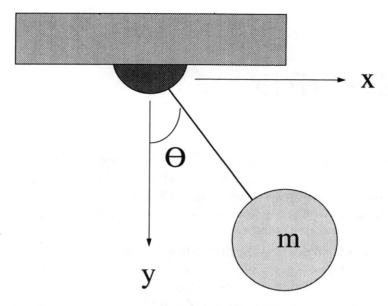

Figure 0.3: A Simple Pendulum

For example, to apply the content knowledge that angular momentum is conserved in the pendulum, we must determine the forces acting on the system. Drawing a free-body diagram makes this easier to do. There is a procedure for drawing free-body diagrams correctly. If we do not possess such procedural knowledge, it would be difficult for us to apply content knowledge.

The third type of knowledge contained in this course is *conditional*. We use conditional knowledge to decide when various methodologies are applicable. For example, if the pendulum is fixed to the Earth, we must decide if the Earth should be assumed fixed, rotating about its axis, or hurling through space.

The focus of the previous courses of this series was using content and procedural knowledge (the conservation laws and strategies for applying the conservation laws) to analyze simple engineering systems. This course builds upon these concepts by increasing the complexity of the systems that we analyze. Thus, the content knowledge that we use is the same as that introduced in the previous courses. The only difference is that we introduce new strategies (procedural and conditional knowledge) that make applying this content knowledge easier. We emphasize these strategies through examples that involve the analysis and design of complex systems. These examples introduce important concepts that are necessary for clarifying system behavior. Learning these methods will allow you to gain some of the experience necessary to make important engineering decisions. This knowledge will be beneficial later when you have to make accurate design decisions to produce a desired behavior.

**THE TEXT ITSELF**

We have arranged the overall structure of the text around the problem solving method. We begin with an introduction of the method and then develop the specific strategies that we need to solve electronics, mechanics, and thermodynamics problems. Then, we apply the concepts learned in these parts to solve complex multidisciplinary problems.

Throughout the text you will notice "conservation boxes." These are boxed environments that introduce the accounting and conservation laws in the text. The purpose of these boxes is to emphasize the type of law that we are applying and to label the terms as input, output, generation, consumption or accumulation. As you progress through the course try to wean yourself off of the information in these boxes. You want to get to a level of understanding that the information in the boxes are second nature. We view these boxes as a "half way point" between the tables used in previous courses and the ability to formulate equations in your head.

You will also notice "objectives" which are short statements that follow the major division headings. The purpose of the objectives is to tell you what you should know when you finish reading a section. If you cannot answer or do what is listed in the objectives, you should seek help from your instructor.

# Chapter 1

# FORMULATION CONCEPTS

Learning to model systems is like a chicken and egg problem. Once you can write models of simple systems, it is helpful to discuss general concepts. On the other hand if you know the general concepts before you begin to model, you will have a greater chance of developing good habits. So what should this text do first? A real dilemma. What we have decided is to present general concepts first then fill in the details later. What this means is that many of the details you find in this chapter may seem premature. Please try to look past the details for now and try to understand the basic concepts. As a minimum, you should consider looking at the [NIF] and [NIA] discussion in this chapter before proceeding.

This chapter discusses the methods we use to formulate mathematical models of systems. The first part of the chapter discusses the accounting and conservation laws. It begins by reviewing the general form of the laws. Then, it discusses the following considerations, which have to be made when applying the laws:

1. what is the objective of the analysis?

2. what is important and what is not?

3. what system should we choose?

4. what is the time period of the analysis?

The later part the chapter introduces several general system concepts. These concepts are simply evaluation tools that make formulating the mathematical model of a system easier. They help us consider the parameters and variables of our problem. They assist us in determining when a problem is not properly defined. They assist us in determining system extent. They tell us the number and type of equations to expect in the formulation process, and they tell us when we have completed the formulation process.

The purpose of this chapter is to develop a solid approach for formulating mathematical models of engineering systems. The chapter first reviews the concepts taught in earlier courses. The scope of this review is not all inclusive. Readers who want a more in-depth review can find references at the end of the text that have this information. Then, the chapter introduces new concepts. Thus, the introduction of the material does not reflect the order in which the formulation process should proceed. In fact, the purpose of the material introduced in the second half of the chapter is to make applying the concepts in the first part easier. Therefore, it is advantageous to apply these concepts first. The scope of the second part of the chapter is also limited. It discusses only general system concepts. There are other system concepts that make applying the conservation laws easier, but they are related to specific types of engineering problems. Therefore, rather than introducing them now, we discuss these concepts in the parts of the text devoted to these specific types of problems.

## 1.1   ACCOUNTING AND CONSERVATION

**Objectives**
*When you complete this section, you should be able to*

1. *List the four steps involved in applying the accounting and conservation laws.*

2. *Explain the difference between accounting and conservation.*

3. *List some conserved properties and explain why they are conserved.*

4. *List some properties that are not conserved and explain why.*

5. *Explain how charge, mass, angular momentum, linear momentum, energy, and entropy can enter and leave a system.*

6. *Distinguish between the two forms of the accounting and conservation equations and explain the difference between them.*

This section reviews the accounting and conservation principles introduced in the **Conservation Principles** [12] course. Accounting principles allow us to keep track of a system's extensive properties. Thus, we can use accounting as a tool to analyze systems. To account an extensive property, we begin by defining the system we want to analyze. Then, we define the time interval that we want to analyze. Then, we apply the accounting and conservation principles. By following these steps, we can analyze the extensive property's interaction with the system and its surroundings. The concept box that follows summarizes these steps (compare these steps to those used in previous classes).

**Concept 1.1** _____

### Accounting Steps

1. *We define the extensive property that we want to account.*
   This step may seem obvious. However, when analyzing systems with many properties, it is not always obvious what we should account. Some properties may be easier to account than others depending on what we want to know about a system. Section 1.1.2 discusses the importance of choosing what to account when analyzing a system's behavior.

2. *We define the system that we want to analyze.*
   The system that we choose determines how we account the extensive property that we selected. Therefore, how we define the system will determine the information that we can get out of our analysis. Furthermore, we not only have to choose a system. We also have to consider what extent of the system to include

in our analysis. Some systems will provide the information that we want in ways that are easier to apply than others. Other systems will give us the wrong information. Therefore, choosing an appropriate system is a very important consideration.

3. *We define the time period of the analysis.*
   Accounting statements differ depending on the time period that we choose. Therefore, we might not get the information that we want if we do not choose the appropriate time interval. Section 1.1.4 explains the importance of defining a time period when analyzing systems.

4. *We apply the accounting principle.*
   $input - output + generation - consumption = accumulation$. This is the only equation that we need to know. It is always true. The only additional information that we need is how the extensive properties that we want to analyze enter, leave, are generated, are consumed, and accumulate in a system.

**End 1.1** ─────────────────────────────────────────────

An important aspect of accounting is conservation. Mass (total and elemental), charge (total), momentum (angular and linear), and energy (total) are *conserved* properties.[1] A property is *conserved* if it cannot be generated or consumed within *any*[2] system. The only way the amount of a conserved property within a system can change is through a direct one to one exchange with the system's surroundings. This is equivalent to saying that the amount of a conserved property within the universe never changes. A modified accounting statement for conserved properties simply drops the generation and consumption terms present in a normal accounting statement (because a conserved property cannot be generated or consumed within any system).

The following box summarizes the accounting and conservation principles introduced in this section:

### Accounting and Conservation Principles

┌─────────────────────────────────────────────────────────────────────┐
│ **Accounting Principle:**                                             │
│ $input - output + generation - consumption = accumulation$            │
│                                                                       │
│ **Conservation Principle:**                                           │
│ $input - output = accumulation$                                       │
└─────────────────────────────────────────────────────────────────────┘

To apply these principles, we only have to know how what we want to account interacts with a system and its surroundings. The following list summarizes the ways that mass, charge, linear momentum, angular momentum, entropy, and energy can interact with a system.

─────────────────────────────

[1] Mass can be converted into energy in a nuclear process. Thus, nuclear processes do not strictly obey mass and energy conservation. However, if we consider mass and energy as a unified property, even nuclear processes obey the conservation laws.

[2] A conserved property is conserved no matter how we define the system. It is important to note that just because a property is not generated or consumed within *a* system does not mean that the property is conserved. It is just not generated or consumed within that particular system. To be conserved, a property must not be generated or consumed within *any* system. This is an important distinction.

1. **mass**

    (a) mass can flow into or out of a system

    (b) mass can accumulate within a system

2. **charge**

    (a) charge can flow into or out of a system

    (b) charge can accumulate within a system

3. **linear momentum**

    (a) mass flow can carry linear momentum into or out of a system

    (b) external forces can transfer linear momentum between a system and its surroundings

    (c) linear momentum can accumulate within a system

4. **angular momentum**

    (a) mass flow can carry angular momentum into or out of a system

    (b) external torques can transfer angular momentum between a system and its surroundings

    (c) angular momentum can accumulate within a system

The conservation of angular momentum has to be applied about a point. For simplicity, you should always choose one of the following points:

(a) the system's center of mass

(b) a stationary point

(c) a point with constant acceleration

(d) a point that accelerates through the system's center of mass

This text will only use the first two types of points.

5. **entropy**

    (a) mass flow can carry entropy into or out of a system

    (b) heat can carry entropy into or out of a system

    (c) entropy can be generated

    (d) entropy can accumulate within a system

6. **energy**

    (a) mass flow can carry energy into or out of a system

    (b) work done by external forces can contribute to the energy of a system

    (c) heat can carry energy into or out of a system

    (d) energy can accumulate within a system

Energy can be difficult to quantify because it can have many different forms. Some of the most common forms are the following:

**[kinetic]** $\frac{1}{2}mv^2$.

**[gravitational potential]** $mgh$, in a constant gravity field.

**[internal]** the energy within the particles of a system ($u$).

**[electrical]** Electrical power is current times voltage, $iV$.

The text will discuss other forms of energy and how to quantify them. As you encounter these additional forms of energy, make a list of them and keep the list in your notes for future reference. Then, whenever you have to apply the conservation of energy, check your notes to see which forms of energy affect the system you are analyzing.

There are two basic forms of the accounting and conservation laws – rate and accumulation. What we want to know about a system dictates which form we apply. If we have a differential time interval for the system, we have to use the rate form of the laws. If we have a finite time interval, we can use the accumulation or rate form. Most often, the rate form produces differential equations, the accumulation form produces algebraic equations. Although differential equations are more difficult to solve, they allow the solution to be obtained as a function of time. The accumulation form is usually much easier to solve but often only provides solutions at a single instance in time.

## 1.1.1 SPECIFYING THE ANALYSIS OBJECTIVE

**Objectives**
*When you complete this section, you should be able to*

1. *Explain why people sometimes solve the wrong problem.*

2. *Distinguish what a problem statement is really asking you to do.*

This section discusses the importance of clearly defining what we want to learn about a system before we analyze it with the accounting laws. In this text, we will deal with processes consisting of many distinct components. The first step in analyzing these processes is to choose the proper system. Before we can do this, we must determine what we know and what we want to determine. This may seem like an elementary concept; however, it can be tricky. It is sometimes difficult to recognize the essence of a problem. As a result, the solution may remedy a symptom rather than solve the real problem directly. Engineers need to be able to see the *total picture* without losing the ability to work with individual problems.

**Example 1.1**

### Cattle Production

The following is the excerpt from a story told to me as a child by my grandfather – Red Fitzgerald. Although it may not be true, he told it with such a straight face, I feel compelled to pass it along.

**QUESTION:**

Jack Burns, a cattle rancher, has noticed that the amount of beef delivered to market has slipped this year. He has also noticed several coyotes roaming his ranch and eating his cattle, what should he do?

Figure 1.1: The cattle on Jack's ranch.

## ANSWER:

An accounting of the cattle on Jack's ranch would produce the following result:

**General  Accounting**

system[ **the cattle population fig.1.1** ]      time period [**finite**]

*input − output + generation − consumption = accumulation*

| **input/output:** | cattle entering  −  cattle leaving |
|---|---|
| **generation:** | cattle births |
| **consumption:** | cattle deaths  +  cattle eaten |
| **accumulation:** | cattle stock |

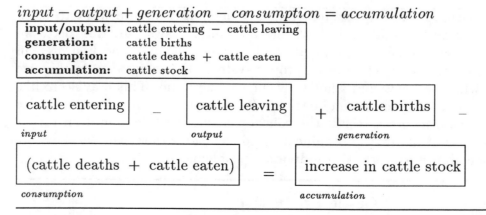

If Jack decides to counteract the affect of the coyotes, he has to increase the input or decrease the output of cattle to his ranch. He could do this by buying cattle or stealing them. Neither of which is a good solution. He could also reduce the number of cattle that leave his ranch. In other words, he could make sure that none are lost or stolen. He already does this. He could try to get the cattle to breed more, but he knows that this occurs at a natural rate that would be difficult to increase. Alternatively, Jack could reduce the number of cattle that die naturally by making sure they stay healthy. He already does this, as well. The only remaining solution is to reduce the number of cattle that are eaten. If Jack does not consider how coyotes impact his ranch's ecosystem, the obvious way to do this is to kill coyotes.

After this analysis, Jack decided to kill coyotes to maintain his cattle population. The decrease in coyotes produced a large increase in rabbits and deer on his ranch. You see, cattle was not the major food source of the coyotes! The deer and rabbit population increase affected the cattle's food supply by eating grass. Their overabundance caused the cattle to have less total weight than before the coyotes were killed. In effect, Jack had more cattle than before, but they weighed less. This analysis failed partially because Jack answered the wrong question. The question he should have answered, was how to maintain the total weight of his cattle.

**End 1.1** ───────────────────────────────────────

This example demonstrates the importance of specifying the objective of an analysis. If the objective were to maximize the cattle population, this could have been accomplished by decreasing the coyote population. However, if the objective was to increase the weight of beef supplied to market, a better solution might have been to raise the cattle in a stockyard where they could be protected from the coyotes. The example also shows that accounting is a useful procedure that we can use to analyze a wide variety of systems (not just bank accounts).

It also shows that we should never take the calculations from a model as being unquestionably correct. There are so many incidental factors contributing to complex systems that we can never be totally confident of any model's results. We must treat all modeling results as subject to verification.

The factors involved in modeling complex systems are certainly not easy to deal with. We have to make many decisions based on intuition, insight, and experience. Right now, do not worry if you do not always make the right choices. You will develop a sense of what's important as you progress through your career. It is important, however, that you learn to question the results and approach of your analysis.

The next example further demonstrates the importance of determining what the overall objective of solving a problem is before solving the problem.

**Example 1.2** ───────────────────────────────────────

### Designing a Production Line

**QUESTION:**

Tom Anderson, the chairman of a personal computer company looked over the books and discovered that the profit was much too low. What steps does he need to take to achieve a better result? What factors does he need to consider?

**ANSWER:**

If Tom simply decides to expand the company by duplicating existing processes at a new plant, he will not really accomplish anything. If the basic processes are inefficient, multiplying them by two or three simply requires two or three times more resources to achieve the desired profit. What Tom wants is to increase the production rates without significantly increasing the cost of manufacturing. Although increasing the size of the company would produce increased production rates, it would also produce increased costs.

The first step in solving this problem is to determine what really needs to be done. If he wants a more efficient process, Tom has to determine what is causing the poor production rates. He cannot adequately solve the problem until he locates its cause. Contrary to intuition, the problem area is not necessarily the process that takes the most time to complete.[3]

**End 1.2** ───────────────────────────────────────────────

In many of the problems in this text, you have to discern what is given and what needs to be determined.

## 1.1.2  DECIDING WHAT TO ACCOUNT

**Objectives**
*When you complete this section, you should be able to*

1. *Recognize that some accounting and conservation laws are more applicable for certain systems.*

2. *Distinguish when a system should be modeled by a different conservation law to make the analysis simpler.*

In the real world, no one is going to tell you how to solve a problem. You will have to decide the steps to take. This section discusses the importance of choosing what property to account when analyzing a system.

There is not a formula for determining what system to analyze or what conservation law to apply to a particular problem. At times, expressing the accounting or conservation of one property over another can be impossible. The best way to decide which law to apply is to consider all the applicable equations and then choose the one(s) that require(s) the fewest assumptions. As you gain more problem solving experience, you will find that these choices become easier to make.

**Example 1.3** ───────────────────────────────────────────

### Choosing the Appropriate Conservation Equation

**PROBLEM:**

A cylinder of mass $m$ and centroidal moment of inertia $I$ [4] slides (or rolls) down a ramp leaving it with a horizontal velocity of $v$ and angular velocity $\omega$. See figure 1.2. The cylinder drops a distance $h$ hitting a cart. The cart has mass of $M$ and rolls on very smooth lightweight rollers. There is friction between the cylinder and the cart.[5] The cylinder strikes end A of the cart and eventually stops moving relative to the cart. Estimate the speed of the cylinder and cart after the cylinder comes to rest relative to the cart.

**FORMULATION:**

To solve this problem, we have to relate what we know with what we are asked to determine. We could use an experimental method to do this. However, we would rather use the conservation laws. The conservation principles that provide meaningful results for understanding the motion of the cylinder and cart are energy and momentum (angular and linear).

───────────────────────

[3] Can you think of one example of this? Think about baking cookies. Do we want to decrease the time that the cookies bake or the time that it takes to mix the cookie batter?

[4] The concept of a moment of inertia was introduced in the *Conservation Principles'* [12] chapter on angular momentum.

[5] This friction could be kinetic or static depending on what is happening.

**Analysis Assumptions**

(1) No mass or charge enters or leaves the system.

(2) We are only concerned with the system's horizontal motion.

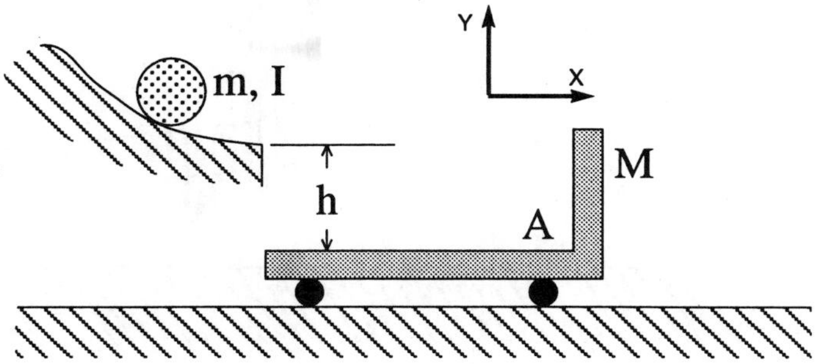


Figure 1.2: (cylhitscart): A Cylinder Sliding (Rolling) Down a Ramp Hitting a Cart

We have to decide which of these laws to apply. To do this, we consider the complications involved in applying the conservation of energy. There are several factors that cause energy to enter and leave the cylinder and the cart. Since potential energy and kinetic energy are easy to quantify, these factors do not overly concern us. The problem is quantifying the more complicated energy transfers that occur when the cylinder and cart collide. Collisions that result in deformation introduce complicated heat, sound, and internal energy terms. These terms are difficult to quantify because the energy transformation is dependent on the material properties of the colliding bodies. There may be no transformation or significant transformation. Since we have no information about the composition of the cylinder and cart, we really cannot account energy. The friction force between the cylinder and cart presents an additional problem. The work done by this force[6] will be very difficult (if not impossible) to calculate. Because of these difficulties, we use the conservation of momentum (our other choice) to model the system rather than the conservation of energy. When applied to the cylinder, the factors that enter the conservation of linear momentum analysis are

1. gravity,
2. the normal force from the cart on the cylinder, and
3. the friction force between the cart and cylinder.

The second and third forces can be very difficult to quantify. Certainly this result is no better than the energy result. We are stuck! What we want is an equation in which these inexpressible forces do not appear. How do we eliminate these forces from our conservation expression?

The answer is to choose the cylinder and cart *together* as the system. This eliminates the forces that are difficult to quantify from the analysis because only external forces appear in the conservation of linear momentum equation. Thus, we only need to express gravity and the vertical reaction forces (from the ground onto the bottom of the cart) to analyze this system adequately. Furthermore, since the vertical motion does not concern us, we can ignore the vertical reaction forces altogether. As a result, we only need to express the horizontal component of the linear momentum[7] equation

[6] In order to do work, a force has to move over a distance. Thus, the frictional force on the cylinder does work only if the cylinder slides at some point in its landing on the cart. Calculating the work done by this force would be very difficult.

[7] The conservation of angular and linear momentum are vector equations.

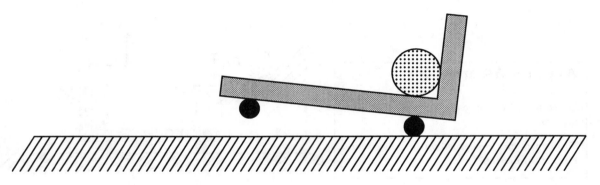

Figure 1.3: The Cart Balanced on One Wheel

to model the cart and cylinder system. In fact, since we are only interested in the final velocity, we can express the conservation of linear momentum over a finite time period. Doing this, we obtain the following:

**Conservation of Linear Momentum (Horizontal Only)**

system[**cart and cylinder fig.1.2**]        time period [**finite**]

*input − output = accumulation*

| **input/output:** | mass flow external forces gravity force | |
|---|---|---|
| **accumulation:** | | $(M+m)v_f - mv\,\vec{i}$ |

$x$:  | 0 |  =  | $(M+m)v_f - mv$ |
       *external forces*        *change in accum.*

The time period we define is from the time the cylinder starts to roll (slide) to the time when the cylinder hits the cart and stops rolling. Notice that $v_f$ is the velocity of the cart and cylinder system *after* the cylinder is at rest relative to the cart. Thus, we can quickly rearrange this result to estimate $v_f$ (and answer the question posed to us by this example). Solving for $v_f$, we obtain the following:

$$v_f = \frac{mv}{m+M}$$

**OTHER CONSIDERATIONS:**

Since we have an answer, we do not need to apply any more accounting or conservation laws. However, we could verify the accuracy of our assumptions by applying the conservation of angular momentum to the cylinder and cart system. In this analysis, we would have to consider the torque applied to the cart from the floor. This torque would result from the difference in the forces acting on the front and rear wheels of the cart. If the cart was balanced on one wheel (see figure 1.3), it could tip over because the wheel may not support the torque induced by the cylinder. The conservation of angular momentum result would allow us to determine whether or not the cart would tip over.

**End 1.3**

This example demonstrates the importance of deciding what to account. In this case, applying the conservation of linear momentum was not only the best choice it was the only choice. In other examples, we will have the ability to choose what property we want to account. In these cases, we can decide what property will be the easiest to account by evaluating what we know about the problem and what we want to determine. This example also demonstrates the significance that choosing a system has in the analysis process. We could not solve the problem until we chose the system of the cylinder and cart together. The next section discusses the considerations that we have to make when defining a system in the analysis process.

## 1.1.3  DEFINING THE SYSTEM OF ANALYSIS

### Objectives
*When you complete this section, you should be able to*

1. *Determine what information to include and what information to ignore in the model of a system.*

2. *List the extensive properties required to model an object's behavior.*

3. *Explain how the properties in objective*

4. *influence the object's behavior.*

5. *Identify systems that will behave differently when isolated from their surroundings.*

6. *Write assumptions that would simplify the analysis of the systems in objective 5.*

---

This section discusses the importance of defining a system when applying the accounting and conservation laws. There are three considerations that we have to make when we define a system. We have to determine

1. what we want to analyze,

2. what factors are important to our analysis, and

3. what extent of the system to include in our analysis.

### DETERMINING WHAT FACTORS ARE IMPORTANT

In engineering practice, we generally know more details than are necessary. Thus, we must be able to discern which factors are important and which are not, when we attempt to model a engineering system's behavior.

For example, suppose that we want to know how much horsepower it takes to propel a family automobile at modest speed. To model the automobile's behavior correctly, we must decide what factors are significant. Do we include tire pressure? Collectively, our answer to questions like this one will determine the type of model we develop. If our decisions are not correct, we will have a model that is either useless for describing the relevant system behavior or is more complex than it needs to be. In some cases, the wrong decision will produce results that are simply wrong!

In modeling a family automobile, we probably would not spend much time determining an optimal tire pressure. However, if we had to model the horsepower requirements for a race car, tire pressure may be very important. Experience has shown that the inflation pressure and tire style play very significant roles in an automobile's performance at high speeds. Successful race car teams pay very close attention to the types of tires they use.

Determining what is important to a system's behavior can be complex. Even systems with similar properties like family automobiles and race cars have different aspects that are significant to their behavior. We must focus on both a system's properties and its purpose when modeling its behavior.

**Example 1.4** ────────────────────────────────────────

### A Falling Rock

**PROBLEM:**

Determine the time it takes for a rock falling from a height of 20 [ft] to hit the ground. Explain your reasoning for including or excluding any factors from this analysis.

**FORMULATION:**

The first step in solving this problem is to develop a relationship between what we know (or can easily measure) and what we want to determine. If it is easy to measure time and distance, we want to

1. measure the height (20 [ft]);

2. drop the rock; and

3. measure the time it takes for the rock to hit the ground.

There is nothing wrong with this experimental approach. However, if we do not have a twenty foot ladder or a stopwatch, then we have to use another approach. With these restrictions, we become more resourceful and realize that force and motion can be related through the conservation laws.

To apply the conservation laws, we generally use strategies and tools specific to the type of system that we are analyzing to make the task easier. For mechanics problems, one of the tools that we use is the free body diagram. Thus, to analyze this system we begin by drawing a freebody diagram (see figure 1.4).

Notice that in the diagram we only show two forces on the rock. One is its weight. The other is air resistance.[8] This ignores the effects of a cross wind because over the twenty foot drop, we do not anticipate these effects to be significant. If there was a strong cross wind, we would have to include its force in our analysis. Furthermore, notice that if the rock has a large density, we could probably ignore the air resistance as well.[9] It cannot be too significant for a rock. We will keep it for the moment just to prove a point.

To determine the weight of the rock, we must know both the local gravitational attraction and the mass. Notice that the problem statement did not say anything about

---

[8] Contact between the rock and the air produces a force that acts to resist the motion of the rock. We call this force air resistance. In this case, air resistance acts upward attempting to slow the rock's downward motion.

[9] The resistance is often proportional to the frontal area times the velocity of the object squared ($resistance \propto Av^2$). If you do not believe that the resistance increases with area and velocity, think about moving your hands in a swimming pool. If you make your hands small and *slice* the water, you feel less resistance than if you make your hands big. Additionally, if you try to move your hands quickly, you encounter more resistance than if you move them slowly.

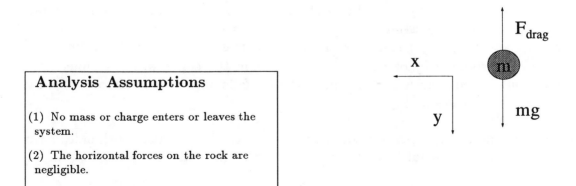

Analysis Assumptions

(1) No mass or charge enters or leaves the system.

(2) The horizontal forces on the rock are negligible.

Figure 1.4: (fbdrock): Free Body Diagram of a Falling Rock

either of these quantities. We call this type of quantity a *parameter*. Parameters[10] are generally assumed to be known. They consist of known constants, tabulated data, or easy to measure quantities (like height or mass).

Applying conservation of linear momentum[11] in the $y$ direction (notice from the figure that we define the $y$ direction as being zero at the point of release and increasing downward), we obtain the following:[12]

**Conservation of Linear Momentum (Vertical Only)**

system[**rock  fig.1.4**]      time period [**differential**]

*input − output = accumulation*

| input/output: | mass flow | |
|---|---|---|
| | external forces | $-f_{\mathrm{drag}}$ |
| | gravity force | $mg$ |
| accumulation: | | $\frac{d}{dt}(m\dot{y})$ |

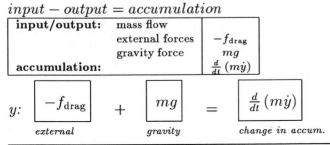

$$y: \quad \boxed{-f_{\mathrm{drag}}} \; + \; \boxed{mg} \; = \; \boxed{\frac{d}{dt}(m\dot{y})}$$

*external*          *gravity*          *change in accum.*

If we apply the product rule for differentiation, this result simplifies to

$$-f_{\mathrm{drag}} + mg = m\ddot{y} + \dot{m}\dot{y} = m\ddot{y} + (0)\dot{y}$$

Thus,

$$-\frac{f_{\mathrm{drag}}}{m} + g = \ddot{y}$$

Notice that if we ignore the drag force, mass has nothing to do with the motion. In fact, we can argue that if the drag force is significantly smaller than the mass, we

[10] Parameters are extra variables in a problem that we can generally assume to be known and constant.

[11] Conservation of linear momentum and conservation of angular momentum produce vector equations (which differ depending on the coordinate system we choose). In this problem, we are using a rectangular Cartesian coordinate system and only one direction (defined as $y$) produces non-trivial results. Thus, we write only the non-trivial component equation to model the system. We ignore the trivial component equations. We handle the vector equations for conservation of momentum in this way throughout this text.

[12] The mathematical notation used in this conservation law is consistent with the rest of the text. Notice that since we defined $y$ as the downward direction from the point of release we are using $\dot{y}$ to represent velocity and $\ddot{y}$ to represent acceleration (where $\dot{y} \equiv \frac{d}{dt}(y)$ and *et cetera*).

can justify ignoring it. What does this imply about a feather or a balloon? Why does physics tell us that all objects fall at the same rate in a vacuum, yet our ordinary experience tells us differently?[13]

Since we are dropping a rock, the effects of air resistance are probably minimal. Thus, we can probably ignore the $-\frac{f_{drag}}{m}$ term from the conservation of linear momentum results. This leaves $g = \ddot{y}$. Since the rock falls only a short distance, we can assume that $g$ is constant.[14]

**SOLUTION:**

With one side of the equation constant, we can easily integrate both sides twice to determine the time it takes for the rock to hit the ground.

$$\int g\, dt = \int \ddot{y}\, dt \quad \overset{reduces}{\leadsto} \quad gt + C_1 = \dot{y}$$

$$\int (gt\, dt + C_1) = \int \ddot{y}\, dt$$

$$\frac{1}{2}gt^2 + (C_1)t + C_2 = y$$

However, since the initial velocity $\frac{d}{dt}(y)$ and initial $y$ position are zero at $t = 0$, our result is $y = gt^2$. Thus,

$$t = \sqrt{\frac{2y}{g}}. \tag{1.1}$$

To find the time when the mass hits, we simply substitute $y = 20$ [ft], into this equation. Thus, $t = \sqrt{\frac{40}{g}}$. Since our answer is in terms of $g$, we have to find its value. In this example, gravity ($g$) is a system parameter. We can implicitly assume that we know its value. We could consult almost any introductory physics or engineering text to find that $g \approx 32.2$ [ft/s$^2$] at the Earth's surface. Therefore, it takes the rock

$$t = \sqrt{\frac{40}{32.2}} = 1.24 \text{ [s]}$$

to hit the ground.

**VERIFICATION:**

We can verify this solution by checking the units of the result.

$$\left( \frac{\text{[ft]}}{\text{[ft/s}^2]} \right)^{\frac{1}{2}} = \text{[s]}.$$

This gives us confidence in our answer because we get the correct units for the quantity that we wanted to find. Furthermore, the magnitude of the answer satisfies common sense. It should take the rock around a second to fall a distance of 20 [ft] (approximately the height of a one story house).

**OTHER CONSIDERATIONS:**

---

[13]This occurs because in a vacuum there are no drag forces. However, we do not live in a vacuum, therefore our experience is that objects fall at different rates (*i.e.* air resistance affects light objects like balloons and feathers more than it affects small heavy objects like rocks). What does this say about our physical intuition?

[14]This is an accurate assumption, but do not over trivialize gravity as always being constant. If we wanted to propel an object into outer space (or pump water up a mountain), gravity would vary with the distance of the object from the center of the Earth. For most cases, we can safely assume that gravity is constant, however, be cautious. In some cases, gravity is not constant. If we misrepresent gravity, we may not model the system's behavior correctly.

What if we are not on the Earth? The problem could now ask us to determine the acceleration of gravity because it is not implicitly known. In this case, equation 1.1 still provides the relationship we need. This time, however, we would calculate $g$ assuming that we know or can measure $y$ and $t$.

**End 1.4** ━━━━━━━━━━━━━━━━━━━━━━━━━━━━━━━━━━━━━━━━━━━━━━━

This example demonstrates the importance of determining what factors are important in the analysis of a system. Real systems interact with many things; however, depending on what we want to know and the type of system that we have, these factors may or may not be important. This is what we have to distinguish. In the previous example, we determined that we could ignore air resistance when modeling the falling motion of a rock. What if we were modeling a Ping-Pong ball? In this case, we would probably want to include air resistance in our model.

## DETERMINING SYSTEM EXTENT

All systems are connected to some sort of surroundings. This section discusses the importance of determining the extent to which these interconnections need to be included in the model of a system's behavior. Ignoring factors of a system that drastically affect its behavior can make a model useless. This can have serious consequences because we may think that we know how the system will behave when we really have no idea.

The Earth in its orbit affects the motion of the Sun. The moon in its orbit affects the motion of the Earth. As you turn the pages of this text, you are affecting the motion of the Earth (although an insignificant amount). As an engineer, it is sometimes necessary to predict the outcome of some event. To do this it is necessary to isolate a system of interest from the rest of the universe. When we isolate a system, we have to include all parts of the universe that will be significantly affected and ignore the remainder. Deciding what to include and what to leave out can be a tough decision. We generally have to rely on our experience to make these decisions. You will gain this experience as you advance through your career. For now, you should at least consider what you left out of an analysis and why, whenever you solve a problem.

In the coyote population problem discussed in example 1.1, we had to decide what aspects of the ranch's ecosystem to include in our model. As it turned out, the parts of the ecosystem that we left out were important. This example demonstrates the complexity that is involved in determining system extent. When analyzing a system's behavior, we have to consider how the system will interact with its surroundings. This means we have to determine which environmental factors are important and which are not in accordance with our objectives.

**Example 1.5** ━━━━━━━━━━━━━━━━━━━━━━━━━━━━━━━━━━━━━━━━━━━━━━━

### Shipping a Delicate Instrument

We have been asked to design a package for shipping a delicate instrument. For simplicity, we model the instrument as a solid mass. To isolate the instrument from bumps, we mount it on a spring. See figure 1.5. We design and test the spring packaging so that it protects the instrument from the worst possible shock.

What would happen if we ship the instrument on a truck? A simplified schematic of the package/truck system appears in figure 1.6. The truck's suspension system

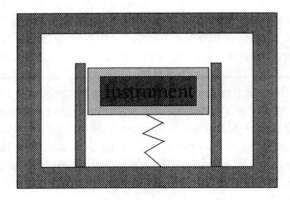

Figure 1.5: A Spring for Isolating the Instrument

will modify the effective spring that isolates the instrument from the ground. The instrument's response to the shock caused by the roadway will be different from that accounted for in the initial model. We may find that the instrument becomes damaged during shipment. The point of this is, if we design the packaging based on the model in figure 1.5, but ship the package as figure 1.6 shows, we may be surprised with the result. Failing to recognize the interaction between the package and the truck could have devastating results on our instrument.

**End 1.5** _____

We have to be aware of the environment in which a system of interest is present. Failing to consider interactions between our system and its environment may produce unexpected and unwanted results.

**Example 1.6** _____

## A Truss Bridge

The bridge in figure 1.7. has heavy bolts that connect the bridge sections together. Because of these connections, it is possible for the joints to transmit moment (twist) from one link to another. Engineers call a structure with this capability a frame. When we analyze the performance of a bridge of this type, we can ignore most of the detail and model the joints as if they transmit no moment. This simplifies the analysis considerably. Since only the ends hold the links together and since the links support no moments,[15] the forces in the link have to be axial. Engineers call these links trusses. In bridges of this type,[16] components withstand far greater axial loads than transverse (bending) loads. The bridge is designed to remain in static equilibrium with zero transverse load in each member.

**End 1.6** _____

---

[15] This is an assumption, but it is not far from reality in most systems.

[16] Recall, the terms truss and frame from the *Conservation Principles'* [12] chapter on statics. A truss is a support device that supports forces only at its ends and cannot support bending moments. A frame is a support device that supports forces and bending moments.

Figure 1.6: The Instrument During Shipment

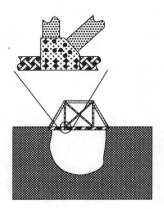

Figure 1.7: A Truss Bridge

This section demonstrated the importance of considering the extent to which environmental factors influence the behavior of a system. If we make our system too small, we may leave out important factors that will affect our objective. This could undermine the usefulness of our analysis and waste time. However, including everything in the model of a system is not practical either. If the system becomes cluttered with superfluous factors, we may never be able to reach a conclusion.

## 1.1.4   DEFINING THE TIME PERIOD OF ANALYSIS

**Objectives**
*When you complete this section, you should be able to*

1. *Explain the importance of specifying the time period of analysis when modeling a system's behavior.*

This section discusses the importance of defining a time period of interest when analyzing the behavior of an occurring process. The importance of defining the time period of interest may not be self-evident, but it is an essential part of modeling a system's behavior. What if someone asked, "How many people are in your engineering class?" We probably would tell them the number of people enrolled in the class for the *entire year*. The point is that our answer could vary depending on the time period we choose to consider when answering the question. For example, our answer could vary if we considered a time when our class does not meet or a day when many of our classmates are sick. If we explicitly define the time period, however, we can accurately account the number of people who enter and leave the class and answer the person's question.

Like class attendance, many time variant factors affect engineering processes. Thus, to analyze an engineering system properly we must define the period of time that we are going to look at that system. As an example, consider the wheel in figure 1.8. For some values of time, the wheel is rolling on a flat plane. For others, the wheel is moving over a circular surface. The forces and velocity on the wheel are different depending on which surface it is on. For example, if the wheel is on the flat plane, the center of the wheel is a constant distance above the ground. Whereas, on the circular surface, the center is a constant distance from point O.

The time period also has implications on the type of modeling that we perform. For example, if we model the motion of a simple pendulum over small time periods (see figure 1.9), we arrive at equations predicting planar motion. However, as the time period increases, the pendulum does not obey this planar model. If we watch a large pendulum swing for several minutes, we will see that it does not have a strictly planar motion. Several museums across the United States have exhibits where we can observe this type of behavior. Our simple planar model is unable to predict this *out of plane* motion. To obtain a satisfactory model, we would have to include the Earth's rotation. Therefore, depending on what we want to determine and the time period of interest, the simple planar pendulum model may or may not be satisfactory.

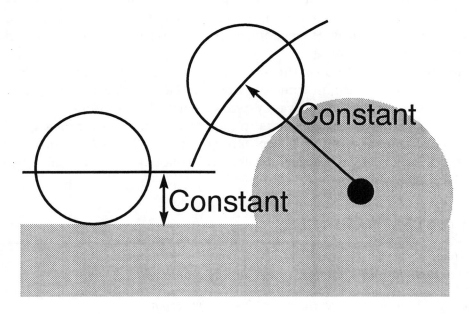

Figure 1.8: A System May Have Multiple Descriptions Depending on When We Observe It

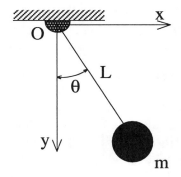

Figure 1.9: A Simple Pendulum

## 1.2    GENERAL SYSTEM CONCEPTS

There are several system concepts that help guide us through the formulation process. Some of these concepts are general. We can apply them regardless of the type of problem being evaluated. This section explains these concepts. This discussion is a prelude to later chapters, which introduce system concepts that are specific to the type of system being evaluated.

In this section, we will apply the accounting and conservation laws to formulate mathematical models of engineering systems. If the formulations seem too difficult to follow, try to focus on the general system concepts being discussed. The text will demonstrate how to use these concepts to guide the formulation process. Remember the purpose of this course is to help you apply our knowledge of the accounting and conservation laws to model complex systems. Do not be discouraged if you do not already have this ability.

### 1.2.1   SYSTEM VARIABLES

**Objectives**
*When you complete this section, you should be able to*

   *1. Recognize and define system variables in problems.*

A system variable quantifies the important properties that describe a system. System variables can be constant. They do not have to change. Thus, the term variable can be misleading. Quantifier would be a better term because it does not suggest variation. However, because the term variable is so popular, we use it throughout this text. Just remember that variables can be constant.

As a general rule, we choose just enough variables to quantify all the important properties entering, exiting, or accumulating in a system (*i.e.* all the quantities that pertain to the conservation laws). The most common mistake that people make when analyzing systems is that they do not use enough variables. We can *always* write equations to model a system's behavior with too many variables, but we can *never* write them with too few.

**Example 1.7**

### Throwing a Baseball

If we wanted to model the motion of a baseball, one of the system variables that we would need is the baseball's velocity. This variable quantifies both the baseball's momentum ($m\vec{v}$) and its kinetic energy ($\frac{1}{2}mv^2$). Notice the presence of $m$ in both the momentum and kinetic energy terms. It is perfectly correct to think of $m$ as another variable in addition to $\vec{v}$ because it quantifies the property of mass. For most problems, however, $m$ is known or given. As a result, most people do not consider $m$ to be a variable.

As the baseball falls to the ground, it moves in the Earth's gravitational field. Thus, another system variable that we would need is the baseball's height (vertical position). Height quantifies the gravitational potential energy of the baseball.

**End 1.7**

As the example above shows, system variables are necessary to quantify all the important quantities that pertain to a system, and that the same variable can describe more than one important quantity. Recognizing the variables necessary to describe a system is an important part of the modeling process. The discussion that follows introduces and explains the system variables we will need to quantify systems in a variety of subjects. As the example above indicated, if we want to model a mechanical system, we may need velocities and positions as system variables. To model an electrical system, we may need voltages and currents. Voltages and currents quantify the energy accumulation that occurs in capacitors and inductors as well as the energy dissipation that occurs in resistors (the typical components of electrical systems). Modeling hydraulic systems generally involves accounting mass. Therefore, we may need pressure and fluid flow rate variables to analyze these systems. Modeling thermodynamic systems generally requires accounting energy and entropy. Thus, we may need temperature, entropy, and internal energy as system variables when modeling these systems.

## 1.2.2 VARIABLE TYPES

**Objectives**

*When you complete this section, you should be able to*

1. *Recognize the two types of variables and give examples of each in various disciplines.*

In all *energetic* systems,[17] there are two basic types of variables. These variable types have different names depending on the discipline. Some disciplines refer to them as effort and flow. Others call them across and through. However, the primary issue is not what we call them but how to recognize them.

Generally, an effort is applied across something and a flow occurs through something. Typically, a flow results because of an effort or an effort results because of a flow. Functions of effort and flow variables are also efforts and flows. In *many* systems, the dot product of an effort and a flow is a function of power. This function can be power or a simple variation of power. The examples that follow list some effort and flow variables of variousfields of study.

In electrical systems, a set of effort and flow variables is voltage ($V$) and current ($i$), respectively. Voltage is an effort applied across something. Current is the flow of charge through something. An applied voltage causes a current to occur or *vice versa*. Charge is the integral of current (a flow) and, therefore, is also a flow. The product of current and voltage is power.

$$iV = \dot{E}$$

In dynamics, a related set of effort and flow variables is force ($\vec{F}$) and velocity ($\vec{v}$), respectively. Linear momentum ($m\vec{v}$) and position ($x$) are both functions of velocity and, therefore, are also flow variables. The dot product of force and velocity is power.

$$\dot{E} = \vec{F} \cdot \vec{v}$$

---

[17]Energetic systems are systems in which energy is defined.

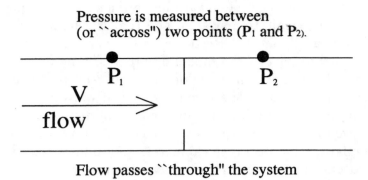

Figure 1.10: Effort (Across) and Flow (Through) Variables in a Hydraulic System

If we use momentum as the flow variable instead of velocity, the effort/flow dot product is simply the product of mass and power. $(\vec{F} \cdot m\vec{v} = m\dot{E})$.

We can extend this discussion to other fields of study. The table that follows lists some common effort and flow variables for hydraulics, thermodynamics, mass transfer, and material science.

| discipline | effort | flow |
|---|---|---|
| hydraulics | pressure | fluid flow |
| thermodynamics | temperature | heat |
| mass transfer | concentration or pressure | diffusion or reaction rate |
| material science | stress | strain |

The underlying point of this discussion is that there are effort and flow variables in every discipline. We can call them whatever we want – effort and flow, across and through, driving force and response, or A and B. It does not matter what we call them.[18] What matters is that we have the ability to recognize and relate these variable types.

**Example 1.8** ━━━━━━━━━━━━━━━━━━━━━━━━━━━━━━━━━━━━━━━━

### Variable Types in Hydraulic Systems

We want to analyze the flow of a constant density fluid through an orifice in a pipe of cross sectional area $A$ (see figure 1.10). The average fluid velocity and pressure at some cross section of the pipe are $v$ and $P$ respectively. The pressure is the *effort* variable of the system because it is what drives the flow variable $(v)$. We can also call pressure the *across* variable. This label across originates from the method used to measure this variable type. For example, when fluid flows through an orifice, the upstream and downstream pressures differ. If we want to measure the pressure, we measure the pressure drop across the orifice. The mass flow rate $(\dot{m})$ is the *through* variable of the system because it is what flows through the orifice. Multiplying the effort variable (pressure, $P$) and the flow variable (mass flow rate, $\dot{m}$) gives a function of power.

$$P\dot{m} = P\left(\rho A v\right) \quad \overset{\text{reduces}}{\rightsquigarrow} \quad PAv\rho = Fv\rho \quad \overset{\text{reduces}}{\rightsquigarrow} \quad \dot{E}\rho$$

**End 1.8** ━━━━━━━━━━━━━━━━━━━━━━━━━━━━━━━━━━━━━━━━━━━━

---

[18]Understanding the general concepts will help us understand the various disciplinary terminology. Thus, the importance is not to learn how all the various disciplines name things, but to learn how to apply a general knowledge to all the various disciplines.

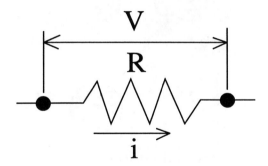

Figure 1.11: Effort (Across) and Flow (Through) Variables in an Electrical System

**Example 1.9**

### Variable Types in Electrical Systems

For the electrical system in figure 1.11 , the voltage drop *across* the resistor is $V$. The current flowing *through* the resistor is $i$. Depending on the terminology that we use, voltage $(V)$ can be called an across (effort) variable and current $(i)$ a through (flow) variable. The product of $V$ and $i$ is power.

**End 1.9**

## 1.2.3 PARAMETERS

**Objectives**
*When you complete this section, you should be able to*

1. *Distinguish between parameters and variables in accounting and conservation equations even when they are not explicitly specified.*

This subsection discusses the concept of parameters. Parameters are extra quantifiers in a problem that are known. When we model systems using the conservation laws, terms other than system variables arise. These additional terms are parameters.[19]

Parameters are important because they often relate effort and flow variables in the conservation laws (*i.e.* they constitute the relation between a set of effort and flow variables). For example, in a resistor, the parameter ($R$ resistance) relates the effort $(V)$ and flow $(i)$ variables as $V = iR$. Parameters are the connection between the system variables we determine and the conservation laws. Without parameters, we often cannot calculate effort and flow variables.

Despite their importance, parameters are excluded from most problem statements. This may sound contradictory, however, the reason is clear. Even though a system's behavior changes as its parameters change - the structure of the conservation equations governing the system do not. Thus, the concept of parameters is important for relating efforts and flows to the laws, but the exact values of the parameters are not

---

[19]Resistance, capacitance, mass, the gravitational constant, are examples of parameters in the conservation laws. Voltage, current, acceleration, and force are generally system variables as we learned earlier.

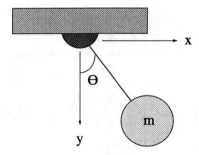

Figure 1.12: A Mass at the End of a String (A Pendulum)

important. We can generally assume that the parameters of a system are known. This assumption allows us to relate a system's efforts and flows within the conservation laws in a general way. With the parameters specified as undetermined variables in these general relations, we do not have to worry about their values. Thus, we achieve a general result that is applicable to a variety of systems with the same behavior pattern.

Parameters are not always known. Sometimes problem statements specify both an effort and a flow and ask us to determine an unknown parameter. For example, if $i = 2$ [amps] and $V = 4$ [V] in the system of figure 1.11, we can find $R$ through the constitutive relation $R = \frac{V}{i}$.[20]

Likewise, parameters do not have to be constants. For example, it is possible to construct a resistor whose resistance changes with temperature or external forces.

**Example 1.10** _____

### How Parameters Relate to the Conservation Laws

**PROBLEM:**

Analyze why the pendulum in a grandfather clock moves the way it does. Develop the mathematical formulation that models the pendulum's motion.

**FORMULATION:**

A grandfather clock's pendulum works exactly like the simple pendulum in figure 1.12. The motion (flow) of the pendulum is the primary system variable. The pendulum mass and length are the system parameters. Since our objective is to model the pendulum's motion, these parameters are of little or no concern to us. Our concern is how to relate the forces acting on the pendulum with its motion. Treating the parameters as known and constant values, we can easily relate the forces (efforts) to the motion (flow) using the conservation laws. If we apply conservation of angular momentum to the mass, we formulate the following relationship:

---

[20]This relationship is Ohm's law. Notice that it has the form resistance=

$$\frac{\text{effort}}{\text{flow}}$$

## Conservation of Angular Momentum (Out of the Page) [@O]

system[ **pendulum f.1.12** ]     time period [**differential**]

*input − output = accumulation*

| input/output: | mass flow | |
|---|---|---|
| | external moment | |
| | gravity moment | $-mgL\sin\theta\ \vec{k}$ |
| accumulation: | | $mL^2\ddot{\theta}\ \vec{k}$ |

*z (CCW - Counter Clock Wise):*     $\boxed{-mgL\sin\theta}$  =  $\boxed{mL^2\ddot{\theta}}$

        *gravity*                    *accumulation*

$$-mgL\sin\theta = mL^2\ddot{\theta}$$

The effort variable is the torque due to the pendulum's weight. The flow variable is $(\theta)$, and the parameters $(g, l$ and $m)$ relate these two variables together.

**OTHER CONSIDERATIONS:**

Alternatively, if we had to calibrate the clock (*i.e.* make sure the motion of the pendulum under the given forces makes one swing per second), the parameters would be our main concern. To calibrate the clock, we would use the known effort and flow variables to calculate the parameters required to cause that type of motion to occur. We can use the same equation as before.

**End 1.10**

## 1.2.4  SYSTEM IMPEDANCE

### Objectives

*When you complete this section, you should be able to*

1. *Explain what impedance means.*

2. *Determine the impedance of simple systems.*

3. *Recognize when problems are well defined.*

Impedance is a fundamental characteristic of a system. If we plot an effort variable on the ordinate and its related flow on the abscissa of a graph, impedance is the quotient of the effort and flow. Impedance characterizes how large an effort has to be to achieve a desired flow[21] and *vice versa*. Calculating impedance for realistic systems can be very difficult. Thus, we will only deal with it qualitatively. We will refer to a system's impedance as large or small, constant or variant, but we will seldom resort to specific quantitative calculations.

Although impedance is difficult to calculate, the concept is helpful for

1. determining when problems are properly defined

2. determining the extent of a system

3. qualitatively describing the behavior of a system.

---

[21] How difficult it is to achieve a desired flow.

**Analysis Assumptions**

(1) The tank is cylindrical and has a very large cross sectional area $A$.

(2) The water momentum in the tank is negligible.

(3) The water pressure at the exit is directly related to the weight of water in the tank (and thus to the height of water in the tank).

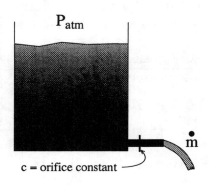

$P_{atm}$

$\dot{m}$

c = orifice constant

Figure 1.13: (atankoffluid): The Farmer's Irrigation Tank

Given an arbitrary system, it is impossible to achieve desired values of both effort and flow variables simultaneously. For example, in the electrical resistor of figure 1.11 with $R = 3\,[\Omega]$, we can achieve a current of 2 [amps] through the resistor by applying a voltage of 2(3) [volts] across the resistor. Provided the resistor does not melt and that we have a 6 [volt] power supply, we will get what we want. On the other hand, if we desire to have 9 [volts] across the resistor, we will have to place $\frac{9}{3}$ [amps] of current through the resistor. Provided that we have the equipment to do this and that the resistor does not melt, we will again receive what we want. If we desire to have both 9 [volts] across the resistor and 2 [amps] through the resistor simultaneously, we will not get what we want. It is impossible! We cannot specify both an arbitrary effort and flow simultaneously and expect to get what we want.

As we will see in many problems, when an effort is specified, the corresponding flow is not and *vice versa*. When a problem specifies both, it is **either** poorly defined **or** its objective is to determine the system's impedance or a parameter. For example, we can specify both an effort of 9 [volts] and a flow of 2 [amps] and ask what resistor allows such a combination.

**Example 1.11** ──────────────────────────────

### Draining a Tank of Fluid

**PROBLEM:**

Henry Swantek, a farmer, wants to irrigate his entire corn crop in one day. Henry has a tank in his field (see figure 1.13) with just enough water to irrigate his entire crop. Develop the mathematical formulation required to determine if the tank will drain fast enough to irrigate Henry's entire crop in one day (it takes all of the water in the tank to irrigate his crop). Then, solve the formulation for height and evaluate the farmer's options. The tank is cylindrical and has a very large cross sectional area, $A$. The exhaust pipe on the tank allows water to escape at a flow rate of approximately $C\sqrt{P}$ (where $P$ is pressure inside the tank at the exhaust pipe and $C$ is a constant).[22] The initial water height in the tank is three feet.

**FORMULATION:**

---

[22] We commonly call this constant the orifice constant.

The first step in solving this problem is to decide what conservation law to apply. Since mass is leaving the system, we decide to apply the conservation of mass to the tank.

## Conservation of Mass

system[**water in the tank f.1.13**]      time period [**differential**]

*input − output = accumulation*

| input/output: | mass flow | $-C\sqrt{P}$ |
| accumulation: | | $\frac{d}{dt}(m)$ |

$$\boxed{-C\sqrt{P}} \quad = \quad \boxed{\frac{d}{dt}(m)}$$

*output*             *change in accum.*

Basically, what we need to determine is if it is possible to provide a certain flow rate for given a water height. However, we know both the water height and the flow rate (because the farmer wants it to drain the tank in a certain time). Furthermore, we know the pressure (a function of gravity, liquid density, and height) in the tank because it is directly proportional to the water's height.[23] Since the problem specifies both the effort and the flow at the same time, we are not optimistic about meeting these conditions without changing the existing parameters.

Despite this negative outlook, we will try inserting these values in the conservation result anyway. We start by replacing the pressure term in the equation. With the water in the tank as the system we can write the conservation of linear momentum as follows:

## Conservation of Linear Momentum (in Vertical)

system[**water f.1.13**]      time period [**differential**]

*input − output = accumulation*

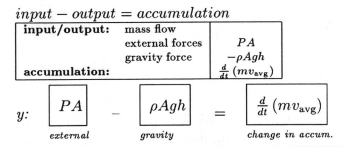

| input/output: | mass flow | |
| | external forces | $PA$ |
| | gravity force | $-\rho Agh$ |
| accumulation: | | $\frac{d}{dt}(mv_{\text{avg}})$ |

$$y: \quad \boxed{PA} \quad - \quad \boxed{\rho Agh} \quad = \quad \boxed{\frac{d}{dt}(mv_{\text{avg}})}$$

*external*     *gravity*     *change in accum.*

Here $PA$ is the pressure force on the water from the tank bottom (hence it pushes upward) and $\rho Ahg$ is the water's weight (hence it is downward). Given that the area of the tank is large and the flow rate is small, we can assume that $\frac{d}{dt}(mv_{\text{avg}})$ is negligibly small. In other words, take the derivative to find: $\frac{d}{dt}(mv_{\text{avg}}) = \frac{d}{dt}(m)v_{\text{avg}} + m\frac{d}{dt}(v_{\text{avg}})$, now if the mass flow rate is small and the area is large, then the change in mass is small and the average velocity $v_{avg}$ is small. If we multiply two small numbers we get an even smaller number so the first term is going to be very small. Now think about the second term. Since the area is large and the mass is being pulled out slowly the average $v$ is small so how much could it change? Not much, we plan to ignore the

---

[23]Pressure is Force over Area. $P = \frac{F}{A}$. The force is the fluid weight which will be mass times g. Fluid mass is density times volume. Therefore $P = \frac{F}{A} = \frac{\rho Ahg}{A} = \underbrace{\rho g}_{\text{parameters}} \ h.$

second term also. This allows us to simplify the conservation result to

$$PA = \rho Agh$$

Thus,

$$P = \rho gh$$

Replacing pressure and mass with $\rho gh$ and $\rho Ah$ in the conservation of mass, gives

$$-C\sqrt{\rho hg} = \frac{d}{dt}(\rho Ah)$$

So why don't we ignore the accumulation term? Well if you think about it, how can we ignore this accumulation? If we did, it would say that $C\sqrt{\rho hg}$ is zero which it isn't. The accumulation starts out very large and given enough time it becomes zero. This is a heck of a change! Think back on the other terms we ignored. How much did the accumulations change? They didn't start out large so how much could the change be? Now this logic doesn't always work, some things that are small have large changes.

**SOLUTION:**

Since $\rho$ and $A$ are constant, the above equation simplifies to

$$-C\sqrt{\rho gh} = \rho A\frac{d}{dt}(h)$$

This differential equation is separable. All we have to do to solve it is

1. separate the variables, and

2. integrate both sides of the separated result.

Doing this, we find that

$$\frac{-C}{A}\sqrt{\frac{g}{\rho}}dt = h^{-\frac{1}{2}}dh$$

$$\frac{-C}{A}\sqrt{\frac{g}{\rho}}t = 2h^{\frac{1}{2}} + \mathcal{C}_1$$

where $\mathcal{C}_1$ is an arbitrary constant of integration. To solve for $\mathcal{C}_1$, we need one initial condition. The initial condition for this problem is the initial height of water in the tank. At t=0, the tank has three feet of water. Therefore,

$$0 = 2(3)^{\frac{1}{2}} + \mathcal{C}_1$$

$$\mathcal{C}_1 = -2\sqrt{3}$$

$$\frac{-C}{A}\sqrt{\frac{g}{\rho}}t = 2h^{\frac{1}{2}} - 2\sqrt{3} \qquad (1.2)$$

Now to find the time when the tank is empty, we solve equation 1.2 for the time when the height of water in the tank is zero.

$$-\frac{C}{A}\sqrt{\frac{g}{\rho}}t_{\text{empty}} = 0 + (-2\sqrt{3})$$

$$t_{\text{empty}} = \frac{2A\sqrt{3\rho}}{C\sqrt{g}}$$

Notice that only if

$$\frac{2A}{C}\sqrt{\frac{3\rho}{g}} < 24 \text{ hours}$$

will the farmer have what he wants. The problem is that the farmer has arbitrarily and simultaneously specified both the effort and the flow. The only way to guarantee achieving his specifications is to size the parameters of the tank exactly right.

The farmer has two choices.

1. He can pressurize the tank. Thereby making the pressure (effort) whatever it needs to be to achieve the desired flow.

2. He can buy a special orifice with the right orifice constant. If he sizes the constant $C$ correctly, the desired effort and flow will meet his specifications.

Either of these two alternatives can produce the results that the farmer desires.
**End 1.11**

## 1.2.5  INPUT/OUTPUT IMPEDANCE

**Objectives**
*When you complete this section, you should be able to*

*1. Determine the inputs and outputs of a system isolated from its surroundings.*

*2. Define the input/output efforts and flows of isolated systems.*

*3. Identify the driver and load in interconnected systems.*

This subsection explains the concepts of input and output impedance. We can use these concepts to determine how connected systems will interact. It often makes sense to view interconnected systems as drivers (sources) and loads (sinks). The driver (source) provides a property and the load (sink) receives the property. Generally, the driver tries to make something happen, and the load responds to the driver. For example, when we ride a bicycle, our body is the driver and the bike is the load. If it is not obvious which system is the driver and which is the load, do not worry about it. We can generally choose either one as the driver or the load.

A driver and load typically exchange properties with each other. Thus, when separated, they often respond much differently than when connected. When interconnected systems interact, they often exchange properties at discrete locations. These locations are called inputs on the loads and outputs on the drivers. For example, consider an AC voltage connected to a resistor. Figure 1.14 shows the system. We view the voltage source as one system and the resistance as a second system. The voltage is obviously the driver and R is the load. The systems connect at terminals AC and BD. Points A and B belong to the source and CD belong to the load. Points A and B are the outputs of the driver; C and D are the inputs to the load.

There is a effort (across or voltage) variable across terminals A and B, this is the output effort. There is also a flow (current or through) variable at terminals A and

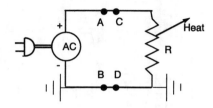

Figure 1.14: An AC Voltage Applied to a Resistor

B. This is the output flow. Note that output flow does not necessarily mean the flow comes out. Output flow refers to the flow in the driver's (the output's) connection. Likewise the effort and flow across and through terminals C and D are the input effort and flow. Notice that the output effort and flow equal the input effort and flow.

The ratio of the output effort to the output flow is the output impedance of the source. A similar ratio for the input is the input impedance of the load.

Also note in figure 1.14 that the resistor is connected to ambient air and giving thermal energy to the air. This is another interconnection. For the resistor/air interface, the resistor is the source and the air is the load. The temperature of the resistor is the output effort and the heat flow is the output flow.

**Example 1.12** _____

### A Bicycle and Rider

**QUESTION:**
Express the dynamics of riding a bicycle up a hill in terms of input and output impedance.

**ANSWER:**
We can think of this process as consisting of three basic elements:

1. the rider,

2. the bicycle, and

3. the hill.

Between the rider and bicycle, the rider is the driver (source) and the bicycle is the load (sink). In producing the forward motion of the bicycle, the rider's feet interact with the bicycle's pedals.[24] The feet are the output of the rider system and the pedals are the input of the bicycle system. The output impedance of the human is the ratio of leg force to leg speed. The input impedance of the bicycle is the ratio of pedal force to pedal speed.

A similar relationship exists between the bicycle and the hill. In this case, the bicycle is the driver and the hill is the load. The bicycle's tires interact with the hill's surface. The output impedance of the bicycle is the ratio of the torque (or force) on the rear wheel to the speed of the rear wheel. The input impedance of the hill is the ratio of the force required to climb the hill to the speed of the climb.

**End 1.12** _____

When a driver and load are connected, the load will typically demand more effort as its flow increases (see figure 1.15). This generally means that the load demands

---

[24]Alternatively, if we wanted to study steering, we would consider how the rider's arms interact with the bicycle's handle bars.

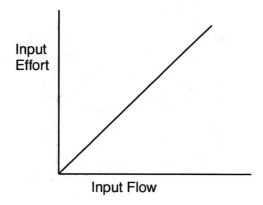

Figure 1.15: A Typical Impedance Relation for a Load

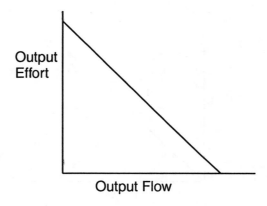

Figure 1.16: A Typical Impedance Relation for a Driver

more power as the flow increases. In contrast, a typical driver will provide decreasing effort as its flow is increased. See figure 1.16. What this means is that a typical driver can only provide a finite amount of power. Thus, the product of effort and flow must be finite. For this stipulation to be met, effort must decrease to zero as flow increases to infinity. For example, someone riding a bike can push hard on slow moving pedals, but as the pedal speed increases, their ability to push hard will decline. What advantage does this give multispeed bicycles?[25]

Each driver/load will have an impedance curve that characterizes it. These curves can be dependent on many factors (*e.g.* temperature and rates of change of effort or flow). Thus, it may not be possible to plot them without making some simplifying assumptions. For example, if the throttle of an automobile engine is opened, the engine can deliver more power than if it is partially closed. An open throttle allows more chemical energy (the air fuel mixture) to enter the engine. The engine produces work (power) by converting this chemical energy into mechanical. Thus, if it has more chemical energy to convert, it can produce more power. Figure 1.17 shows a hypothetical impedance curve for an automobile engine with its throttle in two different positions.

---

[25] The advantage of multispeed bicycles is that the rider can change the pedal speed. Thus, if they want to push harder but the pedals are moving too fast or if they want to push easier but the pedals are moving too slow, all they have to do is change gears. Changing gears modifies the load's input impedance. Thus, the effort/flow required from the driver changes.

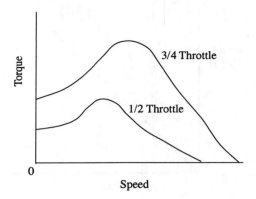

Figure 1.17: A Hypothetical Output Impedance Relation for An Engine at Different Throttle Positions

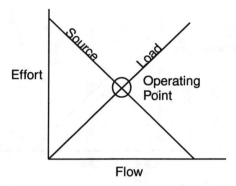

Figure 1.18: A Typical Operating Point

## 1.2.6  OPERATING POINTS

**Objectives**

*When you complete this section, you should be able to*

1. *Explain what an operating point is.*

2. *Find operating points graphically and analytically.*

3. *Explain the effect that high input/output impedance has on a system and its surroundings.*

4. *Explain the effect that low input/output impedance has on a system and its surroundings.*

When connected to a load, a driver must provide the effort and flow required by the load. The point where the effort and flow of a driver matches the effort and flow of a load is the point where a connected system will operate continuously (*i.e.* the steady state operating point). We can determine the operating point of a connected system either by plotting the driver and load's impedance curves on the same graph (see figure 1.18) or by solving the impedance relations of the connected systems together as a system of equations.

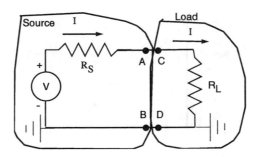

Figure 1.19: A DC Source Connected to a Load

## Example 1.13

### Operating Point of a DC Voltage and Resistor

**PROBLEM:**

Express the behavior of the circuit in figure 1.19 in terms of the input and output impedances of the source and load and find the operating point of the connected circuit.

**FORMULATION:**

The first step in solving this problem is to express what the source's output effort and flow variables are. For this circuit, the source output effort is $V_A$ and the source output flow is $I$. The next step is to develop an impedance relationship between these variables. The general impedance relationship is

$$\text{impedance(resistance)} = \frac{\text{effort}}{\text{flow}}.$$

Therefore, we want to relate the effort and flow variables of the source to the resistance, $R_S$. We can do this by taking $R_S$ as a system and accounting electrical energy

**Electrical Energy Accounting**

system[ $R_S$ **f.1.19** ]      time period [**differential**]

$input - output + generation - consumption = accumulation$

| input/output: | $IV - IV_A$ |
| consumption (thermal): | $I^2 R_S$ |
| generation: | |
| accumulation: | 0 |

$$\boxed{IV} \;-\; \boxed{IV_A} \;-\; \boxed{I^2 R_S} \;=\; \boxed{0}$$

$\quad\;\; input \qquad\quad\; output \qquad\quad consumption \qquad\quad accumulation$

$$VI - V_A I - I^2 R_S = 0$$

This simplifies to

$$\underbrace{V - V_A}_{\text{effort}} = \overbrace{I}^{\text{flow}} \; \underbrace{R_s}_{\text{parameter}} \tag{1.3}$$

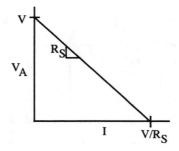

Figure 1.20: Output Impedance of the Source

the impedance relation for the source. Notice that this is the equation of a straight line $\mathcal{Y} = m\mathcal{X} + b$.

$$\underbrace{V_A}_{\mathcal{Y}} = \overbrace{-R_s}^{\text{slope=m}} \underbrace{I}_{\mathcal{X}} + \underbrace{V}_{\text{y intercept=b}}$$

Figure 1.20 shows a plot of this equation. Notice that figure 1.20 has the typical characteristics of a source, when $I$ is large $V_A$ is small. When the load demands a high flow, the source produces a low effort.

Using the same approach that we used for the source, we can relate the input effort of the load ($V_C$) to the input flow of the load ($I$). With $R_L$ as our system, we account electrical energy and get

**Electrical Energy  Accounting**

system[ $R_L$ **f.1.19** ]      time period [**differential**]

$input - output + generation - consumption = accumulation$

| input/output: | $IV_C$ |
|---|---|
| consumption (thermal): | $I^2 R_L$ |
| generation: | |
| accumulation: | 0 |

$$\boxed{IV_C} \quad - \quad \boxed{I^2 R_L} \quad = \quad \boxed{0}$$

$\quad$ input $\qquad\qquad$ consumption $\qquad\qquad$ accumulation

$$IV_C - R_L I^2 = 0$$

This simplifies to

$$V_C = I R_L \qquad\qquad (1.4)$$

the impedance relation of the load. Notice that this is the equation of a straight line $\mathcal{Y} = m\mathcal{X} + b$. Figure 1.21 shows a plot of this equation. Notice that figure 1.21 has typical load impedance. If $V_C$ is large, then $I$ is large.

When we connect the systems, the output effort and output flow has to equal the input effort and input flow. The point where this occurs is the operating point of the system. We can find the operating point graphically by finding the point where the source and load graphs intersect. Figure 1.22 shows how to do this for this system. We can also find the operating point by equating the impedance relations for the source and load. We can do this for this system in the following way:

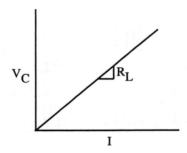

Figure 1.21: Input Impedance of the Load

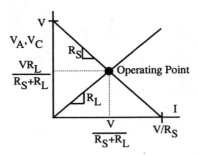

Figure 1.22: Locating the Operating Point Graphically

1. We set $V_A$ in equation 1.3 equal to $V_C$ in equation 1.4.

$$V - R_s I = I R_L$$

2. Then, we solve for $I$.

$$I = \frac{V}{R_s + R_L}$$

3. We use this result in equation 1.4 to solve for $V_C$ (and $V_A$ since they are identical). Thus,

$$V_C = \frac{V R_L}{R_s + R_L}.$$

**End 1.13** ————————————————————————————

**Example 1.14** ————————————————————————————

## A Flashlight

**PROBLEM:**

Determine the operating point of the flashlight circuit in figure 1.23.

**FORMULATION:**

In a flashlight, the battery is the driver (source) and the light bulb is the load. Individual impedance diagrams for the light bulb and the battery are similar to those in figures 1.15 and 1.16. The combined driver/load operating diagram is similar to figure 1.18. When the light bulb and battery are connected, the flashlight will work at the operating point of this diagram. We can find the operating point graphically or mathematically. To find the operating point graphically, we plot the driver/load

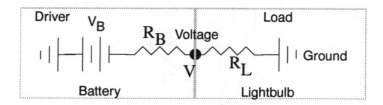

Figure 1.23: The Simplified Electrical Schematic of a Flashlight

impedance diagrams together and determine where they intersect. We did this in figure 1.18. To find the operating point mathematically, we model the individual driver/load systems and solve the resulting equations systematically.

Accounting electrical energy to model the battery and light bulb systems of the flashlight, we get the following:

## Electrical Energy  Accounting

system[ **battery f.1.23** ]      time period [**differential**]

$input - output + generation - consumption = accumulation$

| input/output: | $iV_B - iV$ |
|---|---|
| **consumption (thermal):** | $i^2 R_B$ |
| **generation:** | |
| **accumulation:** | 0 |

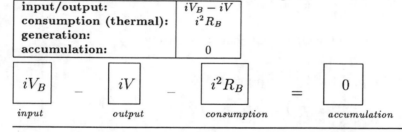

## Electrical Energy  Accounting

system[ **light bulb f.1.23** ]      time period [**differential**]

$input - output + generation - consumption = accumulation$

| input/output: | $iV$ |
|---|---|
| **consumption (thermal):** | $i^2 R_L$ |
| **generation:** | |
| **accumulation:** | 0 |

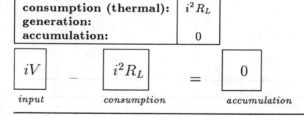

1. the battery (the driver):

$$iV = iV_B - i^2 R_B \quad \overset{\text{reduces}}{\rightsquigarrow} \quad V = V_B - iR_B \tag{1.5}$$

2. the light bulb (the load):

$$iV - i^2 R_L = 0 \quad \overset{\text{reduces}}{\rightsquigarrow} \quad V = iR_L \tag{1.6}$$

Substituting 1.6 into 1.5 gives

$$V_B - iR_B = iR_L \quad \overset{\text{reduces}}{\rightsquigarrow} \quad V_B = i(R_B + R_L)$$

Thus, the operating point of the flashlight is

$$i = \frac{V_B}{R_B + R_L}$$

**End 1.14** ━━━━━━━━━━━━━━━━━━━━━━━━━━━━━━━━━━━━━━━━━━━━

Review the results obtained for the driver and load in example 1.14. If the driver has a low output impedance (small $R_B$), then it can supply a nearly constant output voltage ($V$) for almost any current ($i$) that the load demands. Conversely, if the driver has a high output impedance (large $R_B$), then its output voltage will decrease as the load demands more current. The descriptions below summarize these results in general terms.

[**low output impedance**] allows a driver to provide nearly constant effort regardless of the flow demanded by its load

[**high output impedance**] causes the effort provided by a driver to decrease as the flow demanded by its load increases

A good general purpose driver will have a low output impedance so that it can supply effort to a variety of loads. The electrical power company generates a low output impedance in power lines so that homes will have constant voltage at their wall sockets. This allows devices with different current requirements to operate on the electricity from the same socket.

If the load has a high input impedance (large $R_L$), then it requires little current ($i$) for a large range of input voltages ($V$). If the load has a low input impedance (small $R_L$), then its current demand will increase as the input voltage increases. The descriptions below summarize these results in general terms.

[**low input impedance**] causes the flow required by a load to increase as the effort provided by its driver increases.

[**high input impedance**] causes a load to demand a small flow for a large range of input efforts.

Electrical motors generally have low input impedances. Applying twelve volts to a direct current motor can easily require over five amps of current.

Two disconnected systems that have a very low output impedance and a very high input impedance, respectively, will tend to behave the same even when they are connected together. The following example demonstrates this.

**Example 1.15** ━━━━━━━━━━━━━━━━━━━━━━━━━━━━━━━━━━━━━━━━━━━━

### Using A Volt Meter

**QUESTION:**

Figure 1.24 shows a circuit with a twelve volt battery. The resistor ($R_c$) enclosed by the dashed lines represent the circuit. The voltage source and small resistor ($R_o$) represent the battery. The circuit is not working properly and has to be fixed. Matt Matthews, an electrician is sent to fix it. Matt decides to measure the voltage in the circuit at node $a$. He uses the volt meter in figure 1.25 to do this. Explain where

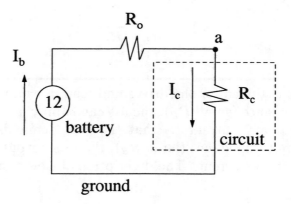

Figure 1.24:

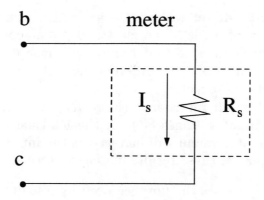

Figure 1.25:

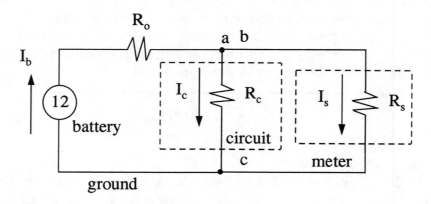

Figure 1.26:

Matt should place the meter. What should the impedance of the meter be for Matt to accurately measure the voltage?

**ANSWER:**

First, we will determine what the voltage at $a$ should be. Accounting the electrical energy in the resistor $R_o$, we find that

**Electrical Energy Accounting**

system[$R_o$ **f.1.24**]    time period [**differential**]

$input - output + generation - consumption = accumulation$

| input/output: | $12I_b - V_aI_b$ |
|---|---|
| **generation:** | |
| **consumption:** | $I_b^2 R_o$ |
| **accumulation:** | |

$$\boxed{12I_b - V_aI_b} \quad - \quad \boxed{I_b^2 R_o} \quad = \quad \boxed{0}$$

*input/output*        *consumption*        *change in accum.*

$$V_a = 12 - I_b R_o$$

Now, taking $R_c$ as a system and accounting for electrical energy, we find that

**Electrical Energy Accounting**

system[$R_c$ **f.1.24**]    time period [**differential**]

$input - output + generation - consumption = accumulation$

| input/output: | $V_aI_c$ |
|---|---|
| **generation:** | |
| **consumption:** | $I_c^2 R_c$ |
| **accumulation:** | |

$$\boxed{V_aI_c} \quad - \quad \boxed{I_c^2 R_c} \quad = \quad \boxed{0}$$

*input/output*        *consumption*        *change in accum.*

$$V_aI_c - I_c^2 R_c = 0$$

Since $I_b = I_c$, this result simplifies to

$$V_aI_b - I_b^2 R_c = 0$$

$$\frac{V_a}{R_c} = I_b$$

Using this result in the equation for $R_o$ gives

$$V_a = 12 - \frac{V_a R_o}{R_c}$$

which simplifies to

$$V_a = \frac{12R_c}{R_c + R_o} \tag{1.7}$$

To determine the voltage at node $a$, Matt places terminal $b$ of the meter on node $a$ and terminal $c$ on ground. Therefore, we will calculate what the meter will read when the circuit and meter are connected in this way (see figure 1.26).

Accounting the electrical energy in $R_o$ gives

---

**Electrical Energy Accounting**

system[$R_o$ **f.1.26**]      time period [**differential**]

*input $-$ output $+$ generation $-$ consumption $=$ accumulation*

| input/output: | $12I_b - V_a I_b$ |
|---|---|
| generation: | |
| consumption: | $I_b^2 R_o$ |
| accumulation: | |

$$\boxed{12I_b - V_a I_b} \quad - \quad \boxed{I_b^2 R_o} \quad = \quad \boxed{0}$$

*input/output*              *consumption*              *change in accum.*

---

$$12I_b - V_a I_b - R_o I_b^2 = 0$$

Accounting the electrical energy in $R_c$, we determine that

---

**Electrical Energy Accounting**

system[$R_c$ **f.1.26**]      time period [**differential**]

*input $-$ output $+$ generation $-$ consumption $=$ accumulation*

| input/output: | $V_a I_c$ |
|---|---|
| generation: | |
| consumption: | $I_c^2 R_c$ |
| accumulation: | |

$$\boxed{V_a I_c} \quad - \quad \boxed{I_c^2 R_c} \quad = \quad \boxed{0}$$

*input/output*              *consumption*              *change in accum.*

---

$$V_a I_c - I_c^2 R_c = 0$$

Accounting the electrical energy in $R_s$, we find that

---

**Electrical Energy Accounting**

system[$R_s$ **f.1.26**]      time period [**differential**]

*input $-$ output $+$ generation $-$ consumption $=$ accumulation*

| input/output: | $V_a I_s$ |
|---|---|
| generation: | |
| consumption: | $I_s^2 R_s$ |
| accumulation: | |

$$\boxed{V_a I_s} \quad - \quad \boxed{I_s^2 R_s} \quad = \quad \boxed{0}$$

*input/output*              *consumption*              *change in accum.*

---

$$V_a I_s - R_s I_s^2 = 0$$

Accounting charge at node $a$, we find that

## Conservation of Charge

system[**node a f.1.26**]  time period [**differential**]

$input - output = accumulation$

| input/output: | $I_b - I_s - I_c$ |
|---|---|
| **accumulation:** | |

| $I_b - I_s - I_c$ | $=$ | $0$ |
|---|---|---|
| *input/output* | | *change in accum.* |

$$I_b - I_s - I_c = 0$$

Combining these equations, it can be determined that

$$V_a = \frac{12R_cR_s}{R_cR_o + R_cR_s + R_oR_s} \tag{1.8}$$

What we want to know is the percent error between what Matt will measure and the actual voltage at node $a$. To determine this, we subtract equation 1.7 from equation 1.8, divide by equation 1.7, and multiply by hundred.

$$\% \text{ error} = 100\frac{\text{ideal} - \text{actual}}{\text{actual}}$$

$$\% \text{ error} = -\frac{100R_cR_o}{R_sR_c + R_sR_o + R_cR_o}$$

To simplify the analysis, we let $R_s = kR_o$. Thus,

$$\% \text{ error} = -\frac{100}{k\left(1 + \frac{R_o}{R_c}\right) + 1}$$

Now notice, as $k$ gets large this expression approaches zero. Matt will measure the true voltage at node $a$ when the input impedance of the meter is large. As $k$ approaches zero, however, the error approaches 100%. In other words, if the input impedance of the meter is small, then Matt's measurement will be completely wrong.

**End 1.15**

The point of the last example is the following. If we have two systems that are not connected and then connect the systems, the systems will disturb the behavior of each other in some way. The amount of disturbance depends on the input/output impedances of the connected systems.

Not all systems exhibit constant responses. Figure 1.27 shows the output torque produced by an electrical motor. Figure 1.28 shows the torque requirements of the load connected to the motor. If we connect these two systems without applying power to the motor, they will have no motion. If we then suddenly apply power to the motor, it will initially have zero (or very small) speed and a large torque. At the same time, the load demands low torque for the low speed. Thus, there is a torque imbalance between the systems. What happens? The extra torque goes to increase the speed of the motor/load combination. The motor and load are not at their operating point. They are exhibiting a transient response.

**Example 1.16**

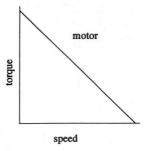

Figure 1.27:

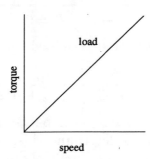

Figure 1.28:

## Multiple Operating Points

**QUESTION:**

Figure 1.29 shows the impedance graphs of a source/load system plotted together. The driver's output impedance is similar to that of a typical AC motor. The torque increases from zero (at low speeds) through a maximum (at intermediate speeds) and returns to zero (at high speeds). The load is similar to a frictional resistance. Its torque increases from zero (at low speeds) until it reaches a constant value (at higher speeds). Locate the system's operating points and discuss what will happen if the load deviates from the operating point.

**ANSWER:**

Using figure 1.29, we see that the system has two operating points (if we ignore the trivial operating point at zero effort and flow). The first operating point occurs at a speed of 0.5 and the second operating point occurs at a speed of 2.6.

The first operating point is unstable. If the system were operating at this point

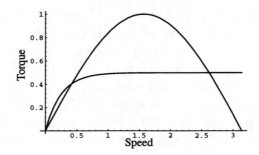

Figure 1.29: An AC Motor Connected to a Load

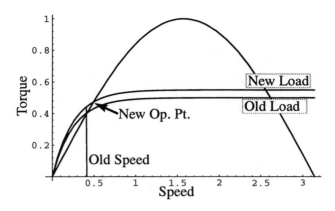

Figure 1.30: Effect of an Increased Load

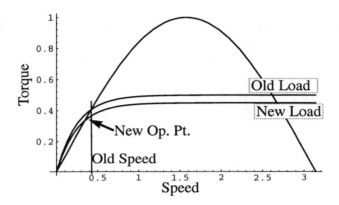

Figure 1.31: Effect of a Decreased Load

and something caused the load to change, the system would no longer operate at this point. For example, if we increase the load on the system (figure 1.30), the motor will slow down to zero. This occurs because the operating speed of the motor does not have enough torque to push the new load. Therefore, the motor slows down. As the motor slows, the torque that the motor provides gets even further from what the new load requires. Thus, the motor continues to slow down until it stops completely. The result is similar to riding a bike uphill. If we use all our effort to make it up the hill and the hill gets steeper (the load increases), our bike speed will begin to drop until it stops completely. The steady state for the bicycle is zero speed. The time period while the bike is slowing is the transient.

On the other hand, if we decrease the load on the system (figure 1.31), the motor will speed up until it reaches the second operating point. This occurs because the operating speed of the motor produces more torque than the load requires. This extra torque causes the motor to speed up. As the motor speeds up, the torque that the motor provides increases. Thus, the motor continues to speed up until the system reaches the second operating point.

The motor stops at the second operating point because at speeds greater than the second operating point the motor does not have enough torque to drive the load and at speeds lower than the second operating point a greater than required torque availability causes the motor to speed up. This is similar to riding a bike uphill when the grade drops. If we continue cycling with the same effort, the bike speeds up.

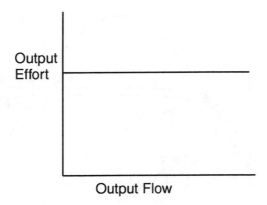

Figure 1.32: An Ideal Effort Source

The right operating point is stable.  Even if we do not start the driver exactly at the right speed, the driver will eventually *return* [26] to the stable operating speed. This happens because at speeds slightly slower than the stable operating speed the driver provides more torque than is necessary to drive the load.  This extra torque increases the driver's speed until it reaches the operating point.  Likewise, at speeds greater than the stable operating speed the driver does not provide enough torque to drive the load.  As a result, the driver slows down until enough torque is available (at the stable operating point).

Describe how we can start the system in this example from rest and so that it operates continuously.

**End 1.16**

Input and output impedances can be extremely helpful when we are trying to determine a system's extent.  Impedance is useful because it provides information about the extent to which two systems interact. As the input impedance of a system increases and the output impedance of its connected pair decreases, the two systems approach isolation. Isolation means that the systems' behavior when interconnected is the same as when separated.

When dealing with systems in isolation, a typical approximation is to assume that a source is ideal.  There are two types of ideal sources – effort and flow. Figure 1.32 shows an ideal effort source. Figure 1.33 shows an ideal flow source.

If an ideal source is connected to a load, the effect of the load on the source is negligible.  We can determine the load's behavior without worrying about what happens to the source.[27]  An ideal effort source produces a known effort (voltage or force) regardless of the load attached to it. Likewise, an ideal flow produces a known flow (current or speed) regardless of the load attached to it.

A typical approximation that we make when determining impedances is to assume that certain loads are ideal.  An ideal effort load has an infinite input impedance. This means it will not absorb any flow from the source (*e.g.* a forcemeter or voltmeter). An ideal flow load has a zero input impedance (*e.g.* a speedometer or ammeter). Just as in ideal source assumptions, when an ideal load is connected to a system, the behavior of the system is independent of the load.

---

[26] In practice the driver will not slow down or speed up exactly to the stable operating speed.  It will more likely overshoot this speed and its torque will push it back towards the operating speed again in a continuous process with the overshoot becoming less and less each time until finally the device operates at the operating speed.

[27] If the system is isolated, its behavior is the same regardless of whether it is connected to another system or not.

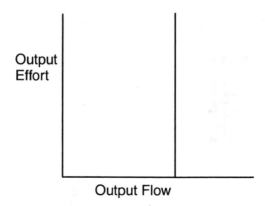

Figure 1.33: An Ideal Flow Source

The following example demonstrates how a system impedance can be adjusted to maximize power delivery.

**Example 1.17** —————————————————————————————————————

### Matching A Motor to a Load

After we formulate a solution to a system's behavior, we have to solve the formulation. For most of the problems in this course, we generate systems of algebraic and differential equations, which can be difficult to solve. This example shows how to use a general equation solver to solve one of these formulations.

**PROBLEM:**

GRAPE JUICE Incorporated (a producer of grape juice) needs a machine that will produce juice from grapes. They have an old electric motor with a torque versus speed curve that resembles a quarter cosine function. The stalling torque of the motor (torque at zero speed) is $\tau$. The no-load speed of the motor (speed at zero torque) is $\omega_o$ [radians/second]. Experimental data indicates that the force required to squeeze grapes is directly proportional to the speed with which the machine smashes them.

Design a machine for GRAPE JUICE that maximizes the amount of grapes that can be juiced with the given motor. We have the ability to adjust the constant of proportionality during the design process by changing the size of the exit orifice of the squashing unit.

**FORMULATION:**

We begin by making the assumption that the power input that we apply to the juicer goes totally toward producing juice. Therefore, it follows that the amount of juice produced increases with the power input. Thus, we can meet the design objective (of maximizing the amount of grapes juiced) by maximizing the power delivered to the juicer.

Figure 1.34 shows the torque curve for the motor and figure 1.35 shows the torque speed curve for a generic juicer. When we use the motor to drive the juicer, the operating point will occur when the delivered torque and speed of the motor exactly balance the requirements of the juicer. This point can be found by superimposing the two torque speed curves onto each other as figure 1.36 shows. To maximize the power delivered to the juicer, we have to place the operating point such that the power (torque times speed) is maximum. To do this we have to find a value of $k$ (through proper design of the juicer) that produces a maximum $T\omega$ at the operating

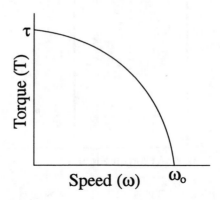

Figure 1.34: The Torque Speed Curve for the Motor

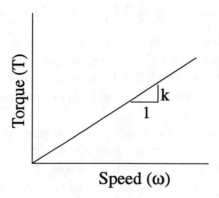

Figure 1.35: The Torque Speed Curve for the Juicer

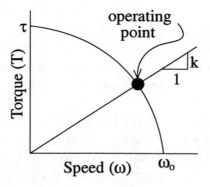

Figure 1.36: The Operating Point of the Motor/Juicer Pair

point. To determine the operating point, we must solve for the intersection of the two torque/speed curves. The mathematical relation for the juicer curve is

$$T_{\text{juicer}} = k\omega_{\text{juicer}}$$

The mathematical relation for the motor curve is

$$T_{\text{motor}} = \tau \cos(\frac{\omega\pi}{2\omega_o})$$

At the operating point, the torque and speed of the juicer equal the torque and speed of the motor, therefore the operating point is

$$k\omega = \tau \cos\left(\frac{\pi\omega}{2\omega_o}\right) \tag{1.9}$$

This equation defines the value of $\omega$ at the operating point in terms of $k$ and the motor parameters $\tau$ and $\omega_o$. Once we find $\omega$, we can express power as follows:

$$P = T_{\text{juicer}}\omega_{\text{juicer}} = k\omega_{\text{juicer}}\omega_{\text{juicer}} = k\omega^2 \tag{1.10}$$

Now if we can solve equation 1.9 for $\omega$. We can substitute $\omega$ into equation 1.10 and maximize power by differentiating equation 1.10 and setting the result to equal to zero.

A significant problem with the approach described is that there are many variables in the equation for power. To reduce these, we will write the equation in a dimensionless form. To accomplish this we will "hide" $\omega_o$ by defining the dimensionless variable $\Omega = \frac{\omega}{\omega_o}$. To "hide" $\tau$, define $\mathcal{T} = \frac{T}{\tau}$. By substituting these into the equations for the juicer and motor:

$$\frac{T_{\text{juicer}}}{\tau} = \mathcal{T}_{\text{juicer}} = \left(\frac{k\omega_o}{\tau}\right)\frac{\omega_{\text{juicer}}}{\omega_o} = K\Omega_{\text{juicer}}$$

Note that we have implicitly defined dimensionless constant $K = \frac{k\omega_o}{\tau}$. Continuing for the motor:

$$\frac{T_{\text{motor}}}{\tau} = \mathcal{T}_{\text{motor}} = \cos\left(\frac{\pi}{2}\frac{\omega}{\omega_o}\right) = \cos\left(\frac{\pi}{2}\Omega_{\text{motor}}\right)$$

Now at the operating point $\mathcal{T}_{\text{motor}} = \mathcal{T}_{\text{juicer}} = \mathcal{T}$ and $\Omega_{\text{motor}} = \Omega_{\text{juicer}} = \Omega$ so:

$$K\Omega = \cos\left(\frac{\pi}{2}\Omega\right) \tag{1.11}$$

$$\frac{P}{\tau\omega_o} = P' = \frac{k\omega_o}{\tau}\Omega^2 = K\Omega^2$$

Note that we have again defined $P' = \frac{P}{\tau\omega_o}$. What we will do now is maximize:

$$P' = K\Omega^2 \tag{1.12}$$

while making sure equation 1.11 is satisfied.

**SOLUTION:**

The solution to this problem was performed with Maple program 1.8.1. First use equation 1.11 to eliminate $K$ from the power (equation 1.12) to find:

$$P' = \Omega \cos\left(\frac{1}{2}\pi\Omega\right) \tag{1.13}$$

Now differentiate this with respect to $\Omega$ to find:

$$\frac{dP'}{d\Omega} = -\frac{\sin(\frac{\pi\Omega}{2})\pi\Omega}{2} + \cos(\frac{\pi\Omega}{2})$$

If we numerically solve for $\Omega$ when this derivative is zero, we find:

$$\Omega = 0.5477$$

with this it follows that:

$$\omega = 0.5477\omega_o$$

Using this in the power, we find:

$$P' = 0.3572$$

so:

$$P = 0.3572\tau\omega_o$$

Substituting these into the constraint equation gives:

$$K = 1.1908$$

and:

$$k = \frac{1.1908\tau}{\omega_o}$$

**End 1.17**  ———————————————————————————

# 1.3   INITIAL AND BOUNDARY CONDITIONS

**Objectives**
*When you complete this section, you should be able to*

1. *Recognize and distinguish between initial and boundary conditions.*

2. *Write simple initial and boundary conditions.*

———————————————————————————

This section discusses the difference between initial conditions and boundary conditions. Many of the processes that we deal with in engineering practice involve systems whose outputs (responses) change with time. Many of these systems can be predicted or modeled with equations. The equations, however, generally predict the response to within an arbitrary constant. We often can determine these arbitrary constants from initial conditions. Initial conditions are values of the response variable (or of a function of the response variable) that we know at some time, $t$. The following concept box discusses initial conditions.

**Concept 1.2**  ———————————————————————————

### Initial Conditions

The fluid filled tank in figure 1.37 has an orifice at its bottom that allows fluid to exit. If we want to determine the height of the liquid in the tank after three minutes, we will have to know the initial level of fluid in the tank. The accounting and conservation

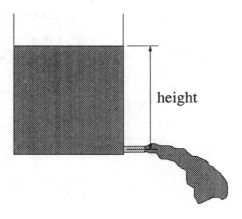

Figure 1.37: A Fluid Filled Tank

laws reflect this because we account for the initial amount of a quantity. Although we often define initial conditions at time zero, this is not a necessity. Generally speaking, we will need the initial condition of every important quantity that our system accumulates.

**End 1.2** ──────────────────────────────────

Boundary conditions are analogous to initial conditions. The difference is that a boundary condition specifies variables as a function of something other than time. The following concept box discusses boundary conditions,

**Concept 1.3** ──────────────────────────────

### Boundary Conditions

We have been asked to determine the temperature distribution in the wall in figure 1.38. The left surface of the wall has a constant elevated temperature maintained with an electric heating unit[28]. The right side of the wall has a lower temperature. Because of this temperature gradient, convective heat transfer (the heat transferred from one molecule to the next) will transport heat from the hot wall to the cold wall. The actual temperature distribution at all points inside the wall is dependent on the surface temperatures. These surface temperatures are boundary conditions.

**End 1.3** ──────────────────────────────────

Although we often specify boundary conditions as a function of position, this is not a necessity. For example, initial conditions can be thought of as boundary conditions where the boundary is with respect to time rather than position.

For most of the problems in this course, initial conditions will be very important whereas boundary conditions will be less important. The reason for this is that the processes we analyze are generally discrete in position and continuous in time. Therefore, it makes sense to have conditions specified for *boundaries* of time and not position.

───────────────────────

[28]We will see in later chapters that this can simply be a cluster of resistors connected to a voltage source.

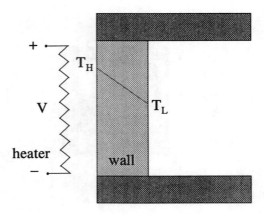

Figure 1.38: A Thick Wall with an Internal Temperature Distribution

# 1.4   NUMBER OF INDEPENDENT FLOWS [NIF]

**Objectives**

*When you complete this section, you should be able to*

1. *Identify flow variables that are related through differentiation.*

2. *Explain what it means to know a variable for all time.*

3. *Identify dependent flow variables.*

4. *Distinguish between independent and dependent flow variables.*

5. *Determine the [NIF] of a system.*

6. *Count equations and unknowns to determine if a model is complete.*

---

This section discusses the concept of independent flows [NIF]. When modeling the behavior of a system, it can be difficult to know when all the necessary equations have been written and if these equations are correct. To ease this difficulty, we employ preliminary analysis techniques. These techniques provide an indication of what to expect from the conservation laws. Thus, they make it easier for us to determine when we have written enough applicable equations.

Determining the number of independent flows [NIF] of a system is a preliminary analysis technique. A system's number of independent flows [NIF] indicates the minimum number of flow histories (*e.g.* flows as a function of time) needed to model the system's behavior. This means that if we know the time history of at least a minimum number of flow variables, we can determine any other flow.[29] For example, a mass moving in one direction has only one [NIF]. Thus, all we need to describe its

---

[29]Because of the relationships that exist between effort and flow variables, a similar concept involving the number of independent efforts is also possible. However, this is not widely used by engineers so we will not discuss it further.

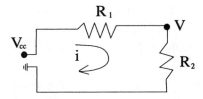

Figure 1.39: A One [NIF] Electrical System

behavior is a single flow variable[30] (*e.g.* $x$ as a function of time). To find momentum, we simply differentiate $x$[31] and multiply it with mass. Similar arguments apply to all the other terms applicable to the behavior of the mass.

## Example 1.18

### An Electrical One [NIF] System

**PROBLEM:**

Formulate a mathematical model of the resistive circuit in figure 1.39.

**FORMULATION:**

The first step in solving this problem would normally be to determine the [NIF] of the system. However, the problem already tells us that the system is one [NIF]. Therefore, we can proceed with the formulation process. The [NIF] of the system tells us that we only need one flow variable to model this system.

Accounting electrical energy for the complete system in figure 1.39 gives

**Electrical Energy   Accounting**

system[ **circuit f.1.39** ]      time period [**differential**]

*input − output + generation − consumption = accumulation*

| input/output: | $iV_{cc}$ |
|---|---|
| consumption (thermal): | $i^2(R_1 + R_2)$ |
| generation: | |
| accumulation: | 0 |

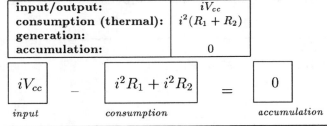

$$iV_{cc} - i^2R_1 - i^2R_2 - \overbrace{(0)i}^{\text{ground}} = 0$$

which reduces to

$$V_{cc} - i\left(R_1 + R_2\right) = 0 \qquad (1.14)$$

Equation 1.14 is an algebraic expression for the single variable $i$ in terms of the input $V_{cc}$ and the parameters $R_1$ and $R_2$.

## End 1.18

Reevaluating the last example, we find that a single algebraic equation is sufficient to determine the current $i$ because the conservation equation has only one unknown.[32]

---

[30]This ignores heating and other environmentally related effects.

[31]Since we know the history of $x$, we can differentiate it.

[32]Its important that you understand that only $i$ is unknown. $R_1$ and $R_2$ are parameters so we assume they are known.

However, if we want to find the voltage $V$ in the circuit we would need another equation.[33] This expresses the fact that the [NIF] of a system is merely an indication of the minimum number of flow variables required to understand a system not the minimum number of equations.

To determine [NIF], we analyze the *maximum* number of independent flows that *can* occur within a system. Each engineering discipline has its own mathematical methods for computing [NIF] and they have different names for this concept. Degree of freedom is the term used most often. However, it is easy to confuse the term degree of freedom because it means something entirely different in thermodynamics. This is why we use the acronym [NIF] instead of degree of freedom. Most electrical engineering textbooks discuss the concept of counting loop equations to determine the minimum number of equations needed to solve a circuit [36]. Counting loop equations is basically the same concept as finding [NIF]. Similarly, most mechanical engineering textbooks use several special methods to determine the [NIF] of complex systems [14]. Most of these special methods are specifically designed to handle a relatively small set of systems. In this text, we will concentrate mainly on systems that are well-behaved and will leave the special cases to more advanced classes.

An easy way to determine the [NIF] of a system is to determine the number of flows that have to be stopped until there is no longer a possibility for flow. If a system has the possibility for flow, [34] then it has at least one [NIF]. Now, imagine that we somehow stop this flow. We could prevent the motion of an object in a mechanical system or we could cut a wire in an electrical system. If another flow is still possible[35], the system has an additional [NIF]. By systematically repeating this process of stopping flows, counting, and looking for more until there are no more possible, we can find the [NIF] of a system.

As a preliminary analysis technique, [NIF] offers a prediction of what the model of a system's behavior should look like. Nevertheless, determining a system's [NIF] is not crucial to the analysis process. [NIF] is merely a tool to help us organize our thoughts.

**Example 1.19** ───────────────────────────────────────────

## [NIF] of Mechanical Systems

**QUESTION:**

Determine the [NIF] of the block in figure 1.40. Assume the block can move and rotate only in the plane of the page.

**ANSWER:**

Analyzing the [NIF] of the block, we see that the mass can move in the $x$ direction. Thus, we begin our count at one [NIF]. If we prevent the block's motion in the positive and negative $x$ direction (with our hand), the block can still move in the $y$ direction. This means there is an additional [NIF] – our count rises to two. If we prevent the block's motion in the $x$ and $y$ directions, it can move by rotating. Thus, the block has yet another [NIF] – our count rises to three. If we prevent the mass from rotating, it

---

[33] We need two equations to solve for $V$ because the electrical energy accounting result for $R_1$ that we use to find $V$ has two unknowns ($i$ and $V$). Two unknowns means we need two equations to solve the system. Thus, we need equation 1.14 and the energy equation for the resistor ($V_{cc} - V = iR_1$) to find this voltage.

[34] This could be a velocity in mechanical systems or a current in electrical systems.

[35] It is important to distinguish possible flows from actual flows. A car traveling straight down a road has one actual flow and three possible flows. To determine [NIF], we count the number of possible flows.

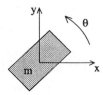

Figure 1.40: A Single Mass Moving Freely on a Plane

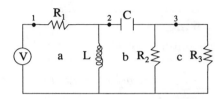

Figure 1.41: A Three [NIF] Electrical Circuit

cannot move – our count ends at three [NIF].

**End 1.19** ⎯⎯⎯⎯⎯⎯⎯⎯⎯⎯⎯⎯⎯⎯⎯⎯⎯⎯⎯⎯⎯⎯⎯⎯⎯⎯⎯

**Example 1.20** ⎯⎯⎯⎯⎯⎯⎯⎯⎯⎯⎯⎯⎯⎯⎯⎯⎯⎯⎯⎯⎯⎯⎯⎯⎯

### [NIF] of Electrical Systems

**QUESTION:**

Determine the [NIF] of

1. the circuit in figure 1.39, and

2. the circuit in figure 1.41.

**ANSWER:**

To determine the [NIF] of these circuits, we count the minimum number of wires that we would have to cut to eliminate all the closed loops in the circuit.

1. **the circuit in figure 1.39:** Current can flow through the circuit therefore it has at least one [NIF]. If we cut a wire somewhere in the circuit, there is no longer a path for current to flow. Thus, the circuit has only one [NIF].

2. **the circuit in figure 1.41:** Current can flow through the circuit therefore we begin with one [NIF]. If we cut the wire at location 1, current can still flow through loops b and c, therefore we have at least one more [NIF]. If we cut the wire at location 2, current can still flow in loop c. Since there are no storage elements in loop c, flow cannot actually occur. However, we are not concerned with whether or not there will be a flow, only if there is a loop through which a flow *could* occur. Our count rises to three [NIF]. If we cut the wire at 3, no loops remain. Flow can no longer occur. Our system has three [NIF].

**End 1.20** ⎯⎯⎯⎯⎯⎯⎯⎯⎯⎯⎯⎯⎯⎯⎯⎯⎯⎯⎯⎯⎯⎯⎯⎯⎯⎯⎯

**Example 1.21** ⎯⎯⎯⎯⎯⎯⎯⎯⎯⎯⎯⎯⎯⎯⎯⎯⎯⎯⎯⎯⎯⎯⎯⎯⎯

### Using Extra Variables

This example is a continuation of example 1.18. It discusses an alternative way to formulate the problem in this example.

**FORMULATION:**

We can formulate the mathematical model of the system in figure 1.39 using an alternative approach that uses more variables than the [NIF] count says the problem requires. The [NIF] of the system is one. This tells us that we need at least one flow variable to formulate the problem. Accounting electrical energy for the entire system, we find that

**Electrical Energy   Accounting**

system[ **circuit f.1.39** ]      time period [**differential**]

$input - output + generation - consumption = accumulation$

| input/output: | $iV_{cc}$ |
|---|---|
| consumption (thermal): | $i^2(R_1 + R_2)$ |
| generation: | |
| accumulation: | $0$ |

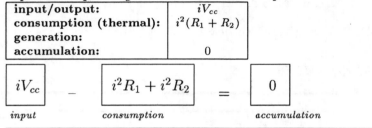

$$V_{cc}i - i^2(R_1 + R_2) = 0$$

If we assume that we know the parameters ($R_1$ and $R_2$) and the effort ($V_{cc}$), then there is only one unknown variable ($i$) in the algebraic equation. Since we have one equation and one unknown, we know that we can solve this equation for the unknown variable. Thus, we know our model is complete with this single equation. We do not need to express any other conservation equations.

However, we may be able to simplify our mathematical model if we apply the conservation laws to different parts of the circuit. Choosing the resistor $R_1$ as our system and accounting electrical energy gives

**Electrical Energy   Accounting**

system[ $R_1$ **f.1.39** ]      time period [**differential**]

$input - output + generation - consumption = accumulation$

| input/output: | $iV_{cc} - iV$ |
|---|---|
| consumption (thermal): | $i^2R_1)$ |
| generation: | |
| accumulation: | $0$ |

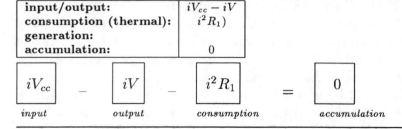

$$(V_{cc} - V)\,i - i^2R_1 = 0 \quad \overset{\text{reduces}}{\leadsto} \quad (V_{cc} - V) - iR_1 = 0$$

This equation contains two unknown variables, $V$ and $i$. Since we cannot solve one equation for two unknowns, we recognize that we need another equation. Choosing the second resistor as our system and accounting electrical energy, we obtain

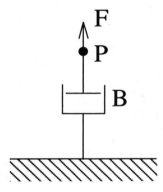

Figure 1.42: An Ideal Viscous Dashpot. Ideal means no thermal effects and no mass.

---

## Electrical Energy Accounting

system[ $R_2$ **f.1.39** ]     time period [**differential**]

$input - output + generation - consumption = accumulation$

| input/output: | $iV$ |
|---|---|
| consumption (thermal): | $i^2 R_2$) |
| generation: | |
| accumulation: | 0 |

$$\boxed{iV} \quad - \quad \boxed{i^2 R_2} \quad = \quad \boxed{0}$$

*input*        *consumption*        *accumulation*

---

$$iV - i^2 R_2 = 0 \quad \overset{\text{reduces}}{\rightsquigarrow} \quad V - iR_2 = 0$$

With this equation, we now have two equations that we can use to solve for the two unknowns.

Notice that if we leave the latter model as is, we require two equations and two variables. However, only one of the variables is a flow variable (as predicted by the [NIF] of the system).

**End 1.21**  ——————————————————————————

**Example 1.22**  ——————————————————————————

### A Mechanical One [NIF] System

**PROBLEM:**

Formulate a mathematical model of the ideal viscous dashpot in figure 1.42. An ideal dashpot dissipates only mechanical energy and does not accumulate energy of any kind.

**FORMULATION:**

The dashpot in the figure has one [NIF] because we can cause point P to have a velocity ($\vec{v}$) through the application of a force ($\vec{F}$). We can determine the relation between this force and velocity by accounting the mechanical energy of the dashpot.[36]

---

[36] $\vec{F} \cdot \vec{v}$ represents the power (the time derivative of work)($\frac{d}{dt}(W) \equiv \dot{W}$) associated with moving the damper. Work equals force times distance therefore power equals force times velocity. A damper of constant damping $B$ dissipates a power of $B\vec{v} \cdot \vec{v}$. We will discuss this in a later chapter.

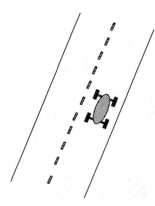

Figure 1.43: An Automobile Traveling in a Straight Path

---

**Mechanical Energy   Accounting**

system[ **dashpot f.1.42** ]      time period [**differential**]

*input − output = accumulation*

| input/output: | $\vec{F} \cdot \vec{v}$ |
|---|---|
| consumption (thermal): | $B\vec{v} \cdot \vec{v}$ |
| generation: | |
| accumulation: | 0 |

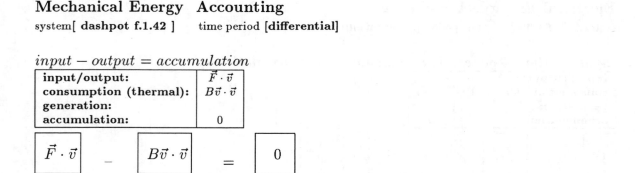

$$\vec{F} \cdot \vec{v} - B\vec{v} \cdot \vec{v} = 0 \tag{1.15}$$

As we expect, this is a single equation relating the input $\vec{F}$ to the single flow variable $\vec{v}$.

**End 1.22**

In addition to its importance as a preliminary analysis technique, the [NIF] of a system can also be useful in the design of new products. As an example, consider the design of a differential gear in an automobile.

**AN AUTOMOBILE DIFFERENTIAL GEAR**

To prevent excessive tire wear, and provide better controllability of an auto, it is necessary to make the tires roll without slip on the roadway. In rear wheel drive automobiles, a gear box connects the engine to the rear wheels. This enables the tires to push the auto forward. When the auto is traveling straight (see figure 1.43), the rear tires rotate identical amounts (assuming both tires have equal diameter) because the distance traveled by each tire is identical. When the auto turns, however, as depicted in figure 1.44, the outer tire must travel a greater distance than the inner. This means the outer tire must rotate faster than the inner one. If it doesn't, either slip will occur between one or more tires or the auto will not travel around the corner.

Since the two rear wheels must rotate at rates that vary depending on the radius of the corner, there must be two independent flows (two independent rotation rates of the tires). This will allow the two rear wheels to spin at different rates under

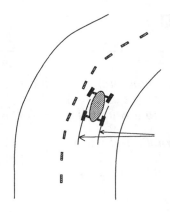

Figure 1.44: An Automobile Traveling in a Turn

a single engine speed. To achieve two independent flows in this manner requires a two [NIF] gearbox. Knowing that the problem requires two [NIF] eliminates a large number of gear boxes as not feasible. This helps to simplify a difficult design process.

Incidentally, because a conventional differential has two [NIF], it is easy to get stuck in mud. For example, if only one rear tire of a conventional differential auto sits on a slick surface, the auto can become stuck. Since counting the number of independent flows is equivalent to counting the independent efforts,[37] a two [NIF] gearbox allows two arbitrary torques to be applied. The two torques on the two [NIF] differential gear come from friction under the tire on the rough surface and the engine. The output is motion of the tire on the slick surface. If the gearbox were only one [NIF], only one flow (rotation) would be possible. Simultaneous engine rotation and zero tire rotation could not occur. Hence, with a single [NIF] differential gear, the engine would force both rear tires to turn. The tire in mud would give no traction, but the tire on the hard surface would push normally.

When an automobile with a conventional two [NIF] differential is traveling normally, its two inputs consist of the engine rotation (engine torque) and an unbalanced torque coming from friction between the roadway and the right and left tires. As the auto takes a corner, friction under the inner tire causes it to rotate slower than the outer tire. The differential allows this slower rotation because it has two [NIF].

---

[37]Perhaps the easiest way to recognize this is to realize that there would be an effort causing every flow. Hence, if there can be $n$ arbitrary flows, there must be the ability to apply $n$ arbitrary efforts.

## 1.5  SYSTEM STATE AND [NIA]

**Objectives**

*When you complete this section, you should be able to*

1. *Explain what it means to know a variable at one instant in time.*

2. *Explain what knowing a single history means.*

3. *Explain how a system accumulates properties.*

4. *Identify dependent accumulations.*

5. *Distinguish between independent and dependent accumulations.*

6. *Determine the [NIA] of a system.*

7. *Define suitable state variables for a system.*

8. *Explain the relationship between [NIA] and differential equations.*

9. *Explain the relationship between [NIA] and initial conditions.*

10. *Distinguish between [NIF] and NIA.*

---

This section explains how to determine the number of independent accumulations [NIA] of a state determined system. It also describes how to determine the number of state variables required to model a state determined system.

The accumulation term in the general accounting and conservation equations represents the change in the amount of a property inside a system over a given time period. For example, the accumulation term for the conservation of mass equation in rate form is $\frac{d}{dt}(m)$. This term describes the rate of change of the system's mass. When the conservation of mass is applied over a finite time period, the accumulation term is $m_{\text{end}} - m_{\text{begin}}$. This term describes how the system's mass changed during the time period.

An accumulation term can be positive, negative, or zero. This means that a system's properties can increase, decrease, or remain the same, respectively. For a system with multiple accumulations, it is possible for some of the accumulations to be related. These are called dependent accumulations. If the accumulations are not related, then they are independent.

To solve an accounting or conservation equation with non-zero independent accumulations of property $P$, we have to know how much $P$ the system contained at some specific time. For example, to solve the finite form of the conservation of mass equation, we have to know $m_{\text{end}}$ or $m_{\text{begin}}$. If it is not clear to you why we need to know the amount of a property contained in a system at some time in order to solve an accounting or conservation equation, consider the following example. There is a bag of marbles sitting on a table. I remove five red marbles from the bag and put six green ones into the bag. How many marbles are in the bag? You cannot answer this question unless I tell you how many marbles were in the bag initially or finally.

To solve the rate form of the equation, we have to know the system's mass at some specific time. In general, we need one value of a property at a specific time for each independent accumulation of the property. Engineers sometimes call knowing a system's property at some time knowing a single piece of history about the system. Knowing a single piece of history is not the same as knowing the complete history of a variable. When we know the complete history, we know the variable over all time. Therefore, can differentiate and integrate the variable. When we only know a single piece of history, we cannot differentiate or integrate the variable.

The variables used to quantify accumulated properties are called state variables. A system can have many state variables. However, most systems have a maximum number of independent state variables. Systems that can be modeled with a finite number of independent state variables are called state determined systems.

We define the [NIA] of a system as the minimum number of state variables required to model the system. For example, if the system requires one state variable, its [NIA] is one. If the system requires two state variables, its [NIA] is two *et cetera*. Because of the relationship between state variables and differential equations, the [NIA] of a system equals the minimum number of first order differential equations required to model the rate of change in the system. A system with three [NIA] requires at least

1. three first order differential equations, or

2. one second order and one first order differential equation, or

3. one third order differential equation

to describe its behavior.

It is possible to convert high order differential equations into a set of first order equations and since each first order differential equation requires one initial condition, the [NIA] also equals the minimum number of initial conditions that must be known to solve a system when it is modeled with the rate form of the conservation and accounting equations.

Like [NIF], [NIA] is a useful preliminary analysis technique. When we have trouble modeling a system, we can use the system's [NIA] as a way of organizing our thoughts. However, we do not have to determine the [NIA] of a system to correctly model its behavior. [NIA] is just a tool that often makes modeling a system's behavior easier.

What kind of information does a system's [NIA] provide? A system's [NIA] equals the minimum number of first order differential equations and initial conditions that we need to determine a system's behavior when modeled with the rate form of the conservation and accounting equations. Also, knowing a system's [NIA] enables us to make quick predictions about a system's response. This is especially beneficial to system's design. [NIA] gives us a quick way to check if we have all the accounting and conservation equations and initial conditions that we need to solve a system. [NIA] also tells us the minimum number of state variables that we need to model a system. Thus, a system's [NIA] gives us an indication of what we need to solve the equations that result from the accounting and conservation laws.

Even if the system is not modeled using a rate form, the [NIA]can help. For example, by determining the [NIA]we will be thinking about what quantities are important in the system response. These then become important terms to consider when the conservation/accounting equations are written in accumulation form.

How do we determine a system's [NIA]? We could write the conservation equations and count the number of differential equations that result to determine [NIA].

However, [NIA] is of greatest advantage when we use it before we apply the accounting/conservation laws. An easy way to determine the [NIA] of a system in advance is to count the number of independent accumulations in the system.[38]

**Example 1.23** _____

### State Variables in a Simple Hydraulic System

**PROBLEM:**

Formulate a mathematical model of the height of fluid in the tank in figure 1.45. We know (or can measure) all the parameters of the tank

1. area ($A$),

2. the flow constant ($C$), flow out is $C\sqrt{P}$, with $P = $ pressure,

3. density ($\rho$), and

4. gravity ($g$).

To demonstrate that we also need to know the history of a variable[39] to solve this problem, we ask the question, "How much fluid will be in the tank an hour from now?" What's the answer? No matter what we do, we cannot answer the question correctly *until* we know some history of a variable or an initial condition (*e.g.* an hour ago the volume was 10 [m³]). The problem could also specify an initial condition by telling us the present condition of the tank.

**FORMULATION:**

The first step in solving this problem is to determine the [NIF] of the system. We determine the [NIA] of the system by counting the *minimum* number of variables needed to express the important accumulations within the tank. The tank has the ability to accumulate

1. mass[40]

2. gravitational potential energy

3. the linear momentum of the water in the tank (provided the velocity of the fluid in the tank is significant)

4. kinetic energy (same comment as in momentum)

5. internal (thermal) energy (if we expect the temperature of the fluid to change significantly).

For simplicity, let's say that $A$ is very large, $C$ is very small (little flow), and that the fluid is at the same temperature as the surrounding air. Under these conditions, we anticipate that the momentum, kinetic energy, and internal energy accumulations will be small compared to the mass and potential energy accumulations. Therefore, the only important accumulations that we need to quantify are mass and potential energy. For the mass accumulation, we could choose mass itself as the state variable. If we did, our conservation of mass equation would appear as follows:

---

[38] This is the origin of the acronym [NIA]. The term used for [NIA] in most engineering disciplines is *order*. However, the term order can be confusing because not all disciplines define it in the same way. This is why we use the acronym.

[39] Remember initial condition is in the definition of a state variable.

[40] We must not think of accumulation as an increase in a substance. It is a *change* in the amount of the substance present in the system. Thus, an accumulation can be an increase or a decrease. For example, we should think of a bucket as *having the ability* to accumulate water regardless of whether water is actually entering or leaving.

**Analysis Assumptions**

(1) The tank is cylindrical and has a very large cross sectional area $A$.

(2) The tank's orifice constant $C$ is small.

(3) No temperature gradient exists between the fluid in the tank and the surrounding air.

(4) The fluid's momentum, $KE$, and internal energy accumulations are negligible.

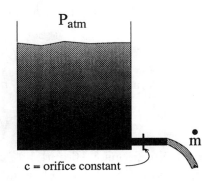

$c$ = orifice constant

Figure 1.45: (atankoffluid): A Simple Tank of Fluid

## Conservation of Mass

system[**tank f.1.45**]     time period [**differential**]

*input − output = accumulation*

| input/output: | mass flow | $-C\sqrt{P}$ |
|---|---|---|
| accumulation: | | $\frac{d}{dt}(m)$ |

$$\boxed{-C\sqrt{P}} \quad = \quad \boxed{\frac{d}{dt}(m)}$$

*output*          *change in accum.*

where $C$ has appropriate units. For this result to be of value, we will have to express the variable $P$ in terms of our state variable $m$. For simplicity, we assume that $A$ is so large and $C$ so small, so that the fluid in the tank drops very slowly and the fluid momentum is negligible. As a result, the fluid pressure is approximately what it would be for a motionless fluid. Therefore, conservation of linear momentum gives us

## Conservation of Linear Momentum (Vector)

system[ **tank f.1.45** ]     time period [ **differential**]

*input − output = accumulation*

| input/output: | mass flow | |
|---|---|---|
| | external forces | $PA$ |
| | gravity force | $mg$ |
| accumulation: | | 0 |

$$y: \quad \boxed{PA} \quad - \quad \boxed{mg} \quad = \quad \boxed{0}$$

*external*          *gravity*          *change in accum.*

$$PA \approx mg$$

Continuing, we obtain

$$-C\sqrt{\frac{mg}{A}} \approx \frac{d}{dt}(m) \tag{1.16}$$

To demonstrate that the choice of state variable is not unique, we could also choose the height of fluid in the tank, $h$, to quantify the accumulation of mass. If we do, we get

$$-C\sqrt{P} = \frac{d}{dt}(\rho A h)$$

This time we relate the pressure to $h$ and get that

$$-C\sqrt{\rho g h} \approx \frac{d}{dt}(\rho A h) \qquad (1.17)$$

Notice that equations 1.16 and 1.17 are identical to each other if we substitute $m = \rho A h$ [41] into equation 1.16. This indicates $h$ and $m$ are mutually satisfactory state variables.

## OTHER CONSIDERATIONS:

Now consider the gravity potential energy accumulation. We can express the potential energy as the product of the fluid weight and the position of the center of mass above the tank bottom, PE$= \frac{\rho A h^2}{2g}$.[42] Since we can express potential energy as a function of $h$, the same state variable used for mass accumulation, we need only one state variable to quantify both accumulations. In fact, if we chose to write the conservation of energy equation for the tank, we would obtain an equation identical to equation 1.17. Because the system requires a minimum of one state variable (based on our decision to throw away kinetic energy and heat accumulations), its [NIA] is one. Based on what we know about [NIA], we expect to obtain only one first order differential equation from the conservation principles.

**End 1.23** _____

Because we express accumulation in the conservation equations in terms of state variables, the minimum number of state variables we need to describe any system is equal to the number of independent accumulations (storage terms) within the system.[43] If it is possible to calculate several accumulated quantities given a single variable, these accumulations are dependent. We can describe dependent accumulations with a combination of other state variables. Not all accumulations in a system contribute to the [NIA]. [NIA] is a count of only *independent* accumulations. Thus, to accurately count [NIA] it is important for us to be able to tell whether a system's accumulations are dependent or independent.

### DEPENDENT ACCUMULATIONS

Dependent accumulations are those that we can describe with the same state variable. The previous example demonstrated two such accumulations. The systems in figure 1.46 all have some form of dependent accumulation elements.

1. The two capacitors accumulate electrical energy. The amount of energy stored is $\frac{1}{2}C_1 V^2$, and $\frac{1}{2}C_2 V^2$ [44] where $V$ is the voltage across the capacitors. Since the voltage across each capacitor is the same, knowing one state variable, $V$,

---

[41] m=(density)(volume)=$\rho V$ = p(area)(height) = $pAh$.

[42] This relation for potential energy is derived from PE= $mgh_{cm} = (\rho V)g(h/2) = (\rho A h)g(h/2) = \rho A h^2/2g$.

[43] For example, a tank of fluid can accumulate fluid. A mass can accumulate kinetic energy and momentum. A mechanical spring can accumulate potential energy. An electrical inductor and capacitor accumulate current, charge, and electrical energy.

[44] The energy stored in a capacitor is equal to the $\frac{1}{2}CV^2$. We discuss this in depth in the chapter on electrical systems.

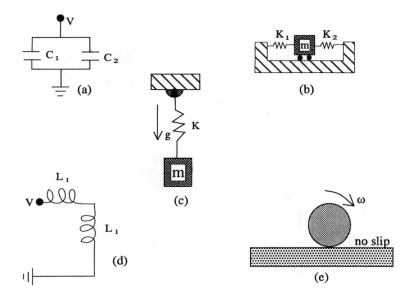

Figure 1.46: Systems with Dependent Energy Storage Elements

allows us to calculate both accumulations. Likewise, the capacitors store charge. Since the stored charge is proportional to $V$, the charge stored and energy are dependent. The system's [NIA] is one. There is one state variable, which can be chosen as $V$.

2. The potential energy stored in each of the two springs is dependent. If we know the deflection in the left spring, we know the deflection in the right spring (as long as we know all of the parameters).[45] We can express the deflection of the springs in terms of the position of the mass, $x$.[46] The mass accumulates both linear momentum and kinetic energy. We can express both of these accumulations using the same variable (the horizontal velocity of the mass$\equiv \dot{x}$). We express the kinetic energy as $\frac{1}{2}m\dot{x}^2$ and the linear momentum as $m\dot{x}$. The state variables in these relations are $x$ and $\dot{x}$. $x$ and $\dot{x}$ are independent because even if we knew $\dot{x}$ for all time, and we could not integrate it with respect to time to find $x$. Instead, we would obtain: $x = \int \dot{x}\,dt + \mathcal{C}_1$ where $\mathcal{C}_\infty$ is an undetermined constant (an initial condition). Knowing $\dot{x}$ does not exclusively allow us to determine $x$. We have to have an extra piece of history (*i.e.* the initial condition of $x$). Thus, the variables $x$ and $\dot{x}$ are independent and therefore so are the accumulations they describe. The system's [NIA] is two. One [NIA] from the kinetic energy (and momentum) in the mass and one from potential energy in the springs. If the mass was storing a significant amount of thermal energy, we would need another variable to describe the thermal energy accumulation, and the [NIA] would be three.

---

[45] The parameters of the springs include the free lengths of the springs and the spring constants.

[46] This is sometimes a point of contention with bond graph methods [31]. Bond graph methods define order (*i.e.* [NIA]) as the number of *independent* energies in a system. The freelengths of springs are not considered to be known. Thus, the spring potentials of this example would depend on unknown freelengths and could be independent. In this text, we are concerned with the more general methodology the conservation principles. As a result, we define [NIA] slightly different than bond graphs methods. Keep in mind that the reason we determine [NIA] is to predict the results of our modeling process. As a result, the slight deviation from specific ideas developed for specific problems is inconsequential, provided we are consistent with the use of our definitions.

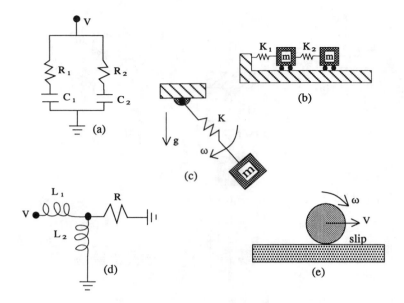

Figure 1.47: Systems with Independent Energy Storage Elements

3. The spring accumulates potential energy and the mass accumulates momentum, kinetic energy and gravity potential energy. The gravity and spring potential energies are dependent. Knowing one (and the system parameters like free length of the spring) allows us to determine the other. The kinetic energy and linear momentum are also dependent, assuming the mass only moves vertically. The system's [NIA] is two.

4. The two inductors accumulate energy. The energy stored in the inductors is $\frac{1}{2}L_1 i^2$ and $\frac{1}{2}L_2 i^2$. Knowing the single state variable (the flow $i$) determines all the accumulations. The system's [NIA] is one.

5. The mass accumulates both translational and rotational kinetic energies and linear and angular momentum. Describing the energy requires two variables such as $\dot{x}$ and $\omega$ (rotation rate). We can quantify the momentums with the same two variables. Since there is no slip between the disk and the ground, we can relate these two variables as $|\dot{x}| = |r\omega|$ (we discuss the derivation of this relation later. Thus, the accumulations are dependent. The system's [NIA] is one.

## INDEPENDENT ACCUMULATIONS

Figure 1.47 shows systems, similar to those in figure 1.46. The difference is that we have slightly modified these systems to remove some of the dependent storage elements.

1. The addition of the resistors allows the voltages across each capacitor to be different from each other. Knowing the energy (or the charge) in one capacitor does not determine the energy (or charge) in the other because the voltages across the capacitors may differ. The system's [NIA] is two. Resistors do not accumulate anything. Thus, they do not enter into the calculation of [NIA] (other than the fact that they cause capacitor accumulations to be independent.

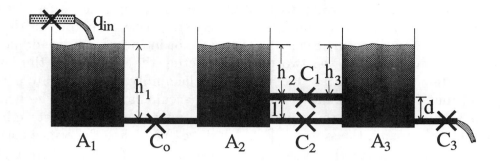

Figure 1.48: A Three Tank System

2. Knowing the deflection in the left spring does not determine the deflection in the right spring. The spring potential energy accumulations are independent. The system is fourth [NIA].

3. The mass can accumulate linear momentum in two directions, angular momentum as well as kinetic energy and gravity potential energy. Knowing the gravity potential does not determine the spring potential, but knowing the two linear momentums can determine the angular momentum and kinetic energies. The system's [NIA] is four.

4. Knowing the current through inductor 1 does not determine the current through inductor 2. The system's [NIA] is two.

5. Because there is slipping between the cylinder and floor, the translational and rotational speeds are not dependent (they can differ by an unknown constant). We cannot relate the linear momentum and angular momentum accumulations. The system's [NIA] is two.

**Example 1.24** _____

## A Three Tank Fluid System

### PROBLEM:

Formulate a mathematical model of the three tank system in figure 1.48. Let's choose the "system" to be the liquid in the three tanks themselves. Hence the fluid in the piping is NOT part of the system. We can make the following assumptions about the system:

1. The tanks can accumulate potential energy and mass. We can quantify these accumulations using the fluid height in the tanks. Actually, the tanks accumulate mass and the mass holds the potential energy.

2. The tanks have such large cross sectional areas that the fluid levels change very slowly. This allows us to ignore the momentum of fluid in each tank.

### ANSWER:
### FORMULATION:

The first step in solving this problem is to determine the [NIF] and [NIA] of the system. If we hold the fluid level constant (stop the flow) in the first tank, flow can still occur in the second or third. If we hold the level constant in the second (stop the flow), flow can still occur. If we hold the level constant (stop the flow) in the third tank, flow can no longer occur. Since the piping is not in the system, we do not care about possible circulation in the two pipes connecting the second and third tanks. Thus, the system's [NIF] is three. Each tank has the ability to accumulate potential energy and mass. We can quantify both of these accumulations with the height of fluid. Since knowing the height in one tank does not automatically determine the height in any other tank, the system's [NIA] is three. We define the orifice constants ($C_0$, $C_1$, $C_2$, and $C_3$) such that flow through the pipes is $C\sqrt{\Delta h}$, where $C$ is the constant for the pipe, and $\Delta h$ is the difference in fluid heights of the tanks connected by the pipe. This is equivalent to saying flow is proportional to the square root of the pressure drop across the pipe.

Let's develop a model of the system by expressing conservation of mass in each tank. Let's assume the orifice constants are different for flow to the left and right. After all, unless the orifice is symmetric, flowrate to the left and right will differ. Constants for flow to the left will be primed values. Constants for flow right will be unprimed. For tank 1, the mass flow rate in is $q_{\text{in}}$. The flow rate out, however, depends on the heights $h_1$ and $h_2$. If neither height is above orifice 0, there is no flow out and the conservation of mass for tank 1 is as follows:

**Conservation of Mass**

system[**tank 1 f.1.48**]      time period [**differential**]

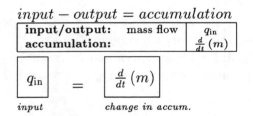

$input - output = accumulation$

| input/output: | mass flow | $q_{\text{in}}$ |
| accumulation: | | $\frac{d}{dt}(m)$ |

$$\boxed{q_{\text{in}}} \quad = \quad \boxed{\frac{d}{dt}(m)}$$

*input*              *change in accum.*

If $h_1$ rises above the orifice, the flow out is $C_0\sqrt{h_1}$. In this case, the conservation equation gives the following relation:

$$q_{\text{in}} - C_0\sqrt{h_1} = \frac{d}{dt}(m_1) = \frac{d}{dt}(\rho A_1 h_1)$$

Provided $h_2 > -l$, and $h_1$ is above the level in tank two, the flow through orifice 0 is $C_0\sqrt{h_1 - h_2 - l}$. If the level in tank two is above the level in tank one, the flow through the orifice is $C_0'\sqrt{h_2 + l - h_1}$. Depending on the height of the fluid in the tank, we use one of the following conservation equations.

$$q_{\text{in}} - C_0\sqrt{h_1 - h_2 - l} = \frac{d}{dt}(\rho A_1 h_1) \quad \text{if } h_1 > h_2 + l \qquad (1.18)$$

or

$$q_{\text{in}} + C_0'\sqrt{h_2 + l - h_1} = \frac{d}{dt}(\rho A_1 h_1) \quad \text{if } h_2 + l > h_1$$

When solving these equations, we must constantly check the fluid levels of the tanks. Because if the fluid levels covers or uncovers an inlet/outlet pipe, we must change the equations that we are using.

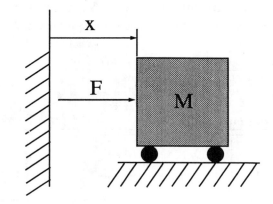

Analysis Assumptions

(1) No mass or charge enters or leaves the system.

(2) The only forces acting on the system are in the $x$ direction.

Figure 1.49: (singlemassonrollers): A Mass on Frictionless Rollers

The remaining tanks also have many variations depending on the fluid levels. For simplicity of the discussion however, lets assume $h_1 > h_2 + l > h_3 + d > d$. The conservation of mass for tank two becomes the following:

$$C_0\sqrt{h_1 - h_2 - l} - C_1\sqrt{h_2 - h_3} - C_2\sqrt{h_2 + l - h_3 - l} = \frac{d}{dt}(\rho A_2 h_2)$$

Finally, for the third tank, we get that

$$C_1\sqrt{h_2 - h_3} + C_2\sqrt{h_2 + l - h_3 - l} - C_3\sqrt{h_3 + d} = \frac{d}{dt}(\rho A_3 h_3)$$

Notice that one selection of state variables for this problem is $h_1$, $h_2$, and $h_3$ and that the system's model consists of three first order differential equations. This is consistent with the system's [NIA] of three. Also, note that there are only three flow variables () which is consistent with the system's [NIF] of three.

**End 1.24**

**Example 1.25**

## A Simple Mass on Rollers

**PROBLEM:**

Formulate a mathematical model of the mass $M$ in figure 1.49. The mass is mounted on frictionless rollers and subjected to an applied force ($F$).

**ANSWER:**

**FORMULATION:**

The first step in solving this problem is to determine the [NIA] of the system. The mass has the ability to accumulate linear momentum in the $x$ direction. We account this momentum using the state variable $\dot{x}$ (velocity). We can quantify the only other accumulation in the system (kinetic energy) using the same state variable $\dot{x}$. Thus, the system has one [NIA]. Applying conservation of linear momentum produces the following expression:

## Conservation of Linear Momentum (Vector)

system[**mass f.1.49**]        time period [**differential**]

$input - output = accumulation$

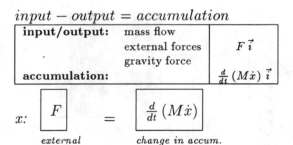

| input/output: | mass flow external forces gravity force | $F\,\vec{i}$ |
|---|---|---|
| accumulation: | | $\frac{d}{dt}\left(M\dot{x}\right)\,\vec{i}$ |

$$x: \quad \boxed{F} \quad = \quad \boxed{\frac{d}{dt}\left(M\dot{x}\right)}$$

$\quad\quad\quad$ *external* $\quad\quad\quad\quad$ *change in accum.*

$$F = M\ddot{x} \tag{1.19}$$

This looks like a second order equation describing the physics of the system, but we can perform a simple variable substitution and make it first order. For example, we can define $v = \dot{x}$. Using this in equation 1.19 gives us

$$F = M\dot{v} \tag{1.20}$$

This equation is clearly first order and completely describes the physics of the system. To solve for the position of the mass, we can use equation 1.20 and $\dot{x} = v$.

## OTHER CONSIDERATIONS:

This example emphasizes an important point. The mass has only one [NIA] because it only has one independent accumulation. As a result, the mass requires only one state variable to model its physics (its properties). We may choose to use higher order equations to model the system if more than its physics interests us, but this does not mean the system actually has a higher [NIA]. This simply means that we are interested in more than the physics of the system, and another equation is necessary to find what we want. In this example, the position of the mass does not quantify an independent accumulation. Therefore, we do not need $x$ to model the physics of the system.

**End 1.25**

A system's [NIF] and [NIA] are *not* the same. The [NIF] tells us the number of flow variables (as a function of time) that we will need to model the system. The [NIA] tells us the number of initial values (histories) we will need. As an example, consider the mass and spring system of figure 1.50. There is only one [NIF] in the system. If we knew $x$ as a *function of time*,[47] we could calculate all flow terms that would appear in either the energy or momentum equations. Notice that when counting the [NIF], we determine the number of functions of time of different variables that are needed.

Now think about [NIA]. [NIA] deals with initial values (histories). This means the minimum number of conditions that we need to know right now to quantify all the system's accumulations. For example, if we knew that at this moment $x = 3$ (this is not a function of time only a single value), we could calculate the potential energy accumulated in the spring.[48] However, if the only information that we know

---

[47]To know something as a function of time means that we know it at all times. For example, to know $x$ as a function of time means that we can plot the function $x$ versus time. To know $x$ as a function of time means that we also know (or can calculate) all of $x$'s derivatives.

[48]This assumes that we know the sizes of things (*e.g.* the freelengths of the springs *et cetera*).

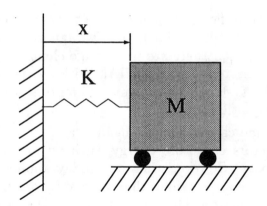

Figure 1.50: A Two [NIA] One [NIF] System

is that $x = 3$ right now, then there is no way for us to calculate the kinetic energy or momentum accumulated in the mass. To know these accumulations, we would need another initial condition (history) such as $\dot{x} = 0$ right now. Thus, the system's [NIA] is two.

Notice that we count $x$ and $\dot{x}$ as one [NIF] because if we knew $x$ as a function of time, we would also know (or could easily calculate) $\dot{x}$. We count $x$ and $\dot{x}$ as distinct [NIA] because if we only know one initial value ($x$) we would not automatically know $\dot{x}$.

Unlike [NIA], which indicates the number of first order differential equations needed for modeling a system, the [NIF] predicts differential and algebraic equations alike. Therefore, when a system's [NIF] is greater than its [NIA], their difference ([NIF]-[NIA]) equals the number of variables that are algebraic functions of the system's state variables, parameters, and known excitations (inputs). When the [NIA] is greater than its [NIF], the numerical difference ([NIA]- [NIF]) equals the number of state variables that are derivatives of other state variables. The state variables that are derivatives of other state variables cannot be directly modified by excitations applied to the system. They can be affected only through their derivatives.

**Example 1.26** ───────────────────────────

## [NIF] and [NIA]

**QUESTION:**
Determine the [NIF] and [NIA] of the systems in

1. example 1.18 (figure 1.39)

2. example 1.19 (figure 1.40)

3. example 1.22 (figure 1.42).

Then, based on their [NIF] and [NIA], discuss the number and type of equations (algebraic or differential) that we will need to model the systems.

**ANSWER:**

1. Example 1.18 (figure 1.39):
   The circuit has only one [NIF] because it has only one independent flow (if we

cut the circuit, no other flows are possible). The resistor cannot store any energy other than thermal (it can get hot), so if we ignore thermal energy (if we are not interested in it), then there are no storage elements in the circuit (thus no accumulation). The system has zero [NIA]. With a [NIA] of zero, we expect that we can describe the system without any differential equations. This is exactly what we did (see equation 1.14).

There is always some capacitance and inductance in any circuit and since capacitors and inductors can store energy, such an idealized zero [NIA] system will never really exist. However, there are many applications where the inductance and capacitance are so small that we can ignore them. Thus, we can use the model given in equation 1.14 to approximate the behavior of the circuit in figure 1.39 if we assume that the circuit has ideal components.

If we have a system with few (if any) accumulation terms, the [NIA] will be small. However, the system may still require several variables depending on the [NIF] of the system. These idealized systems often occur when we study the motion of negligible mass machines, the current flow in resistor circuits, and the fluid flow in pipes.

2. Example 1.19 (figure 1.40):
As we determined earlier, the system has three [NIF]. If the plane containing the motion is horizontal, there is no accumulation of gravity potential energy. With this assumption, the mass has three [NIA] because we need three variables to describe the linear and angular momentums accumulated by the mass. We can quantify the kinetic energy accumulation using the same three variables that we use to quantify these momentums. Thus, it does not introduce any additional [NIA].

The system's [NIF] and [NIA] tell us that we will need three first order differential equations (conservation of linear momentum in the x and y and the conservation of angular momentum about the z axis) and at least three flow variables $(\ddot{x}, \ddot{y}, \ddot{\theta})$ to model this system.

3. Example 1.22 (figure 1.42):
The system has only one [NIF] because it contains only one independent flow (if we hold the motion of the damper in one direction, no other motion is possible). Furthermore, since the system does not accumulate anything, it is zero [NIA]. We will see later that an ideal damper does not store anything except thermal energy which we choose to ignore here. This tells us that we only need algebraic equations and one flow variable to model the system's behavior. Notice from the results of the formulation process (see equation 1.15) that the model consists of a single equation relating the effort $F$ to the single flow variable $v$.[49]

**End 1.26** ──────────────────────────────────────────────

**Example 1.27** ──────────────────────────────────────────────

### [NIF] and [NIA] of a Mass and Damper

──────────

[49]This is exactly what the equation introduced earlier predicts. Since the [NIF] is greater than the [NIA], [NIF]-[NIA] equals the number of variables that are algebraic equations of the system's other variables.

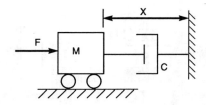

Figure 1.51: A Mass and Damper.

**PROBLEM:**
Figure 1.51 shows a mass connected to a damper. The damper (*e.g.* a shock absorber) connects the mass with the ground. We will discuss dampers in more detail later in the text, but for now all you need to know is that an ideal damper cannot store any energy. In fact, an ideal damper's sole purpose is to convert mechanical energy into thermal energy. Find the [NIF] and [NIA] for the mass and damper.

**FORMULATION:**
What is the [NIF]? If we assume the mass can only move left and right, then the [NIF] is one, and $x$ is a reasonable flow variable. What is the [NIA]? If we ignore any thermal effects, the only independent accumulations are the

1. linear momentum of the mass in the horizontal direction, and

2. the kinetic energy of the mass.

$\dot{x}$ quantifies both of these accumulations. Therefore, these accumulations are not independent. We only need a single state variable ($\dot{x}$) to quantify all the accumulations occurring within the system. The system's [NIA] is one.

**FORMULATION:**
If you are willing to accept that the amount of mechanical power converted by an ideal damper equals $\dot{E}_{\text{damper}} = C\dot{x}^2$, then we can derive an equation for predicting all the extensive properties for the system by accounting mechanical energy.

**Mechanical Energy Accounting**
system[**Mass and Damper**]      time period [**differential**]

*input − output + generation − consumption = accumulation*

| input/output: | mass flow | 0 |
|---|---|---|
|  | work | $-F\dot{x}$ |
| generation: |  | 0 |
| consumption: |  | $C\dot{x}^2$ |
| accumulation: |  | $\frac{1}{2}m\dot{x}^2$ |

$$\boxed{-F\dot{x}} \quad - \quad \boxed{C\dot{x}^2} \quad = \quad \boxed{\frac{d}{dt}\left(\frac{1}{2}m\dot{x}^2\right)}$$

*Power Out by Force*        *Lost by Conversion*         *Accumulation*

Taking the derivative gives us:

$$-F\dot{x} - C\dot{x}^2 = m\dot{x}\ddot{x}$$

which simplifies to:

$$-F - C\dot{x} = m\ddot{x} \tag{1.21}$$

**INTERPRETATION:**

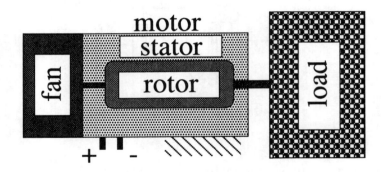

Figure 1.52: A Fan Cooled Permanent Magnet DC Motor

Is equation 1.21 really first order, as the [NIA] predicts it should be? No, not as written, but what if we substitute $v = \dot{x}$ into the equation. In this case,

$$-F - Cv = m\dot{v}$$

This equation is first order. Is it always possible to reduce the order of a differential equation? No, but since the [NIA] of this system was one, we knew that equation 1.21 could be reduced.

**End 1.27** ─────────────────────────────────────────────

**Example 1.28** ───────────────────────────────────────────

### A Fan Cooled DC Motor

**PROBLEM:**

Discuss the approach required to formulate a mathematical model of the permanent magnet DC motor in Figure 1.52. The motor spins a load and a fan that blows cooling air across the motor housing. Begin by determining the [NIF] and [NIA] of the motor. Then, discuss the parameters and variables required to model the motor.

A permanent magnet DC motor consists of a stationary magnetic field generated by permanent magnets, and a rotating magnetic field generated by electromagnetism. We call the stationary magnetic field the Stator and the rotating magnet the Rotor. Since an electromagnet is merely a coil of wire, it is often satisfactory to model the motor as simply a resistor and inductor in series. See figure 1.53. The resistance and inductance are not really separate. The resistance comes from the resistivity of the wire used for the electromagnet. It is uniformly distributed along the coil, but the effect of the distributed resistance is approximately equivalent to a lumped resistor (see figure 1.53).

The rotor consists of a shaft of metal with a coil of wire (the electromagnet) wrapped around it. Bearings at its two ends, mount the shaft in place. When the motor operates, the rotor rotates. The load can be anything we want the motor to turn and is normally rigidly connected to the motor's shaft. The load, in this example, is simply a mass that is sitting in an oil bath.[50] The cooling fan is also rigidly attached to the motor shaft and rotates with the motor. The motor's rotor turns when a voltage develops across the terminals of the electromagnet (a and b).

---

[50]This might occur for example in an oil pump.

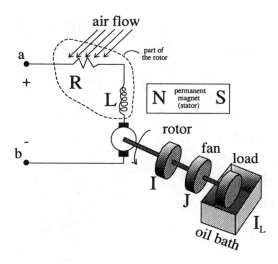

Figure 1.53: A Schematic Diagram of a DC Permanent Magnet Motor

## FORMULATION:

To formulate a model that describes the operation of this system, we begin by determining the [NIA] and [NIF] of the system.

The first step is to choose variables that are capable of defining the state of the system. Figure 1.54 shows some of these variables. After we define the necessary variables, we can search for the accumulation elements and determine which are independent. The inductor accumulates electrical energy and the rotor shaft accumulates both rotational kinetic energy and angular momentum. The fan and load accumulate kinetic energy and angular momentum as well. Since the motor coolant is air, it is safe to assume that it is heating up. We would probably want to count the storage of thermal energy by the rotor if this is true. Since temperature accounts thermal energy, we could define $T$ as the temperature of the motor. If we can treat all the heating components as if they are at a uniform temperature, a single temperature will suffice. However, if each component is at different temperatures, individual temperatures are necessary to account for each item's thermal energy. Consequently, the [NIA] would increase. If we assume the air flow is so large that we can ignore its energy storage capacity (we will in this problem), we need not count it.

Of these accumulations, the kinetic energy and angular momentum accumulated by the rotor, the fan, and the load are dependent. We can define these accumulations in terms of the same variable ($\dot{\theta}$). The energy accumulation in the inductor and the thermal energy accumulation in the rotor are additional independent accumulations that we can define in terms of $i$ (the motor current flow) and $T$ (the rotor temperature) respectively. Thus, the system has three [NIA]. Acceptable variables for describing the state of the system are $i$, $\dot{\theta}$, and $T$.

To determine the [NIF], we count the number of arbitrary flow variables. The load mass can rotate, so we start with one [NIF]. If we hold the load mass so it cannot rotate, none of the other masses can rotate either because they are rigidly connected to each other. Additionally, if the only air flow comes from the fan, the air flow rate is zero when we hold the load. Continuing with the load held, it is possible to pass a current through the circuit so there is now two [NIF]. If we cut the circuit to prevent

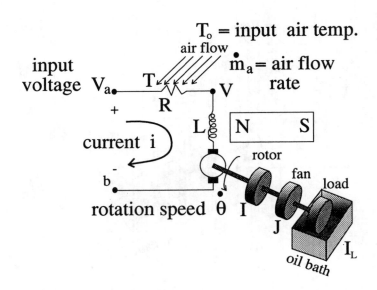

Figure 1.54: Some Suitable Variables for the Fan Cooled Motor

current from flowing, there is still the possibility for heat to flow from the hot rotor to the air. Thus, there are at least three [NIF]. If we prevent the heat from flowing, there is nothing else that can flow so we stop counting. The system has three [NIF].

There are other parameters and variables of interest in this problem. For example, we would most likely know the voltage applied to the electromagnet ($V_a$); the temperature of the ambient air ($T_o$); the resistance ($R$) and inductance ($L$) of the motor; the inertias of the rotor, fan, and load ($I$, $J$, $I_L$); and the damping of the oil bath ($b$).

**End 1.28** ───────────────────────────────────────────────

It is possible to use information about [NIF] to help calculate the [NIA]. We know that elements that accumulate properties have state variables. These state variables can be either an effort or a flow. For example, a capacitor accumulates electrical energy. The amount of energy stored in a capacitor is proportional to the voltage (effort variable) across the capacitor. An inductor also accumulates electrical energy. The amount of energy in an inductor is proportional to the current (flow variable) through the inductor. Since a system can only have [NIF] independent flows, it can only have [NIF] independent accumulations described by flow state variables. In addition to the [NIF] independent flow state variables, the system may also have several effort state variables.

**Example 1.29** ───────────────────────────────────────────

### Using [NIF] to Estimate [NIA]

**QUESTION:**
Determine the [NIF] and [NIA] of the four bar mechanism in figure 1.55.
**ANSWER:**
Each member (A, B, and C) of the mechanism has mass. We would like to know the [NIA] of the system. Member A accumulates angular momentum, gravitational potential energy, kinetic energy, and linear momentum. However, because it rotates about a fixed point, we can quantify all of these accumulations using the variables ($\theta$

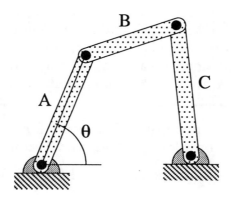

Figure 1.55: A Four Bar Mechanism (We Count Ground As a Bar)

and $\dot{\theta}$). Member C's accumulations like A's require an angle and its derivative. The question is are these accumulations independent? Member B's accumulations are not as obvious as A's and C's. Member B seems to move in a plane – both translating and rotating. How many independent accumulations are present there? The fourth bar of the mechanism is ground it has no accumulations. Notice that all the system's accumulations are related to flow variables.

To help alleviate the confusion, consider counting the number of [NIF] in the mechanism. First the mechanism can move so we begin with one [NIF]. Since member A has the ability to rotate let's hold it and prevent it from rotating. If A is held, neither member B nor C has the ability to move, unless the machine breaks apart which we assume does not happen. The system has only one [NIF]. Since there is only one [NIF], we can specify the motion of the entire system using a single motion variable such as $\dot{\theta}$. This implies that we can express the momentums (linear and angular) and kinetic energies of all three bodies as functions of a single variable. Likewise, we can calculate the position of all the members once we know a single position (such as $\theta$). Hence, the [NIA] must be two. The state variables can be $\theta$ and $\dot{\theta}$.

**End 1.29**

## 1.5.1 VARIABLES DETERMINED BY DIFFERENTIAL EQUATIONS

**Objectives**

*When you complete this section, you should be able to*

1. *Identify the variables of a differential equation can be determined when the equation is solved.*

2. *Determine whether or not there are enough equations to uniquely determine all the variables in a set of independent algebraic and differential equations containing several unknowns.*

This subsection discusses ways to determine the number of variables that we can solve for with a given differential equation. When solving a set of algebraic equations, we can solve for only one unknown per equation. A single differential equation, however, may be solved for more than one variable (counting a variable and its derivatives separately). We can solve a single differential equation for a single variable (say $x$) and all its derivatives, up to its highest derivative in the equation. If two or more variables appear in the equation, we can usually choose to solve for either one. All of this discussion assumes that we have the appropriate initial conditions.

**Concept 1.4**

### Determining the Number of Differential Equations Required To Completely Model a System

The application of a conservation equation produces the following equation:

$$\ddot{x} + \dot{y} + 3\dot{x} + 4x + 5y = 6$$

We can use this equation to find $x$, $\dot{x}$, and $\ddot{x}$, or $y$ and $\dot{y}$ as a function of time. However, because we cannot use this equation to find all the variables that are unknown, we know that this is not a complete model of our system. We have to find another expression. We apply another conservation equation and find that

$$\dot{x} + 3\dot{y} + 2x + 4y = 0$$

We can use this equation to find $y$ and $\dot{y}$ or $x$ and $\dot{x}$. Thus, combining this equation with the previous equation gives us the ability to calculate all the unknown variables.

A similar way to look at the same problem is to say there are two unknowns ($\ddot{x}$) and ($\dot{y}$). Notice that we count only the highest derivatives as unknowns. Since each equation gives us the ability to solve for only one unknown, we have two equations and two unknowns. We can generally solve this type of mathematical problem (*i.e.* a system of $n$ equations and $n$ unknowns) for a unique answer.

**End 1.4**

If we only write accounting and conservation equations for independent accumulations, then the resulting differential equations will always be independent.

## 1.6  CONSTRAINTS

**Objectives**
*When you complete this section, you should be able to*

1. *Explain the concept of a constraint.*

2. *Distinguish between constraint equations and accounting/conservation equations.*

3. *Identify the effort constraint associated with a flow constraint.*

4. *List the constraints of a system.*

This section discusses the concept of constraint equations and their relationship to the conservation laws. Constraint equations are relations between the effort and flow variables of a system. Try touching your right elbow with your right hand. Under most conditions, no matter how hard you try you will not be able to do this. Try pushing a wall with your hand. Under normal conditions, no matter how hard you push, your hand will not move. What if we wanted to analyze the motion of our hand under these conditions? How could we express the limitations on its motion? When analyzing engineering systems, similar restrictions to motion arise. Constraints are the relations that we use to express these restrictions.

Not surprisingly, considering the relation between effort and flow, efforts and flows have constraints associated with them. Saying that a book is resting on a table is a simple flow constraint. The book has 0 flow. However, for this 0 flow to occur a constraining effort must be acting on the book. Similarly, if we constrain a circuit to have a certain current (the constrained flow), there must be some driving device capable of providing the required voltage (the constraining effort).

These relations between constraining efforts and constrained flows (and *vice versa*) produce constraint equations. Constraint equations can be algebraic or differential. When constraint equations are differential, we need an appropriate number of initial conditions to solve them. Constraint equations are generally not obtained directly from the conservation laws. More often, they are expressions of geometry or trigonometry.[51]

Constraint equations are especially important in problems where we use more than the minimum number of variables to model a system. In such cases, we need constraint equations to relate the extra variables. A common mistake is to overlook constraints.[52] When this happens, the description of the system is incomplete. We should suspect that a model is incomplete when

1. the number of equations does not match the number of variables

2. the equations have more flows than [NIF] predicts

3. the equations are higher order than [NIA] predicts.

Whenever these inconsistencies arise, we have to reevaluate the problem carefully to determine where we made a mistake. We may have miscalculated the [NIA], [NIF], or overlooked constraints.

Although [NIA] is the theoretical minimum number of state variables required to model a system, the practical number may be larger than the [NIA]. For example, if two state variables are constrained together through an algebraic relation, one can theoretically solve the constraint to write one of the variables in terms of the other. By expressing one of the variables in terms of the other in the set of differential equations that models the system, we can reduce the number of required variables. However, it is not always practical to model a system with the *minimum* number of variables. There is nothing wrong with using additional variables if it makes solving a system easier. Using extra variables is particularly useful when a constraint equation is difficult to solve. In these cases, we just add the constraint equations to the set of equations that model the system. By leaving the extra variables in the equations, we have additional unknowns. However, by using the constraint equations, we can determine these additional unknowns.

**Example 1.30** _____

### [NIA] and Constraints in a Simple Pendulum

---

[51] For example, the relation between the height and volume of fluid in a tank is a simple constraint relation.

[52] Although most constraints are simple and easy to recognize, in some problems, constraint equations can be very complicated. In fact, the study of positional constraints dominates kinematics (the mathematics that governs the physics of motion).

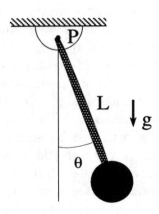

**Analysis Assumptions**

(1) No mass or charge enters or leaves the system.

(2) Pendulum material is strong enough to maintain constraints.

(3) T is the tension in the string.

(4) The pendulum moves only in the plane of the paper.

Figure 1.56: (pend1): A Single Mass Pendulum Moving in a Constant Gravitational Field

## PROBLEM:

We want to build a clock using the swing of a pendulum as the fundamental time period. The clock will calculate time by counting the oscillations of the pendulum.[53] Formulate a mathematical model of the motion of the pendulum in figure 1.56 so that we can use its fundamental time period to build our clock. The pendulum (in figure 1.56) has a mass of $M$ and a length of $L$. It is free to move and gravity acts on it as the figure shows.

## FORMULATION:

To formulate a mathematical model of the pendulum, we first have to consider what type of constraints the pendulum will have when it is in the clock. Constraining the pendulum allows us to simplify our analysis and achieve a more realistic design.

For our clock design, we can safely assume that point $P$ is fixed. This constraint allows us to ignore the motion of the Earth. We can also assume that the shaft of the pendulum constrains its motion to the plane of the paper. Constructing the pendulum of a solid material makes these constraints physically possible. Although guaranteeing the constraint may require external constraint forces to keep the mass in the plane, we will assume the solid material comprising the pendulum is strong enough to supply these forces.[54]

Because its motion is planar, we can assume the pendulum moves in a circular path. This means we have introduced another constraint that says the pendulum's mass must remain a fixed distance $L$ from point $P$ (see figure 1.56). Again, this constraint requires forces to maintain it, and we assume the pendulum material is strong enough to supply them. If the material cannot, our constraint is not valid and the problem is more complex. For example, if the pendulum weighs 50 [lbs] and only a thin thread supports it, the thread will surely break and the mass will no longer be a constant distance from $P$. The thread would not be strong enough to maintain the force of constraint. We could still model the motion of the pendulum mass even if the thread breaks, but its [NIA] (and consequently the required number of variables)

---

[53] Counting oscillations is a common method for accounting time. In fact, modern time pieces operate on the same concept but have replaced the mechanical oscillator (a pendulum) with more accurate electronic oscillators (such as quartz crystals and atomic vibrations).

[54] If we use a string incapable of supplying these constraint forces to support the pendulum, the mass *would* swing out of the plane.

would increase. However, this does not necessarily mean that this motion would be more difficult to model.

Since the [NIA] of the pendulum we are using is two, we anticipate two state variables in our model. Since the pendulum is moving in a circular motion, satisfactory state variables are $\theta$ and $\dot{\theta}$. To model the motion of the pendulum we have to apply a conservation law. Since conservation of angular momentum seems to be the easiest to apply, we will use it.

---

**Conservation of Angular Momentum (CCW - Counter Clockwise) (Vector) [@ P]**

system[pendulum f.1.56]    time period [**differential**]

*input − output = accumulation*

| input/output: | mass flow | |
|---|---|---|
| | external moment | |
| | gravity moment | $-mgL\sin(\theta)\,\vec{k}$ |
| accumulation: | | $\frac{d}{dt}\left(mL^2\dot{\theta}\right)\vec{k}$ |

$$z: \quad \boxed{-mgL\sin(\theta)} \quad = \quad \boxed{\frac{d}{dt}\left(mL^2\dot{\theta}\right)}$$

    *gravity*               *change in accum.*

---

This gives us the following model for the motion of the pendulum:

$$-mgL\sin(\theta) = mL^2\ddot{\theta} \tag{1.22}$$

Notice that equation 1.22 is a second order differential equation with two state variables. This is exactly what the system [NIA] predicted. If we wanted to find the tension (the force of constraint) in the pendulum shaft, we could apply linear momentum in a direction parallel to the string. Note that this is rather difficult to do so we will just write it down for now, later in the text we will show how to compute this expression.

$$-T + mg\cos(\theta) = -mL\dot{\theta}^2 \tag{1.23}$$

here $\theta$ is the angle that appears in equation 1.22.

**End 1.30**

---

# 1.7   SUMMARY AND PROBLEMS

## SUMMARY

The following list reviews the approach that we use to formulate a mathematical model of a system's behavior using the accounting and conservation laws:

1. *Specify the objective of the analysis*

2. *Define the system of interest.* Pay particular attention to the extent that the system reacts with its surroundings, and determine whether these factors are important in modeling the behavior of the system. Determine what factors in the system are significant and what factors can be ignored.

3. *Define a time period of interest.* You may have to define multiple time periods to model a system's behavior correctly.

4. *Determine the [NIF] of the system.* [NIF] is useful because it tells us the minimum number of flow variables needed to model the system's behavior correctly.

5. *Determine the [NIA] of the system.* [NIA] is useful because it tells us the minimum number of first order differential equations and the minimum number of initial conditions needed to model the system's behavior correctly when the conservation/accounting equations are written in rate form.

6. *Apply the accounting and conservation principles (the laws).* Try to determine the law(s) that can describe the system's behavior in the simplest possible way.

7. *Define any necessary initial and boundary conditions.* If initial and boundary conditions are necessary (if a differential equation results from the laws), they must be determined independent of the model. They must be known or measured.

# QUESTIONS

To answer the following questions, call on the information from this chapter and your own life experience. Most of these questions are subjective; however, they are very useful in demonstrating the key points introduced in this chapter. Namely, they focus on jargon, the conservation laws, and system modeling (with emphasis on defining the system and time period, choosing the most applicable conservation law, and determining the factors that are necessary to model the system's behavior accurately).

1. Describe the steps of the problem solving method. Explain what is involved in each step and why the step is important.

2. Based on what you know about thermodynamics, heat transfer, and house construction; make a list of items you believe should be included in a model of the heating and air conditioning system for a house. Give a brief reason for each item you include in your list. (Hint: think about climates, building materials, and house design).

3. If we were going to put a satellite in orbit, would the Earth's spin about its axis matter? (Explain). What about the Sun's spin about its axis? (Explain).

4. If we were studying the dynamic characteristics of a semitrailer, would a collision with a bird in the radiator be significant? (Explain). Would this collision be significant if we were studying a jet aircraft? (Explain).

5. We want to model a human respiratory system, will the model be significantly different for a person 6'5" tall relative to a person 5'6" tall? (Explain). Is the difference significant?

6. If we are studying heat transfer from human beings (*e.g.* we might be working for an outdoor clothing company) is height more important than surface area? (Explain).

7. What kinds of items should we include in an insect population model?

8. If we were accounting for significant radiation sources in a person's life, which sources of radiation should we include? (Explain).

9. Is the amount of paint on a commercial airplane significant? Should we include its weight in a model of the airplane's behavior? (Explain).

10. Are there any advantages to using double pane windows in a house in the tropics? (Explain).

11. Why are automobiles becoming more streamlined? What does this mean in terms of the importance of shape in design? (Explain).

12. We own an engineering consulting firm. A waste disposal company contracts us to retrofit their equipment (trucks, clothes, machinery, *et cetera.*) so that they can compete in the lucrative toxic waste market. What equipment of theirs should we retrofit for use with toxic waste? (Explain).

13. We have to design an incinerator for a hospital. Should we let the combustion products (smoke) exhaust directly to the air? (Explain).

14. We want to study the temperature distribution of a skillet on a stove. Are initial conditions or boundary conditions (or both) important for this analysis? (Explain).

15. What factors influence the ride quality of an automobile, an airplane, a motorcycle, and a horse? (Explain).

16. The chairman of a potato chip company wants to improve the production volume of his company's new potato chip. What steps does he need to take to achieve this goal? What factors does he need to consider?

17. We cannot necessarily increase a process' production rate just by increasing the process' most time consuming step. Think of an example of when this is true?

18. If we ignore the drag force acting on an object, then the object's motion is independent of its mass. What does this imply about a feather or a balloon versus an anvil or a cannonball? Why does physics tell us that all objects fall at the same rate in a vacuum, yet our ordinary experience tells us differently?

19. List as many effort and flow variables as you can.

20. List the effort and flow variables that we would find in the following disciplines:

    (a) Chemical Engineering,

    (b) Nuclear Engineering,

    (c) Mechanical Engineering,

    (d) Civil Engineering,

    (e) Aerospace Engineering,

    (f) Electrical Engineering,

    (g) Computer Engineering,

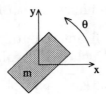

Figure 1.57:

(h) Industrial Engineering,

(i) Petroleum Engineering.

21. How could we design the tank in example 1.11 so that we still achieve the stated objective without modifying the orifice constant $C$? (Hint: Speak generally. *e.g.* Should we make the tank larger, shorter, *et cetera*?)

22. Explain why a ten speed bicycle can be ridden at higher speeds for longer periods of time over varying terrain than a conventional bicycle. Also explain why bicycle racers who race on a smooth level oval track use bikes with only one speed.

23. What is the effect on input impedance of a gear change in a bicycle?

24. Discuss how we could experimentally determine the following:

    (a) an orifice constant,

    (b) the inertia of a body,

    (c) specific heat of a fluid, and

    (d) electrical capacitance.

25. What type of input and output impedance would be best for a voltmeter? (Explain).

26. What type of input and output impedance would be best for an ammeter? (Explain).

27. What type of input and output impedance would be best for a automobile engine? (Explain).

28. How many [NIF] does a block of wood have?

29. How many motors would we need on a robot if it must move a block of wood in all possible directions and rotations?

    [NIF] [NIA]

30. Determine the [NIA] of the block in figure 1.57. The block can move and rotate only in the plane of the page.

31. Determine the [NIF] of the circuits in

    (a) figure 1.58

    (b) figure 1.59

    (c) figure 1.60

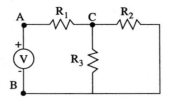

Figure 1.58:

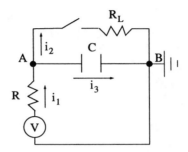

Figure 1.59:

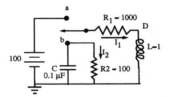

Figure 1.60:

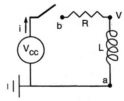

Figure 1.61:

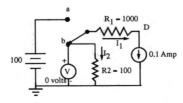

Figure 1.62:

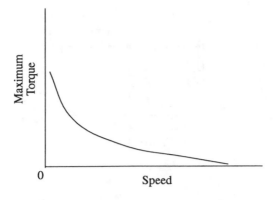

Figure 1.63:

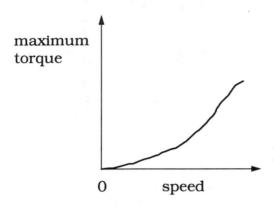

Figure 1.64:

(d) figure 1.61

(e) figure 1.62

32. We have a standard motor pump system that we use to pump water from a well. Explain how the pump works. Sketches of the input and output impedance relations for the motor and pump appear in figures 1.63 and 1.64. Explain what happens as the flow variable increases for these relationships. Is this the type of behavior that we expect from a typical source and load? Furthermore, use the input and output impedance sketches to determine

   (a) the relationship between the amount of water pumped and the motor speed

   (b) the relationship between the kinetic energy required to pump the water and the pump speed (hint KE$= \frac{1}{2}mv^2$)

   (c) the motor response required to achieve increased water flow

   (d) the operating point of the pump.

33. Express the dynamics of running up a hill in terms of input and output impedance.

34. A bike rider can only provide a finite amount of power. Therefore, since power equals the product of effort and flow, we know that as the flow of the bike increases to infinity, the effort of the bike rider must decrease to zero. This means that a rider can push hard on slow moving pedals, but as the pedal speed

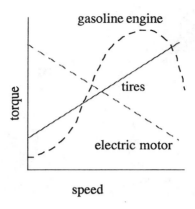

Figure 1.65:

increases to infinity, the rider's ability to push hard will decrease to zero. Explain the advantage this gives to multispeed bicycles?

35. What would we need to know in order to determine the amount of water in a tank two hours from now?

36. Try pushing a wall with your hand. Under normal conditions, no matter how hard you push, your hand will not move. What if we wanted to analyze the motion of our hand under these conditions? How could we express the limitations on its motion?

37. A farmer is having a problem with insects eating his wheat, what should he do? What factors should he include in the analysis of his farm?

38. A fisherman is having a problem with sharks eating all the fish in his nets, what should he do? What factors should he include in the analysis of his nets?

39. Explain why motors are designed with low input impedances.

40. Explain would a sensor be designed with high input impedance?

41. List real world examples of two interconnected systems. For each, explain which system is the driver and which is the load. For each driver, discuss whether it would work best if it had a high or a low output impedance. Do the same for each load.

42. Find two real world examples of devices that are designed with a high output impedance.

43. Explain the conditions when we would want a system with an

    (a) infinite input impedance and
    (b) zero input impedance.

44. Figure 1.65 shows load diagrams of

    (a) an automobile,
    (b) a gasoline engine, and
    (c) an electric motor.

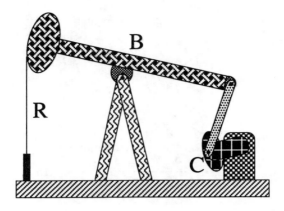

Figure 1.66:

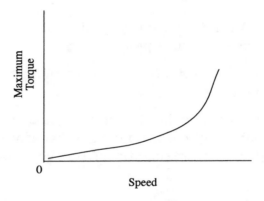

Figure 1.67:

Explain how the automobile will act when starting from rest with both drivers. If any driver fails to operate at low speeds, explain how we could fix the system to allow start up.

45. Consider the oil well pump of figure 1.66. Notice that the torque/speed relationship of figure 1.67 has the absolute value of the maximum torque as the ordinate. Without resorting to quantitative means, sketch a graph of the torque the motor must produce to pump the oil, versus crank (member $C$) angle. Let the starting crank position be the position where the pump rod (member R) is at its lowest position. Label the significant points on the graph.

46. Determine the [NIA] and define the state variables of the systems depicted in figure 1.68.

47. Determine the [NIA] of the circuits in figure 1.69 and figure 1.70.

48. Determine the [NIA] and define state variables for the system in figure 1.71.

49. Analytically determine the operating point of the basic light producing circuit in figure 1.72.

50. For the circuit in figure 1.73,

    (a) determine the [NIF] of the circuit

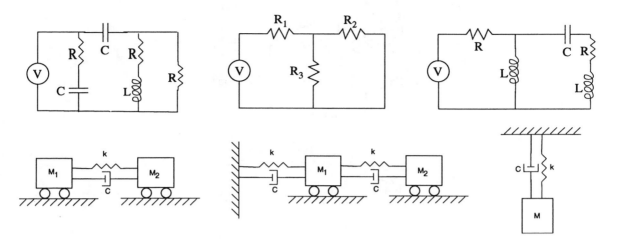

Figure 1.68:  Problems

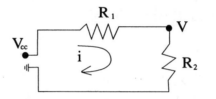

Figure 1.69:

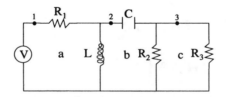

Figure 1.70:

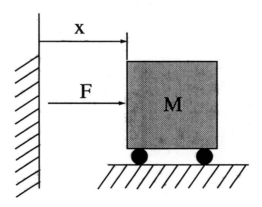

Figure 1.71:

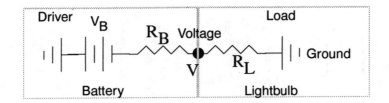

Figure 1.72:

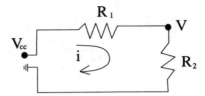

Figure 1.73:

(b) determine the [NIA] of the circuit

(c) formulate a mathematical model of the circuit.

51. For the system in figure 1.74,

    (a) determine the [NIF] of the system

    (b) determine the [NIA] of the system

52. For the tank in figure 1.75,

    (a) determine the [NIF] of the tank

    (b) determine the [NIA] of the tank

## 1.8   MAPLE PROGRAMS

This section includes the Maple files referenced in the text. Plots generated by Maple are displayed in the text, not in this section.

### 1.8.1   File: mapjuice.ms - The Maple Program for Example 1.17.

```
>  c := k*w = cos(Pi*w/2);
```

$$c := k\,w = \cos(\frac{1}{2}\,\pi\,w)$$

```
>  p := k*w^2;
```

$$p := k\,w^2$$

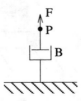

Figure 1.74:

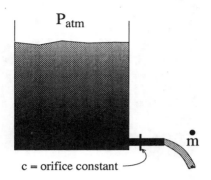

$P_{atm}$

c = orifice constant

Figure 1.75:

```
>  power := subs(k=solve(c,k),p);
```
$$power := \cos(\frac{1}{2}\pi w)\, w$$

```
>  ex := diff(power,w);
```
$$ex := -\frac{1}{2}\sin(\frac{1}{2}\pi w)\,\pi\, w + \cos(\frac{1}{2}\pi w)$$

```
>  wbest := fsolve(ex,w=0..1);
```
$$wbest := .5477053736$$

```
>  evalf(subs(w=wbest,power));
```
$$.3572050230$$

```
>  kbest := solve(subs(w=wbest,c),k);
```
$$kbest := 1.190758125$$

# Chapter 2

# ELECTRICAL SYSTEM CONCEPTS

The focus of this part of the text is electrical system concepts. The properties of most importance in electronics are energy and charge. In this chapter, we emphasize the concepts that we use to model electrical systems. When dealing with electrical systems, we will encounter resistors, capacitors, inductors, mutual inductors, diodes, switches, transistors, and amplifiers. To solve electrical problems, we must know what these items are and how to deal with them. This chapter explains these items and introduces the methods we need to analyze electrical systems.

## 2.1  VOLTAGE AND CURRENT

**Objectives**
*When you complete this section, you should be able to*

1. *Explain the difference between conventional current and electron current.*

2. *Be able to recognize perfect wires in a circuit.*

3. *Be able to explain how voltage varies along a perfect wire.*

4. *Locate and label all of the nodes in a circuit.*

5. *Explain what a ground node is.*

This section discusses the concepts of voltage and current – the effort and flow of electrical systems. The flow variable is called current. There are two types of current that we will encounter in electrical circuits – conventional and electron. In the early years before people understood the nature of electricity, scientists believed electricity consisted of a flow. They imagined that the flow consisted of positive charges moving from the + side of a battery (or other source) toward the − side. However, this is incorrect. As we now know and have measurements to verify, current is a flow of negative charges (electrons). Unfortunately, people are so accustomed to thinking of current as a flow of + charges, it would be counterproductive to switch all textbooks and try to reeducate everyone. Changing the way people think about current is not necessary for most applications anyway. We can explain most concepts in electrical systems with either positive or negative particle flow. There are some devices, however, whose behavior we can only explain after recognizing that current is a flow of negative particles. As a result, we will discuss two types of current, conventional and electron. By default, current, means conventional current is a flow of positive charges from the + terminal to the −.

The most common effort variable in electrical circuits is voltage. If you understand the concept of a voltage, wonderful. If you do not consider the following analogy. Suppose water flows in a pipe.

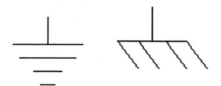

Figure 2.1: Two Types of Ground Symbols.

The water would be analogous to current (water flow is the movement of $H_2O$ molecules and current is the flow of electrons). To make the water flow, you must apply a pressure. You have to hook the hose to a water faucet that has a water pressure. The pressure is analogous to voltage. The voltage "forces" the current to flow.

The simplest item you will find in an electrical circuit is an ideal wire. The voltage along a perfect wire is constant. Consider the water analogy again. A hose is like a wire, it carries flow. There is a pressure (a voltage) on one end of the hose (the wire). Because the hose converts kinetic energy of the water flow into other forms, the water loses energy as it flows in the pipe. For example, friction between the pipe and water removes some energy. Because of this energy loss, the pipe must have a higher pressure on one end than on the other. As an experiment, take a long garden hose and fill it with water, then try to blow the water out. Its really hard to do. This is because of the losses in the hose. Now imagine as the hose gets more perfect, it would be easier to move the water and you would not have to blow nearly as hard. If the hose were nearly perfect, if there were only minute energy losses along the hose, the water would flow without much pressure change at all. Take the limit as the hose becomes ideal, flow could occur without any pressure change at all. This is the same thing that happens with a perfect wire. If the wire is perfect, flow (current) can occur without any push (voltage) difference from one end of the wire to the other. Of course there are no perfect wires but we will discuss the differences in the section on resistance. A perfect wire is drawn as a line.

Perfect wires are used to connect other "components" together. When components in a circuit are connected together, nodes are formed. A node is a junction between two or morecomponents. To locate all the nodes in a problem, sequentially consider all the components. We will find a node on each end of eachcomponent. For simplicity, all nodes which are separated by perfect wire (drawn as straight lines on diagrams) are consolidated into a singlenode. Nodes are drawn on diagrams as dots. We will discuss voltages at nodes.

Except for special semiconductor devices, we can treat voltage as a relative quantity. As a result, it is common practice to pick an origin for electrical voltage by choosing a ground node. Basically, with few exceptions, we can arbitrarily pick any one node in a circuit and assume its voltage is zero. When the ground node is chosen, we typically mark it with a ground symbol. There are several types of ground symbols, figure 2.1 shows two common ground symbols. In most circuits, current does not flow out of the circuit at the ground node, the ground symbol merely "indicates" the node that is defined to have zero voltage. Sometimes you will see a circuit with more than one ground symbol. This occurs in wiring diagrams of automobiles, clothes washers and other large objects. Figure 2.2 shows a simple wiring diagram from an automobile. The components will be discussed later but focus on the ground symbols. When the ground symbols are the same shape, as they are in figure 2.2 it means the "ground" nodes are actually connected. For example in the figure, points A and B are electrically connected. Typically, this is accomplished by using the metal car body or washing machine cabinet as a wire. For example, point A might be connected to the metal car chassis inside the engine compartment and point B is connected to the car chassis near the tail light. Since both are connected to the same metal, they are electrically connected.

Occasionally you will see a circuit with two different looking ground symbols. Usually this means the grounds are not electrically connected. You will often see this type of diagram in devices that are in some way isolated. For example, in the case of a computer, signals can be transmitted to other computers by applying voltages. These signals may have their own ground. The power connection that you plug into the wall may have its own ground. The signal ground and power ground may not be electrically connected.

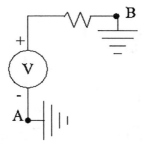

Figure 2.2: A Wiring Diagram Typical of That Used in Automobile Documentation.

Voltage is measured in Volts and current is measured in Amps. One Volt times one Amp equals 1 Watt or 1V times 1Amp = 1 Watt = 1 Newton Meter/Second.

## 2.2  SOURCES AND RESISTORS

**Objectives**
*When you complete this section, you should be able to*

1. *Determine the [NIF] of electrical circuits.*

2. *Identify ideal wires in a circuit diagram and explain what voltage does along the length of the wire.*

3. *Identify ideal resistors in a circuit diagram and calculate the energy dissipation in terms of voltage and current.*

4. *Identify ideal sources in a circuit diagram and explain how they act. For example, whether voltage/current is known across/through them.*

5. *Label a ground node on a circuit diagram.*

6. *Account energy and charge in a circuit of perfect resistors, wires and sources then solve the equations for currents and voltages.*

7. *Determine the voltage across a resistor given current through it.*

8. *Determine the current through a resistor given voltage across it.*

There are four types of perfect sources that we will consider. A voltage source, provides a known voltage across itself regardless of the flow through it. It can provide an infinite flow if necessary. An ideal current source provides a known current regardless of the voltage across it. Figure 2.3 shows the schematic diagrams of these sources. We will also consider dependent voltage and dependent current sources. A dependent source produces a voltage or current as a function of some circuit variable. Figure 2.4 shows the schematic diagrams of dependent sources. A dependent source is usually constructed using some sensor or transducer of some kind. Sources provide energy to a circuit. Therefore, they need to receive energy from something else. For example, a source may "plug into" a wall outlet, or as in a battery, obtain its supply from a chemical reaction. Typically, we do not concern ourselves with where the energy comes from so we do not show the connection into the source from the energy provider. When we account for electrical energy in a circuit, we normally leave the source out of the system. If you do not leave the source out of the energy formulation, you will need to account for the energy supply that "powers" the source.

There are of course no real world "ideal" sources but there are devices that get close. For example, an automobile battery is a pretty good emulator of a voltage source (especially at low currents). Ideal current sources have to be constructed from many other components and really

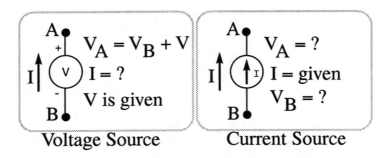

Figure 2.3: Voltage and Current Sources

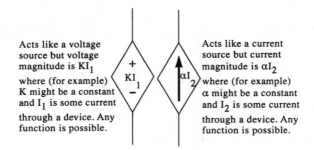

Figure 2.4: Dependent Voltage and Current Sources

are an idealization. They do however simulate part of the behavior of some semiconductor devices therefore it is helpful to learn to deal with them in "simple" circuits before confronting them in more complex problems. It is possible to model real sources by connecting ideal sources with resistors and other components so if you know how to handle ideal devices, you will be able to handle real devices too.

Ideally resistive circuits consist of perfect wires, perfect sources, and perfect resistors. An ideal (perfect) resistor provides a voltage across it proportional to the current through it. Figure 2.5 shows the current and voltage relations for a perfect resistor $R$. If current $I$ flows in the direction shown, the left side voltage is higher than the right. The following expression relates the voltage difference to the current for the circuit:

$$V_{\text{high}} - V_{\text{low}} = RI \tag{2.1}$$

Notice that whenever we write voltage/current relationships for any element, we will always draw a current arrow. Then we subtract the voltage at the arrowhead from the voltage at the arrow tail (remember tail minus head, its always tail minus head for the current as drawn with an arrow). If we consistently express voltage and currents this way, we will have less trouble keeping the + and − signs consistent in our equations.

Real wires unlike ideal wires have a voltage drop along them. Often you can model a real wire as an ideal wire with a resistor. For example, a real wire will have a resistance per length. Multiply the wire length by the resistance per length and draw an ideal resistor with that value. This is a simplification to the real case but it works for simple applications. Keep in mind however that the

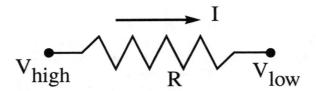

Figure 2.5: Current and Voltage for a Perfect Resistor

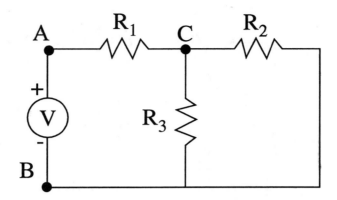

Figure 2.6: A Two Loop Resistor Network

resistance per length of most wire is very small therefore for most reasonably sized applications, real wires are pretty dog gone close to ideal wires so why worry.

The amount of electrical energy dissipated (transformed into other forms) by a resistor is

**Electrical Energy Accounting**
system[**resistor fig.2.5**]      time period [**differential**]

$input - output + generation - consumption = accumulation$

| input/output: | $V_{high}I - V_{low}I$ |
|---|---|
| generation: | |
| consumption: | $P_{diss}$ |
| accumulation: | |

| $V_{high}I - V_{low}I$ | $-$ | $P_{diss}$ | $=$ | $0$ |
|---|---|---|---|---|
| *input/output* | | *consumption* | | *change in accum.* |

$$P_{diss} = V_{high}I - V_{low}I = (V_{high} - V_{low})I$$

Using equation 2.1 gives

$$P_{diss} = I^2 R = \frac{(V_{high} - V_{low})^2}{R}$$

Resistive circuits have no accumulation elements except perhaps for thermal energy (they can get hot). If we ignore thermal effects, resistive circuits are always zero [NIA]. Because they are zero [NIA], the mathematical models that we develop will be algebraic equations. This means that the changes in voltage and currents in resistive circuits can happen instantaneously.

The [NIF] of an electrical circuit is equal to the number of closed circuits. An easy way to determine the [NIF] is to imagine cutting wires with a wire cutter until there are no more closed loops.

The following example demonstrates [NIF] and the choice of ground. The example accounts for electrical energy. Electrical energy accounting is not very convenient because it results in quadratic equations. We will demonstrate another method for obtaining equations a bit later but this example starts with something we trust is more familiar.

**Example 2.1** ————————————————————————————

## A Simple Resistor Circuit

**PROBLEM:**
Determine the electrical energy dissipated in the circuit in figure 2.6.
**FORMULATION:**

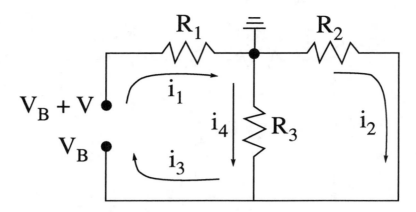

Figure 2.7: A System for Determining Currents

Since there is a closed loop in the circuit, we have at least one [NIF]. Imagine cutting a wire to prevent the loop. Cut the wire between C and $R_3$ and there is still a closed loop. Therefore, the circuit has another [NIF]. When we cut a second wire (say between $R_1$ and $R_2$), there are no other closed loops, therefore the circuit has two [NIF]. Because the circuit contains only resistors, nothing accumulates (zero [NIA]). Thus all equations describing the system should be algebraic.

To determine the nodes, consider each element. Nodes A and B surround the source element. Nodes A and C surround $R_1$. Notice that if we defined another node on the left of $R_1$ we would then consolidate it with A since they would be separated by only wire. Nodes C and B surround both $R_2$ and $R_3$.

One of the early steps in analyzing electrical systems is to choose an origin for measuring voltages (choosing ground). Except in rare circumstances, such as the analysis of some solid state devices, the choice of ground is arbitrary. To demonstrate this, we will choose node C as ground.

As our system, we will take the circuit without a power supply (see figure 2.7). There is no accumulation. Note that $V_B$ represents the voltage of node B. This will be the typical notation in the text. Also note that we utilize the definition of the voltage source to express $V_A = V_B + V$. We have also labeled four flows (currents) in the circuit. Only two of these are independent because the system is two [NIF]. We could treat $I_1$ and $I_2$ as independent. Therefore, there must be a relationship between these and the other two. $I_1$ and $I_3$ are not independent. Explain why. Are $I_1$ and $I_4$ independent?

Applying conservation of charge for the complete system of figure 2.7 (remember the system has no accumulation), we get the following: (since $I_1$ enters and $I_3$ exits and nothing goes in or out of the ground node)

---

**Conservation of Charge**
system[**circuit fig.2.7**]      time period [**differential**]

*input − output = accumulation*

| input/output: | $I_1 - I_3$ |
| **accumulation:** | |

| $I_1 - I_3$ | $=$ | $0$ |
|---|---|---|
| *input/output* | | *change in accum.* |

---

Summing what enters, subtracting what exits, and setting this equal to the time rate of change of the accumulation, we get the following:

$$I_1 - I_3 = 0$$

Now accounting for electrical energy, we find (find these terms on your own, this is important, we are making a point and you need to find the terms yourself, do it!)

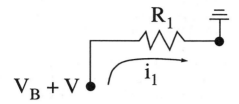

Figure 2.8: An Extra System Chosen for Our Analysis

**Electrical Energy Accounting**

system[ciruit fig.2.7]     time period [**differential**]

*input − output + generation − consumption = accumulation*

| input/output: | $(V_B + V)\, I_1 - V_B I_3$ |
|---|---|
| consumption (thermal): | $\frac{(V_B+V)^2}{R_1} + \frac{V_B^2}{R_3} + \frac{V_B^2}{R_2}$ |
| generation: | |
| accumulation: | |

$$
\boxed{(V_B + V)\, I_1 - V_B I_3} \;-\; \boxed{\frac{(V_B+V)^2}{R_1} + \frac{V_B^2}{R_3} + \frac{V_B^2}{R_2}} \;=\; \boxed{0}
$$

*input/output*          *consumption*          *change in accum.*

$$
(V_B + V)\, I_1 - V_B I_3 - \frac{(V_B + V)^2}{R_1} - \frac{V_B^2}{R_3} - \frac{V_B^2}{R_2} = 0
$$

We now have two equations but three unknowns ($V_B$, $I_1$, and $I_3$). We clearly need another equation. To obtain another, we will choose another system. We will choose the system in figure 2.8. Again accounting for electrical energy, we find (note that this time we use the current relationship for power dissipation, just for kicks)

**Electrical Energy Accounting**

system[**circuit fig.2.8**]     time period [**differential**]

*input − output + generation − consumption = accumulation*

| input/output: | $(V_B + V)\, I_1$ |
|---|---|
| consumption (thermal): | $I_1^2 R_1$ |
| generation: | |
| accumulation: | |

$$
\boxed{(V_B + V)\, I_1} \;-\; \boxed{I_1^2 R_1} \;=\; \boxed{0}
$$

*input/output*          *consumption*          *change in accum.*

$$
(V_B + V)\, I_1 - I_1^2 R_1 = 0
$$

**SOLUTION:**

The solution to these equations was determined using Maple (see Maple session 2.12.1). Maple determined three possible solutions. Solution 1 is:

$$
\left\{ V_B = 0, I_1 = \frac{V}{R_1}, I_3 = \frac{V}{R_1} \right\}
$$

The second solution is:

$$
\left\{ V_B = -\frac{R_3\, R_2\, V}{R_3\, R_2 + R_1\, R_2 + R_1\, R_3}, I_1 = \frac{(R_2 + R_3)\, V}{R_3\, R_2 + R_1\, R_2 + R_1\, R_3}, I_3 = \frac{(R_2 + R_3)\, V}{R_3\, R_2 + R_1\, R_2 + R_1\, R_3} \right\}
$$

The third solution is much longer and is a complex number which is meaningless in this context hence we know it is incorrect so it will not be printed. Now the problem is to identify which of the above two solutions is valid.

The correct result is the second one. Note that although the first solution solves the equations, it is not physically possible. If $V_B$ were really zero, then $I_2$ and $I_4$ would also be zero (write an energy equation to prove it) and $I_1$ nonzero. If this really were the case, then the conservation of charge into the ground terminal (*e.g.* $I_1 - I_2 - I_4 = 0 = I_1 \neq 0$) would be violated. This tells us that

1. we cannot blindly trust the results of an equation solver just because it gives us an answer (we have to verify that the answer makes physical sense)

2. the electrical energy equation can often be difficult to deal with because it is quadratic and therefore gives us "extra" solutions.

We can determine a correct solution by dividing the first equation by $V_B$ (this prevents $V_B = 0$ from being a valid solution).

$$\frac{(V_B + V) I_1}{V_B} - \frac{V_B I_3}{V_B} - \frac{(V_B + V)^2}{R_1 V_B} - \frac{V_B}{R_3} - \frac{V_B}{R_2} = 0$$

## OTHER CONSIDERATIONS:

An easier way to solve the last example is to apply conservation of charge on the ground node. Doing this we find (again, do yourself a favor, find these terms yourself)

**Conservation of Charge**
system[**ground fig.2.7**]      time period [**differential**]

*input $-$ output $=$ accumulation*

| **input/output:** ground | $\frac{V_B+V}{R_1} + \frac{V_B}{R_3} + \frac{V_B}{R_2}$ |
|---|---|
| **accumulation:** | |

| $\frac{V_B+V}{R_1} + \frac{V_B}{R_3} + \frac{V_B}{R_2}$ | $=$ | $0$ |
|---|---|---|
| *input/output* | | *change in accum.* |

Summing what enters, subtracting what exits, and setting this equal to the time rate of change of the accumulation, we get the following:

$$\frac{V_B + V}{R_1} + \frac{V_B}{R_3} + \frac{V_B}{R_2} = 0$$

This one equation allows us to solve for $V_B$. Then we can find $I_1$ from the system in figure 2.8.
**End 2.1**

The points to get out of the previous example include:

1. that the accounting of electrical energy is not the easiest method for modeling electrical systems,

2. current does not flow into a ground node,

3. when using electrical energy accounting, avoid including the sources.

The next section derives a method that is easier to use than electrical energy.

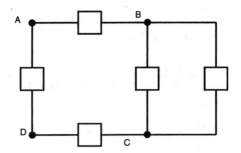

Figure 2.9: A General Five Element, Two [NIF] System

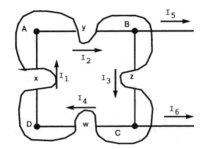

Figure 2.10: System for Energy Accounting

## 2.3 KIRCHHOFF'S VOLTAGE LAW

**Objectives**
*When you complete this section, you should be able to*

1. *Explain the relationship between Kirchhoff's voltage law and the conservation/accounting principles.*

2. *List the steps for applying Kirchhoff's voltage law to a circuit.*

3. *Apply KVL to circuits with perfect wires, resistors, and sources.*

4. *Relate flow variables (currents) together using conservation of charge at a node.*

In this section we will introduce the general method we will use to solve circuits. Many students have developed their own method for solving circuits. If you are such a student, we recommend you skip ahead to example 2.6 and see if your method allows you to correctly solve the circuit problem. If it does then maybe you can skip most of this chapter. If it doesn't then you should try to modify your method to incorporate what we are demonstrating here.

There is a general method for solving circuits which is easier than accounting electrical energy. The method based on what is known as Kirchhoff's Voltage Law. We can derive the law as a combination of conservation of charge and accounting of electrical energy. We will not provide a detailed derivation of the law but will demonstrate how the law fits in the conservation framework.

The two [NIF] system in figure 2.9 has five elements. The empty boxes represent any element that you want to install. The analysis we plan does not depend on what the elements are so they are shown as blocks. Consider choosing the system shown in figure 2.10. We have chosen a system consisting of only the wires which form a single loop. Now accounting for electrical energy, we find that

**Electrical Energy Accounting**

system[**single loop fig.2.10**]      time period [**differential**]

*input − output + generation − consumption = accumulation*

| input/output:<br>consumption (thermal):<br>generation:<br>accumulation: | $V_A I_1 + V_B I_2 + V_C I_3 + V_D I_4 - V_A I_2 - V_B I_5 - V_B I_3 - V_C I_6 - V_C I_4 - V_D I_1$ |
|---|---|

$$V_A I_1 + V_B I_2 + V_C I_3 + V_D I_4 - V_A I_2 - V_B I_5 - V_B I_3 - V_C I_6 - V_C I_4 - V_D I_1 \qquad =$$

*input/output*

$$\boxed{0}$$

*change in accum.*

Collecting terms, we find that

$$(V_A - V_D)\, I_1 + (V_B - V_A)\, I_2 + (V_C - V_B)\, I_3 - V_B I_5 - V_C I_6 + (V_D - V_C)\, I_4 = 0$$

From conservation of charge into each node, one can easily show that

**Conservation of Charge**

system[**the individual nodes fig.2.10**]      time period [**differential**]

*input − output = accumulation*

| input/output: node A<br>accumulation: | $I_1 - I_2$ |     | input/output: node B<br>accumulation: | $I_2 - I_5 - I_3$ |
|---|---|---|---|---|
| input/output: node C<br>accumulation: | $I_3 - I_6 - I_4$ | input/output: node D<br>accumulation: | $I_4 - I_1$ |  |

$$\boxed{I_1 - I_2} \quad = \quad \boxed{0}$$

*input/output*                *change in accum.*

$$\boxed{I_2 - I_5 - I_3} \quad = \quad \boxed{0}$$

*input/output*                *change in accum.*

$$\boxed{I_3 - I_6 - I_4} \quad = \quad \boxed{0}$$

*input/output*                *change in accum.*

$$\boxed{I_4 - I_1} \quad = \quad \boxed{0}$$

*input/output*                *change in accum.*

From these equations, we find that

$$I_1 = I_2 = I_4$$

and

$$I_5 = -I_6$$

Using these relations, we can factor the currents out of our energy result. Doing this, we obtain the following:

$$(V_A - V_D) + (V_B - V_A) + (V_C - V_B) + (V_D - V_C) = 0$$

The following general steps summarize the application of Kirchhoff's Voltage Law.

1. Count the [NIF]. You should expect to write [NIF] KVL equations.

2. Label nodes and currents. To label current, draw an arrow that symbolizes current beside each element. If there are 5 elements draw 5 arrows. Don't worry about drawing the arrows in any particular direction. It will work out regardless of the direction you choose.

3. Choose ground.

4. Choose a loop. Basically you are going to "walk" around the circuit until you return to the starting point. A loop is the path you plan to "walk". Some loops may produce equations that are easier to solve than others but any loop is valid. In fact it isn't even necessary that the loop consist of a closed circuit. Expect to choose [NIF] loops. When you apply more than one KVL make each loop different from the previous and be sure you walk past each element in at least one of the loops. Don't make any element jealous by ignoring it!

5. Arbitrarily choose a starting point on your loop and choose a "walking" direction.

6. Pretend to "walk" around the loop starting at your beginning and proceed in your chosen direction. For each element (resistor, source, or other device) you "step over" do the following:

   (a) Express the magnitude of voltage change.

   (b) Imagine that the expression of step 6a is positive.

   (c) Determine which side of the component has a higher voltage based on the results from step 6b.

   (d) Determine what you would see (a rise or a fall) if you moved across the component (in the direction that you chose to walk) based on the results from step 6c.

   (e) If the voltage change goes up as you walk, put a + in front of the term; if the voltage change goes down, put a - sign.

7. When you arrive back at the starting point sum all the changes (rises add, falls subtract) and set the sum equal to 0.

Table 2.1: Steps For Applying Kirchhoff's Voltage Law.

This final equation tells us that if we add the voltage rise across each element in a closed loop, the sum will be zero. This is a simple statement of Kirchhoff's law.

When applying Kirchhoff's law to a circuit, use the steps summarized in Table 2.1. After choosing a loop, decide the direction to move around the loop and add all voltage rises, subtract voltage falls as you walk around the loop, stepping over elements.

For practice, we will apply Kirchhoff's law to example 2.1. We begin by determining that the [NIF] is 2 (expect 2 KVL equations), the [NIA] is 0 (expect algebra equations only). Next label the nodes and currents of the circuit (see figure 2.11). Notice that there are currents (arrows) beside all the elements. Current (arrow) 3 is the current that flows through the voltage source. Also note that a conservation of charge at node A will tell you that current 3 is the same as current 1. If you get good at doing these simple calculations in your head you can save a little time by not drawing too many currents. For example, if you **know** that current 3 is the same as 1, don't draw 3 (or better yet draw 3 but put it in the same direction as 1 and label it 1). Until you get really good, draw all the arrows you can, it will help you keep things straight. Now we choose ground, see the figure. By the way, most people would choose node B as ground but we are trying to prove a point by choosing C. The loop we choose first is ACBA. We will move counter clockwise around this loop adding voltage rises and subtracting voltage falls as we go. Again most students would choose to walk clockwise but we are doing the opposite to prove that it can be done.

Stepping over the voltage source from A to B, the voltage changes by $V$.

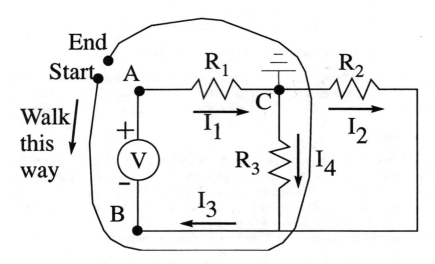

Figure 2.11: A Loop to Walk Around

| (?)            |           | (?)             |
|----------------|-----------|-----------------|
| **Step Over:** | **Magnitude** | **Rise or Fall?** |
| Voltage source | $V$       | (see discussion) |

According to the figure, A has a higher voltage than B (see the + and - signs on the source). Therefore, if we move from A to B, the voltage falls as we step over the source.

| (−)            |           | (−)             |
|----------------|-----------|-----------------|
| **Step Over:** | **Magnitude** | **Rise or Fall?** |
| Voltage source | $V$       | Fall            |

Walking along the wire, there is no voltage change because its a perfect wire. Stepping over $R_3$, the voltage changes by $R_3 I_4$ (see equation 2.1). The result is

| (?)            |           | (?)             |
|----------------|-----------|-----------------|
| **Step Over:** | **Magnitude** | **Rise or Fall?** |
| $R_3$          | $R_3 I_4$ | (see discussion) |

To determine whether to put a + or a - sign in front of this term, determine whether it rises or falls as we step over. To do this, play an imaginary game, imagine $I_4$ is positive.[1] If $I_4 > 0$ (look at the figure with $I_4$) then it flows in the direction drawn, which is downward. Notice this last point, if we compute positive values for $I$, the arrows drawn are correct, if the $I$ is found to be negative, we drew it backward. Okay so we imagine $I_4$ is downward, if the current flows downward, the voltage at C is higher than the voltage at B. You will discover that for most elements,[2] the higher voltage side is on the tail of the arrow you drew on the diagram. This is true for resistors. For this example, C is on the tail of the current so it is higher than B, therefore, the voltage appears to rise as we walk.

| (+)            |           | (+)             |
|----------------|-----------|-----------------|
| **Step Over:** | **Magnitude** | **Rise or Fall?** |
| $R_3$          | $R_3 I_4$ | Rise            |

Similarly, as we walk from C to A, the voltage changes by $R_1 I_1$. The voltage expression is

| (?)            |           | (?)             |
|----------------|-----------|-----------------|
| **Step Over:** | **Magnitude** | **Rise or Fall?** |
| $R_1$          | $R_1 I_1$ | (see discussion) |

---

[1]Imagine that the expression for the magnitude change is positive. In this case, imagine that $R_3 I_4$ is positive. Since $R_3$ is always positive, we imagine $I_4 > 0$.

[2]The exceptions are elements that put energy into a circuit like a source.

If $I_1$ is positive, it flows from left to right, as the figure shows, hence the tail (left) is higher than the head (right). Therefore, $V_A$ is higher than $V_C$ so it appears to rise.

| (+) | | (+) |
|---|---|---|
| **Step Over:** | **Magnitude** | **Rise or Fall?** |
| $R_1$ | $R_1 I_1$ | Rise |

Since we are finally back to our starting point, we sum the rises, subtract the falls and set the expression to zero. This gives

$$-V + R_3 I_4 + R_1 I_1 = 0 \qquad (2.2)$$

In this one equation we have two unknowns, $I_4$ and $I_1$. We need another equation. Since [NIF] is two, we expect to write another KVL.

We can obtain a second equation by walking around a loop between CBC (across $R_3$ and $R_2$). Notice that this loop is different from the first and once we finish walking, we will have stepped over all the elements at least once. If we start at C and walk clockwise. We step over $R_2$

| (?) | | (?) |
|---|---|---|
| **Step Over:** | **Magnitude** | **Rise or Fall?** |
| $R_2$ | $R_2 I_2$ | (see discussion) |

If $I_2 > 0$ the voltage on the left of $R_2$ is higher than the right side. Since we are walking from left to right, it looks like a drop

| (−) | | (−) |
|---|---|---|
| **Step Over:** | **Magnitude** | **Rise or Fall?** |
| $R_2$ | $R_2 I_2$ | Fall |

Walking over $R_3$, we write that

| (?) | | (?) |
|---|---|---|
| **Step Over:** | **Magnitude** | **Rise or Fall?** |
| $R_3$ | $R_3 I_4$ | (see discussion) |

Due to the direction drawn for $I_4$, if $I_4 > 0$ the voltage on top of $R_3$ (node C) is higher than on the bottom. As a result, it looks like a rise as we walk. The expression is

| (+) | | (+) |
|---|---|---|
| **Step Over:** | **Magnitude** | **Rise or Fall?** |
| $R_3$ | $R_3 I_4$ | Rise |

Since we are back where we started, we sum rises subtract falls and set the expression to zero.

$$-R_2 I_2 + R_3 I_4 = 0 \qquad (2.3)$$

In equations 2.2 and 2.3 there are three unknowns ($I_1$, $I_2$ and $I_4$). All three of the unknowns are flow variables. Since there should only be two flow variables (there are two [NIF]) we need to relate one flow to the others. The way to relate flows in an electrical circuit is to apply conservation of charge. To do this look for a system in which $I_1$, $I_2$ and $I_4$ are the only flows present. Node C is just such a location. Note that if you chose node A, you could relate $I_3$ and $I_1$ but we don't want to do that. Applying conservation of charge for node C, we find that

**Conservation of Charge**

system[**node C fig.2.11**]      time period [**differential**]

*input − output = accumulation*

| **input/output:** node C | $I_1 - I_2 - I_4$ |
|---|---|
| **accumulation:** | |

| $I_1 - I_2 - I_4$ | = | $0$ |
|---|---|---|
| *input/output* | | *change in accum.* |

$$I_1 - I_2 - I_4 = 0 \qquad\qquad (2.4)$$

This with the previous two equations gives us everything we need.

Before leaving this example, determine how equation 2.3 would change if we drew $I_4$ in the opposite direction. Could we determine the value of the currents in the circuit even if we drew them in the wrong direction? Explain how the results would differ. Do it before you read the answer. What you would find is the + or - sign in front of the terms $R_3 I_4$ flip. When you solve the equations your answer to $I_4$ would also be the opposite + or - from the equations in the book give. For example if you solve our equations as written you will get a + answer for $I_4$ which means you drew the $I_4$ arrow correctly. Try this yourself, let $V = 10$, $R_1 = 1$, $R_2 = 2$ and $R_3 = 3$ to find $I_4 = +1.82$. Current flows from C to B through $R_3$. If you drew $I_4$ up instead of down and wrote the equations correctly then solved them you would compute $I_4 = -1.82$, the negative meaning you drew the $I_4$ arrow upside down, current flows downward from C to B through $R_3$!

One of the difficulties many students have with circuits is that they try to mentally calculate too much. For example, in a problem like we just completed, it is common for students to try analyzing the circuit without labeling $I_4$. Essentially, some students develop a tendency to mentally calculate things like $I_4$ rather than labeling it and explicitly computing it. Now in the current example, it is not difficult to do this, but there will be times when the mental calculation will be either too formidable to handle, or slightly tricky causing a error. A safe practice is for you to label all quantities then explicitly calculate them.

In most situations, you will write the maximum number of KVL equations only to discover that you need more equations. It is therefore important to understand where to find the extra equations. When dealing with a complex circuit first determine the [NIF] and [NIA]. The [NIF] is most important in the present discussion. Now to obtain a sufficient number of equations, first write [NIF] KVL equations (or apply conservation of energy whichever you prefer). After writing these equations, make a list of the unknowns. Also count the number of unknown flow variables. If the number of independent equations matches the number of unknowns, attempt a solution. If however, there are too many unknowns then either the problem is underconstrained (not well defined) or you must find extra equations. To find extra equations, first look at the number of unknown flows. It is always possible to write your KVL equations so they only contain [NIF] flows. If your equations contain more than [NIF] flows, then you can relate the flows. To relate the flows in electrical circuits, choose a node and express conservation of charge in terms of currents. This is what equation 2.4 did. If there are current sources in a circuit, then these sources often specify some of the independent flows. When this happens, the number of independent node equations you can write is increased by the number of independent current sources. Table 2.2 summarizes this strategy.

Occasionally, after writing [NIF] KVL equations and the maximum possible conservation of charge at a node equations, you still need more equations. When this happens, it is often easiest to choose a single element as a system and account for electrical energy.

## Example 2.2

### KVL with Voltage and Current Sources.

**PROBLEM:**
Just for kicks, determine the currents through all the elements shown in figure 2.12 and all the node voltages. Let $V = 10$ volts, $I = 2$ amps, and $R_i = i$ (eg. $R_3 = 3$).
**SOLUTION:**
First determine the [NIF] by counting the number of wires that must be cut to prevent any closed loops. The [NIF] is 3. Therefore we expect to write 3 KVL equations. Next find all the nodes and label one ground. Then draw arrows representing currents through each element. There are 3 nodes as figure 2.13 shows. Notice that there are six flows one of which is known ($I = 2$ amps). The way to determine the number of charge equations is to count the number of unknown flows (5) subtract the [NIF] and add the number of current sources. The number of conservation of charge equations we expect is = 5 - 3 + 1 = 3.[3]

---
[3]Some students like to take all the flows including the current sources and subtract the [NIF]. This is the same thing as the steps say. You decide how you want to do it and then be consistent. Do it the same each time.

To find equations for an electrical system, the following steps often help.

1. Express [NIF] KVL equations.

2. Count the number of unknowns. If they match, attempt solution.

3. Count the number of unknown flow variables. Call this $U_f$. Since the current through a current source is a given, don't label it (don't count it) as an unknown flow.

4. Count the number of current sources. Call this number $C_s$. Only count nonredundant current sources, for example, if two 3 amp current sources are placed in the same wire, one of them is redundant because once one fixes the current at 3 amps, its fixed, you don't need the other source. This would happen so rarely, you probably would never see it except on an exam! We hope we haven't given your instructor nasty ideas.

5. Write $U_f - [NIF] + C_s$ conservation of charge at node equations.

6. If there are still insufficient equations then either the problem is underconstrained (poorly defined) or you should account for energy on single elements until you reach the number of equations needed. This should be done after step 5.

Table 2.2: General Procedure for Finding Equations.

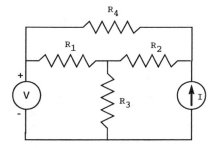

Figure 2.12: A Resistor Network with a Voltage and Current Source.

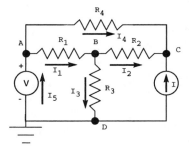

Figure 2.13: Nodes and Element Currents.

For the first of our three KVL equations start at D then go DABD. Stepping over the voltage source gives:

| (?) | | (?) |
|---|---|---|
| **Step Over:** | **Magnitude** | **Rise or Fall?** |
| Voltage source | $V$ | (see discussion) |

According to the figure, A has a higher voltage than D (see the + and - signs on the source). Therefore, if we step from D to A, the voltage rises.

| (+) | | (+) |
|---|---|---|
| **Step Over:** | **Magnitude** | **Rise or Fall?** |
| Voltage source | $V$ | Rise |

Stepping over $R_1$, the voltage changes by $R_1 I_1$ (see equation 2.1). The result is

| (?) | | (?) |
|---|---|---|
| **Step Over:** | **Magnitude** | **Rise or Fall?** |
| $R_1$ | $R_1 I_1$ | (see discussion) |

To determine whether to put a + or a - sign in front of this term, determine whether it rises or falls as we step over. Imagine $I_1$ is positive. If $I_1 > 0$ (look at figure 2.13) the tail of the arrow is higher voltage than the head so the voltage falls as we walk over $R_1$.

| (−) | | (−) |
|---|---|---|
| **Step Over:** | **Magnitude** | **Rise or Fall?** |
| $R_1$ | $R_1 I_1$ | Fall |

Similarly, as we walk from B to D, the voltage changes by $R_3 I_3$. The voltage expression is

| (?) | | (?) |
|---|---|---|
| **Step Over:** | **Magnitude** | **Rise or Fall?** |
| $R_3$ | $R_3 I_3$ | (see discussion) |

If $I_3$ is positive, the tail is higher voltage so B is higher than D so it falls as we step over:

| (−) | | (−) |
|---|---|---|
| **Step Over:** | **Magnitude** | **Rise or Fall?** |
| $R_3$ | $R_3 I_3$ | Fall |

Since we are back to our starting point, we sum the rises, subtract the falls and set the expression to zero. This gives

$$V - R_1 I_1 - R_3 I_3 = 0 \qquad (2.5)$$

For the second KVL start at A step over $R_4$ to make the loop ACBA. Now without much detail we have:

| (−) | | (−) |
|---|---|---|
| **Step Over:** | **Magnitude** | **Rise or Fall?** |
| $R_4$ | $R_4 I_4$ | Fall |

It falls over $R_4$ because the tail is high. Next we have:

| (+) | | (+) |
|---|---|---|
| **Step Over:** | **Magnitude** | **Rise or Fall?** |
| $R_2$ | $R_2 I_2$ | Rise |

It rises over $R_2$ because the tail is high but we are walking from the right to the left (from head which is low to the tail which is high). Now finally we have:

| (+) | | (+) |
|---|---|---|
| **Step Over:** | **Magnitude** | **Rise or Fall?** |
| $R_1$ | $R_1 I_1$ | Rise |

Again it rises because we are walking from head to tail, head is low, tail is high so it rises. Some students get confused with this so if you are having trouble put a little x on the tails of all the $I$ arrows you draw. The x marks the high side then as you walk over the resistors ask yourself "am I stepping up to the x (a rise) or stepping down from the x (a fall)". Also note that in equation 2.5 the $R_1I_1$ term was negative (we were walking left to right then) and now its positive (we are walking right to left now). This is exactly correct. What you write in one KVL may differ from what you write in another, it all depends on which way you are walking in the loop and that is arbitrary! Now that we are back to the start, we have:

$$-R_4I_4 + R_2I_2 + R_1I_1 = 0 \tag{2.6}$$

Now for the third and final loop we have to walk over the current source since we haven't done that yet so choose the loop DCBD. Of course another valid loop would be DCBAD but, what the heck, use DCBD. **Now here is a common mistake that students make, and I'll bet your teacher will test this so pay close attention!** The voltage drop across a current source is **NOT** necessarily zero, it might be but it probably isn't. Remember, if you **KNOW a flow** (current) you **do not usually know the effort** (voltage) and when you KNOW the effort (like in the voltage source) you do not usually know the flow. For some reason many students have no problem with the fact that they know $V$ across the voltage source but do not know the $I_5$ through it. But for some reason when they know the current (like $I$ through the current source) they want to also say they know the voltage across it is zero. WRONG! Okay enough said, back to the problem. The voltage change across the current source is:

| (?) | | (?) |
|---|---|---|
| **Step Over:** | **Magnitude** | **Rise or Fall?** |
| Current Source | $V_C - V_D$ | (see discussion) |

First of all notice the rise or fall is the difference in voltage between C and D hence we subtract them. Second, it doesn't really matter whether we write C minus D or D minus C because we are going to use our imagination to get the sign correct. In fact, after we do it this way, we'll repeat it another way to show that its the same thing. Play the imaginary game, pretend that the expression we wrote $(V_C - V_D)$ is positive. This means C is above D, so put a pencil mark x on C. Now we are walking from D to C so we step up on the x, hence this is a rise:

| (+) | | (+) |
|---|---|---|
| **Step Over:** | **Magnitude** | **Rise or Fall?** |
| Current Source | $V_C - V_D$ | Rise |

Now since D is ground, it is zero and this reduces to:

| (+) | | (+) |
|---|---|---|
| **Step Over:** | **Magnitude** | **Rise or Fall?** |
| Current Source | $V_C$ | Rise |

Okay now as we promised, let's do the same term the other way. First express the voltage change:

| (?) | | (?) |
|---|---|---|
| **Step Over:** | **Magnitude** | **Rise or Fall?** |
| Current Source | $V_D - V_C$ | (see discussion) |

Now pretend that the expression is positive, so D is higher than C. Erase the x on C and put it on D. Now we are still walking from D to C therefore we step down from the x so this is a fall:

| (−) | | (−) |
|---|---|---|
| **Step Over:** | **Magnitude** | **Rise or Fall?** |
| Current Source | $V_D - V_C$ | Fall |

Since D is ground, it is zero and we have:

| (−) | | (−) |
|---|---|---|
| **Step Over:** | **Magnitude** | **Rise or Fall?** |
| Current Source | $-V_C$ | Fall |

Notice that when we put all the terms together the first expression ($V_C$) is a rise so we add it, hence we have a $+V_C$ in the equation. The second expression ($-V_C$) is a fall so we will subtract it to obtain $-(-V_C) = +V_C$ (the first negative is due to the fact that it is a fall, the second comes from the expression for the voltage change). This is the same thing. You see, the equation will come out the same regardless of whether you take D minus C or C minus D.

Okay, back to the rest of the walking. Step over $R_2$ to find:

| (+) | | (+) |
|---|---|---|
| **Step Over:** | **Magnitude** | **Rise or Fall?** |
| $R_2$ | $R_2 I_2$ | Rise |

now step over $R_3$:

| (−) | | (−) |
|---|---|---|
| **Step Over:** | **Magnitude** | **Rise or Fall?** |
| $R_3$ | $R_3 I_3$ | Fall |

Our walk is finished so add the rises and subtract the falls to find:

$$V_C + R_2 I_2 - R_3 I_3 = 0 \tag{2.7}$$

In the three equations, we have five unknowns, $I_1$, $I_2$, $I_3$, $I_4$ and $V_C$. There is 3 [NIF] so there should be only 3 currents (one of which is $I$) so we can keep only two of the four flows in the equations. We have to relate some of the flows. We expected this already when we estimated the number of conservation of charge equations we needed. Let's write them. Choose node B to relate currents 1, 3 and 2 as:

$$I_1 - I - 3 - I_2 = 0 \tag{2.8}$$

Now node C to relate $I$, $I_2$ and $I_4$:

$$I_2 + I + I_4 = 0 \tag{2.9}$$

Now notice that we have 5 equations with 5 unknowns but we have not used the third charge equation we predicted to need. Why? Well notice that $I_5$ does not appear in any of the equations, if we don't care about $I_5$ we are done. Solve the equations. If we do want to find $I_5$ then write one more charge equation. This time use node A:

$$I_5 - I_1 - I_4 = 0 \tag{2.10}$$

Now the only thing left is to solve the six equations for the six unknowns. Maple gives us the following:

$$I_2 = -1.56, I_3 = 2.89, V_C = 11.8, I_1 = 1.33, I_4 = -0.444, I_5 = 0.889$$

Now this takes care of one requirement which is to determine the currents flowing in each element. What we still do not know are the node voltages.

Once we know the currents in each element, the node voltages are relatively easy to determine. One method to find the voltages is to take individual elements as a system and write conservation of energy or sometimes just the impedance relationship for each. For example take $R_2$ as a system and write the impedance relationship:

$$V_{\text{tail}} - V_{\text{head}} = IR$$

which is:

$$V_B - V_C = I_2 R_2 \rightarrow V_B - 11.8 = -1.56(2) \rightarrow V_B = 8.67$$

Now do the same for $R_4$:

$$V_{\text{tail}} - V_{\text{head}} = IR \rightarrow V_A - V_C = I_4 R_4 \rightarrow V_A = 11.8 + (-0.444)(4) = 10.02$$

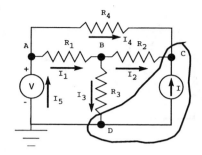

Figure 2.14: The Circle Shows a System of Two Nodes.

Now we have all the calculations we wanted to find.

Now look at the results. Notice that $I_1 > 0$ this means the arrow drawn for resistor 1 is correct. The current flows from A toward B. The value of $I_2 < 0$ which means the current is backwards to what we drew. The current flows from C toward B. Current 4 is negative, so the flow is from C to A. We know the current flows from B to D because the value found for $I_3 > 0$. Current 5 is positive so flow travels from D to A, hence the source is adding electrical energy to the circuit. If current flowed into the high voltage side of the voltage source, the circuit would be "charging" the source. If it were a battery and current went into the positive terminal, the battery is "charging". Write a conservation of energy on the voltage source to prove this to yourself. Is the current source putting energy into the circuit, or is the circuit putting energy into the current source? How do you know.[4]

**End 2.2** ─────────────────────────────────────────────────
**Example 2.3** ─────────────────────────────────────────────

### Conservation of Charge at Two Nodes

**PROBLEM:**
Show that it is possible to express conservation of charge on a system that has more than 1 node.
**SOLUTION:**
Consider the system circled in figure 2.14 which consists of two nodes and a current source. We will apply the conservation of charge to the system and use the results from example 2.2 to check our results.

Now remember, the purpose of the circle is to make you focus on the input/output to the system. You only have to locate places where charge crosses the circle and you will know where charge enters or exits. Charge at a rate of $I_5$ exits the circle travelling through the voltage source. Charge at the rate of $I_3$ enters at the wire connected to $R_3$. An amount $I_2$ enters from $R_2$ and $I_4$ enters through $R_4$. The quantity $I$ does not enter nor exit since it does not cross the circle. If you draw the circle differently, you can have different terms but for the one shown these are the terms that enter and exit. Summarizing the terms we have:

$$-I_5 + I_3 + I_2 + I_4 = 0 \tag{2.11}$$

Now we will check this by plugging in the values determined from example 2.2. Using the numbers we have:

$$-0.889 + 2.89 + (-1.56) + (-0.444) = -0.003 \approx 0$$

Actually, if you did the algebra exactly (as Maple would do) you will find that equation 2.11 is exactly zero.

What this example is supposed to show (which you probably already know) is that you can apply the conservation of charge on all types of systems not just single nodes. We will almost always use

---

[4]The current source is putting energy into the circuit because current comes out the more positive side of the source.

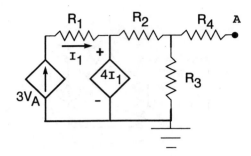

Figure 2.15: A Circuit With a Dependent Voltage and Current Source.

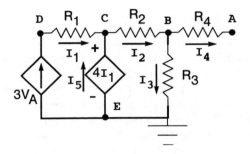

Figure 2.16: Nodes and Currents in the Circuit.

single nodes in this text but you can do whatever you find useful.

**End 2.3** ────────────────────────────────────────────

**Example 2.4** ──────────────────────────────────────

## A Dependent Voltage and Current Source.

**PROBLEM:**

In this example, we will demonstrate handling dependent sources and wierd things like incomplete circuits. The objective is to determine the element currents and node voltages in the circuit shown in figure 2.15. Let the resistance values match the subscript. For example $R_3 = 3$ ohms.

**SOLUTION:**

The circuit has 2 [NIF], 5 nodes (see figure 2.16) and six currents (counting the dependent current source). We expect to write two KVL equations and $5 - 2 + 1 = 4$ conservation of charge equations.

Without much ado, write a KVL around the loop EDCE. Remember, there may be a voltage across the current source from E to D. The KVL is (put in the steps if you need to):

$$V_D - R_1 I_1 - 4I_1 = 0 \tag{2.12}$$

Note that the dependent source has a voltage change of $4I_1$. Now for the loop ECBE. Again write the steps if you need to:

$$4I_1 - R_2 I_2 - R_3 I_3 = 0 \tag{2.13}$$

There are 5 unknowns, $I_1$, $I_2$, $I_3$, $V_A$ and $V_D$ but only 2 equations. Since there are too many unknowns, we write conservation of charge equations. First start with node D:

$$3V_A - I_1 = 0 \tag{2.14}$$

now take node C:

$$I_1 + I_5 - I_2 = 0 \tag{2.15}$$

now for node B:

$$I_2 - I_3 - I_4 = 0 \tag{2.16}$$

finally node A:

$$I_4 = 0 \qquad (2.17)$$

Another two unknowns ($I_4$ and $I_5$) have been introduced in these equations so there are now seven unknowns and 6 equations. We need one more. What would happen if you wrote the conservation of charge at node E? The equation would be dependent with equations 2.14 to 2.17. How do we know? Look at how many flows are involved in the node D, C, B, A and E equations. There are six currents 1, 2, 3, 4, 5 and the dependent flow. Only five of these are unknown which means if all five node equations (for nodes D, C, B, A and E) are independent then we could find all the unknown flows by relating them one to another. But this cannot be. Since there is 2 [NIF], there must be 2 flows that cannot be related. If all of this is confusing, write out the node E equation and try to solve the problem using it. You cannot do it.

Well the situation now is there are seven unknowns but only six equations. If we solve it now we find:

$$I_1 = 3V_A, I_4 = 0, V_D = 15V_A, I_3 = \frac{12}{5}V_A, I_2 = \frac{12}{5}V_A, I_5 = -3/5V_A$$

Note that everything can be found in terms of $V_A$. Can you find $V_A$ since we have used all the KVL equations and all the node equations? Sure. Remember you were warned that sometimes you need extra equations after you write all the expected equations. You know the missing one is not a node equation so pick another system and apply energy or impedance relationships. Take $R_4$ as the system to find:

$$V_B - V_A = R_4 I_4$$

Since $I_4 = 0$, we find $V_A = V_B$ (you may have known this by looking at the circuit). Now we need to find $V_B$ so take $R_3$:

$$V_B - V_E = R_3 I_3$$

Since $V_E = 0$ we can simplify this to:

$$V_A = R_3 I_3$$

Using the value for $I_3$ we have:

$$V_A = R_3 \frac{12}{5}V_A \rightarrow 0 = \left[\frac{36}{5} - 1\right] V_A \rightarrow V_A = 0$$

Wow! What a shock! The voltage at A is zero which means all the currents are zero! This is a result of this particular circuit configuration and is not a general result. Hey after all, this is just an example!

**End 2.4** _____

**Example 2.5** _____

## Another Dependent Source Problem.

**PROBLEM:**

As a final example consider a circuit similar to the one figure 2.15 shows. The only exception is to replace the dependent voltage source with a constant voltage source of 10 volts.

**SOLUTION:**

Without too many words, use the same KVL and node equations as previously. Start with KVL loop EDCE:

$$V_D - R_1 I_1 - 10 = 0$$

Now KVL loop ECBE:

$$10 - R_2 I_2 - R_3 I_3 = 0$$

Now node D:

$$3V_A - I_1 = 0$$

Node C:

$$I_1 + I_5 - I_2 = 0$$

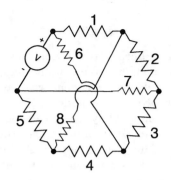

Figure 2.17: A Challenging Circuit.

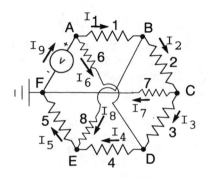

Figure 2.18: Nodes in the Challenging Circuit.

Node B:
$$I_2 - I_3 - I_4 = 0$$

Node A:
$$I_4 = 0$$

Element $R_4$:
$$V_B - V_A = R_4 I_4$$

Element $R_3$:
$$V_B - 0 = R_3 I_3$$

Now solve to find:

$$I_4 = 0, V_D = 28, I_5 = -16, I_3 = 2, V_B = 6, V_A = 6, I_1 = 18, I_2 = 2$$

Notice that $I_5 < 0$ so the current flows into the positive terminal of the voltage source. If the source were a battery, it would be "charging".

**End 2.5** ━━━━━━━━━━━━━━━━━━━━━━━━━━━━━━━━━━━━━━━━━━━━━

**Example 2.6** ━━━━━━━━━━━━━━━━━━━━━━━━━━━━━━━━━━━━━━━━━

## A Nonplanar Circuit

**PROBLEM:**
    Determine all the node voltages and currents present in the circuit shown in figure 2.17.
**SOLUTION:**
    The circuit has 4 [NIF], and 6 nodes. Figure 2.18 shows the 6 nodes, ground and the definitions of the element currents.
    For the first KVL take loop FADEF:

$$10 - 6I_6 - 4I_4 - 5I_5 = 0$$

The second KVL take loop FABEF:

$$10 - 1I_1 - 8I_8 - 5I_5 = 0$$

For the next KVL, take loop FABCF:

$$10 - 1I_1 - 2I_2 - 7I_7 = 0$$

For the final, KVL use loop FABCDEF:

$$10 - 1I_1 - 2I_2 - 3I_3 - 4I_4 - 5I_5 = 0$$

There are 8 flows in these equations plus 1 through the voltage source making 9 flows - 4 [NIF] = 5 conservation of charge equations. Take node A first:

$$I_9 - I_1 - I_6 = 0$$

Now for node B:

$$I_1 - I_2 - I_8 = 0$$

Now node C:

$$I_2 - I_3 - I_7 = 0$$

Now node D:

$$I_3 + I_6 - I_4 = 0$$

Finally node E:

$$I_8 + I_4 - I_5 = 0$$

This gives us 9 equations for 9 unknowns. Maple can solve these easily to obtain:

$$I_1 = 1.401, I_4 = 0.498, I_6 = 0.528, I_2 = 0.931, I_5 = 0.968, I_3 = -0.0309, I_8 = 0.470, I_7 = 0.962, I_9 = 1.93$$

Note that all the current arrows are drawn correctly except $I_3$ which flows from D to C, pretty weird.
Having all the element currents makes it easy to find the node voltages. Take element 5 as:

$$V_E - V_F = 5I_5 \rightarrow V_E = 4.838$$

Now take element 7:

$$V_C - V_F = 7I_7 \rightarrow V_C = 6.737$$

Now element 8:

$$V_B - V_E = 8I_8 \rightarrow V_B = 8.599$$

Now take element 4:

$$V_D - V_E = 4I_4 \rightarrow V_D = 6.829$$

From the voltage source:

$$V_A = 10$$

**End 2.6**

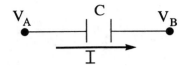

Figure 2.19: A Diagram of Electrical Capacitance

## 2.4   ELECTRIC FIELDS AND CAPACITANCE

**Objectives**
*When you complete this section, you should be able to*

1. *Recognize ideal capacitors when drawn in a circuit.*

2. *Determine the [NIF] of a circuit having capacitors.*

3. *Explain why current can flow in a capacitor even though its circuit diagram looks like a break.*

4. *Recognize realistic voltage as a function of time graphs for voltage across a capacitor.*

5. *Calculate the electrical energy stored in a capacitor.*

6. *Determine the [NIA] in capacitive circuits.*

7. *Calculate the voltages and currents in capacitive circuits.*

8. *Recognize the voltage/current (across/through) relationships for capacitors.*

A capacitor is formed by placing two conductors very close to each other. Normally, for large capacitance, the manufacturer uses a large area on the two conductors (often called plates) and places them very close together. To prevent electrical contact between the two plates, the manufacturer places an isolator (dielectric) between them. A capacitor operates when placed in a closed circuit and negative voltage on one plate forces electrons to collect in it driving electrons out of the other.

Capacitance is very common. In fact, circuits can have capacitance even without explicitly designing it in them. For example, power lines, which are generally several miles long, can have a significant capacitance. Power lines always have at least two wires, one high voltage and one ground. These wires form part of a circuit and are separated by an insulator. Even though the wires may be separated a large distance (decreasing capacitance per unit length), their length causes them to possess considerable capacitance. Likewise, integrated circuits[5] can posses considerable capacitance due to the close proximity of components.

Figure 2.19 shows electrical capacitance schematically. The capacitor has the ability to store electrical energy [12]. We can write the electrical energy stored as a function of the voltage across the capacitor $(V)$ as $E = \frac{1}{2} C \left( V_A - V_B \right)^2$. The capacitor has a voltage/current (impedance ) relationship, which we can derive by accounting the electrical energy of the system. Taking the single capacitor of figure 2.19 as our system and accounting electrical energy, we get

---

[5]An integrated circuit is a collection of elements built into a very small volume. ICs typically require magnifying lenses to observe. They are very common and are usually packaged inside black plastic or ceramic with electrical connections protruding.

**Electrical Energy Accounting**
system[**capacitor fig.2.19**]       time period [**differential**]

*input − output + generation − consumption = accumulation*

| input/output: | $V_A I - V_B I$ |
|---|---|
| consumption (thermal): | |
| generation: | |
| accumulation: | $\frac{1}{2} C (V_A - V_B)^2$ |

| $V_A I - V_B I$ | = | $\frac{d}{dt} \left( \frac{1}{2} C (V_A - V_B)^2 \right)$ |
|---|---|---|
| *input/output* | | *change in accum.* |

Summing what enters, subtracting what exits, and setting this equal to the time rate of change of the accumulation, we get the following:

$$V_A I - V_B I = \frac{d}{dt} \left( \frac{1}{2} C (V_A - V_B)^2 \right) = C (V_A - V_B) \left( \dot{V}_A - \dot{V}_B \right)$$

or if we divide by $(V_A - V_B)$

$$I = C \left( \dot{V}_A - \dot{V}_B \right) \tag{2.18}$$

Capacitance is measured in Farads. Real capacitors are typically in sizes much smaller than 1 Farad. Micro or Pico Farad capacitors are common. According to equation 2.18 we can relate Farads to other units as follows. 1 Amp = 1 Farad Volt/Second. Or using power we have, 1 Farad times Volt squared per Second = 1 Watt. This leads to: 1 Farad times Volt Squared = 1 Newton Meter.

If we account for net charge in a complete capacitor, we will find no charge accumulation. For every electron entering one side of the capacitor, an electron leaves the other side. In fact, the movement of electrons is the current. If we take a single plate of the capacitor as a system, charge accumulation can occur. This accumulation would equal $CV$ where $C$ is the capacitance of the capacitor and $V$ is the voltage across the capacitor.

When counting the [NIA] in a circuit with capacitors, each capacitor has the ability to accumulate energy. It can also accumulate charge if the system consists of only one plate. Since the charge in a plate and the energy are both given as functions of voltage, the ideal capacitor cannot provide more than one [NIA]. Since accumulation in a capacitor is a function of the voltage across the capacitor, we look at the voltage to determine if multiple capacitors provide independent accumulations. For example if knowing the voltage across one capacitor at time $t$ automatically determines the voltage across another capacitor then they are dependent.

Consider the impedance relationship given in equation 2.18. If the voltage across the capacitor is constant, the number of electrons in each plate remains constant being in force[6] equilibrium. Only when the voltage changes do the electrons move. When they move, they move instantaneously (actually they move very quickly but not instantaneously) until they regain force equilibrium. Hence, only if the voltage continues to change will electrons (current) continue to flow.

Although electricity is very different from water, we will use an analogy to help explain why current flows in a capacitor. Now this is a simplified view of the phenomenon so do not go too far with the analogy. Suppose you have a pipe with water in it. You cut the pipe and on the two open ends just formed you put rubber balloons each partially filled with water. As water is pumped around (through) the pipe, water fills one balloon and empties from the other. As one balloon fills, the rubber stretches increasing the pressure inside. As the other empties, the rubber relaxes and its pressure drops. Maybe the balloon even gets "sucked into" the pipe creating a negative pressure. Eventually (assuming the balloons do not break) the pressure in one balloon is so large and in the other it is so small that pump cannot handle such a pressure difference that it stops moving water. Flow stops, but one balloon remains very full and the other remains very empty. If the pump's pressure handling ability changes (increases or decreases), then flow will resume. If the pump's pressure handling ability increases, a little more water will enter the full balloon (increasing

---

[6]The forces acting on the electrons are electrostatic (likes repel, opposites attract).

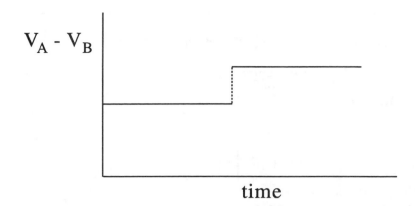

Figure 2.20: Impossible Voltage Condition on a Capacitor

pressure) and the same amount will leave the empty balloon (decreasing pressure) until the pump cannot handle the difference and flow stops again. If the pressure decreases, water will leave the full balloon and move to the empty balloon until the balloon pressures balance what the pump can handle. This is similar to a capacitor. Voltage is like the pressure and current is like the water flow. Voltage across the capacitor is like the pump pressure. If current flows in a capacitor, electrons flow in one side (into one balloon) and different electrons flow out the other (out of the other balloon).

When computing the [NIF] of a circuit having capacitors, students sometimes get confused by the capacitors. For example, to find the [NIF] one attempts to determine the minimum number of wires to cut so there are no closed loops in the circuit. When they get to a loop having a capacitor, due to its symbol looking like a cut wire, they often think the loop is already cut and fail to count it. Since it is possible for current to flow in a capacitor, it is not the same as a cut wire.

Consider the possibility that the voltage across a capacitor might instantly jump as figure 2.20 shows. If the voltage does jump, there will be an infinite derivative of voltage (vertical change) implying from equation 2.18 that the current flow is infinite. This is clearly unrealistic. Only in special idealized circuits will the voltage across a capacitor jump. In fact, a rule of thumb is that capacitors attempt to prevent sudden voltage changes. The next example explores the possibility of sudden voltage changes across a capacitor.

**Example 2.7** ────────────────────────────────────────────

## Discharging a Capacitor

**PROBLEM:**

Dr. Zap does research on high power laser beams. His laser beam requires very large current for a short period of time. He believes he can obtain this by using the circuit in figure 2.21. His concept is to open the switch, let the voltage and currents settle, and then suddenly close the switch. He hopes that this will cause a sudden change in $V_A$ (e.g. $\dot{V}_A = \infty$) that will produce an infinite current through the capacitor (according to equation 2.18). Determine the actual current that Dr. Zap's circuit will provide. Dr. Zap closes the switch at time zero and then reopens it after 0.02 [sec].

**ANSWER:**

The resistor $R_L$ represents the laser beam and $R$ represents the output impedance of the power source. Ideally, Dr. Zap wants an ideal source in which $R = 0$. Such a source would give him as much current as his laser demands, but unfortunately, he cannot get the current he needs.[7] The resistor in line with the ideal voltage supply is a common method for modeling real sources. For example an automobile battery is intended to provide a constant (often 12 volts) voltage to a circuit. It cannot provide a constant voltage with infinite current. The electrical power (voltage times current) comes

────────────────────────────────────────────

[7]Incidentally, this is not unrealistic. There are often times when the flow demanded from a power company exceeds its ability. When this happens the supply voltage drops and the power users experience a brown out. Usually brown outs occur because many people are demanding current. The way a person "demands" a lot of current is to connect a small resistor between the terminals of the power outlet. Since voltage magnitude is supposed to be constant, the smaller the resistance, the more the current.

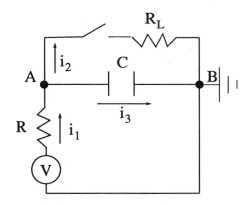

Figure 2.21: A Circuit that Quickly Discharges a Capacitor

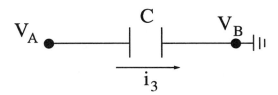

Figure 2.22: A System Consisting of a Capacitor

from a chemical reaction inside the battery and the current is limited by the speed at which the reaction occurs (among other things). A good strong battery would have a small output impedance ($R$).

The basic idea of Dr. Zap is not too radical, essentially he is using the capacitor to store a supply of energy, which he will collect slowly (governed by $R$). After collecting the energy, he plans to dump it quickly through $R_L$. What we will do is calculate the current through $R_L$. For this analysis, we will use the accounting of electrical energy just for practice. Later however, we will use Kirchhoff's voltage law (KVL) almost exclusively.

First, we determine the [NIF] and [NIA] of Dr. Zap's circuit. With the switch closed, there is two [NIF]. When it is open, there is one [NIF]. The two possibilities are summarized in the table

| | Condition | [NIF] | Comments |
|---|---|---|---|
| below. | Switch open | 1 | Occurs when $t > 0.02$, $i_2 = 0$ |
| | Switch closed | 2 | Occurs when $0 < t < 0.02$, $i_2$ is unknown |

Accounting for the electrical energy in the capacitor, we find that

**Electrical Energy Accounting**

system[**capacitor fig.2.22**]      time period [**differential**]

*input − output + generation − consumption = accumulation*

| input/output: | $V_A i_3 - V_B i_3$ |
|---|---|
| consumption (thermal): | |
| generation: | |
| accumulation: | $\frac{1}{2}C\left(V_A - V_B\right)^2$ |

| $V_A i_3 - V_B i_3$ | $=$ | $\frac{d}{dt}\left(\frac{1}{2}C\left(V_A - V_B\right)^2\right)$ |
|---|---|---|
| *input/output* | | *change in accum.* |

Since $V_B$ is on the grounded wire, it has zero voltage. Thus, the accounting statement simplifies to

$$V_A i_3 = \frac{d}{dt}\left(\frac{1}{2}CV_A^2\right)$$

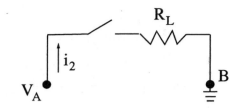

Figure 2.23: A System for Determining $i_2$

or

$$i_3 = C\dot{V}_A \tag{2.19}$$

We could have written this immediately using the impedance relation for a capacitor. Accounting for the electrical energy in the resistor $R$, we find that

**Electrical Energy Accounting**

system[**resistor $R$**]      time period [**differential**]

*input − output + generation − consumption = accumulation*

| | |
|---|---|
| **input/output:** | $Vi_1 - V_Ai_1$ |
| **consumption (thermal):** | $i_1^2 R$ |
| **generation:** | |
| **accumulation:** | |

$$\boxed{Vi_1 - V_Ai_1} \quad - \quad \boxed{i_1^2 R} \quad = \quad \boxed{0}$$

*input/output*        *consumption*        *change in accum.*

$$Vi_1 - V_Ai_1 - i_i^2 R = 0$$

which simplifies to

$$i_1 = \frac{V - V_A}{R} \tag{2.20}$$

This could have been determined immediately using the impedance relation for a resistor. Finally, we can determine $i_2$ by choosing the system in figure 2.23.

If the switch is open, $i_2 = 0$. In this case, the electrical energy accounting statement for figure 2.23 would be

**Electrical Energy Accounting**

system[**circuit fig.2.23**]      time period [**differential**]

*input − output + generation − consumption = accumulation*

| | |
|---|---|
| **input/output:** | $V_Ai_2 - V_Bi_2$ |
| **consumption (thermal):** | $i_2^2 R_L$ |
| **generation:** | |
| **accumulation:** | |

$$\boxed{V_Ai_2 - V_Bi_2} \quad - \quad \boxed{i_2^2 R_L} \quad = \quad \boxed{0}$$

*input/output*        *consumption*        *change in accum.*

Since $V_B = 0$, this simplifies to

$$V_Ai_2 - i_2^2 R_L = 0$$

$$i_2 = \frac{V_A}{R_L} \tag{2.21}$$

We now have three equations containing four unknowns ($i_1$, $i_2$, $i_3$, and $V_A$). One of our equations is a first order differential equation (as [NIA] predicts). There are three unknown flows, and there

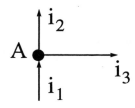

Figure 2.24: Node A Chosen as a System

should be only one or two (depending on whether the switch is open or closed). Thus, we need to relate flows. This is done by applying the conservation of charge at node A. Node A has all three flows entering or leaving so it is a good choice. Applying the conservation of charge, we find that

**Conservation of Charge**

system[**node A fig.2.24**]    time period [**differential**]

*input − output = accumulation*

| input/output: node A accumulation: | $i_1 - i_2 - i_3$ |
|---|---|

| $i_1 - i_2 - i_3$ | $=$ | $0$ |
|---|---|---|
| *input/output* | | *change in accum.* |

$$i_1 - i_2 - i_3 = 0 \tag{2.22}$$

Substituting equations 2.19, 2.20 and 2.21 into 2.22 gives

$$\frac{V - V_A}{R} - C\dot{V_A} - \left\{ \begin{array}{ll} 0 & \text{if the switch is open} \\ \frac{V_A}{R_L} & \text{if the switch is closed} \end{array} \right. = 0 \tag{2.23}$$

**SOLUTION:**

Because our equation for $V_A$ (equation 2.23) is one [NIA], we will need one initial condition before we can solve for the response. Later we will discuss in detail how to determine initial conditions, but for the moment lets assume the switch has been left open a long time and that $V_A = V$. At this point, we have to recognize that this initial condition is indeed possible. To recognize this, consider whether or not it satisfies Dr. Zap's equations. The initial condition means $i_1 = 0$,[8] $i_2 = 0$ because the switch is open and $i_3 = 0$.[9] Since equation 2.22 is obviously satisfied by zero currents, the initial condition is possible. Later in the text, we will show why $V_A = V$ is the only possible initial condition for a constant $V$ if the switch is open for a very long time.

At time zero, Dr. Zap closes the switch. Then, after 0.02 seconds, he reopens it (he obviously has fast hands). Because we want to determine the response of the circuit, we must solve for $V_A$ and the associated currents. We can solve equation 2.23 in a number ways. We can find the analytical solution by integrating the equation piece by piece (once when the switch is open, then again when the switch is closed). Or, we can find the numerical solution. For this problem, we used maple to find an analytical solution. For further details, see the Maple listing 2.12.2. When the switch is closed, the solution for $V_A$ is

$$V_A(t) = \frac{V}{(R_l + R)} \left( e^{-\frac{t(R_l + R)}{CRR_l}} R + R_l \right) \tag{2.24}$$

When the switch is open, the solution for $V_A$ is

$$V_A(t) = -\frac{V}{(R_l + R)} \left( -1.0R_l - 1.0R + Re^{\frac{-1.0t + 0.02}{CR}} - 1.0Re^{\frac{-1.0R_l t - 0.02}{CRR_l}} \right) \tag{2.25}$$

---

[8]$i_1$ is zero because $V_A = V$, see equation 2.20.

[9]$i_3$ is zero because $V_A$ is constant, see equation 2.19.

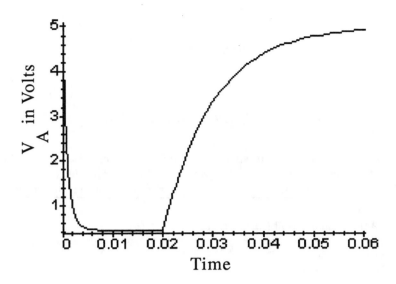

Figure 2.25: $V_A$ Versus Time for Zap's Circuit.

To make sense of this, let $C = 0.0001$ [F], $R_l = 10$ [$\Omega$], $R = 100$ [$\Omega$] and $V = 5$ [V]. With these values, figure 2.25 shows a plot of $V_A$ versus time. To find currents, we use $V_A$ with equations 2.20 and 2.21 to find $i_1$ and $i_2$. Then, we use equation 2.20 to solve for $i_3$. Figure 2.26 shows the current versus time relationship of the circuit.

**Discussion:**

The figure shows that as the switch is suddenly closed, we get a sharp jump in $i_2$, but the voltage $V_A$ does not jump. We should have expected this because the capacitor attempts to prevent sudden changes in voltage. As the switch suddenly opens, $i_2$ drops to zero, but again the voltage does not jump. The reason the voltage cannot jump is because a jump would imply an infinite current flow through the capacitor. If an infinite current did flow, there would have to be an infinite current coming into node A through one or both of the resistors. An infinite current across either resistor would require an infinite voltage drop. An infinite voltage drop cannot exist for finite $V$. As a result, the current is limited. Basically, the resistors are choking off the large current.

Notice that for a short time, Dr. Zap is getting more current than he would have without the capacitor. The next question is why is this happening? When you study differential equations, you will learn that differential equations like equation 2.23 have two parts to their solution, a transient (a part that usually disappears) and a steady state (part that "hangs around"). Equation 2.23 is no exception. The transient is the part with the decaying exponential. The steady state is everything left after the exponential becomes very small. Figure 2.25 shows the transient as the part that falls from 5 volts. The steady state is the part of the curve that is flat.[10] Following the constant voltage segment, the switch is thrown and another transient ensues as the voltage builds toward 5. The figure quits before showing the second steady state region. Transient periods normally occur after a change. In this case, the change is the position of the switch.

Basically, the circuit has a transient period when $V_A$ is falling from its initial value of $V$ down to its steady state value (where it levels off at approximately 0.008 seconds in figure 2.25). The extra current comes from the fact that $V_A$ is elevated above its steady state value for a short time. With a larger than normal $V_A$ across the laser, there is a larger than normal current.

To see how the capacitor operates, consider the circuit without the capacitor. This circuit would be zero [NIA]. The transient state of this circuit would disappear instantaneously. Opening the switch would cause $V_A$ to rise to $V$, but as soon as the switch is closed, $V_A$ would immediately fall to its steady state value, and $i_2 < \frac{V}{R_l}$. The circuit with the capacitor in it has a [NIA] of one. You

---

[10] The steady state solution is not always constant, but it is persistent. You will learn more about this in a differential equations class.

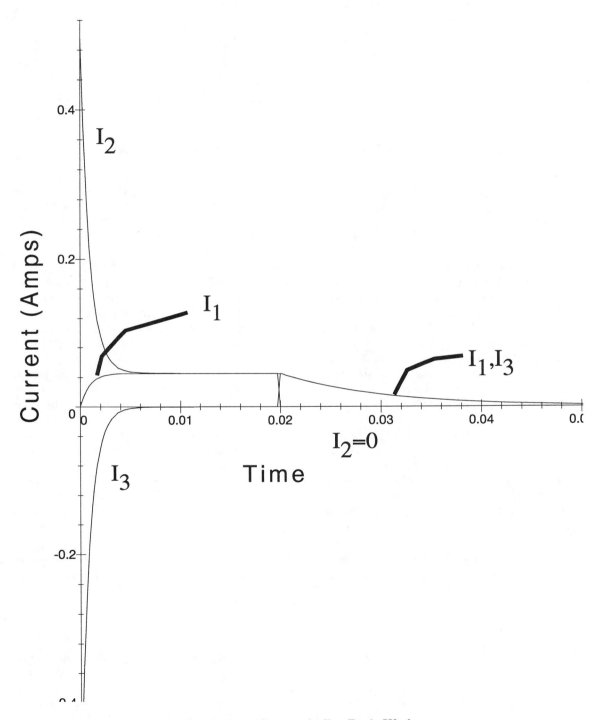

Figure 2.26: Current in Dr. Zap's Work

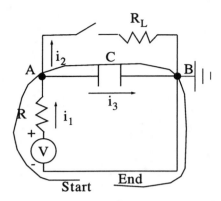

Figure 2.27: The Path We Will Walk to Find One KVL Equation for Zap.

see one effect of [NIA] is to produce a transient. This circuit has a transient state because it can accumulate charge, because it has a [NIA]. The longer the transient of the circuit, (the less negative the power on the exponential terms in the solution) the longer it takes for $V_A$ to fall to its steady state value and the longer we will have an elevated current. Since [NIA] is related to the transient, what would produce a longer transient, a large capacitance, or a smaller?

Another way to visualize this situation is to consider the output impedance of the source and the capacitor. The output impedance of the source ($R$) chokes off the output current available from the supply causing the supply voltage at node A to drop well below five. Since the capacitor does not have an output impedance resistor choking its current off, it is able to supply a larger voltage to the laser than the source. Unfortunately, the capacitor's voltage drops with the amount of current so it cannot sustain the voltage forever. With this way of thinking, a larger capacitance would hold a large current on the laser for a longer time (same conclusion as before).

A third way to understand this situation is through energy. The capacitor has a storage of electrical energy, which it dumps into the laser when the switch is closed. Eventually the energy stored in the capacitor is exhausted and its contribution to the increased flow through the laser drops off.

If Dr. Zap wanted the circuit to provide an elevated current to the laser for a longer time period, he should increase $C$. This would increase the transient state. To see this, consider the exponential terms in equations 2.24 and 2.25, increasing $C$ makes the power in these terms a smaller negative number, hence they "hang around" longer. From an energy standpoint, increasing $C$ increases the energy stored in the capacitor so that there is more to dump through the laser. Note, however, the maximum current Dr. Zap can get at any time does not change when $C$ is changed. The maximum current is governed by the impedance of the laser ($R_l$) and the maximum voltage in the circuit ($V$).

**End 2.7** ━━━━━━━━━━━━━━━━━━━━━━━━━━━━━━━━━━━━━━━━━━━━━━━

**Example 2.8** ━━━━━━━━━━━━━━━━━━━━━━━━━━━━━━━━━━━━━━━━━━━

## Dr. Zap Revisited

**PROBLEM:**

What we desire to do now is solve Dr. Zap's problem using the more convenient method of Kirchhoff's Voltage Law.

**FORMULATION**

Zap's circuit has two [NIF] and one [NIA] therefore we expect to write two KVL equations and expect to find one first order differential equation. Figure 2.27 shows the path we will take for the first loop. The first thing we step over is the voltage source, which looks like a rise:

| (+) | | (+) |
|---|---|---|
| **Step Over:** | **Magnitude** | **Rise or Fall?** |
| Voltage Source | $V$ | Rise |

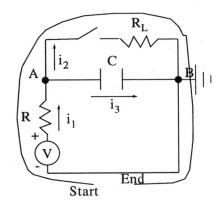

Figure 2.28: The Second Path for KVL.

Next we step over the resistor $R$:

| (?) | | (?) |
|---|---|---|
| **Step Over:** | **Magnitude** | **Rise or Fall?** |
| $R$ | $I_1R$ | (see discussion) |

The tail of the current always has the higher voltage so if $I_1$ is positive (look at the picture) then as we walk from bottom to top, the voltage appears to fall, hence:

| (−) | | (−) |
|---|---|---|
| **Step Over:** | **Magnitude** | **Rise or Fall?** |
| $R$ | $I_1R$ | Fall |

Next we step over the capacitor and write:

| (?) | | (?) |
|---|---|---|
| **Step Over:** | **Magnitude** | **Rise or Fall?** |
| $C$ | $V_A - V_B = V_A$ | (see discussion) |

Notice that we avoid writing the expression that contains $C$ and current because it is an integral and we prefer to postpone using an integral because many people have difficulty understanding the integral term. Every time we cheat like this, we will have to write an extra equation for the capacitor's impedance, more on this later. To determine whether this expression is a rise or a fall, we plan a little game. We'll do the same game throughout the text so it will benefit thinking about it carefully now. Pretend conditions are just right in the circuit so that the expression we wrote $(V_A - V_B)$ is positive, in other words, pretend $V_A > V_B$. Now we are not saying this is true, we are simply looking for consistency. If this were true, would it be a rise or a fall? Note that if $V_A$ is positive, then the voltage change looks like a fall so:

| (−) | | (−) |
|---|---|---|
| **Step Over:** | **Magnitude** | **Rise or Fall?** |
| $C$ | $V_A$ | Fall |

After stepping over the capacitor, we are back at the start therefore we collect all the voltage changes and equate them to zero to find:

$$V - RI_1 - V_A = 0 \qquad (2.26)$$

For the second loop, we will walk the loop shown in figure 2.28 First we step across the voltage supply:

| (+) | | (+) |
|---|---|---|
| **Step Over:** | **Magnitude** | **Rise or Fall?** |
| Voltage Source | $V$ | Rise |

Next we step across the resistor $R$:

| (−) | | (−) |
|---|---|---|
| **Step Over:** | **Magnitude** | **Rise or Fall?** |
| $R$ | $I_1 R$ | Fall |

Now step across the laser $R_L$:

| (?) | | (?) |
|---|---|---|
| **Step Over:** | **Magnitude** | **Rise or Fall?** |
| $R_L$ | $R_L I_2$ | (see discussion) |

If $I_2$ is positive, then based on our label for $I_2$ the voltage on the left (the tail) is larger than the right and the voltage change appears to fall, hence:

| (−) | | (−) |
|---|---|---|
| **Step Over:** | **Magnitude** | **Rise or Fall?** |
| $R_L$ | $R_L I_2$ | Fall |

Since we are back at the starting location, collect the rises and falls to find:

$$V - RI_1 - R_L I_2 = 0 \qquad (2.27)$$

Now, count the unknowns in the equations. They are: $I_1$, $V_A$, and $I_2$; two flows and an effort. Since there are more unknowns than equations, we need another equation. Should we write conservation of charge at a node to relate one flow to the other? No. Why? Because there are two [NIF] implying there should be two flows which is exactly what we have. We need something besides relating flows. This situation happened since we ignored the capacitor while performing KVL, we should go back and write an equation for the capacitor. You can expect this to happen. For every capacitor you avoid during the KVL (for every one you "cheat" on), you can expect to have to write an equation eventually which is what we do next.

Choose the capacitor and account for energy as:

---

**Electrical Energy Accounting**
system[**the capacitor alone**]     time period [**differential**]

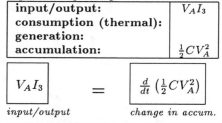

*input − output + generation − consumption = accumulation*

| | |
|---|---|
| **input/output:** | $V_A I_3$ |
| **consumption (thermal):** | |
| **generation:** | |
| **accumulation:** | $\frac{1}{2} C V_A^2$ |

| $V_A I_3$ | $=$ | $\frac{d}{dt}\left(\frac{1}{2} C V_A^2\right)$ |
|---|---|---|
| *input/output* | | *change in accum.* |

---

This gives us:

$$I_3 = C\dot{V}_A \qquad (2.28)$$

In actuality, we should have simply written the impedance relationship for the capacitor (which is exactly equation 2.28 and not wasted our time with the energy formulation which is what we will do from now on).

Now we have a single first order equation which matches the [NIA] but now we have four unknowns: ($I_1$, $I_2$, $I_3$ and $V_A$). There are three flows but there should be two ([NIF] is two), so write a conservation of charge at a node. Pick node $A$ and write:

**Conservation of Charge**

system[**Node** $A$]     time period [**differential**]

$input - output = accumulation$

| input/output: | $I_1 - I_2 - I_3$ |
|---|---|
| accumulation: | 0 |

$$\boxed{I_1 - I_2 - I_3} \quad = \quad \boxed{0}$$

input/output          change in accum.

This last equation:

$$I_1 - I_2 - I_3 = 0 \tag{2.29}$$

gives us everything we need.

**Solution**

The solution process requires solving two parts, one with the switch closed and the other with it open. If the switch is closed, solve equations 2.26, 2.27 and 2.29 for the three currents as:

$$I_1 = \frac{V - V_A}{R}$$

$$I_2 = \frac{V_A}{R_L}$$

and:

$$I_3 = \frac{-RV_A + R_L V - R_L V_A}{R_L R}$$

now use $I_3$ in equation 2.28:

$$C\dot{V}_A = \frac{R_L V - R_L V_A - R V_A}{R_L R}$$

If the switch is open, then $I_2 = 0$ so solve equations 2.29 and 2.26 for $I_1$ and $I_3$ as:

$$I_1 = \frac{V - V_A}{R}$$

$$I_3 = \frac{V - V_A}{R}$$

again use $I_3$ in equation 2.28 to find:

$$C\dot{V}_A = \frac{V - V_A}{R}$$

**End 2.8**

As you can see from the previous examples, switching capacitive loads can sometimes generate very large currents. This can be a problem if large currents are undesirable. Consider the circuit in figure 2.29.

We will discuss transistors in detail later, but for now think of the transistor as a switch. When $V_{\text{in}}$ is ground, the transistor allows very little current $i$ to flow. As the voltage $V_{\text{in}}$ increases,[11] the resistance to $i$ decreases causing the voltage at A to drop to roughly 0.7 [volts] above ground. If $V_{\text{in}}$ rises sharply,[12] then $V_A$ could drop suddenly to 0.7 [volts]. This sudden drop in voltage causes the capacitor to dump charge. This produces a large current $i$, which flows through the transistor to ground. By accounting for electrical energy, we can calculate the amount of energy that the transistor dissipates (converts mostly to thermal energy).

---

[11]Actually its the current into the wire not the voltage. More on that later.
[12]This might happen if there is small capacitance in the circuit connected to $V_{\text{in}}$.

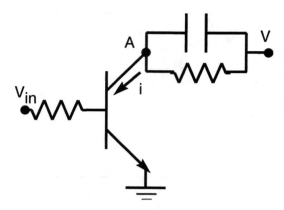

Figure 2.29: A Transistor Switching a Capacitive Circuit

---

**Electrical Energy Accounting**
system[transitor fig.2.29]          time period [**differential**]

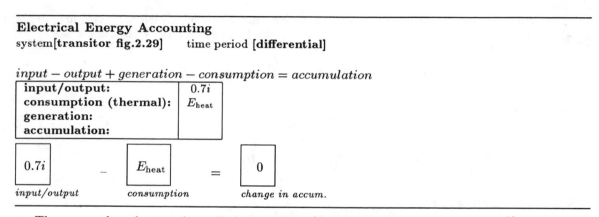

$input - output + generation - consumption = accumulation$

| input/output: | $0.7i$ |
|---|---|
| consumption (thermal): | $E_{\text{heat}}$ |
| generation: | |
| accumulation: | |

$$\boxed{0.7i} \quad - \quad \boxed{E_{\text{heat}}} \quad = \quad \boxed{0}$$

input/output        consumption              change in accum.

---

The power that the transistor dissipates, $E_{\text{heat}}$ (in this situation) is roughly $0.7i$.[13] If $i$ is large, and it will be if the capacitance is large, then the transistor will dissipate a large amount of electrical energy. Most of this electrical energy is converted into thermal energy. If the temperature increase caused by this conversion to thermal energy is too severe (that is if the transistor cannot store much or if it cannot conduct it away to someone else), it will ruin the transistor. To protect the transistor, what you want to do is either:

1. provide something to accumulate the thermal energy. For example this might be done by making the transistor large so the energy can be "spread out" and stored in the mass of the transistor.

2. provide a way for the thermal energy to be transferred from the transistor. For example you could attach the device to a large piece of metal which will conduct the thermal energy away. Of course the metal would have to give the energy to something so maybe you should blow air across it. This is called a heat sink.

What do you think "power" transistors (devices for switching large power) look like? Are they big or small? Where would you expect them to be located? They are much bigger than others and are located on heat sinks near cooling fans.

So far these examples have demonstrated a few uses for capacitors. We will discuss these uses more in section 2.6 but for now, we will simply solve some examples of circuits with capacitors.

## Example 2.9 ────────────────────────────────

### Two Capacitors With a Voltage Source.

**PROBLEM:**

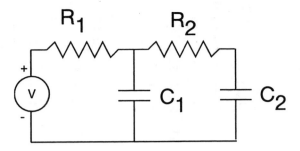

Figure 2.30: A Practice Circuit with 2 Capacitors.

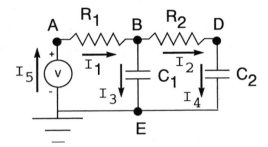

Figure 2.31: Nodes and Currents for the Practice Circuit.

Figure 2.30 shows a circuit with two capacitors. Determine the voltages across each capacitor and the current through the voltage supply. Let $R_i = i$ Ohms and $C_i = i$ microFarads. First let the voltage $V = 10$ then let it be $V = 10\sin(10^6 t)$ where $t$ is time. Later we will discuss how to find initial conditions but for now we take them to be zero.

**FORMULATION:**

The circuit has two [NIF] and two [NIA]. Figure 2.31 shows ground, the nodes and currents for the circuit (note there is no node C to avoid confusion with a capacitor). We expect 2 KVL and 5-2= 3 charge equations. Since the circuit is two [NIA] we expect to need two initial conditions. The initial conditions we need are the initial accumulations in the capacitors. The problem says to assume these values to be zero. Since accumulation in a capacitor is directly related to the voltage, this means to take the voltage across each capacitor to be zero at the initial condition.

For the first KVL take loop EABE. You should be familiar with all the steps except perhaps stepping over the capacitor so we will show the detail of that step. First we have:

| (?) | | (?) |
|---|---|---|
| **Step Over:** | **Magnitude** | **Rise or Fall?** |
| $C_1$ | $V_B - V_E$ | (see discussion) |

Notice that we are "cheating" by saying the voltage change is the difference in voltages. Now pretend that the expression is positive, that $V_B > V_E$. If so, this would be a fall as we walk from B to E so:

| (−) | | (−) |
|---|---|---|
| **Step Over:** | **Magnitude** | **Rise or Fall?** |
| $C_1$ | $V_B - V_E$ | Fall |

The complete KVL is:

$$V - R_1 I_1 - (V_B - V_E) = 0 \qquad (2.30)$$

Take loop EBDE for the second. This is actually insane since walking EABDE would involve only one capacitor and would result in an easier equation, but this **IS** practice. Again for the first capacitor:

---

[13]Notice that the wasted energy for a device that does not accumulate energy is simply the voltage drop across the device times the current through it.

| (?) | | (?) |
|---|---|---|
| **Step Over:** $C_1$ | **Magnitude** $V_B - V_E$ | **Rise or Fall?** (see discussion) |

Pretend B is above E but this time we are walking from E to B so this will look like a rise.

| (+) | | (+) |
|---|---|---|
| **Step Over:** $C_1$ | **Magnitude** $V_B - V_E$ | **Rise or Fall?** Rise |

For the second capacitor:

| (?) | | (?) |
|---|---|---|
| **Step Over:** $C_2$ | **Magnitude** $V_D - V_E$ | **Rise or Fall?** (see discussion) |

Pretend that $V_D > V_E$ and since we are walking from D to E, this looks like a fall.

| (−) | | (−) |
|---|---|---|
| **Step Over:** $C_2$ | **Magnitude** $V_D - V_E$ | **Rise or Fall?** Fall |

The complete KVL is:

$$+ (V_B - V_E) - R_2 I_2 - (V_D - V_E) = 0 \tag{2.31}$$

Now, before we forget, since we cheated on two capacitors we write their impedance equations as:

$$I = C \frac{d}{dt} \left( V_{\text{tail}} - V_{\text{head}} \right)$$

so:

$$I_3 = C_1 \frac{d}{dt} \left( V_B - V_E \right) \tag{2.32}$$

and:

$$I_4 = C_2 \frac{d}{dt} \left( V_D - V_E \right) \tag{2.33}$$

Now write the 3 charge equations. First node A:

$$I_5 - I_1 = 0 \tag{2.34}$$

Now node B:

$$I_1 - I_3 - I_2 = 0 \tag{2.35}$$

Finally node D:

$$I_2 - I_4 = 0 \tag{2.36}$$

Of course $V_E = 0$ so we have seven equations (equations 2.30 through 2.36) for seven unknowns $(I_1, I_2, I_3, I_4, I_5, V_B, V_D)$. There are 2 first order differential equations (equations 2.32 and 2.33) which matches what we expect from the [NIA]. We should be able to solve.

**SOLUTION/DISCUSSION:**

We solved for the unknowns when $V = 10$ using maple file 2.12.5. Figure 2.32 shows the results for the two capacitor voltages. The maple listing also gives all the currents. You could solve these equations by hand, but we will leave that for your differential equations class. The expression for the current through the voltage supply is:

$$I_5 = 4.13 \, e^{-157,000 \, t} + 5.87 \, e^{-1,590,000 \, t} \tag{2.37}$$

Note a couple of things from the results.

1. Notice that there is a transient part of the solution. You can see this in figure 2.32 in the region where the voltages are changing from 0 to 10 volts. Also note in equation 2.37 the transient is evident from the exponential terms with negative exponents, they disappear before long.

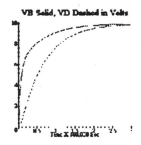

Figure 2.32: Voltage $V_B$ and $V_D$ Versus Time X $10^5$.

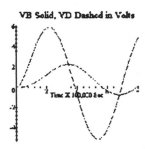

Figure 2.33: Voltage Response With Sinusoidal Source.

2. Notice that there is a steady state response which you can see where the voltages are leveling off. What is the steady state value of the current through the source? It is zero, look at equation 2.37 there is nothing left after the transients become very small.

3. Note that the capacitor $C_2$ takes longer to reach steady state than does $C_1$ this is due to the fact that it accumulates more energy (it has a larger $C$, its bigger and holds more) and because there are two resistors "choking back" some of the current trying to come from the source.

4. Note that the time it takes to respond (to reach steady state) is very short. This is typical of electrical circuits, they respond quickly.

5. Note that there are two exponential terms in the current. This is due to the fact that the circuit has two [NIA]. The number of exponentials will not always equal the [NIA] but quite often they do. They result from the fact that the accumulations "fill" at different rates.

Now for the solution when the source is sinusoidal. The same maple program solved this case as well. Figure 2.33 shows the response with the sinusoidal source. The source current is given by:

$$I_5 = -0.632e^{-157000t} - 2.64e^{-1590000t} + 3.276\cos(1000000.0t) + 5.69\sin(1000000.0t)$$

Notice the following:

1. It is not easy to see the transient in the figure, but it is evident in the current. There are still exponentials with negative exponents.

2. Notice that the exponents are the same as they were before. The reason for this is the time it takes the accumulations to fill depends on the size of the accumulation (the value of $C$, how much room the accumulation or capacitor has) and the amount being "choked back" by the resistors. This has nothing to do with the source.

3. Notice that the sine terms have the same frequency (the term multiplying time in the sine and cosine) as the source. Basically they are trying to follow the source. They try to fill and/or empty when the source goes high then low.

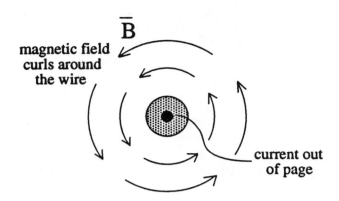

Figure 2.34: Magnetic Field, $\vec{B}$, Generated by Current in a Wire Pointed Straight Out of the Page

4. The capacitors never make it up to the source voltage. This is because when the source is high, they begin to fill but before they can fill completely reaching the source, the source drops and becomes negative. The accumulations then try to respond by emptying but before they can empty, the source has changed again. Notice that $C_2$ has more trouble keeping up with the source level because it has filled very little before the source has changed. What would you expect the voltages to look like if the source equals $10 \sin(1t)$? They would look very much like $10 \sin(1t)$. Since the source is changing slowly, the capacitors have time to fill up and empty almost completely before the source changes significantly.

5. Explain why the $C_2$ voltage in the figure is "out of phase" with the $C_1$. Out of phase means the voltage on the $C_2$ peaks and valleys at a different time than $C_1$.

**End 2.9** _____

All of the things we noted in the previous example will be discussed in detail in your differential equations class but even before you take that class you can make predictions and explain why things should and do behave as they do by discussing the [NIA], accumulations, and rates of accumulation. Whether or not you are "expected" to find solutions to these equations is up to your instructor. We will demonstrate a few solutions throughout the text.

If you need more examples of capacitor circuits. Look for them in the following sections. We will continue to solve circuits with capacitors but we should move on to another topic.

## 2.5   MAGNETIC FIELDS AND INDUCTANCE

**Objectives**
*When you complete this section, you should be able to*

1. *Explain what a magnetic field is.*

2. *Determine the relationship between current and magnetic fields.*

_____

When current flows in a wire, it generates a magnetic field. The strength of the magnetic field is proportional to the current.

$$\left|\vec{B}\right| = \mu i \tag{2.38}$$

Fields are distributed over space and therefore are beyond the scope of this course, however we will discuss the effects of fields in a qualitative manner. Figure 2.34 shows the orientation of the magnetic field generated by a current flowing in a wire. If we point our right thumb in the direction of current flow, the magnetic field ($\vec{B}$) curls in the direction of our fingers. The tip of our fingers is the north pole. The finger base is the south pole. Ordinary magnets produce $\vec{B}$ fields also. Figure 2.35 shows

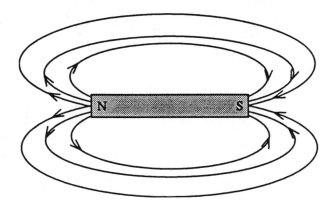

Figure 2.35: Magnetic Field Produced by a Standard Bar Magnet

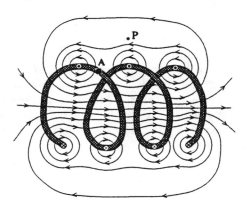

Figure 2.36: A Coil of Wire for Generating a Large Magnetic Field

a typical $\vec{B}$ field generated by a bar magnet. Notice that the $\vec{B}$ is always a closed loop. It is however not necessarily circular.

The magnetic field generated by a single straight wire is relatively small. However, we can amplify the field by coiling the wire. Coiled wires that amplify a magnetic field are called inductors. Figure 2.36 shows a wire that has been coiled into a cylinder. The magnetic field generated by each element of the coil add to each other producing a strong concentrated magnetic field in the center of the coil.

When iron or other magnetic material[14] is placed in a magnetic field, the iron attracts the field so to speak. Essentially, the field is concentrated and focused by the magnetic material. This is why electromagnets (coiled wire) are typically built around iron cores.

## 2.5.1  FORCE GENERATION

**Objectives**
*When you complete this section, you should be able to*

1. *Determine the approximate force (magnitude and direction) acting on a conductor when it conducts a current in the presence of an external magnetic field.*

If a current flows in the presence of a magnetic field, interaction between the current and magnetic field can generate a force. We can find the force ($\vec{F}$) using the following equation:

$$\vec{F} = \vec{\imath} \times \vec{B} \tag{2.39}$$

[14] Formally what we are referring to is material with a high magnetic permeability.

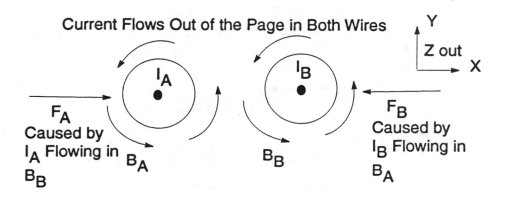

Figure 2.37: Two Parallel Wires with Current

where $\vec{i}$ points in the direction of conventional current and $\vec{B}$ is in the direction of the magnetic field. The current is in [amps] and the magnetic field strength is in [Tesla]. One [Tesla] equals one [Newton/amp m]. To understand this better, consider figure 2.37. The two wires, A and B, carry current in the direction that the figure shows. Because the magnetic field caused by current in wire A curls completely around wire A, its generated magnetic field does not generate an unbalanced force on wire A. However, if we expose the current in wire B to the magnetic field caused by wire A, a force on wire B will result. Likewise, the field caused by B exerts force on A.

The forces acting on the wires in figure 2.37 are

$$\vec{F}_A = \vec{i}_A \times \vec{B}_B$$

$$\vec{F}_B = \vec{i}_B \times \vec{B}_A$$

Since the magnetic fields are proportional to the current, we have

$$\left|\vec{B}_A\right| = \mu \left|i_A\right|$$

$$\left|\vec{B}_B\right| = \mu \left|i_B\right|$$

For wire A, the current vector points out of the page (in $\vec{z}$) and the magnetic field caused by B (at least on the average near A) points downward (in $-\vec{y}$). The cross product between a vector out and one down is to the right. We can calculate this using the right hand rule or with vector math.

$$\vec{z} \times (-\vec{y}) = -(\vec{z} \times \vec{y}) = -(-\vec{x}) = \vec{x}$$

Thus, $F_A$ points to the right. Its magnitude is approximately $\mu \left|\vec{i}_B\right| \left|\vec{i}_A\right|$.

For wire B, the current vector points out of the page ($\vec{z}$) and the magnetic field caused by A (at least that near wire B) points upward ($\vec{y}$). The cross product between a vector out and one up is to the left. Thus, $F_B$ points to the left. We can calculate the cross product with the right hand rule or with vector math.

$$\vec{z} \times \vec{y} = -\vec{x}$$

Its magnitude is $\mu \left|i_A\right| \left|i_B\right|$. Notice that the forces are equal and opposite. Can you think of an application for the phenomenon described in this section? Are there forces acting on a coiled wire (an inductor) that has a current flowing? Yes, but for ordinary magnitudes of current, the forces are very small. If the coil is a superconductor with a huge magnitude of current, the force can become important.

## 2.5.2 VOLTAGE INDUCTION

**Objectives**

*When you complete this section, you should be able to*

1. *Explain the concept of flux.*

2. *Describe how magnetic fields can induce a voltage.*

3. *Calculate the electrical energy stored in an inductor.*

4. *Determine the impedance relationship for an inductor.*

5. *Explain what a transformer is and how to make one.*

6. *Calculate the voltages and currents in inductive circuits.*

7. *Recognize the voltage/current relationships of inductors and transformers.*

---

Converse to the generation of force, if a conductor is in the presence of an external changing magnetic field, a current will be generated in the conductor. This is called induction. The induced voltage is proportional to the negative rate of change of the magnetic field's flux.

Flux $(\phi)$ quantifies the interaction between a magnetic field and an object. Formally, we define flux (measured in [Webers]) as

$$\phi = \int_S \vec{B} \cdot d\vec{S}$$

where $d\vec{s}$ is a unit of area, its a vector normal to a surface. As the name implies,[15] magnetic flux is the amount of magnetic field which cuts through a surface.

The following is a simplified way of visualizing flux. If we have a magnetic field (*e.g.* a bar magnet), then we can visualize the field by imagining that the field is a spray of water. Where the field curves, so does the spray.[16] Strong regions of the field represent high fluid speed. If we place the magnet into a fishing net, the magnetic field (water) will flow through the net. We would see some areas on the net where strong fields (fast water flow) enters and exits. Whereas, other areas would have weaker fields (slower water flow). The field (flow streams) would also cut through the net at various angles. If we summed up the total amount of field (water flow) exiting the net, we would find the flux through the net. If we summed the field (water) exiting the net through a subset of the net, then we would have the flux for that subset of net. Flux is dependent not only on the shape and strength of the field, but also on the size and shape of the area.

If we place a conductor in a changing flux,[17] a voltage will be generated in the conductor. We can determine the voltage generated using the following equation:

$$V = -\frac{d\phi}{dt}$$

Here, $V$ is the voltage induced in the conductor, and $\phi$ is the flux through the area spanned by the conductor. What this means is that the voltage generated in a conductor (a wire for example) is related to the change in flux through the area enclosed by the conductor. If the flux changes quickly, the voltage is large. This phenomenon is called induction. The voltage is induced. Can you think of an application for such a phenomenon?

**Concept 2.1** ─────────────────────────────────────────────

### Induced Voltage in A Wire Loop

For the rectangular wire in figure 2.38, the magnetic field generated by the stationary magnet poles

---

[15] In general, flux is the rate of transfer of a substance across a given surface. For example, if water flows in a pipe, having cross sectional area $S$, with a speed $V$; the flux of water is $\int_S V \, dS$ or $VS$.

[16] Here is one of the problems with the analogy between water and fields. The field can easily curve whereas a water spray will not.

[17] The flux can change due to a change in the strength or direction of $\vec{B}$ or due to a change in area $S$.

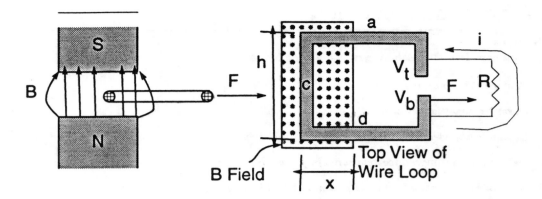

Figure 2.38: A Wire Loop Placed Between Two Magnet Poles

is a constant $B$ and points in the direction indicated in the figure. The flux cutting through the area enclosed by the wire loop is approximately $Bxh$.[18] Since both $h$ and $B$ are constant, the rate of change of the flux is dependent on $\dot{x}$.

$$\frac{d\phi}{dt} = Bh\dot{x}$$

If $x$ increases, the area increases therefore so does the flux. A voltage is induced such that $V_b$ is $Bh\dot{x}$ [volts] above $V_t$. In the remainder, we will discuss why the voltage at b is higher than at t.

First, the motion of the loop requires that we apply a force $F$, to the loop. If we remove $F$, the motion stops. This means, we must always work to make the loop move. To understand why this is true, consider the electrical energy in the loop. If we join the two ends, t and b, the induced voltage (whatever the polarity) will cause a current to flow. Suppose the wire is real and therefore has a small resistance. As the current flows through the real wire, electrical energy will be dissipated by the wire's resistance. If we remove $F$ and the wire continues to move, the current will continue to convert electrical energy into thermal. This thermal energy must come from something and the only thing available is the kinetic energy of the moving loop. Eventually, the kinetic energy is wasted by converting it to thermal and the wire's motion stops. Hence, we must apply $F$ and move the wire (we must be adding energy to the wire) so the wire can convert motion into electrical flow which turns the energy into thermal energy. The wire heats up very slightly.

Now, we will deduce the polarity of induced voltage. We join the ends t and b with a resistor so that the induced voltage causes a current to flow. If the voltage at b is greater than at t when we pull the loop from the field, then the current will flow counterclockwise (from bottom to top through the resistor). Since this current would flow in the presence of the external magnetic field (generated by the magnet), there will be forces on the loop. On the loop leg labeled a, current flows left so the force on that segment (from equation 2.39) points outward (up). On segment c, the current flows down so the force points to the left. The force on leg d is outward (down). Since the force on segment c points to the left, it opposes the force ($F$) attempting to withdraw the wire. If we made a mistake, and thought the current was clockwise, the forces would reverse direction. Therefore, the force on leg c would help drive the loop out of the field. Clearly this would mean that once we initiate the motion, $F$ could be removed and the wire would continue to move due to the force on leg c. This, however, violates the conservation of energy statement discussed in the last paragraph.

How would this discussion change (or would it), if the wire was a superconducting material with zero resistance?

**End 2.1**

---

[18] This is only approximate because:

1. the field is disrupted slightly by a change in material (the wire),

2. fringe effects at the edge of the field (the curved $\vec{B}$ in the figure), and

3. many other complicating effects.

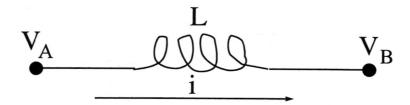

Figure 2.39: Variables Used to Define the Impedance of an Inductor

Consider what happens when current flows in a coil of wire (an inductor) as figure 2.36 shows. When the current flows in the coil, a magnetic field develops. Thus, every section of the coil has a current flowing in the presence of a magnetic field caused by the current at other sections. For example, consider a point A along the wire. Since current flows through point A, it produces a magnetic field and other points along the wire are in A's field. Likewise A is in the fields produced by other points too. Now if the current flowing in the wire changes (let it increase), then the magnetic field (generated by A and all other points of the wire) changes (it increases). Since point A is a conductor in a changing external magnetic field, an induced voltage occurs in point A. In fact, the voltage drops at A since the induced voltage is opposite magnetic field change. The positive increase in field (caused by increase in current) induces a negative change in voltage.

The voltage induction process in a single coil is called self induction and the coil is called an inductor. The magnitude of induced voltage drop for a given current change depends on the magnetic field which, in turn, depends on the coil geometry and magnetic properties. The magnitude of self induction can also vary along the coil but in this text, we will treat the coil as a single item, thus only the overall average relation interests us. In most cases, we can approximate the electrical energy stored in the magnetic field [12] of a perfect inductor as $E = \frac{1}{2}LI^2$ where $L$ is the coil's inductance measured in [Henry],[19] and $I$ is the current. Using the system in figure 2.39, we can derive the impedance relation for self inductance. Accounting for electrical energy, we find that

**Electrical Energy Accounting**
system[**inductor fig.2.39**]    time period [**differential**]

*input − output + generation − consumption = accumulation*

| | |
|---|---|
| **input/output:** | $V_A I - V_B I$ |
| **consumption (thermal):** | |
| **generation:** | |
| **accumulation:** | $\frac{1}{2}LI^2$ |

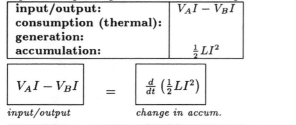

This will simplify to

$$V_A - V_B = L\frac{d}{dt}(I) \tag{2.40}$$

It is possible for an inductor to transform some electrical energy into other forms. For example, if an iron core is placed inside an inductor, the changing flux of the inductor will induce voltage in the iron. These voltages cause currents (eddy currents) to flow in a closed path in the iron. Because iron is not a perfect wire, it has a resistance to the current and so these eddy currents convert some electrical energy into thermal. The thermal energy converted in large inductors with large changes in current can be significant so if you work for an electrical distribution company, you would be concerned with it. We will ignore all forms of loss in the inductor. If you want to include it, you could model the "real" inductor as a perfect inductor connected to a resistor. The resistor models the electrical energy losses.

## Example 2.10

[19]Obviously if $\frac{1}{2}LI^2$ is energy, a [Henry] is related to other units as [Henry] = [Watt second/amp$^2$] = [volt second/amp.].

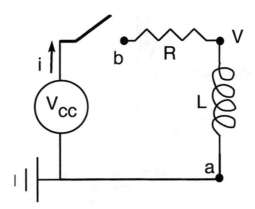

Figure 2.40: A Switched Inductor Circuit

## A Resistor and Inductor in Series

**PROBLEM:**

Determine the current $i$ and voltage $V$ of the circuit in figure 2.40 if we close the switch at t=0 (the switch is open at $t=0^-$). The circuit consists of only an ideal resistor, ideal wire, and ideal inductor. In a real system, these idealized components do not exist. For example, real wire will have a small[20] resistance or even an inductance or capacitance and both the inductor and resistor might have capacitance. As a matter of fact, this circuit could represent a real inductor because the resistance models the energy losses present in the inductor.

**FORMULATION:**

The first step in solving this problem is to determine the [NIF] and [NIA] of the system. When the switch is closed, the idealized circuit has one [NIF] and one [NIA] Therefore, we expect one flow variable and one independent accumulation (for example, one conservation equation should be sufficient). Taking the circuit from points b to a as the system and accounting for electrical energy, we find that

**Electrical Energy Accounting**

system[**circuit fig.2.40**]        time period [**differential**]

*input − output + generation − consumption = accumulation*

| input/output: | $V_{cc}i$ |
|---|---|
| consumption (thermal): | $i^2R$ |
| generation: | |
| accumulation: | $\frac{1}{2}Li^2$ |

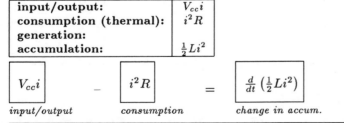

which simplifies to

$$V_{cc}i - i^2R = \frac{d}{dt}\left(\frac{1}{2}Li^2\right)$$

which reduces to

$$V_{cc}i - i^2R = Li\frac{d}{dt}(i)$$

$$V_{cc} - iR = L\frac{d}{dt}(i) \tag{2.41}$$

---

[20]We hope that the real elements and the approximate ideal elements are close. This may or may not be the case. For example, in small integrated circuit components, the undesired capacitance, inductance, resistance of the components can be significant.

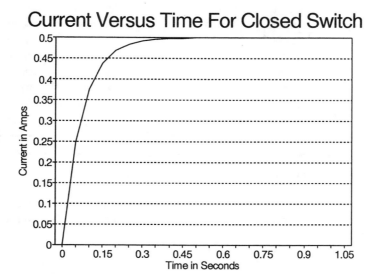

Figure 2.41: Current Versus Time for the Circuit Results

At the beginning, at t=0, the switch is open and $i = 0$. This is our initial condition. Though the current starts at zero, it has an initial positive derivative which means it increases. What happens is the current starts small but builds as equation 2.41 describes once the switch is closed (t>0).

**SOLUTION:**

We can solve equation 2.41 analytically or numerically. Maple program 2.12.3 demonstrates the solution. Figure 2.41 shows a plot of the current versus time. Why does the current build? The inductor is storing energy and it takes a little while for the energy to get into the inductor. Eventually the current stops changing because the resistor "chokes back" the current.

Once we know the current, we can calculate the voltage $V$ by accounting for the electrical energy of the resistor. Doing this we find that

**Electrical Energy Accounting**

system[**resistor fig.2.40**]     time period [**differential**]

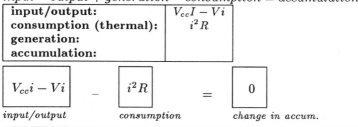

$$V_{cc}i - Vi - i^2R = 0$$

Equivalently, we could use our knowledge of a resistor's impedance to write

$$\frac{V_{cc} - V}{R} = i$$

**OTHER CONSIDERATIONS:**

Now suppose that at time $\tau > 0$, we open the switch. For the idealized system, when the switch opens, current must immediately drop to zero. However, there is electrical energy stored in the inductor. The energy is given by $E = \frac{1}{2}Li^2$ which means that if current goes to zero, the energy must suddenly leave the inductor. This energy **must** go somewhere. In the idealized circuit, the energy has nowhere to go therefore we have a problem. In a real circuit the electrical energy in the inductor will rush out causing the voltage at point b to become more and more negative in an attempt to store energy in the small capacitances (which will exist) and dissipate energy in R (energy turned into thermal in a resistor is $I^2R$ or $(V_{\text{tail}} - V_{\text{head}})^2/R$). Ultimately, all of the energy will be stored and dissipated or else the voltage at b will become so low that a spark jumps across the switch. Since the resistance of an air gap is large, it would not take much spark current to dissipate a large amount of energy (note $E = i^2R$). Describe what happens to the switch if a spark does occur. How could we protect the switch? Connecting an inductor like this is actually a dangerous thing to do so don't do it.

**End 2.10** ━━━━━━━━━━━━━━━━━━━━━━━━━━━━━━━━━━━━━━━━━━━━━━━━━━━

## Example 2.11 ━━━━━━━━━━━━━━━━━━━━━━━━━━━━━━━━━━━━━━━━━━━

### A Resistor and Inductor Revisited

Now let's redo the previous example using KVL instead of energy. Of course there is one [NIF] and one [NIA]. Since there is one [NIF], we will write one KVL. There is only one loop in the problem so walk clockwise around it.

Stepping over the voltage source, we have:

| (+) | | (+) |
|---|---|---|
| **Step Over:** | **Magnitude** | **Rise or Fall?** |
| Voltage Source | $V_{cc}$ | Rise |

stepping over the resistor:

| (−) | | (−) |
|---|---|---|
| **Step Over:** | **Magnitude** | **Rise or Fall?** |
| Resistor | $Ri$ | Fall |

finally stepping over the inductor:

| (?) | | (?) |
|---|---|---|
| **Step Over:** | **Magnitude** | **Rise or Fall?** |
| Inductor | $L\frac{d}{dt}(i)$ | (see discussion) |

To determine whether this is a rise or a fall, pretend that $L\frac{d}{dt}(i) > 0$ so that $i$ is increasing. Based on the diagram in figure 2.40, positive $i$ flows into the top of the inductor, therefore positive $\frac{d}{dt}(i)$ causes a higher voltage at the top of the inductor. Hence, as we walk across the inductor, the voltage appears to fall. Therefore:

| (−) | | (−) |
|---|---|---|
| **Step Over:** | **Magnitude** | **Rise or Fall?** |
| Inductor | $L\frac{d}{dt}(i)$ | Fall |

Altogether we have:

$$V_{cc} - Ri - L\frac{d}{dt}(i) = 0$$

This is one first order differential equation for one unknown flow variable $i$.

Note that when applying KVL to inductors, the tail of the arrows drawn next to the inductor has the high voltage. This is identical to resistors and capacitors.

**End 2.11** ━━━━━━━━━━━━━━━━━━━━━━━━━━━━━━━━━━━━━━━━━━━━━━━━

A transformer (shown schematically in figure 2.42) is an electrical device consisting of two or more coils of wire typically wrapped around a common iron core. The coils are electrically insulated from each other. The iron core serves as a magnetic field conductor or concentrator. The concentrated magnetic fields generated in each coil's center is contained and conducted by the iron into the opposite coil's center, allowing the two magnetic fields to interact. An ordinary two coil transformer has three dominate inductances. The transformer in the figure has two primary (self) inductances, (denoted by $L_l$ and $L_r$) and a single mutual inductance (denoted by M).

The mutual inductance results from the influence of one coil on the other. For example, current $i_l$ flowing in one wrap of the left coil helps generate a magnetic field in the iron core. If the current in

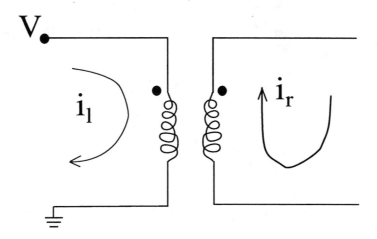

Figure 2.42: An Electrical Transformer

the left coil changes, so does its magnetic field. The changing field induces voltage in any conductor nearby. Since both the left and right coils are in the field, a voltage is induced in both. Mutual inductance is when the left induces voltage in the right and vice versa. Unlike self inductance, mutual inductance can induce a voltage rise or fall depending on the direction of the windings on each side.

In summary, the voltage rise/drop in the left coil is quantitatively

$$V_l = L_l \frac{di_l}{dt} \pm M \frac{di_r}{dt} \tag{2.42}$$

and the rise/drop in the right coil is

$$V_r = L_r \frac{di_r}{dt} \pm M \frac{di_l}{dt}$$

We will discuss how to determine the plus or minus sign shortly. We can approximate the electrical energy stored by a simple transformer as

$$E = \frac{1}{2} L_l i_l^2 + \frac{1}{2} L_r i_r^2 \pm M i_l i_r \tag{2.43}$$

The effect of mutual inductance is that extra electrical energy can be stored. This extra energy can then be transferred back and forth between the two isolated circuits. This electrical energy transfer manifests itself in the form of voltage rises/drops. In general, a transformer will convert some electrical energy into another form - thermal. One source of electrical energy loss is eddy currents in the iron core developed by the changing flux of the coils. Just as in the simple inductor, we can accommodate some forms of energy conversion by using different constants in equation 2.43. We will do this throughout and will ignore all other forms of energy conversion.

Whether the plus or minus appears in equation 2.42 depends on the direction of the magnetic fields which depends on the direction of the coils. To designate the direction of the windings, two dots mark the coil as Figure 2.42 shows. We interpret these polarity markings (the dots) in the following way. Suppose we are writing a KVL equation for a transformer. Draw currents through each of the coils. Based on the drawing of the currents, the voltage change caused by self induction always produces a higher voltage at the tail side of the coil. Hence the self induction of the left coil in figure 2.42 is high on the dotted end. The self induction of the right coil is high on its dotted end also (the tail side). Now for mutual induction. If a coil self induces his dotted side high, it will also mutually induce "his buddy's" dotted side high. If a coil self induces his undotted side high, then it mutually induces his buddy's undotted side high. For the system in figure 2.42, positive changes in the currents mutually induce larger voltages at the dotted sides of the coils. The following steps may help you:

1. Look at the coil for which you want to find the mutual term.

2. Find the current that causes the mutual induction. Determine if the tail of that current lies on the dotted or on the undotted side.

3. If the tail from step 2 is on the dotted side then the mutual term is high on the dotted side. If the tail from step 2 is on the undotted side, then the mutual term is high on the undotted side.

4. Now that you know which side is high, determine whether the term is a rise or fall depending on the direction you are walking.

The authors use the previous process however many students find it too complex to remember. If you agree then perhaps the next technique will be easier for you. Another rule makes it easy to determine the proper sign convention for voltage rises and falls, and for electrical energy storage. If both positive currents (as drawn) enter the inductances on the same side (either the dotted side or the undotted side), then the mutual inductance terms have the same sign as the primary inductance terms. If the positive currents enter opposite ends of the coils, the mutual inductance terms have signs opposite the primary inductance terms. To use this rule do the following:

1. Determine the + - signs of the self inductance terms. Here the tail side of the current arrow are high so depending on which direction you are walking you should be able to determine the proper sign.

2. If both current arrows you drew flow into the dotted end of their coils or if they both flow into the undotted ends they are the same. If one arrow flows into the dotted end while the other flows into the undotted, they are different. Determine if the arrows are the "same" or "different".

3. If the arrows are the same, the + - signs on the mutual induction terms are the same as the self induction. If the arrows are different, the mutual terms are different from the self.

In keeping with our previous examples, the next example demonstrates how to apply the conservation of energy to a transformer. It is a little confusing so if you want to skip it and focus on the examples that use KVL methods, they are easier.

## Example 2.12

### A Transformer Circuit

**PROBLEM:**

Determine the mathematical formulation required to solve for the voltage $V_s$ of the transformer circuit in figure 2.43. A voltage source of $V_p$ excites the circuit.

**FORMULATION:**

The circuit has two [NIF] and two [NIA]. The two state variables are the currents in the two loops. To begin, we choose the subsystem in figure 2.44. Notice we are assuming there is no electrical energy loss in our system. If the iron core, for example, transforms electrical energy into thermal, we exclude it from our system. Accounting for electrical energy, we obtain the following:

**Electrical Energy Accounting**
system[**system fig.2.44**]       time period [**differential**]

$input - output + generation - consumption = accumulation$

| input/output: | $V_p i_p + V_s i_s$ |
|---|---|
| consumption (thermal): | |
| generation: | |
| accumulation: | $\frac{1}{2} L_p i_p^2 + \frac{1}{2} L_s i_s^2 + M i_p i_s$ |

| $V_p i_p + V_s i_s$ | $=$ | $\frac{d}{dt} \left( \frac{1}{2} L_p i_p^2 + \frac{1}{2} L_s i_s^2 + M i_p i_s \right)$ |
|---|---|---|
| *input/output* | | *change in accum.* |

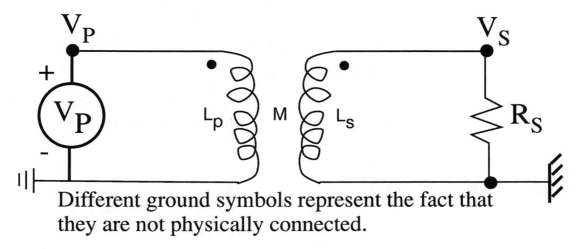

Different ground symbols represent the fact that they are not physically connected.

Figure 2.43: A Simple Transformer Circuit

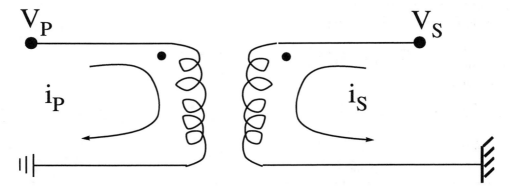

Figure 2.44: First System for Analyzing the Transformer Circuit

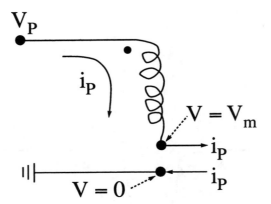

Figure 2.45: Second System For Analyzing the Transformer Circuit

In equation form, this is

$$V_p i_p + V_s i_s = \frac{d}{dt}\left(\frac{1}{2}L_p i_p^2 + \frac{1}{2}L_s i_s^2 + M i_p i_s\right) \tag{2.44}$$

Since the system has two [NIA] and equation 2.44 is first order, we know we need another equation with accumulation. Choose the left coil without the mutual induction and without losses as our system (see figure 2.45). We represent cutting out the mutual induction from the left coil by showing the current leaving the system at the mutually induced voltage $V_m$ then reentering the system at ground voltage. Because positive $i_s$ flows into the dotted end of the coil, the induced voltage in the left will be higher on the end closer to the dot. This means that positive $\frac{d}{dt}(i_s)$ will cause $V_m$ to be higher than 0. The voltage induced in the left coil by the right $(V_m)$ is

$$V_m = +M\frac{d}{dt}(i_s)$$

Using this result in the second system allows us to account for electrical energy as follows:

**Electrical Energy Accounting**
system[**system fig.2.45**]        time period [**differential**]

*input − output + generation − consumption = accumulation*

| input/output: | $V_p i_p - M\frac{d}{dt}(i_s) i_p$ |
|---|---|
| **consumption (thermal):** | |
| **generation:** | |
| **accumulation:** | $\frac{1}{2}L_p i_p^2$ |

| $V_p i_p - M\frac{d}{dt}(i_s) i_p$ | = | $\frac{d}{dt}\left(\frac{1}{2}L_p i_p^2\right)$ |
|---|---|---|
| *input/output* | | *change in accum.* |

$$V_p i_p - M\frac{d}{dt}(i_s) i_p = \frac{d}{dt}\left(\frac{1}{2}L_p i_p^2\right) \tag{2.45}$$

Now, because we have introduced more variables than absolutely necessary (both $V_s$ and $i_s$) we need to eliminate one of them. We can do this easily using the system in figure 2.46. Accounting electrical energy, we find that

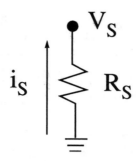

Figure 2.46: System For Eliminating Extra Variables in the Transformer Circuit

**Electrical Energy Accounting**

system[**resistor fig.2.46**]     time period [**differential**]

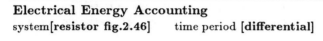

*input − output + generation − consumption = accumulation*

| input/output: | $-V_s i_s$ |
|---|---|
| consumption (thermal): | $i_s^2 R_s$ |
| generation: | |
| accumulation: | |

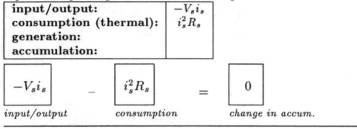

| $-V_s i_s$ | − | $i_s^2 R_s$ | = | 0 |
|---|---|---|---|---|
| *input/output* | | *consumption* | | *change in accum.* |

$$-V_s i_s - i_s^2 R_s = 0$$

We can simplify this to

$$- V_s - i_s R_s = 0 \qquad (2.46)$$

Equations 2.44, 2.45, and 2.46 give us three equations with three unknowns. Therefore, the problem formulation is complete.

**End 2.12**

In general, the self inductances ($L_p$ and $L_s$) and the mutual inductance ($M$) (used in the previous example 2.12) are given quantities for the transformer. In other words, when you buy the transformer, these values are specified. Under ideal conditions there is a relationship between them and these relationships will be discussed in section 2.5.3. If you solve the equations derived in example 2.12 for current and determine the conditions that guarantee a realistic transformer you will find that $M^2 < L_p L_s$.

**Example 2.13**

### A Transformer Circuit Revisited.

This example will solve the previous example using KVL. There are two [NIF], therefore we will write two KVL. For the first loop, walk clockwise around the primary beginning at the ground. We have:

| (+) | | (+) |
|---|---|---|
| **Step Over:** | **Magnitude** | **Rise or Fall?** |
| Voltage Source | $V_p$ | Rise |

Now stepping over the primary coil, voltages changes because of the primary self inductance and because of the mutual induction. First the effect of the primary is:

| (?) | | (?) |
|---|---|---|
| **Step Over:** | **Magnitude** | **Rise or Fall?** |
| Primary Self Induction | $L_p \frac{d}{dt}(i_p)$ | (see discussion) |

If $\frac{d}{dt}(i_p) > 0$ the coil will self induce a high voltage at the tail of the $i_p$ current therefore the top of the coil is high and the voltage change will look like a fall.

| (−) | | (−) |
|---|---|---|
| **Step Over:** | **Magnitude** | **Rise or Fall?** |
| Primary Self Induction | $L_p\frac{d}{dt}(i_p)$ | Fall |

Now for the mutual induction:

| (?) | | (?) |
|---|---|---|
| **Step Over:** | **Magnitude** | **Rise or Fall?** |
| Mutual Induction on Primary | $M\frac{d}{dt}(i_s)$ | (see discussion) |

Note that it is a change in the secondary current that mutually induces voltage in the primary. Now if $M\frac{d}{dt}(i_s) > 0$ then $i_s$ self induces the dotted side of the secondary high. Since $i_s$ self induces its dotted end high, it also mutually induces the dotted end of the primary high. Since the dotted end of the coil is high, this appears as a fall. So:

| (−) | | (−) |
|---|---|---|
| **Step Over:** | **Magnitude** | **Rise or Fall?** |
| Mutual Induction on Primary | $M\frac{d}{dt}(i_s)$ | Fall |

Having arrived back at the start, we collect the terms to find:

$$V_p - L_p\frac{d}{dt}(i_p) - M\frac{d}{dt}(i_s) = 0 \tag{2.47}$$

Look at figure 2.44 showing the currents. Notice that both currents go into the dotted ends. Since they enter the "same" sides of the coils, they have the "same" signs as the self induction. Do they? Look at equation 2.47, yes they do.

Now walk counterclockwise around the secondary loop starting at ground. First, step over the resistor:

| (?) | | (?) |
|---|---|---|
| **Step Over:** | **Magnitude** | **Rise or Fall?** |
| Resistor | $Ri_s$ | (see discussion) |

If $Ri_s > 0$ then current flows up through $R$, hence the bottom of the resistor has the high voltage and this will appear to be a fall when we walk. So:

| (−) | | (−) |
|---|---|---|
| **Step Over:** | **Magnitude** | **Rise or Fall?** |
| Resistor | $R_s i_s$ | Fall |

Next step over the secondary coil. First compute the voltage change caused by self induction:

| (?) | | (?) |
|---|---|---|
| **Step Over:** | **Magnitude** | **Rise or Fall?** |
| Secondary Self Induction | $L_s\frac{d}{dt}(i_s)$ | (see discussion) |

If $L_s\frac{d}{dt}(i_s) > 0$ then $i_s$ self induces the top of the coil high and this will appear as a fall when we walk. Therefore:

| (−) | | (−) |
|---|---|---|
| **Step Over:** | **Magnitude** | **Rise or Fall?** |
| Secondary Self Induction | $L_s\frac{d}{dt}(i_s)$ | Fall |

For the mutual induction in the secondary coil:

| (?) | | (?) |
|---|---|---|
| **Step Over:** | **Magnitude** | **Rise or Fall?** |
| Secondary Mutual Induction | $M\frac{d}{dt}(i_p)$ | (see discussion) |

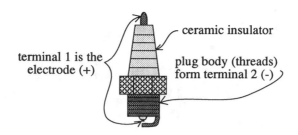

Figure 2.47: A Typical Spark Plug

Now if $M\frac{d}{dt}(i_p) > 0$ then $i_p$ is self inducing the dotted end of the primary high, hence it also mutually induces the dotted end of the secondary high. If the dotted end of the secondary is high, it appears to be a fall as we walk. So:

| (−) | | (−) |
|---|---|---|
| **Step Over:** | **Magnitude** | **Rise or Fall?** |
| Secondary Mutual Induction | $M\frac{d}{dt}(i_p)$ | Fall |

Having arrived back at the start, collect terms to find:

$$-R_s i_s - L_s \frac{d}{dt}(i_s) - M\frac{d}{dt}(i_p) = 0 \qquad (2.48)$$

Again, since the currents go into the same side of their coils, the mutual induction term should have the same sign as the self. Does it? Yes.

Now the only question remaining is: are the two equations derived in this example the same as those derived before? Yes. In fact equation 2.48 is exactly the same as equation 2.45. To demonstrate equation 2.44 also matches, multiply equation 2.47 by $i_s$, multiply equation 2.48 by $i_p$ and add them together. Finally substitute $V_s = -i_s R_s$ (from equation 2.46) to arrive at equation 2.44.

Since this formulation and the previous give the same equations, why use KVL? Because the equations resulting from KVL are often linear (they are here), whereas those from energy tend to be nonlinear. Nonlinear equations are often very difficult to solve.

**End 2.13** ━━━━━━━━━━━━━━━━━━━━━━━━━━━━━━━━━━━━━━━━━━━━━━

**Example 2.14** ━━━━━━━━━━━━━━━━━━━━━━━━━━━━━━━━━━━━━━━━━━━━━━

## A Spark Plug Circuit

**PROBLEM:**

A spark plug is an important component of gasoline powered automobiles. The purpose of the spark plug is to ignite the air/gasoline mixture contained in the cylinders of the engine. A typical spark plug appears in figure 2.47. The plug consists of two electrical terminals. One terminal consists of the plug body which forms electrical ground by contacting (screwing into) the engine body. The second terminal extends down the length of the plug, and is electrically isolated from ground. When the air/fuel mixture is to be ignited, we place a very large (thousands) voltage across the plug terminals causing a spark to jump across the terminals of the plug. The spark ignites the air/fuel mixture. The problem we have is to generate, from the 12 [volt] battery in the automobile, the large voltage required for the spark.

Conventionally, the way we do this is to use an electrical transformer. How can we determine if a transformer is the proper device to use? Does what we learned in example 2.10 help? The schematic of an ideal spark plug circuit appears in figure 2.48. In addition to the transformer, the circuits have some resistance, a switch,[21] and the 12 [volt] automobile battery. The right circuit (in relation to the figure) which uses subscripts s connects to the spark plug. Notice that it is an open circuit because of the plug gap. Formulate a mathematical model of the circuit that we can use to predict the magnitude of spark voltage ($V_s$).

---

[21] When the switch is mechanical, engineers call them points.

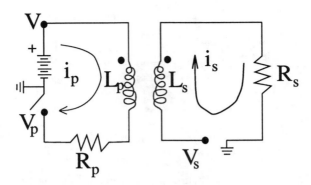

Figure 2.48: A Typical Spark Plug Circuit

## FORMULATION:

Accounting electrical energy (if you don't like to use energy, skip to the next example) for the complete circuit (left and right sides), we obtain the following:

**Electrical Energy Accounting**

system[**complete circuit fig.2.48**]          time period [**differential**]

*input − output + generation − consumption = accumulation*

| input/output: | $Vi_p + V_s i_s - V_p i_p$ |
|---|---|
| consumption (thermal): | $i_p^2 R_p + i_s^2 R_s$ |
| generation: | |
| accumulation: | $\left(\frac{1}{2}L_p i_p^2 + \frac{1}{2}L_s i_s^2 - M i_p i_s\right)$ |

$$\boxed{Vi_p + V_s i_s - V_p i_p} \; - \; \boxed{i_p^2 R_p + i_s^2 R_s} \; = \; \boxed{\frac{d}{dt}\left(\left(\frac{1}{2}L_p i_p^2 + \frac{1}{2}L_s i_s^2 - M i_p i_s\right)\right)}$$

*input/output*               *consumption*                *change in accum.*

$$Vi_p - V_p i_p + V_s i_s - i_p^2 R_p - i_s^2 R_s = \frac{d(\frac{1}{2}L_p i_p^2 + \frac{1}{2}L_s i_s^2 - M i_p i_s)}{dt}$$

which reduces to

$$Vi_p - V_p i_p + V_s i_s - i_p^2 R_p - i_s^2 R_s = L_p i_p \frac{d}{dt}\left(i_p\right) + L_s i_s \frac{d}{dt}\left(i_s\right) - M\frac{d}{dt}\left(i_p\right)i_s - M i_p \frac{d}{dt}\left(i_s\right) \quad (2.49)$$

Notice that the currents flow into different sides of the coils. Current $i_p$ flows into the dotted side but the $i_s$ flows into its undotted side. These are different. Since these are different, the signs on the mutual terms should be different from the self induction. Note that they are different.

Since the system has two [NIA] and this equation is only a first order equation, we recognize we need another equation. To get this equation we must take another system. We will choose the left side as our system with the mutual induction cut out. We show it in figure 2.49. Our system contains only the primary inductance. To account for the induced effects, we have shown the effect of the mutual induction as a break, with one side labeled $V'$ and the other $V' + M\frac{d}{dt}\left(i_s\right)$. The effect of the mutual induction is to generate a voltage in the circuit. Based on the polarity markings on the transformer, $\frac{d}{dt}\left(i_s\right)$ generates a voltage such that the undotted side of the left inductor is slightly higher than it would be without the mutual induction. We show this by indicating the break out of the system at voltage $V'$, and a return to the system at voltage $V' + M\frac{d}{dt}\left(i_s\right)$ (the voltage the mutual induction causes). Accounting for electrical energy for the system in figure 2.49, we find that

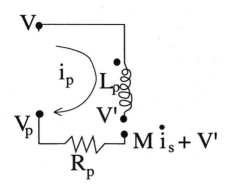

Figure 2.49: The Left Side of the Spark Plug Circuit

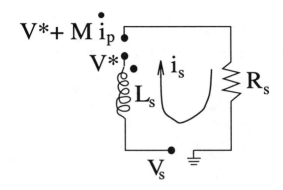

Figure 2.50: Right Hand Side of the Spark Plug Circuit

**Electrical Energy Accounting**

system[**circuit fig.2.49**]    time period [**differential**]

*input − output + generation − consumption = accumulation*

| input/output: | $Vi_p + \left(V' + M\frac{d}{dt}(i_s)\right)i_p - V'i_p - V_pi_p$ |
|---|---|
| **consumption (thermal):** | $i_p^2 R_p$ |
| **generation:** | |
| **accumulation:** | $\frac{1}{2}L_p i_p^2$ |

$$\boxed{Vi_p + \left(V' + M\tfrac{d}{dt}(i_s)\right)i_p - V'i_p + V_pi_p} \quad - \quad \boxed{i_p^2 R_p} \quad = \quad \boxed{\tfrac{d}{dt}\left(\tfrac{1}{2}L_p i_p^2\right)}$$

*input/output*                    *consumption*          *change in accum.*

$$Vi_p + \left(V' + M\frac{d}{dt}(i_s)\right)i_p - V_pi_p - V'i_p - i_p^2 R_p = \frac{d}{dt}\left(\frac{1}{2}L_p i_p^2\right)$$

which reduces to

$$V + M\frac{d}{dt}(i_s) - i_p R_p - V_p = L_p\frac{d}{dt}(i_p) \tag{2.50}$$

We can solve equations 2.49 and 2.50 for the current if we want. However, for practice, we will derive the equations for the right side of the circuit. Also notice that the signs on the mutual term is different from the sign on the self induction (put them on the same side of the equation to verify this).

Consider the system in figure 2.50 which does not contain the mutual induction. Again the effect of the mutual inductance appears as the break in the system. The accounting of electrical energy gives

**Electrical Energy Accounting**
system[circuit fig.2.50]      time period [**differential**]

$$input - output + generation - consumption = accumulation$$

| | |
|---|---|
| **input/output:** | $V_s i_s + \left(V^\star + M\frac{d}{dt}(i_p)\right)i_s - V^\star i_s$ |
| **consumption (thermal):** | $i_s^2 R_s$ |
| **generation:** | |
| **accumulation:** | $\frac{1}{2}L_s i_s^2$ |

$$\boxed{V_s i_s + \left(V^\star + M\frac{d}{dt}(i_p)\right)i_s - V^\star i_s} \quad - \quad \boxed{i_s^2 R_s} \quad = \quad \boxed{\frac{d}{dt}\left(\frac{1}{2}L_s i_s^2\right)}$$

input/output                                 consumption          change in accum.

$$V_s i_s - V^\star i_s + \left(V^\star + M\frac{d}{dt}(i_p)\right)i_s - i_s^2 R_s = \frac{d}{dt}\left(\frac{1}{2}L_s i_s^2\right)$$

Assuming that $V_s$ is sufficient to generate a spark allowing nonzero $i_s$, we can reduce the previous to

$$V_s + M\frac{d}{dt}(i_p) - i_s R_s = L_s \frac{d}{dt}(i_s) \tag{2.51}$$

Just to prove that equations 2.50 and 2.51 are valid we will multiply by appropriate currents and add them to show they are equivalent to equation 2.49.

$$Vi_p + M\frac{d}{dt}(i_s)i_p - i_p^2 R_p - V_p i_p + V_s i_s + M\frac{d}{dt}(i_p)i_s - i_s^2 R_s = L_p i_p \frac{d}{dt}(i_p) + L_s i_s \frac{d}{dt}(i_s)$$

The reader can verify that this is identical to equation 2.49. This makes sense because the system is only two [NIA]. Therefore, one of the three equations that we wrote had to be dependent on the other two. For convenience only, we will proceed with the example using equations 2.50 and 2.51.

Since the current in circuit 2 is zero before the spark occurs, we can simplify the two circuit equations for times before the spark occurs

$$V_s = -M\frac{di_p}{dt} \tag{2.52}$$

and

$$V - L_p \frac{di_p}{dt} - R_p i_p - V_p = 0$$

We can see from equation 2.52 that the method for generating large $V_s$ is to cause a large rate of change of $i_p$. This is the reason for the switch in the left hand circuit. The idea is that by throwing the switch, we can make $i_p$ change dramatically enough to generate a large spark voltage.

Now, we will investigate the effects of the switch a bit more closely. Solving the last equation for the change in $i_p$, we obtain

$$\frac{di_p}{dt} = \frac{V - R_p i_p - V_p}{L_p} \tag{2.53}$$

Now if at $t = 0^-$ (just before we desire the spark) the switch is open, then at initial time, (assuming the left circuit has settled down since the last spark)[22] $i_p = 0$, $V_p = V$, and $\frac{d}{dt}(i_p) = 0$. Now suddenly, we close the switch and $V_p = 0$, $i_p = 0$, and $\frac{d}{dt}(i_p) = \frac{V}{L_p}$. This generates a spark voltage of (using equation 2.52)

$$V_s = -M\frac{V}{L_p}$$

Now since $V$ is roughly 12 [volts], unless $\frac{M}{L_p}$ is on the order of a thousand, we cannot cause a spark.

As an alternative plan, suppose that before the spark we keep the switch closed. At the initial time (again assuming things have settled), $i_p = \frac{V}{R_p}$,[23] $V_p = 0$, and $\frac{d}{dt}(i_p) = 0$. If we suddenly open

---

[22] What we really mean is that enough time has elapsed that the transient component which consists of a decaying exponential, is zero.

[23] We will discuss later precisely how to determine this initial condition. We will leave it for a later exercise.

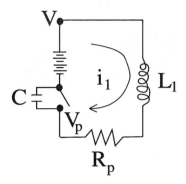

Figure 2.51: A Condenser (Capacitor) Used to Reduce Point Damage

the switch, the inductor $L_p$ will attempt to keep $i_p$ flowing; however, since there is no place for the current to originate, the current stops instantaneously. Opening the switch causes an infinite rate of change of $i_p$ (at least theoretically), thus producing infinite spark voltage $V_s$ (see equation 2.52). Actually, there is some nonideal capacitance in the switch and electrons flow (current flows) into the small capacitance of the switch. Therefore, $i_p$ does not stop instantaneously. The sudden change in $i_p$ causes $V_s$ to rise several hundred [volts] but not to infinity.

The voltage $V_p$ also rises significantly. Consider the voltage $V_p$ predicted in equation 2.53. If $\frac{d}{dt}(i_p)$ is large negative, $V_p$ becomes large positive. In fact, $V_p$ can become so large that in older automobiles which employed a mechanical switch (called points), sparking across the switch was very common. The sparks caused deterioration of the point's surface causing the precise timing of the spark to degrade. To help reduce damage to the points, a capacitor (usually called a condenser) was connected across the gap as figure 2.51 shows. The capacitor prevented $V_p$ from rising too high until the points opened far enough to prevent arcing. There can also be a resistor in line with the capacitor but the capacitor is the key component. Most modern electronic ignition systems still use transformers (coils), but mechanical switches (points) have been replaced with electronic switches.

**End 2.14** ━━━━━━━━━━━━━━━━━━━━━━━━━━━━━━━━━━━━━━━━━━━━━━━━━━━━━━━━━━━━━━━━━━

If arcing does occur across the points (as the previous example described), which side of the switch would be damaged the most? The side connected to $V_p$ is damaged the most. Since real current is the flow of electrons, when an arc occurs it will be a bunch of electrons jumping across the gap. Since point p is the more positive side of the switch, the electrons will jump from the bottom of the battery over to point p. It is the collision of the electrons with the switch that blows little chunks of metal out of the way. There is a manufacturing process call Electric Discharge Machining (EDM) where a graphite rod with a large negative voltage is brought near a grounded metal workpiece. Sparks fly from the graphite rod to the metal, blasting chunks of metal away. The graphite rod is eaten away a little too, but not nearly as much as the metal workpiece.

As the previous example shows, when we attempt to switch a high inductive load, we can develop very high voltages across the switch. Whatever our mechanism is for switching, we must deal with this high voltage. When switching inductive loads using a transistor it is not uncommon for the voltage to exceed the break down voltage and force current through the transistor. This causes significant electrical energy conversion to thermal energy and can possibly damage the device.[24] Basically an inductor attempts to provide continuous and smooth currents. When sudden current changes occur, high voltages typically result.

Air conditioning units typically have high inductive loads (the fan and compressor are ordinarily AC induction motors). Often, switching is performed using a Mercury filled switch as figure 2.52 shows. To turn on, the switch is rotated so the Mercury touches both contacts. To turn off, the Mercury is rotated away from the contacts. When the inductive load causes arcing at the switch, the energy vaporizes the Mercury which ultimately condenses back inside the switch. If you watch the right kind of thermostat when the air conditioning is shutting off (especially in the dark), you

───────────────────

[24] Transistors are discussed in detail later.

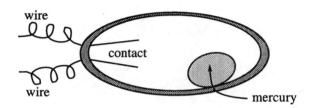

Figure 2.52: A Mercury Filled Switch

will see the spark caused by the arcing.

## Example 2.15

### A Spark Plug Circuit Revisited.

This example will solve the spark plug circuit using KVL. Since there are two loops in the circuit, we will write two KVL equations. Take loop one around the left side clockwise from ground:

**KVL For the Left Loop**

system[**Battery Loop**]      time period [**differential**]

$$\sum^{\text{loop}} V = 0$$

| rise: | $V - L_p \frac{d}{dt}(i_p) + M \frac{d}{dt}(i_s) - R_p i_p - V_p$ |
|---|---|
| fall: | $0$ |

$$\boxed{V - L_p \frac{d}{dt}(i_p) + M \frac{d}{dt}(i_s) - R_p i_p - V_p} \quad - \quad \boxed{0} \quad = \quad \boxed{0}$$

$\quad\;\;$ *rises* $\qquad\qquad\qquad\qquad\qquad\qquad\qquad$ *falls*

For the spark plug loop walking clockwise starting from ground:

**KVL For the Right Loop**

system[**Spark Plug Loop**]      time period [**differential**]

$$\sum^{\text{loop}} V = 0$$

| rise: | $V_s - L_s \frac{d}{dt}(i_s) + M \frac{d}{dt}(i_p) - R_s i_s$ |
|---|---|
| fall: | $0$ |

$$\boxed{V_s - L_s \frac{d}{dt}(i_s) + M \frac{d}{dt}(i_p) - R_s i_s} \quad - \quad \boxed{0} \quad = \quad \boxed{0}$$

$\quad\;\;$ *rises* $\qquad\qquad\qquad\qquad\qquad\qquad\qquad$ *falls*

In summary, our two equations for this system are:

$$V - L_p \frac{d}{dt}(i_p) + M \frac{d}{dt}(i_s) - R_p i_p - V_p = 0$$

$$V_s - L_s \frac{d}{dt}(i_s) + M \frac{d}{dt}(i_p) - R_s i_s = 0$$

**End 2.15**

## 2.5.3   Induction, Mutual Induction and Wire Wraps

Inductance is a function of the number of loops of wire used in a coil. Consider two wraps of wire from a coil, call them A and B. Current in loop A will produce a magnetic field whose strength is proportional to current. This magnetic field will produce a flux in all other loops including B. The flux caused by loop A $(d\phi)$ on loop B is proportional to the magnetic field and the area the field cuts through. Since the field is proportional to current we can say that the flux caused by A on B is $d\phi = ki$. The constant of proportionality $(k)$ depends on things like the size of the wire and the

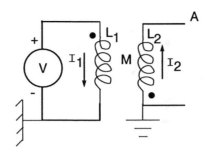

Figure 2.53: An Ideal Transformer Driven by an AC Source.

size of the loop. Now if there are $N$ loops on the coil, and if all of loop A's magnetic field enters all the loops then the total flux caused by loop A will be pretty close to $kNi$. Loop A is not the only loop creating a magnetic field, all the loops generate field therefore there should be pretty close to $N$ terms like $kNi$ hence the total should be pretty close to $kNNi$. Changing current causes a change in flux and a change in flux induces voltage therefore the induced voltage is basically $kN^2 \frac{d}{dt}(i)$. By inspection we notice that the self induction $L$ is given by $L \approx kN^2$. Note that some of the unspoken assumptions are that the loops are *identical*, the coil has a large number of wraps so that we can ignore the edges, and the field of all N wraps generate equal field on the coil.

Many students are familiar with handy transformer formulas that involve the number of wraps on the primary and secondary. We will now relate M to the number of wraps in the coils. Start with two coils A and B, one on coil 1, the other on coil 2. Current in loop 1 ($i_1$) allows A to produce a magnetic field whose strength is proportional to current 1. Assume this magnetic field is **completely** conducted to coil 2 so it will produce a flux in loops of coil 2 including loop B which is approximately $ki_1$. If the entire field from wrap A interacts perfectly with all the loops in coil 2 then the total flux caused by loop A is $kN_2i_1$ where $N_2$ are the number of loops on coil 2. If there are $N_1$ loops identical to A then the total flux on coil 2 caused by loop 1 is $kN_1N_2i_1$. By inspection we have $M \approx kN_1N_2$. Since the self induction values are given by (see previous paragragh): $L_1 = kN_1^2$ and $L_2 = kN_2^2$. It follows that $M^2 = L_1L_2$. Note that it is theoretically possible to have a transformer that does not convert any electrical energy into thermal energy yet not satisfy the simple relationships given here. In other words, if you consider an ideal transformer to be one with no energy losses (energy transformations into other forms) you still cannot assume the idealized equations are valid.

The following example demonstrates how the "handy little formulas" you may have used previously are related to the more general type of analysis this book teaches.

## Example 2.16

### Handy BUT LIMITED Transformer Equations

**PROBLEM:**

Figure 2.53 shows an ideal transformer. Let coil 2 have 100 times the windings as coil 1. Also let the source provide a voltage equal to $\sin(\omega t)$ where $\omega$ is a known constant. Determine the voltage ($V_A$) across coil 2.

**FORMULATION:**

There is one [NIF] in the circuit and one [NIA]. Applying KVL around the left loop gives:

$$V - L_1\frac{d}{dt}(I)_1 - M\frac{d}{dt}(I)_2 = 0$$

Since coil 2 is not connected, $I_2 = 0$. Therefore we can solve for $I_1$ by integrating:

$$I_1 = \int \frac{V}{L_1}dt + C = \frac{-\cos(\omega t)}{\omega L_1} + C$$

The value of $C$ accommodates the initial condition. We will not worry about the value of $C$ at the moment.

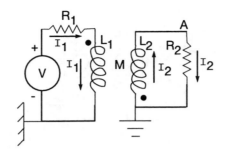

Figure 2.54: A Transformer with a Load.

Next we need to determine the voltage at terminal A. We could do this by writing the impedance relation for a coil with mutual induction but it might be easier to apply a KVL around the right loop. Remember, we told you the circuit need not be completed to write a KVL and here is an example of that. A KVL around the right loop starting at ground is:

$$-L_2 \frac{d}{dt}(I)_2 - M\frac{d}{dt}(I)_1 - V_A = 0$$

Since $I_2 = 0$ we have $V_A = -M\frac{d}{dt}(I)_1$. Differentiating $I_1$ gives:

$$V_A = -M\frac{\sin(\omega t)}{L_1}$$

From our discussion about the number of wraps in a coil, we know that under ideal cases $L_1 = kN_1^2$ and $M = L_1 L_2 = kN_1 N_2$ (where $N_1$ and $N_2$ are the numbers of loops in each coil). Simplifying we have $\frac{M}{L_1} = \frac{kN_1 N_2}{kN_1^2} = 100$ hence:

$$V_A = -100\sin(\omega t)$$

What happens is the transformer boosts the output voltage by the ratio of the numbers of windings. This is perhaps the handy formula that many students have memorized.

**End 2.16** ───────────────────────────────────────────────

The coil that connects to a source is typically called the primary, the remaining coil is called the secondary. If the number of windings on the secondary is larger than those on the primary, the transformer "steps up" (increases) the magnitude of a sinusoidal voltage. If the secondary has fewer windings, the transformer "steps down" the voltage. The transformer does **not** step up power, and the simplified equations are not applicable to all circumstances as the following example demonstrates.

**Example 2.17** ───────────────────────────────────────────

## A Loaded Transformer.

**PROBLEM:**

Figure 2.54 shows a transformer with two resistors. If the voltage source is $\sin(\omega t)$, determine the magnitude of voltage supplied to the load $R_2$. Also discuss the power transformation inside the transformer.

**FORMULATION:**

The circuit has 2 [NIF] and 2 [NIA]. We will apply KVL clockwise around the primary and secondary loops. Without any discussion the resulting equations are:

$$V - R_1 I_1 - L_1\frac{d}{dt}(I)_1 - M\frac{d}{dt}(I)_2 = 0$$

$$-L_2\frac{d}{dt}(I)_2 - M\frac{d}{dt}(I)_1 - R_2 I_2 = 0$$

To find the voltage at node A we write:

$$V_A = R_2 I_2$$

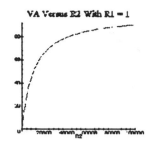

Figure 2.55: The Effect of a Load Resistance on the Transformer.

These equations can be solved. It is not our intention to teach solution methods, we used maple, see program 2.12.6. There are easier ways to finding the solution which you will learn in other courses, for now, we simply need the solution so we can make a few points. The program includes some comments, we will summarize the solution here. If we let inductance be related to the number of windings we can write:

$$L_1 = N_1^2 L$$

$$L_2 = N_2^2 L$$

$$M = N_1 N_2 L$$

where $L$ is a constant of proportionality. Maple tells us the voltage across $R_2$ is:

$$V_A = \frac{\sqrt{L^2 {N_1}^2 \omega^2 {R_2}^2 {N_2}^2 \left({R_2}^2 {R_1}^2 + \omega^2 L^2 {N_2}^4 {R_1}^2 + 2\omega^2 L^2 {N_2}^2 R_1 R\mathit{2} {N_1}^2 + \omega^2 L^2 {R_2}^2 {N_1}^4\right)}}{{R_2}^2 {R_1}^2 + \omega^2 L^2 {N_2}^4 {R_1}^2 + 2\omega^2 L^2 {N_2}^2 R_1 R_2 {N_1}^2 + \omega^2 L^2 {R_2}^2 {N_1}^4}$$

This is very ugly looking however note a couple of things, first if $R_1 = 0$, the expression reduces to the simplified expression:

$$V_A = \frac{N_2}{N_1}$$

If $R_1 \neq 0$ the expression is much larger. To make things a little easier let's put in some numbers. Let $N_1 = 1$, $L = 1$, $N_2 = 100$ and $\omega = 60$Hertz (what is supplied from the wall outlet in the United States). The expression for $V_A$ becomes:

$$V_A = 37700 \frac{\sqrt{{R_2}^2 \left({R_2}^2 {R_1}^2 + 142(10)^{11} {R_1}^2 + 284(10)^7 R_1 R_2 + 142000 {R_2}^2\right)}}{{R_2}^2 {R_1}^2 + 142(10)^{11} {R_1}^2 + 284(10)^7 R_1 R_2 + 142000 {R_2}^2} \tag{2.54}$$

Figure 2.55 shows the relationship between the voltage $V_A$ and the load resistance $R_2$. What the figure shows is that even for small source loads $R_1$ the voltage applied to the load is seriously degraded (reduced) for all but the largest loads. The effect of an increasing $R_1$, see the maple program to see it for yourself, is to make the situation worse. What this example demonstrates is that the simplified transformer equations are too simplified for most cases since even with the smallest loads applied, the voltage conversion degrades.

Another question is what is the effect of $\omega$? Figure 2.56 shows the effect of frequency on the response of the transformer. The figure shows the response when $R_1 = 1$ Ohm and $R_2 = \infty$. At very small frequencies, the system achieves its maximum voltage conversion. Why does the transformer work better for higher frequencies? Take the derivative of $\sin(\omega t)$. You get $-\omega \cos(\omega t)$ which means that if $\omega$ is large, the changes are large and since the inductance relies on change, it responds better.

Next, consider conservation of energy applied to the transformer. Power into the transformer is $V I_1 + 0$, power out is $V_A I_2 + 0$. The transformer accumulates energy but after awhile whatever energy the transformer stores will reach a steady state. In other words, the energy stored in the transformer is a function of the currents. Unless the current becomes infinite, the energy storage is finite. Let's take the time period for our analysis such that when the change in accumulation in the transformer is zero. Hence the energy input during this time equals the energy output. What

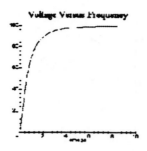

Figure 2.56: Voltage Versus Frequency for a Loaded Transformer.

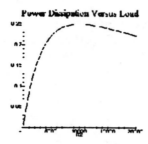

Figure 2.57: Power Dissipated in $R_2$ Versus $R_2$.

this means is the "average" voltage $V$ times "average" current $I_1$ equals "average" voltage $V_A$ times "average" current $I_2$. You see, since $I_1$ is limited (by the resistor $R_1$) the quantity $VI_1$ is limited which limits the product $V_A I_2$ therefore as $V_A$ increases, $I_2$ must decrease and vice versa. So can a transformer step up voltage? Yes. Can it step up current? Yes by stepping down voltage. Can it step up voltage and current at the same time? NO! Note that when $R_2 = \infty$, $I_2 = 0$ nothing flows but $|V_A|$ was as large as it would ever be. When $R_2$ was small, $|V_A|$ was small but the current $|I_2| > 0$.

Finally, is there a load (a value of $R_2$) that maximizes the power converted in the load? Think about it, if $R_2 = \infty$ the current $I_2 = 0$ and $I_2^2 R_2 = 0$ no power is converted. What about if $R_2 = 0$, in that case $V_A = 0$ (look at equation 2.54) and power $(V_A^2 / R_2)$ is also zero. There are values of $R_2$ that allow nonzero power conversion. Since power is zero for two values of $R_2$ and nonzero for at least one value of $R_2$ then there is at least one value of $R_2$ that makes power conversion maximum. If you compute the power dissipated in $R_2$ from equation 2.54 you would find:

$$P = \frac{L^2 N_1{}^2 \omega^2 R_2 N_2{}^2}{R_2{}^2 R_1{}^2 + \omega^2 L^2 N_2{}^4 R_1{}^2 + 2.0\omega^2 L^2 N_2{}^2 R_1 R_2 N_1{}^2 + \omega^2 L^2 R_2{}^2 N_1{}^4}$$

If you use the same values for the constants that we have used previously, this expression reduces to:

$$P = 10000 \frac{\omega^2 R_2}{R_2{}^2 + 100000000\omega^2 + 20000\omega^2 R_2 + \omega^2 R_2{}^2}$$

This expression peaks with relatively small values of $\omega$. Figure 2.57 show the power magnitude versus the load magnitude. Notice there is, as we expected a maximum power dissipation.

This concept is what is behind "impedance matching" transformers. For example suppose you have a stereo modeled as the sinusoidal voltage source and resistor $R_1$. The $R_1$ is similar to the output impedance of your stereo (actually the stereo's impedance is more complex than a single resistor but the concept is similar). The stereo actually generates a voltage with multiple voltage levels and frequencies. These represent the sound that is to be produced. The load $R_2$ is similar to the speaker. The real speaker is of course more complex, but the idea is the same. The objective is to connect the stereo to the speaker in such a way that as much power as possible is dissipated in the speaker ($R_2$) after all you want the system to crank. If you have too many, too few, or the

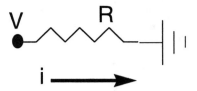

Figure 2.58: A Resistor with Applied Voltage

speakers are not properly sized (wrong value of $R_2$) you may not convert as much power as you could into sound. The purpose of an impedance matching transformer is to match the impedances to maximize the power delivery.

**End 2.17**
_____

## 2.6   IMPEDANCE RELATIONSHIPS

**Objectives**
*When you complete this section, you should be able to*

1. *Calculate the current through a resistor, capacitor and inductor when a sinusoidally varying voltage is applied across it.*

2. *Explain what the equivalent resistance is to a capacitor and inductor at high and low frequencies.*

3. *Use the equivalent resistance concept in the design of simple filters.*

4. *Explain some of the uses for capacitors and inductors.*

5. *Explain the effect of input/output impedance on circuit design.*

Up to this point, the text has focused on the analysis of circuits. Given the circuit diagram you could compute the voltages and currents. This is useful to make sure a circuit acts the way you want it to act. In this section, we discuss some methods to help you decide how to design a circuit. In other words, how to decide to put a capacitor or inductor in a circuit to make a response that you want.

This section discusses impedances of electrical components. When thinking about devices, it is sometimes convenient to think about their impedance relationships.[25] Although it is possible to use energy relationships exclusively to derive circuit equations, impedance relations can be helpful at times. This is especially true when trying to design circuits.

Consider a resistor with a voltage applied to it as figure 2.58 shows. Accounting for electrical energy for the resistor gives us a voltage/current relationship for the resistor.

_____
[25] Remember that the impedance relation is the relationship between the device's effort and flow variables.

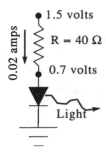

Figure 2.59: A Resistor Impeding the Flow of Current Through a Light Emitting Diode

## Electrical Energy  Accounting

system[ **resistor fig.2.58** ]      time period [**differential**]

$$input - output + generation - consumption = accumulation$$

| input/output: | $iV$ |
|---|---|
| consumption (thermal): | $i^2R$ |
| generation: | |
| accumulation: | $0$ |

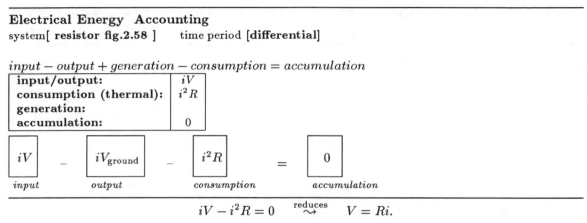

$$iV - i^2R = 0 \quad \overset{\text{reduces}}{\rightsquigarrow} \quad V = Ri.$$

Notice that the simple resistor has a simple (constant) impedance ($R$). Also note that whenever $V$ changes, an immediate and proportional change occurs in $i$. Since the change in current is immediate, the resistor can respond as well to a quickly changing voltage as a slowly changing one. If the voltage across the resistor is sinusoidal (if $V = V_o \sin(\omega t)$) then the current is sinusoidal, ($i = \frac{V_o}{R} \sin(\omega t)$). The resistor has the same response to high and low frequency (high frequency is large $\omega$, low frequency is small $\omega$) signals (voltages).

We commonly use resistors to

1. Impede the flow of current. For example, preventing a short circuit between a source and ground, or choking back flow into a device. Figure 2.59 shows a resistor inline with a light emitting diode (LED). The diode has the characteristic that when current passes through it as shown, it emits light. The more current, the more light. The current that passes through the diode is a nonlinear function of the voltage across it. For example at no current flow, there will be 0 volts across the diode (like a resistor). As current increases, voltage increases the voltage reaches the forward bias voltage (approximately 0.7 [volts] for many diodes). At approximately the forward bias voltage, increasing the current does not significantly increase the voltage. Very strange indeed. Basically, a "generic" LED will conduct approximately 20 [milliamps] of current at the forward bias of 0.7 [volts] and it will emit light. Write a conservation of energy on the diode to determine what will happen if you connect a diode between the terminals of a 1.5 volt battery. Too much current will make the diode heat up and burn out. Too little current and it doesn't emit much light. to burn out, or not emit light. If you connect a 1.5 [volt] battery straight to the diode, there will be more than 0.7 [volts] across the diode and more than 0.02 [amps] of current. This may cause the diode to burn up. Placing a resistor as the figure shows will choke back the current and drop the voltage to the desired value. How could you compute the required value of resistance to use? By the way, a diode (including LEDs) are one way valves for current. Conventional current can flow in the direction of the big black arrow LED symbol but it normally does not flow the other direction. Diodes will be discussed later.

2. Convert current flow into a voltage. For example, if we want to measure current but only have

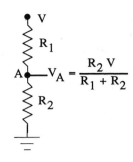

Figure 2.60: A Voltage Divider

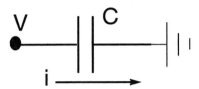

Figure 2.61: A Capacitor with Applied Voltage

a voltmeter, we can place a resistor in the flow, measure the voltage across the resistor and relate voltage to current. In such an application, what are the advantages of using a small resistor versus a large resistor? A small resistor will not impede the flow that you are trying to measure, but it will not create a very large voltage change either. Hence a small resistor will not disturb what you are trying to measure, but you better have a good voltmeter that measures small voltages!

3. Converting electrical energy into another form. Usually the other form is thermal energy. An incandescent light bulb operates on this concept. The filament gets so hot, it glows.

4. Constructing a voltage divider. We have a voltage source at $V$ [volts] but need a different voltage. The circuit shown in figure 2.60 demonstrates how we can create a circuit with a couple of resistors that provides the desired voltage. What happens to the voltage produced by the divider when current is extracted from point A? Maple program 2.12.9 shows the computation. What are the advantages and disadvantages to having large and small values of resistance in the voltage divider? The maple program shows the disadvantage of a large $R_1$. The problem with small resistors is there will be plenty of wasted current flow from the power source to ground straight through the divider.

Now consider a capacitor with an applied voltage as figure 2.61 shows. Accounting for electrical energy in the capacitor, we find that

**Electrical Energy Accounting**

system[ **capacitor fig.2.61** ]      time period [**differential**]

$input - output + generation - consumption = accumulation$

| input/output: | $iV$ |
|---|---|
| **consumption (thermal):** | |
| **generation:** | |
| **accumulation:** | $\frac{d}{dt}\left(\frac{1}{2}CV^2\right)$ |

| $iV$ | $-$ | $iV_{\text{ground}}$ | $=$ | $\frac{d}{dt}\left(\frac{1}{2}CV^2\right)$ |
|---|---|---|---|---|
| *input* | | *output* | | *accumulation* |

$$iV = \frac{d}{dt}\left(\frac{1}{2}CV^2\right)$$

which simplifies to

$$iV = \frac{1}{2}C2V\dot{V} = CV\dot{V}$$

$$i = C\dot{V} \tag{2.55}$$

Here, the impedance relation is more complex because it depends on how $V$ changes. Let's consider one simple case. Suppose the voltage across a capacitor is sinusoidal with a frequency $\omega$.[26] We can write the voltage as

$$V = v\sin(\omega t) \tag{2.56}$$

If we differentiate, we find that

$$\dot{V} = v\omega\cos(\omega t)$$

Using this in equation 2.55 gives

$$Cv\omega\cos(\omega t) = i$$

This says that the current changes in a similar way as the voltage (with the same frequency $\omega$), but it is out of phase with the voltage (voltage is sine, current is cosine) which means that voltage and current are never simultaneous maximums.

If we solve for the voltage $v$ rather than $i$, we find that

$$v = \frac{1}{C\omega\cos(\omega t)}i$$

In this form, the capacitor behavior looks similar to resistor behavior which leads us to the idea of comparing them. If we think of an equivalent resistance for the capacitor[27] the term $\frac{1}{C\omega}$ is quite like the magnitude of resistance. With this thinking, at low frequencies (small $\omega$) and/or small capacitance, the capacitor acts like a large resistance ($\frac{1}{C\omega}$ is large). In fact, at zero frequency the capacitor acts like an infinite resistance, or open circuit. At high frequencies (large $\omega$) and/or large capacitance, the capacitor acts like a small resistance ($\frac{1}{C\omega}$ is small). This is why engineers often use the following rule of thumb. A capacitor is a short circuit (low resistance) at high frequency and an open circuit (high resistance) at low frequency. We can summarize this concept as follows:

1. For high frequency, a capacitor has low resistance. In the limit, high frequency is infinity and the capacitor will look like a short circuit (like a wire).

2. For low frequency, a capacitor has high resistance. In the limit, low frequency is zero and the capacitor will look like an open circuit (like a broken wire).

Keep in mind, however that these are approximations to what the capacitor does after a long time (the steady state response), the capacitor's actual response depends on its initial conditions and the time. For example, a capacitor without any charge differential across its plates will conduct current from a DC (zero frequency) supply for a limited amount of time. Eventually however, it stops conducting.

Before leaving the capacitor, there is another interpretation which comes in handy. Because of the behavior indicated by equation 2.56, we can think of a capacitor as a device that attempts to prevent its voltage drop from changing quickly. For example, if $V$ changes quickly, then $\dot{V}$ will be large thus requiring large current. If the capacitor is a member of a circuit, then there must be another set of components in the circuit which are capable of supplying (or absorbing) the large

---

[26] Actually a sinusoidal voltage is not restrictive. Since it is possible to express all continuous functions as a series of sinusoids, knowing what happens for a single sinusoid gives a good idea of what will happen with more complex functions.

[27] Resistance and capacitance are nothing alike. However, since people are accustomed to thinking about resistance and unaccustomed to thinking about capacitance, it is helpful (especially in design problems) to say capacitance acts like resistance. This allows us to think about a circuit in terms of a resistance. It gives a designer a good idea of what will happen before actually taking the time to analyze it.

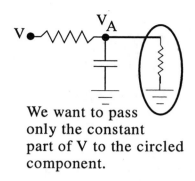

We want to pass
only the constant
part of V to the circled
component.

Figure 2.62: A Smoothing Device

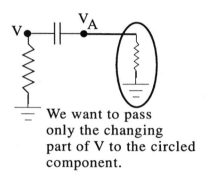

We want to pass
only the changing
part of V to the circled
component.

Figure 2.63: A DC Reducing Device

current flowing in the capacitor. If this current cannot be handled, the voltage change across the capacitor must remain small. For this reason, engineers use large capacitors in systems where a constant voltage is required. Examples of such systems are DC power supplies and DC to DC converters. On the flip side, if a circuit requires large amounts of current, this current might be achieved by forcing a large voltage change across a capacitor. Systems such as high power pulsed lasers use this technique (although Dr. Zap was a made up problem his idea is not that bizzare). As a result of the capacitor's behavior, circuits with large capacitance tend to have smooth voltage changes but may have considerably large current.

We commonly use capacitors to

1. Smooth the voltage changes in a circuit. Figure 2.62 shows a capacitor connected to smooth the voltages at point A. Suppose the voltage $V$ is coming from a AC source such as a 110 [volt] household connection and we desire for the fluctuations in voltage at A to be reduced. The capacitor accomplishes this. For example, if $V_A$ tries to rise, the capacitor conducts some current away from A (due to the changing $V_A$) helping to reduce $V_A$. If $V_A$ begins to fall, the capacitor conducts current into A helping to hold $V_A$ up. The resistor is needed to shelter (choke off) the source current coming from $V$. Without the resistor, the relatively small currents conducted by C would be consumed by the source. The resistor and capacitor connected as shown make up what is called a low pass filter. It passes the lower frequencies to the circled load. Note that what is high and what is low depends on the specific values used for the resistance and capacitance. In a design, you sketch out what you think you want in a circuit (by thinking about impedance) then you test and select sizes by analyzing (computing voltages and currents using KVL and other equations). Maple program 2.12.7 shows how the analysis can be performed.

2. Eliminating (reducing) the effects of constant voltage. Figure 2.63 shows a device which reduces the amount of low frequency components transmitted to $V_A$. Since the ideal resistor conducts all frequencies equally well and the capacitor conducts the high frequencies better than others, the high frequency parts of $V$ travel through the capacitor while the low frequency

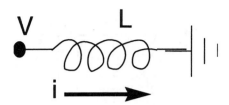

Figure 2.64: An Inductor with an Applied Voltage

parts get dumped through the resistor. Devices like the one in the figure are common in measurement components and is called a high pass filter. Some call is AC coupling. The Maple program 2.12.8 shows the analysis. It also shows that the resistor on the left is not required.

3. Store energy temporarily. We discussed an application of this in example 2.14.

Now consider the single inductor with an applied voltage as figure 2.64 shows. Accounting for electrical energy in the inductor, we find that

**Electrical Energy  Accounting**
system[ **inductor fig.2.64** ]      time period [**differential**]

$input - output + generation - consumption = accumulation$

| input/output: | $iV$ |
| **consumption (thermal):** | |
| **generation:** | |
| **accumulation:** | $\frac{d}{dt}\left(\frac{1}{2}Li^2\right)$ |

$$\boxed{iV}_{\text{input}} - \boxed{iV_{\text{ground}}}_{\text{output}} = \boxed{\frac{d}{dt}\left(\frac{1}{2}Li^2\right)}_{\text{change in accumulation}}$$

$$iV = \frac{d}{dt}\left(\frac{1}{2}Li^2\right)$$

which simplifies to

$$iV = \frac{1}{2}L2i\frac{d}{dt}\left(i\right) = Li\frac{d}{dt}\left(i\right)$$

$$V = L\frac{d}{dt}\left(i\right) \tag{2.57}$$

Again, the impedance relationship is not as simple as a resistor because it depends on the change in current. Suppose we consider just one case where the current $I$ changes sinusoidally[28] as $I = i\sin(\omega t)$. Differentiating the current relation, we find that

$$\dot{I} = i\omega\cos(\omega t)$$

Using this in equation 2.57 gives

$$V = L\omega\cos(\omega t)i$$

Just like the capacitor, the voltage and current change with the same frequency but are out of phase. Just as for the capacitor, we can think of the inductor as having an equivalent resistance with magnitude of $L\omega$. This behavior is just opposite to the capacitor. At low frequencies and low inductance, the equivalent resistance is small; an inductor is a low frequency short circuit. At high frequencies and high inductance, the equivalent resistance is large; an inductor is a high frequency open circuit.

---

[28] Let's not worry how we can make that happen right now.

As a final interpretation, based on the behavior indicated by equation 2.57, an inductor attempts to keep current from changing quickly. For example, if current changes suddenly, then $\frac{d}{dt}(I)$ is large implying a large $V$. If the inductor is a component in a circuit, then something in the circuit must generate or absorb the large voltage. If the large voltage is impossible, then the current cannot change quickly. For this reason, we will find large inductances in circuits that either attempt to hold current constant or generate large voltages. We will also find inductance in systems that develop significant magnetic fields. Systems containing large inductances often have smooth current changes but significant voltage swings.

We commonly use inductors to

1. Generate large voltages. This application was demonstrated in example 2.14.

2. Smooth current changes in a circuit. An automobile radio, for example, is often subject to large voltage/current swings due to the high inductance alternator[29] sourcing energy into the auto's circuit. To help reduce these swings, the radio places an inductor in series with the power connection. The inductor passes the low frequency components and chops off the higher frequencies.

3. To electrically isolate systems. In the form of a transformer, changing voltage/current on one side (the primary) induces voltage/current on the other (the secondary). Because the two sides are not electrically connected, they can have their own grounds. This is especially important in constructing sensors (avoiding ground loops), or for applying power to a circuit at a level referenced above ground. Transformers also provide some protection against lightning strikes. Protection against lightning is difficult to achieve, however a transformer can provide some isolation from parts of circuits subject to lightning.

4. Temporarily storing energy.

5. Changing the voltage of AC power. Transformers are used to produce a rise or drop in AC power. For example, power transmitted over large distances by power companies is transmitted at an elevated voltage. Before it enters a household, the power is run through a transformer to drop the voltage down to 110 [volts]. A transformer is a very efficient method for this conversion.

6. Transformers can also be used to match impedances. For example audio amplifier equipment has an output impedance which is a function of its components. For maximum power delivery, the equipment should be connected to a load (speakers for example) which have an impedance matched to the equipment. This doesn't always happen. To modify the impedance relationships, we can place a transformer between the two systems. The amplifier drives the primary side of the transformer, the load is connected to the secondary. In analogy, the transformer performs the same function as a gearbox. The transformer is electrical whereas the gearbox is mechanical.

7. Used in conjunction with a capacitor, an inductor forms part of a resonant or tuned circuit. Essentially a tuned circuit responds best to frequencies it is tuned for and responds poorly to all other frequencies. Radio receivers and transmitters, radar, lasers and many other devices use resonant circuits. Figure 2.65 shows a circuit tuned to a frequency of $\sqrt{\frac{1}{LC}}$. As an exercise, solve for the voltage at B assuming that $V_A = \sin\left(\frac{kt}{LC}\right)$. Find the solution for several values of $k \neq 1$. Based on the results, what do you think the solution is when $k = 1$? Maple program 2.12.10 shows the solution. Basically what happens is if the voltage at A is sinusoidal at the "tuned" frequency, then there is a very large voltage magnitude at B, otherwise the voltage at B is not extraordinary. Tuned circuits are used when one needs to get large amounts of energy into or out of a circuit.

Inductors can also be used to create low and high pass filters. If you have an option of building a filter with a capacitor or with an inductor, use the capacitor because they are usually cheaper than inductors.

---

[29]An alternator, sometimes a generator, is used for battery charging. It typically has considerable inductance because it operates on the electromagnetic principle of induction.

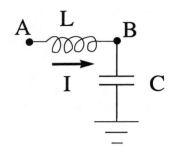

Figure 2.65: A Resonant Circuit

## 2.7    CIRCUITS WITH SWITCHES

**Objectives**

*When you complete this section, you should be able to*

1. *Explain how many and what type of initial conditions are required to complete a circuit formulation.*

2. *Explain why one might need initial conditions that differ from what is predicted.*

3. *Explain why you seldom need the initial conditions for circuits with sinusoidal sources.*

4. *Describe what the steady state response of a capacitor and inductor are when excited by DC sources.*

5. *Describe the continuity conditions that capacitors and inductors impose on circuits and give some exceptions to them.*

6. *Recognize circuits in which inductors or capacitors do not impose continuity on the circuit.*

7. *Determine circuit variables at an initial time for circuits with DC sources and inductors and capacitors at continuity conditions.*

---

This section will discuss how to calculate initial conditions in circuits with switches. When we have a circuit with capacitors or inductors, we have a circuit that can accumulate electrical energy. Therefore, the circuit is nonzero [NIA]. Recall that the [NIA] is the minimum number of first order differential equations required to express the conservation and accounting equations. The [NIA] also equals the minimum number of initial conditions needed to solve the conservation and accounting equations. At this point, we turn our attention to determining these initial conditions. This text will only cover initial conditions for capacitors and inductors subjected to constant sources. When sinusoidal sources are present, one usually desires to find the steady state response of the circuit. Since the steady state response does not depend on the initial conditions, one usually does not bother to determine them. Some previous examples demonstrated how a steady state solution can be obtained and you will learn more about this in a differential equations class so we will not cover it here.

Usually initial time to an engineer means just after some event has taken place. In electrical circuit analysis, an event is usually the opening or closing of a switch.

Consider first the behavior of the capacitor, that figure 2.66 shows. If the capacitor has no electrical energy stored and the switch is closed, electrons from the negative terminal will rush in to occupy the bottom plate of the capacitor. As each electron flows in the bottom, a different electron pops out of the top plate and moves toward the positive battery terminal. This process is current flow, the capacitor flows current. Before long however, the bottom plate is holding as many electrons as the voltage source can force in and flow stops. In summary, a capacitor will flow current, even from a constant source, at least for a little while.

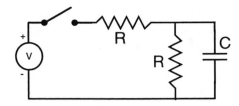

Figure 2.66: A Simple Capacitor, Resistor Circuit

Real components that accumulate properties (like a capacitor and inductor) exhibit two types of response, transient (temporary) and steady state (persisting). Mathematicians often call the types homogeneous and nonhomogeneous (sometimes called homogeneous and particular). An example of transient or homogeneous behavior is the temporary flow through a capacitor. We discovered that the magnitude of the long term (steady state) response of a capacitor, $C$, at low frequency is equivalent to a resistance of $\frac{1}{C\omega}$. For zero frequency (constant sources $\omega = 0$), the capacitor has infinite resistance. Only transient or temporary flow is possible. After long times, the capacitor begins to look like a broken wire.

Now consider the response of a capacitor over small intervals of time. Recall the impedance relationship of a capacitor $C\frac{d}{dt}(V) = I$ where $I$ is the current flowing through the capacitor and $V$ is the voltage across the capacitor. If we integrate the impedance relation over an infinitesimal time period, we find that

$$\int_t^{t+dt} C\, dV = \int_t^{t+dt} I\, dt$$

If the current is finite and $dt$ is infinitesimal, then the right hand side of the equation is zero. Thus, this relation simplifies to

$$\int_t^{t+dt} C\, dV = C\left(V|_{t+dt} - V|_t\right) = 0$$

The voltage at time $t$ equals the voltage at time $t + dt$. This means that the voltage is continuous over time. The capacitor does not like jumps in voltage! This is the continuity relation imposed by the capacitor. If there cannot be an infinite current, then there cannot be a jump in the voltage across the capacitor.

Now consider an inductor. We also discovered that the magnitude of long term response of an inductor, $L$, at low frequency is equivalent to a resistance of $L\omega$. For zero frequency (constant sources $\omega = 0$), the inductor has zero resistance at steady state. After the transient period, the inductor does not inhibit flow of current. After long times, the inductor begins to look like a perfect wire.

To determine the response over small intervals of time, recall the impedance relationship of an inductor $L\frac{d}{dt}(I) = V$ where $I$ is the current flowing through the inductor and $V$ is the voltage across it. If we integrate the impedance relation over an infinitesimal time period, we find that

$$\int_t^{t+dt} L\, dI = \int_t^{t+dt} V\, dt$$

If the voltage is finite and $dt$ is infinitesimal, the right hand side of the equation is zero. Thus, this relation simplifies to

$$\int_t^{t+dt} L\, dI = L\left(I|_{t+dt} - I|_t\right) = 0$$

The current at time $t$ equals the current at time $t + dt$. This means that the current is continuous over time. The inductor does not like jumps in current! This is the continuity relation imposed by inductors, if the voltage is finite, the current through an inductor cannot suddenly change.

Now to put this together. Suppose we have an event taking place (*e.g.* opening or closing a switch), immediately after the event all capacitors/inductors will desire to keep their voltage/current exactly what is was before the event. Before the event, all capacitors/inductors display their steady state behaviors. Concept 2.7 summarizes the steps to use in finding initial conditions for constant sources.

## Concept 2.2 ─────────────────────────────────────────

### Finding Initial Conditions

1. Make sure that there cannot be infinite voltages or infinite currents in the circuit. What this means is:

    (a) Make sure there are no loops with only sources and capacitors (an ideal source can provide infinite current).

    (b) Make sure no inductors exist in loops which get cut by the event. In other words, the spark plug circuit in Example 2.14 fails the test. When the switch opens, the current through the inductor has no place to go and an infinite (theoretically) voltage occurs.

2. Determine what the event is. For example is it the opening of a switch or the closing? If there are multiple event occurring one after another, we must assume there is sufficient time between the events that they all reach their steady state operation. Example 2.7 showed how to handle an series of events when there was insufficient time between the events.

3. Redraw the circuit in its before the event (BTE) state. For example, if the event is the opening of a switch, then before the opening, the switch is closed, draw the switch closed. At the BTE, capacitors and inductors display their steady state behavior.

    (a) Replace all inductors with perfect wire connections.

    (b) Replace all capacitors with broken wires.

4. Analyze the BTE circuit using conservation of charge, accounting of electrical energy, or Kirchhoff's law. Actually most BTE circuits are so simple, you can "recognize" the voltages and currents without writing a bunch of equations. If you get confused, write the equations.

    (a) Determine the currents flowing in all the perfect wires that used to be the inductors.

    (b) Determine the voltages across all the broken wires that used to be capacitors.

5. Redraw the circuit in its after the event (ATE) state. Capacitors have the same voltage and inductors have the same current as BTE.

    (a) Replace all inductors with current sources. The magnitude and direction of the sources match the BTE currents determined in step 4.

    (b) Replace all capacitors with voltage sources. The magnitude and polarity of the sources match the BTE voltages determined in step 4

6. Determine all the initial condition voltages and currents necessary to formulate a complete mathematical model of the system. Unless you have manipulated your equations, the BTE voltages and currents will most likely be what you need.

**End 2.2** ─────────────────────────────────────────

## Example 2.18 ─────────────────────────────────────────

### A Simple Switched Circuit

**PROBLEM:**

At $t=0^-$ (this means just before time zero), the switch in the circuit in figure 2.67 is at a. At $t=0^+$ (this means just after time zero), the switch is at b. At $t = 0$ we move the switch to b. Determine $V_b$, $I_1$, $\frac{d}{dt}(I_1)$ and $\frac{d^2}{dt^2}(I_1)$ immediately after the switch is thrown. Notice that the circuit

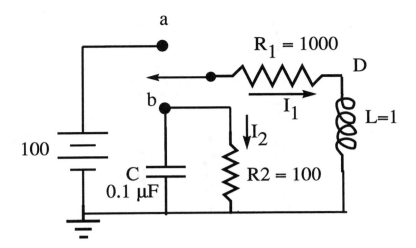

Figure 2.67: A Simple Switched Circuit

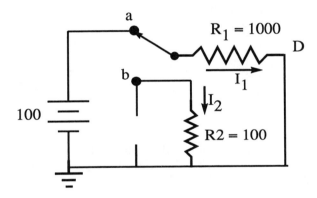

Figure 2.68: The BTE State of the Circuit

has two [NIA] which means you will need only 2 initial conditions. The initial conditions will be the voltage at b and the initial current $I_1$. The initial voltage across the capacitor which tells you how much energy and charge is initially contained in the capacitor and the initial current tells you how much energy is initially contained in the inductor. You will not need the first and second derivatives of $I_1$ to solve the conservation equations. Since this is just an example, we are asking for them to make a point.

**FORMULATION:**

Note that there are no loops (regardless of the switch position) in which there are only sources and capacitors. Also note however that if there is a time when the switch is neither in position a or b then the inductor will be in a loop which is cut by the event and this circuit would not be so easy to analyze! Since the problem says the switch is at one moment in position a then its in position b we will proceed. What do you think would "really" happen? Think about this as we work through the example.

First, we draw the circuit in the BTE state. The capacitor is a broken wire and the inductor is a perfect wire. Figure 2.68 shows this. Applying the conservation of charge at node b, we find that

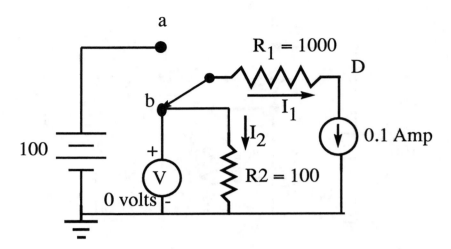

Figure 2.69: The Circuit in its ATE State

---

**Conservation of Charge**
system[**node b fig.2.68**]      time period [**differential**]

*input − output = accumulation*

| input/output: node b | $-I_{2,\text{BTE}}$ |
|---|---|
| **accumulation:** | |

| $-I_{2,\text{BTE}}$ | $=$ | $0$ |
|---|---|---|
| *input/output* | | *change in accum.* |

---

Therefore, $I_{2,\text{BTE}} = 0$ (you could have seen this right off the diagram, no). Since there is no current through $R_2$, there is no voltage across it.[30]  Therefore, $V_{\text{b,BTE}} = 0$. By inspection, $V_{\text{a,BTE}} = 100$ and $V_{\text{D,BTE}} = 0$. Using the impedance relation for $R_1$, we find

$$V_{\text{tail}} - V_{\text{head}} = IR \rightarrow \frac{100}{R_1} = I_{1,\text{BTE}} = \frac{100}{1000} = 0.1$$

Now, draw the circuit in its ATE state. The capacitor is replaced with a voltage source of magnitude $V_{\text{b,BTE}} = 0$.[31]  The inductor is replaced with a current source of magnitude $I_{1,\text{BTE}} = 0.1$. Figure 2.69 shows the circuit in its ATE state. Notice that the arrow in the current source matches the direction of the current through the inductor at BTE. By inspection $V_{\text{b,ATE}} = 0$. Since $V_{\text{b,ATE}} = 0$, there is no voltage across $R_2$. Therefore, $I_{2,\text{ATE}} = 0$. Using the impedance relation for $R_1$ gives

$$\frac{(V_{\text{b,ATE}} - V_{\text{D,ATE}})}{R_1} = I_{1,\text{ATE}}$$

Substituting the known quantities, this reduces to

$$\frac{-V_{\text{D,ATE}}}{1000} = 0.1$$

Solving this relation, we find that $V_{\text{D,ATE}} = -100$. Strange wouldn't you say? The voltage at D jumps from zero at BTE to -100 at ATE. Is this possible? Yes, only the voltage across a capacitor is guaranteed to have no jumps.

Since the BTE and ATE states correspond to specific instances in time, we cannot attempt to differentiate any of our BTE or ATE results to determine the derivatives we need. What we need is a

---

[30] Remember $V_{\text{tail}} - V\text{head} = IR$ for a resistor. If $I = 0$ then $V_{\text{tail}} = V_{\text{head}}$

[31] A voltage source of zero magnitude is a perfect wire.

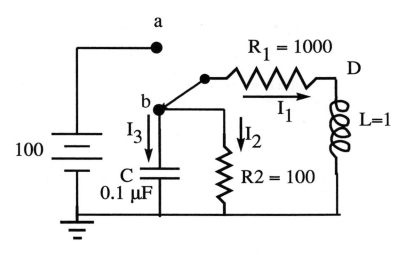

Figure 2.70: The Circuit For General Times After the Event

general relationship for the circuit values so we can differentiate. Since we want the derivatives ATE, we draw the circuit a little time ATE. We redraw the circuit keeping in the inductor and capacitor and also showing the switch at terminal b. You see its just a little while after the event, we have not reached steady state after the event, its just a wee bit later but it is later so the switch has been thrown. This represents the real system a wee bit after the event. We know that the initial values for this circuit correspond to the ATE values. Figure 2.70 shows the circuit. Applying Kirchhoff's law to the loop across C to D and back to ground gives

**Kirchhoff's Voltage Law (KVL)**

system[ **loop CD fig.2.70** ]     time period [**differential**]

$$\sum{}^{\text{loop}} V = 0$$

| rise: | $V_b$ |
|-------|-------|
| fall: | $R_1 I_1 + L\frac{d}{dt}(I_1)$ |

$$\boxed{V_b}_{\,rise} \quad - \quad \boxed{R_1 I_1 + L\frac{d}{dt}(I_1)}_{\,fall} \quad = \quad \boxed{0}$$

$$V_b - R_1 I_1 - L\frac{d}{dt}(I_1) = 0$$

Solving for $\frac{d}{dt}(I_1)$ gives:

$$\frac{V_b - R_1 I_1}{L} = \frac{d}{dt}(I_1)$$

now using the impedance relation for resistor $R_2$ (that is $V_b - 0 = R_2 I_2$) we find that

$$\frac{I_2 R_2 - R_1 I_1}{L} = \frac{d}{dt}(I_1) \tag{2.58}$$

Since we want $\frac{d}{dt}(I_1)$ at ATE, we substitute in known ATE values. This gives

$$\frac{(0)(100) - (1000)(0.1)}{1} = -100 = \frac{d}{dt}(I_{1,\text{ATE}})$$

Because equation 2.58 is valid at all times, we can differentiate it. Differentiating equation 2.58, we find that

$$\frac{\frac{d}{dt}(I_2) R_2 - R_1 \frac{d}{dt}(I_1)}{L} = \frac{d^2}{dt^2}(I_1) \tag{2.59}$$

Equation 2.59 involves two unknowns, $\frac{d^2}{dt^2}(I_1)$ and $\frac{d}{dt}(I_2)$. We need another equation. We need something we can use to find the derivative of $I_2$. Using the impedance relation of $R_2$, we find that

$$V_b = I_2 R_2$$

Differentiating and solving (remember if we are differentiating equations we need equations that are good over a period of time which is why we didn't put in 0 for $V_b$) for $\frac{d}{dt}(I_2)$ gives

$$\frac{d}{dt}(I_2) = \frac{\dot{V}_b}{R_2}$$

Dog gone, its always something, we cannot find $\frac{d}{dt}(I_2)$ until we find the derivative of $V_b$. To determine $\dot{V}_b$, we use the impedance of $C$.

$$I_3 = C\dot{V}_b$$

Solving the last equation for $\dot{V}_b$, we find that

$$\dot{V}_b = \frac{I_3}{C}$$

Well now we need $I_3$. Finally, applying the conservation of charge at node b, we find that

$$-I_1 - I_2 - I_3 = 0$$

which puts $I_3$ in terms of things we know, whew.

Now we can evaluate all this at ATE (we have finished all the differentiation), so we have (remember $I_2$ at ATE is zero)

$$-I_{1,ATE} - I_{2,ATE} - I_{3,ATE} = -0.1 - 0 - I_{3,ATE} = 0$$

so $I_{3,ATE} = -0.1$. Therefore, working back through our equations, we find that

$$\dot{V}_{b,ATE} = \frac{I_{3,ATE}}{C} = \frac{-0.1}{0.0000001} = -10^6$$

$$\frac{d}{dt}(I_{2,ATE}) = \frac{\dot{V}_{b,ATE}}{R_2} = \frac{-10^6}{100} = -10^4$$

$$\frac{\frac{d}{dt}(I_{2,ATE})R_2 - R_1\frac{d}{dt}(I_{1,ATE})}{L} = \frac{d^2}{dt^2}(I_{1,ATE}) = \frac{-10^4(100) - (1000)(-100)}{1} = -10^6 + 10^5$$

Man, finding higher derivatives in circuits is a pain. Don't do it if you don't have to and you will not have to unless your instructor makes you, or if you manipulate your circuit equations after you get them.

**End 2.18**

As a bit of practical knowledge, the type of switch shown in figure 2.67 is a single pole, double throw (SPDT). Single pole means there is a single "common" connection, double throw means that connection can be in one of two locations. In this case, it can be in position a or position b. In an ordinary SPDT switch, there is some time when the common pole is neither in position a or b as we mentioned earlier in this paragraph, but we are ignoring that fact for the moment. You can purchase switches in a variety of poles and throws depending on your needs. In addition to specifying the poles and throws you sometimes specify the normal operating conditions. For example if the switch is a pushbutton you might say it is a SPSTNO (single pole single throw normally open meaning the switch is open until you push it when it makes contact), or DPSTNC (double pole single throw normally closed which means two wires are connected until you hit the button and both open simultaneously). Classify the switches in the next example.

**Example 2.19**

**A Circuit with Several Events.**

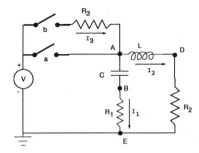

Figure 2.71: A Circuit with Two Switches.

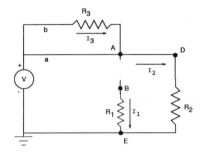

Figure 2.72: Before Switch a Opens.

Figure 2.71 shows a circuit with two switches. Both switches a and b are independent of each other and can be thrown at will. The figure shows the switches open so it is clear where they are but we will look at what happens when the switches start out closed, then we open a, then open b, then close a, then close b. We want to determine all the currents and voltages just after each event. If you wanted to find the minimum number of things, all you need as initial conditions is the current $I_2$ and the voltage difference $V_A - V_B$ just after each event. We will find all the currents and all the voltages for practice. The switches in this circuit are single pole, single throw switches. To make the calculations easy let $R_1 = 1$, $R_2 = 2$ and $R_3 = 3$ Ohms and let $V = 10$ volts.

**EVENT ONE, a OPENS:**

The first event is opening switch a. Notice that opening switch still allows the inductor to be in a loop so it can satisfy its continuity. The capacitor is not in any loop containing only sources and capacitors so it also satisfies its continuity.

Start by drawing a circuit diagram before the event. Both switches are closed, and (assuming there has been plenty of time since the last event)[32] the inductors look like ordinary wire and the capacitors look like breaks in the wire. Figure 2.72 shows this BTE circuit. From inspection, $V_A = V = 10$. From the impedance relation for $R_3$, $V_{\text{tail}} - V_{\text{head}} = 10 - 10 = 0 = R_3 I_3 \rightarrow I_3 = 0$. From a conservation of charge at node B, $0 - I_1 = 0 \rightarrow I_1 = 0$. From the impedance of $R_1$, (you do the details from now on) $V_B = 0$. From inspection, $V_D = V_A = 10$. From the impedance of $R_2$, $I_2 = \frac{V}{R_2} = 5$. In summary we have:

BEFORE EVENT a opens : $V_A = 10$Volts, $V_B = 0$Volts, $V_D = 10$Volts, $I_1 = 0$Amps, $I_2 = 5$Amps, $I_3 = 0$Amps

The current $I_2$ and voltage difference $V_A - V_B$ are the initial conditions you need if you wanted to solve the differential equations (they tell you how much accumulation is in each element). Since the problem asks for the values after the event, we will continue with the ATE.

Now draw the after the event circuit. Here inductors look like current sources with the same current as before the event. Every capacitor looks like a voltage source with the same voltage as what was across the capacitor BTE. Figure 2.73 shows this ATE circuit. By inspection, $I_2 = 5$

---

[32]You have seen the time it takes common circuits to reach their steady states. Look back at Dr. Zap as one example (example 2.7). What our "long time" assumption means is that the circuit has reached its steady state. If we are flipping switches by hand, chances are, the assumption is satisfied.

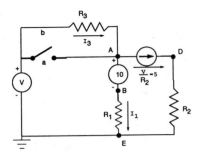

Figure 2.73: Just After Switch a Opens.

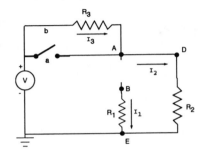

Figure 2.74: The Circuit BTE Switch b Opening.

(its the current through the current source). From the impedance of $R_2$ (do it, write it), $V_D = 10$. Now it gets a little tougher. Don't make the faulty assumption that $V_A = V_D$ it isn't. Remember when you know the flow, you don't usually know the effort. Although you know the flow through the current source you **do not know** the voltage across it. Many students want to say the voltage across the current source is zero. **WRONG!** None of the remaining terms seem obvious to us, so we plan to use KVL. Take the loop EADE to find:

$$R_1 I_1 + 10 - (V_A - V_D) - R_2 I_2 = 0 \rightarrow I_1 + 10 - V_A + 10 - 10 = 0$$

There are two unknowns and one equation so we will take loop EADE across the voltage source and $R_3$ to find:

$$10 - R_3 I_3 - 10 - R_1 I_1 = 0 \rightarrow -3I_3 - I_1 = 0$$

Now a node equation (conservation of charge) at A:

$$I_3 - I_1 - 5 = 0$$

These equations define $I_1$, $I_3$ and $V_A$, to find $V_B$ use the impedance of $R_1$:

$$V_B = R_1 I_1 \rightarrow V_B = I_1$$

Solving for currents and voltage gives:

AFTER EVENT a opens :$V_A = 6.25$Volts, $V_B = -3.75$Volts, $V_D = 10$Volts, $I_1 = -3.75$Amps, $I_2 = 5$Amps, $I_3 = 1.25$Amps

### EVENT TWO, b OPENS

By assuming sufficient time has gone by since the previous event, the circuit has reached steady state and we draw what it looks like Before The Event. Note that after the event the inductor will still be in a closed loop and there will not be a loop with nothing but capacitors and sources, hence the continuity conditions are satisfied. Figure 2.74 shows the BTE for this event. Capacitors are broken wires and inductors are perfect wires. Now to find the voltages and currents. By inspection, $I_1 = 0$. From the impedance of $R_1$, $V_B = 0$. From a KVL around the outer loop:

$$10 - R_3 I_3 - R_2 I_2 = 0$$

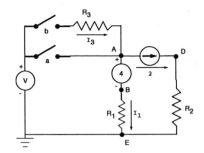

Figure 2.75: The Circuit ATE Switch b Opening.

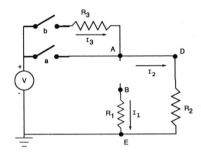

Figure 2.76: The Circuit BTE Switch a Closing.

and charge at node A:

$$I_3 - I_2 = 0$$

solving these two we have: $I_2 = 2$, $I_3 = 2$. From the impedance of $R_2$: $V_D = R_2 I_2 = 4$, from inspection: $V_A = V_D = 4$. In summary, we have:

BEFORE EVENT b opens : $V_A = 4$ Volts, $V_B = 0$ Volts, $V_D = 4$ Volts, $I_1 = 0$ Amps, $I_2 = 2$ Amps, $I_3 = 2$ Amps

If all we wanted was to solve the differential equations for the circuit, the initial conditions for b opening are $I_2 = 2$ and $V_A - V_B = 4$, but the problem asks for everything ATE so we will continue.

Figure 2.75 shows the circuit at ATE. The capacitor is a voltage source with the magnitude from the BTE voltages $V_A - V_B = 4$, the inductor is a current source with magnitude from the BTE current $I_2 = 2$. From inspection $I_3 = 0$. From the impedance of $R_2$: $V_D = R_2 I_2 = 4$. From node A: $-I_1 - 2 = 0 \rightarrow I_1 = -2$. From impedance of $R_1$: $V_B = R_1 I_1 = -2$ and from the source representing the capacitor: $V_A = V_B + 4 = 2$. In summary:

AFTER EVENT b opens : $V_A = 2$ Volts, $V_B = -2$ Volts, $V_D = 4$ Volts, $I_1 = -2$ Amps, $I_2 = 2$ Amps, $I_3 = 0$ Amps

**EVENT THREE, a CLOSES**

Continuity is satisfied for this event so start by drawing the BTE as figure 2.76 shows. From inspection and the impedance for $R_1$ and $R_2$, we have:

BEFORE EVENT a closes : $V_A = 0$ Volts, $V_B = 0$ Volts, $V_D = 0$ Volts, $I_1 = 0$ Amps, $I_2 = 0$ Amps, $I_3 = 0$ Amps

Figure 2.77 shows the ATE diagram. From the impedance of $R_2$, $V_D = 0$. From inspection, $V_A = 10$ and $I_3 = 0$. From the voltage source that is the capacitor, $V_B = V_A - 0 = 10$. From the impedance of $R_1$: $V_B = R_1 I_1 \rightarrow I_1 = V_B = 10$. In summary:

AFTER EVENT a closes : $V_A = 10$ Volts, $V_B = 10$ Volts, $V_D = 0$ Volts, $I_1 = 10$ Amps, $I_2 = 0$ Amps, $I_3 = 0$ Amps

**FINAL EVENT, b CLOSES**

Continuity is satisfied so draw the BTE as shown in figure 2.78. From inspection $V_A = 10$,

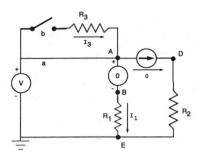

Figure 2.77: The Circuit ATE Switch a Closing.

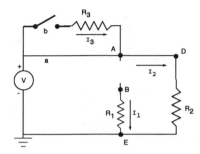

Figure 2.78: The Circuit BTE Switch b Closing.

$V_D = 10$, $I_3 = 0$, $I_1 = 0$. From the impedance of $R_1$ and $R_2$: $V_D = I_2 R_2 \rightarrow I_2 = 5$, $V_B = R_1 I_1 = 0$. In summary:

BEFORE EVENT b closes : $V_A = 10$Volts, $V_B = 0$Volts, $V_D = 10$Volts, $I_1 = 0$Amps, $I_2 = 5$Amps, $I_3 = 0$Amps

Now figure 2.79 shows the ATE circuit. From inspection, $V_A = 10$, $I_2 = 5$. From impedance of $R_3$ and $R_2$: $10 - 10 = R_3 I_3 \rightarrow I_3 = 0$, $V_D = R_2 I_2 = 10$. From the source that was the capacitor: $V_B = V_A - 10 = 0$. From the impedance of $R_1$: $V_B = R_1 I_1 \rightarrow I_1 = 0$. In summary:

AFTER EVENT b closes : $V_A = 10$Volts, $V_B = 0$Volts, $V_D = 10$Volts, $I_1 = 0$Amps, $I_2 = 5$Amps, $I_3 = 0$Amps

**DISCUSSION:**
Note that $I_2$ never jumps, it is continuous at all time. This is the result of the fact that the inductor tries to prevent sudden changes in $I_2$. Also note that since $V_D = R_2 I_2$, it never jumps either. If you notice, voltage $V_A$ and $V_B$ both jump however note that the quantity $V_A - V_B$ never jumps. This is because the capacitor attempts to keep the voltage across itself continuous. Since $I_1 = \frac{V_B}{R_1}$ and since $V_B$ jumps, $I_1$ also jumps. This is fine, the capacitor does not care about changes in current.

**End 2.19**

**Example 2.20**

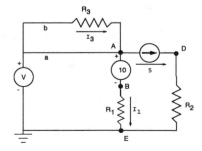

Figure 2.79: The Circuit ATE Switch b Closing.

## Finding First Derivatives.

For practice, let's determine the first derivatives of the voltages and currents in the circuit that figure 2.71 shows right after the event of switch b opening. Assume a was open for a long time when b is suddenly opened.

**FORMULATION:**

First recall the solution for voltage and current ATE b opening from the previous example. The solution was:

AFTER EVENT b opens : $V_A = 2\text{Volts}, V_B = -2\text{Volts}, V_D = 4\text{Volts}, I_1 = -2\text{Amps}, I_2 = 2\text{Amps}, I_3 = 0\text{Amps}$

Unfortunately these are values valid only at a single moment in time so we cannot differentiate them. We must therefore go back to the circuit and derive equations that are valid a some general time after the event and use these equations to determine derivatives. From inspection of $R_3$ for all time after ATE it is true that: $I_3 = 0$. Taking the derivative of this gives: $\frac{d}{dt}(I_3) = 0$.

From figure 2.71 we write a KVL around EBADE for times after ATE:

$$+R_1 I_1 + (V_A - V_B) - L\frac{d}{dt}(I_2) - R_2 I_2 = 0$$

this tells us:

$$\frac{d}{dt}(I_2) = \frac{-2I_2 + I_1 + V_A - V_B}{L}$$

plugging in values at ATE we have:

$$\frac{d}{dt}(I_2) = \frac{-4 - 2 + 2 - (-2)}{L} = \frac{-2}{L}$$

From the impedance of $R_2$: $V_D = R_2 I_2$, taking a derivative:

$$\frac{d}{dt}(V_D) = R_2\frac{d}{dt}(I_2) = 2\frac{-2}{L} = \frac{-4}{L}$$

From the impedance of the capacitor:

$$I_1 = C\frac{d}{dt}(V_A - V_B) \tag{2.60}$$

This has two unknowns so we cannot solve it immediately. How about the charge at node A. In general we have: $I_3 - I_2 - I_1 = 0$. We can solve for $I_1$ as: $I_1 = I_3 - I_2$, now differentiate to find: $\frac{d}{dt}(I_1) = \frac{d}{dt}(I_3) - \frac{d}{dt}(I_1) = 0 - \frac{-2}{L} = \frac{2}{L}$. Now from impedance of $R_1$: $V_B = R_1 I_1$, differentiating:

$$\frac{d}{dt}(V_B) = 1\frac{d}{dt}(I_1) = \frac{2}{L}$$

Now we can use this in equation 2.60 to find:

$$-2 = C\left(\frac{d}{dt}(V_A) - \frac{d}{dt}(V_B)\right) = C\frac{d}{dt}(V_A) - C\frac{2}{L} \rightarrow \frac{d}{dt}(V_A) = \frac{-2}{C} + \frac{2}{L}$$

The values we found here are only valid at ATE since we plugged in values at ATE to get them. If you want voltages and currents at some significant time after ATE, you will need to solve the circuit differential equations.

**End 2.20** ─────────────────────────────────────

If you must find the initial conditions for a capacitor in a circuit when the capacitor has an infinite current flowing through it, we use the following two relationships between charge, current and voltage:

$$V = \frac{q}{C} \tag{2.61}$$

$$\frac{d}{dt}(q) = I$$

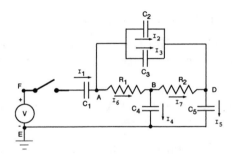

Figure 2.80: A Switched Circuit That Allows an Infinite Current to Flow Through Capacitors.

where $V$ is the voltage across the capacitor, $q$ is the charge difference between the plates of the capacitor, $C$ is the capacitance, $I$ is the current flowing through the capacitor and $t$ is time. Equation 2.61 comes from the definition of capacitance. The second comes from the fact that current is the flow rate of charge. We can integrate the second equation to find:

$$q - q_o = \int_{t=0^-}^{t=0^+} I(t)dt \tag{2.62}$$

where $q_o$ is the charge in the capacitor before the event. The limits of integration are across the infinitesimal time it takes for the event. Ordinarily if $I$ is finite, the integral is zero which means the charge doesn't change, which means the voltage doesn't change. However if the current is infinite, the integral will be nonzero and the charge in the capacitor can change across the time of the event. If the charge jumps, the voltage does too. The following example shows how to use these equations to find the initial conditions for capacitors with infinite current.

## Example 2.21

### A Bad Arrangement of Capacitors

Figure 2.80 shows a circuit that allows an infinite current to flow through several capacitors. You can tell that this is true, because there is a loop that contains only sources and capacitors. For example, loop EFADE is one that contains one voltage source and several capacitors. When you analyze a circuit like this, the capacitors can have a sudden jump in their voltages. To demonstrate how to handle this circuit, determine the voltages across the capacitors when the switch closes. Let $V = 10$, $C_i = i$ (not realistic values but it keeps the numbers pretty), and $R_i = i$.

**FORMULATION:**

Assume that the switch has been open for a long time. If so the voltage across each capacitor is zero before the event. This means the initial charges across each capacitor (see equation 2.61) is zero also. When the switch is closed, an infinite current will flow through capacitor 1, 2, 3, and 5. Because there are resistors between nodes A and B currents $I_6$ and $I_7$ cannot be infinite, therefore capacitor $C_4$ does not have an infinite current hence its voltage cannot jump during the event. The initial voltage across $C_4$ is zero.

Next write a KVL around EFADE over $C_2$. We use equation 2.61 to express the voltage rises, falls over the capacitors the moment after the event as follows:

$$V - \frac{q_1}{C_1} - \frac{q_2}{C_2} - \frac{q_5}{C_5} = 0$$

If one of the capacitors had a nonzero voltage before the event we would write its rise/fall as $\frac{q_o + \delta q}{C}$ where $q_o$ is the initial charge and $\delta q$ is the charge "added" during the event. Anyway, back to the problem. We eliminate the values of charge using equation 2.62, putting charge in terms of current. We therefore have:

$$V - \frac{\int_{t=0^-}^{t=0^+} I_1 dt}{C_1} - \frac{\int_{t=0^-}^{t=0^+} I_2 dt}{C_2} - \frac{\int_{t=0^-}^{t=0^+} I_5 dt}{C_5} = 0$$

(If we were doing this by hand without a word processor with a cut and paste function we would write some simple symbol like $dI$ to represent $\int_{t=0^-}^{t=0^+} I(t)dt$.) So far we have one equation, but three unknowns, the currents through the capacitors. We need more equations. There is another loop (EFADE over $C_3$) so we will write a KVL equation for it.

$$V - \frac{q_1}{C_1} - \frac{q_3}{C_3} - \frac{q_5}{C_5} = 0$$

Replacing charge with current gives:

$$V - \frac{\int_{t=0^-}^{t=0^+} I_1 dt}{C_1} - \frac{\int_{t=0^-}^{t=0^+} I_3 dt}{C_3} - \frac{\int_{t=0^-}^{t=0^+} I_5 dt}{C_5} = 0$$

This does not help us very much because it brings in one new unknown. We have two equations but four unknowns.

When the switch is closed, the circuit has three [NIF], therefore we have too many flows. We will write some node equations. We should be able to write 7-3=4 node equations relating the flows. Let's choose node A:

$$I_1 - I_2 - I_3 - I_6 = 0 \tag{2.63}$$

Now do node D:

$$I_2 + I_3 + I_7 - I_5 = 0 \tag{2.64}$$

Now node B:

$$I_6 - I_7 - I_4 = 0 \tag{2.65}$$

Now for a little reasoning. At the moment the event happens, what currents are infinite? Currents 1, 2, 3, and 5. Look at equation 2.63, the only finite current in the equation is $I_6$. We are adding HUGE currents $I_1$, $I_2$ and $I_3$ to some tiny value (in comparison) $I_6$. Current $I_6$ is practically zero for all it does. Equation 2.63 can be approximated as:

$$I_1 - I_2 - I_3 = 0$$

To make the terms look like what we see in the KVL equations, we can integrate this over the event to say:

$$\int_{t=0^-}^{t=0^+} I_1 dt - \int_{t=0^-}^{t=0^+} I_2 dt - \int_{t=0^-}^{t=0^+} I_3 dt = 0$$

The same argument applies to $I_7$ in equation 2.64 so we can write it as:

$$\int_{t=0^-}^{t=0^+} I_2 dt + \int_{t=0^-}^{t=0^+} I_3 dt - \int_{t=0^-}^{t=0^+} I_5 dt = 0$$

If we take our two KVL equations with the last two node equations we can solve for the current integrals:

$$\int_{t=0^-}^{t=0^+} I_1 dt = q_1 = \frac{50}{7}, \int_{t=0^-}^{t=0^+} I_2 dt = q_2 = \frac{20}{7}, \int_{t=0^-}^{t=0^+} I_3 dt = q_3 = \frac{30}{7}, \int_{t=0^-}^{t=0^+} I_5 dt = q_5 = \frac{50}{7}$$

Having the charges added by the event, we use equation 2.61 to find the voltages right after the event.

$$10 - V_A = \frac{q_1}{C_1} = \frac{50}{7}$$

$$V_A - V_D = \frac{q_2}{C_2} = \frac{10}{7} = \frac{q_3}{C_3} = \frac{10}{7}$$

$$V_D - 0 = \frac{q_5}{C_5} = \frac{10}{7}$$

These answer the question since they are the voltages across the capacitors but if we want the node voltages we solve to find $V_D = \frac{10}{7}$, $V_B = 0$ (because the voltage across $C_4$ does not jump during

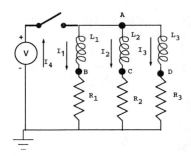

Figure 2.81: A Circuit With Three Inductors That are Forced To Instantly Change Current.

the event and it was zero before), $V_A = \frac{20}{7}$. Having the voltages, we can compute the remaining currents. Using the impedance of $R_1$ gives:

$$V_A - V_B = \frac{20}{7} - 0 = I_6 R_1 \rightarrow I_6 = \frac{20}{7}$$

Using the impedance of $R_2$ gives:

$$V_B - V_D = 0 - \frac{10}{7} = I_7 R_2 \rightarrow I_7 = -\frac{5}{7}$$

Using equation 2.65 gives:

$$I_4 = \frac{25}{7}$$

The other currents are infinite at the moment of the event. If you needed to compute derivatives you would proceed as usual.

If the switch opened rather than closed, would you have a similar problem with infinite currents? No, none of the loops that remain can have an infinite current. Think about the loop around $C_2$ and $C_3$. Convince yourself that infinite current cannot flow out of one capacitor and into the other.

**End 2.21** ▬▬▬▬▬▬▬▬▬▬▬▬▬▬▬▬▬▬▬▬▬▬▬▬▬▬▬▬▬▬▬▬▬▬▬▬▬▬▬▬▬▬▬▬▬▬▬

If you have a circuit in which a switch stops the current through an inductor, you will have a similar problem in which some voltages become infinite and the inductor will be forced to change its flow instantly. Suppose we have an inductor $L$ with a current $I$ flowing in it, we can write:

$$V_{\text{tail}} - V_{\text{head}} = L\frac{d}{dt}(I) \rightarrow \int_{t=0^-}^{t=0^+} (V_{\text{tail}} - V_{\text{head}})\, dt = L(I - I_o)$$

where $I_o$ is the current before the event and $I$ is the flow after the event. Ordinarily, the tail and head voltages are finite and the integral is zero which means the current does not change. However, if the current DOES change, at least one of the voltages becomes infinite and the integral is nonzero. Define $\psi_{\text{tail}} = \int_{t=0^-}^{t=0^+} (V_{\text{tail}})\, dt$ (similarly define $\psi_{\text{head}}$) so:

$$\frac{\psi_{\text{tail}} - \psi_{\text{head}}}{L} = I - I_o \tag{2.66}$$

The next example demonstrates how to use this definition to determine the initial conditions on inductors when they are forced to change their flow instantly.

**Example 2.22** ▬▬▬▬▬▬▬▬▬▬▬▬▬▬▬▬▬▬▬▬▬▬▬▬▬▬▬▬▬▬▬▬▬▬▬▬▬▬▬▬▬▬▬▬

### Initial Conditions When Inductors Change Current Instantly

Determine the node voltages and currents the moment the switch shown in figure 2.81 opens. Take $V = 10$, $L_i = i$, and $R_i = i$.

**FORMULATION:**

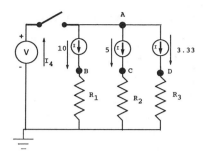

Figure 2.82: What we Might Think is the Correct ATE Diagram.

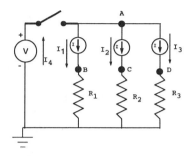

Figure 2.83: The Proper ATE Circuit Once we Realize Some Inductors Change Current.

Quite often, students have trouble recognizing when inductors are forced to change their flow. The reason it happens here is because the only wires connected to node A have inductors and switches. Consider what happens, after a long time, the inductors begin to settle down and they try to keep the current in their leg of the loop constant. The inductors try to make currents $I_1 + I_2 + I_3$ constant and their sum equals $I_4$. Once the switch is opened, $I_4$ suddenly changes which means some or all of the other currents must change.

Suppose you did not recognize that this circuit is special. What should you do? Well, what we will do is pretend that we did not know the circuit was special and we approach it in the standard way, watch how we will suddenly discover that something strange is happening, and what we will do about it.

We will draw the circuit in the BTE state, to determine the currents and voltages at the nodes. What we would find is: $V_A = V_B = V_C = V_D = 10$, $I_1 = 10, I_2 = 5, I_3 = 3.33$, and $I_4 = 18.33$. Next we might draw what we think is the ATE circuit. Figure 2.82 shows what this might look like. Now at this point, we hopefully would consider the [NIF]. If you notice the [NIF] is 2, but we have 3 flows given by the current sources. How many node equations can we write? We can write -1 node equation. Minus 1, what the heck? What is this saying? It says we have specified too many flows. We have given more flows than the circuit can have. This is our indication that at least one of the inductors must suddenly change its flow, because we cannot have all three of them specified.

Figure 2.83 shows the proper way to draw the ATE circuit once we realize that the inductors change their current. From the impedance of each resistor, we can find equations for the voltages at B, C and D:

$$V_B = I_1 R_1, V_C = I_2 R_2, V_D = I_3 R_3 \qquad (2.67)$$

Note that the voltages are finite because the currents are finite. Now use equation 2.66 to relate the voltages across the inductors to their currents:

$$\frac{\psi_A - \psi_B}{L_1} = I_1 - I_{1o}$$

$$\frac{\psi_A - \psi_C}{L_2} = I_2 - I_{2o}$$

$$\frac{\psi_A - \psi_D}{L_3} = I_3 - I_{3o}$$

Since the voltages at B, C, and D are finite, the values of $\psi_B$, $\psi_C$ and $\psi_D$ are zero. Therefore, also using values from the BTE:

$$\frac{\psi_A}{L_1} = I_1 - 10 \tag{2.68}$$

$$\frac{\psi_A}{L_2} = I_2 - 5 \tag{2.69}$$

$$\frac{\psi_A}{L_3} = I_3 - 3.33 \tag{2.70}$$

In equations 2.67 through 2.70, there are seven unknowns, $\psi_A$, three currents and three voltages. Since there is to [NIF] equals 2 we can write one node equation. Choose node A.

$$-I_1 - I_2 - I_3 = 0$$

Now we can solve for voltages and currents to find:

$$I_1 = 0, I_2 = 0, I_3 = 0, V_B = 0, V_C = 0, V_D = 0, \psi_A = -10$$

The fact that everything but $\psi_A$ is zero is a coincidence! Also note that $V_A \neq -10$, the integral of $V_A$ over the infinitesimal time of the event is -10, the value of $V_A$ is negative infinity. To prove that the zeros are coincidental, solve the equations if we change $R_2 = 4$. Now the solution is:

$$I_1 = \frac{15}{11}, I_2 = \frac{-20}{11}, I_3 = \frac{5}{11}, V_B = \frac{15}{11}, V_C = \frac{-80}{11}, V_D = \frac{15}{11}, \psi_A = \frac{-95}{11}$$

**End 2.22**
_____

## 2.8   MAKING ASSUMPTIONS
_____

**Objectives**
*When you complete this section, you should be able to*

1. *Recognize when assumptions must be made.*

2. *Be able to determine what type of assumption should be made.*

3. *Be able to test assumptions.*

4. *Be able to continue with a problem when you find that assumptions are violated.*

_____

Occasionally you may find yourself with a circuit that requires an assumption. For example suppose a circuit has a fuse in it. A fuse is a device that breaks the circuit if an excessive current flows through the fuse. They are used for safety reasons.

**Example 2.23** _____

### A Circuit With a Fuse.

Determine the current and node voltages in the simple circuit shown in figure 2.84 The fuse will burn out if the current through it exceeds 2 amps. We will get one answer if the fuse is burned, and another if it isn't. The solution depends on the current, but we don't know the current until we have the solution. Of course, for this problem, the answer is so simple that is is not difficult to tell in advance whether the fuse is burned or not. If we cannot tell in advance then we will have to make an assumption. For practice, we will pretend that we cannot tell whether the fuse is burned or nor and proceed. Therefore we will make an assumption and test the assumption.

The test that we will use for the assumption involves current. If the current is too large it is blown, if the current is not too large, it isn't blown. Therefore we want to solve our problem for current. If we assume the fuse is burned, then we will "know" the current through the fuse, it would

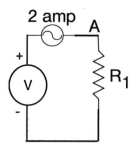

Figure 2.84: A Circuit with a 2 Amp Fuse.

be zero. This will not allow us to solve the circuit for the current. The assumption we should make about the fuse therefore is that it is not burned which allows us to solve for current.

Since we assume the fuse is not burned we can write a KVL equation around the loop. We will find that code:

$$I = \frac{V}{R_1}$$

Let $V = 10$ and $R_1 = 2$ so $I = 5$. Now that we know what the current is, we can test the assumption. In this example, the current is equal to 5 amps. Since 5 amps is greater than 2 amps, the fuse will burn out. This tells us that our assumption is incorrect. Now that we know the fuse is burned out, we could to the problem again. This time we know what the fuse is doing. The current is zero and the voltage at A is zero.

**End 2.23**

## 2.9  EQUIVALENT CIRCUITS

**Objectives**
*When you complete this section, you should be able to*

1. *Explain what one means by an equivalent circuit.*

2. *Explain why the concept of equivalent circuits offers advantages to an investigator.*

3. *Recognize series and parallel elements.*

4. *Reduce series and parallel elements to a single equivalent element.*

This section discusses how we can replace circuit elements with equivalent elements. When dealing with complex circuits, it is sometimes beneficial to replace one circuit (or part of a circuit) with a simpler, equivalent circuit. Circuits (or pieces of circuits) are equivalent when their impedances are the same. If a simplification is to be useful, they should be easy to determine. Therefore, there is an emphasis on developing a bag of tricks that we can apply with ease.

Consider the piece of a circuit in figure 2.85. The resistors are in series. A more descriptive way to refer to the resistors is to say they are common flow resistors. This term says that the resistors have the same flow variable.

What we want to do is replace these common flow (in series) resistors with a single equivalent resistor. The first step is to determine the impedance of the common flow segment. Accounting electrical energy, we find that

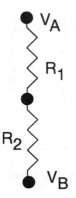

Figure 2.85: Two Resistors in Series (Common Flow)

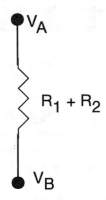

Figure 2.86: Equivalent Resistance of Series Resistors

**Electrical Energy  Accounting**
system[ **$R_1$ and  $R_2$ fig.2.85** ]      time period [**differential**]

*input $-$ output $+$ generation $-$ consumption $=$ accumulation*

| input/output: | $iV_A - iV_B$ |
|---|---|
| **consumption (thermal):** | $i^2R_1 + i^2R_2$ |
| **generation:** | |
| **accumulation:** | 0 |

$$\boxed{V_A i} \quad - \quad \boxed{V_B i} \quad - \quad \boxed{i^2R_1 + i^2R_2} \quad = \quad \boxed{0}$$
   *input*        *output*      *consumption*       *accumulation*

$$V_A i - V_B i = i^2 \left( R_1 + R_2 \right)$$

$$V_A - V_B = i \left( R_1 + R_2 \right)$$

Therefore, the impedance for the resistors is $(R_1 + R_2)$. Therefore, we can replace these resistors with the equivalent $R_e$ such that $R_e = (R_1 + R_2)$. This equivalent element appears in figure 2.86. An easy rule of thumb, whenever two resistances experience the same flow, we can replace them with a single resistor as we show here.

Now consider the common effort resistors (in parallel) in figure 2.87.    Taking resistor 1 as our system, we find (from the impedance relation) that

$$V_A - V_B = i_1 R_1 \tag{2.71}$$

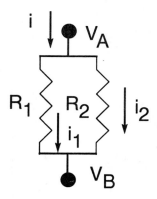

Figure 2.87: Two Resistors in Parallel (Common Effort)

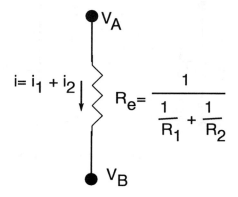

Figure 2.88: Equivalent Resistance for Parallel (Common Effort) Resistors

Likewise, taking resistor 2 as our system, we find (from the impedance relation) that

$$V_A - V_B = i_2 R_2 \qquad (2.72)$$

For the equivalent resistor in figure 2.88, the current is the sum of each individual current. Solving equations 2.71 and 2.72 for current and adding gives

$$\frac{V_A - V_B}{R_1} + \frac{V_A - V_B}{R_2} = i_1 + i_2 = i$$

$$(V_A - V_B)\left(\frac{1}{R_1} + \frac{1}{R_2}\right) = i$$

$$(V_A - V_B)\left(\frac{R_1 + R_2}{R_1 R_2}\right) = i$$

$$V_A - V_B = i\frac{R_1 R_2}{R_1 + R_2} = iR_e \qquad (2.73)$$

Using equation 2.73, we can see that the equivalent resistance is

$$R_e = \frac{R_1 R_2}{R_1 + R_2} = \frac{1}{\frac{1}{R_1} + \frac{1}{R_2}}$$

We can establish equivalence relationships for other elements as easily as we did for resistances. For example, figure 2.89 shows equivalencies for inductors and capacitors.

Consider the voltage and current sources in figure 2.90. We desire to determine the conditions

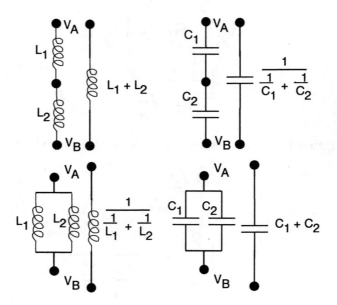

Figure 2.89: Equivalent Impedances for Inductors and Capacitors

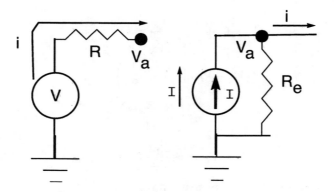

Figure 2.90: A Voltage and Current Source

under which the two sources are equivalent. By equivalent, we mean that the impedance (relation between $V_a$ and $i$) should be identical for each element.

For the voltage source, a system comprised of the resistor gives an impedance relation of

$$V - V_a = iR \tag{2.74}$$

For the current source, we can apply the conservation of charge into node a.

**Conservation of Charge**
system[**node a fig.2.90**]    time period [**differential**]

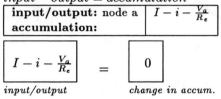

*input − output = accumulation*

| **input/output:** node a | $I - i - \frac{V_a}{R_e}$ |
|---|---|
| **accumulation:** | |

| $I - i - \frac{V_a}{R_e}$ | $=$ | $0$ |
|---|---|---|
| *input/output* | | *change in accum.* |

In equation form, we have

$$I - i - \frac{V_a}{R_e} = 0$$

Solving for $i$ and reducing the fractional terms, we find that

$$IR_e - V_a = R_e i \tag{2.75}$$

If the two sources are to be equivalent, then equation 2.74 should be identical (term by term) with equation 2.75. Therefore, we have the following two requirements:

$$V - V_a \equiv IR_e - V_a$$

and

$$iR \equiv R_e i$$

Thus,

$$R_e = R$$

and

$$I = \frac{V}{R_e} = \frac{V}{R}$$

We can convert a voltage source in a common flow (in series) with a resistor into a current source in a common effort (in parallel) with the same resistor and *vice versa*. There are many other equivalencies that we can determine and store in our bag of tricks for later use. We use equivalent circuits to simplify the application of the concepts we used in this part of the text. We can derive these relations by equating the impedances of the original and equivalent circuits.

## 2.10    SEMICONDUCTOR DEVICES

This section discusses some simple semiconductors. Conductance is a plot of voltage on the abscissa (horizontal) and current on the ordinate (vertical). Impedance is a plot of current on the abscissa and voltage on the ordinate. In the study of semiconductors, it is common practice of electrical engineers to plot the conductance of devices rather than the impedance. This text will continue that practice.

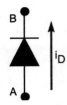

Figure 2.91: A Schematic Diagram of a Diode

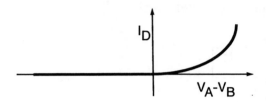

Figure 2.92: Simplified Diagram of Current Versus Bias Voltage for a Diode

## 2.10.1   DIODES

**Objectives**
*When you complete this section, you should be able to*

1. *Explain what conductance is.*

2. *Sketch the conductance diagram for a "typical" diode.*

3. *Solve for voltage and current in circuits containing diodes by approximating the diode with an idealized conduction.*

Figure 2.91 shows a schematic diagram of a diode. In simple terms, a diode acts like a one way valve for current. When the voltage on terminal A is more positive than that on B (this is forward bias), current flows in the direction indicated by $i_D$. When the diode is reverse biased (voltage B is more positive than A), little current flows. A simple diagram showing the current versus bias voltage appears in figure 2.92. A conductance diagram is a plot of current and voltage with current plotted on the vertical axis and voltage along the horizontal. The conductance relationship, over a large range of voltages, for a generic diode appears in figure 2.93. Typically, $V_f$ (forward bias) is on the order of 0.7 [volts] and $V_b$ (reverse breakdown) is roughly 60 [volts].

There are many different types of diodes available, each with different characteristics. For exam-

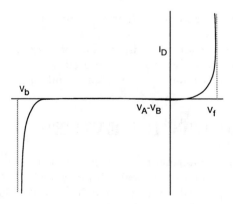

Figure 2.93: The Impedance Relationship of a Generic Diode

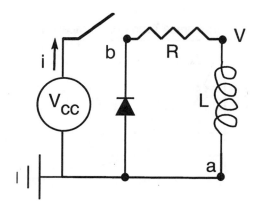

Figure 2.94: A Clamping Diode in a Switched Inductive Circuit

ple Zener diodes have small, precise reverse breakdown voltages because they are used in reverse bias configurations to provide precise voltage levels. High frequency diodes are small so they respond quickly and power handling diodes are large to handle the heating involved. Because diodes are manufactured in batches, it is difficult to guarantee precise relationships from batch to batch.

## Example 2.24

### Clamping Diode

**PROBLEM:**

As example 2.10 demonstrated, switching inductive loads can create voltages large enough to generate sparks. Generally, these sparks must be reduced or eliminated. One method of doing this is to install a clamping[33] diode (see figure 2.94).

Now for our circuit. When we close the switch, current begins to flow through the resistor and inductor because very little current can flow backwards through the diode (provided $V_b$ is below the reverse breakdown). When the switch opens, the voltage at b drops because the inductor tries to keep its current flowing. As a result, it starves b of + charges.[34] Before long $V_b$ drops enough that $V_a - V_b$ is above $V_f$ for the diode ($\approx 0.7$ [volts]) and the diode begins to conduct. If we assume, for this example, that the conductance diagram for forward bias (voltage) greater than $V_f$ is a vertical line (see figure 2.93) then the diode conducts as much current as necessary to guarantee $V_a - V_b = V_f$.[35] Determine how the clamping diode affects the behavior of this circuit.

**FORMULATION:**

Suppose the switch has been closed for a long time and we suddenly open it. When we open the switch, $V_b = -V_f$. Therefore, if we account the electrical energy in the resistor and inductor between b and a, we find that

---

[33]The term clamping refers to the role of the diode not to a specific type.

[34]Actually the inductor tries to keep the electrons flowing (the electron current). Therefore, electrons accumulate at point b. The effect is the same, $V_b$ becomes negative.

[35]A real diode cannot conduct infinite current because it will either burn up or reach a point where the conductance diagram is nonvertical.

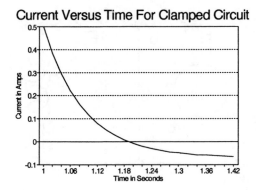

Figure 2.95: Current Versus Time for A Clamped Circuit

---

**Electrical Energy Accounting**

system[circuit fig.2.94]      time period [**differential**]

$input - output + generation - consumption = accumulation$

| input/output: | $-V_f I$ |
|---|---|
| **consumption (thermal):** | $I^2 R$ |
| **generation:** | |
| **accumulation:** | $\frac{d}{dt}\left(\frac{1}{2}LI^2\right)$ |

$$
\boxed{-V_f I} \quad - \quad \boxed{I^2 R} \quad = \quad \boxed{\frac{d}{dt}\left(\frac{1}{2}LI^2\right)}
$$

*input/output*          *consumption*                *change in accum.*

---

Note that the power entering at node b is negative because the voltage at b is below ground due to the diode's bias. Combining all of the terms, we get

$$-V_f I - I^2 R = \frac{d}{dt}\left(\frac{1}{2}LI^2\right)$$

This reduces to

$$-V_f - RI = L\dot{I} \tag{2.76}$$

**SOLUTION:**

We can integrate this differential equation analytically or numerically. The Maple program 2.12.4 shows how it can be solved. Note that to find the initial condition for $I$ we use the Before and After Event material discussed earlier. For this system $I$ at $t = 0$ is $\frac{V_{cc}}{R}$. The results appear in figure 2.95. Notice in figure 2.95 that the current becomes negative. This results from the simple treatment of the diode. Essentially we modeled the diode as if it held $V_b$ at 0.7 [volts], however as the current drops, the diode conductance will reach the rounded corner between 0 and $V_f$ [volts] as figure 2.93 shows. When we reach this point, the voltage at terminal b will begin to rise toward ground. Eventually the current becomes 0 and the voltage at b is ground. To model the circuit more precisely, we would need a better expression for the diode's conductance. If you need extremely good simulations, we suggest you use a software package intended for that. One excellent package is Electronics Workbench [7].
**End 2.24**

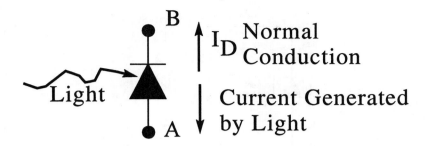

Figure 2.96: The Schematic Diagram of a Photodiode

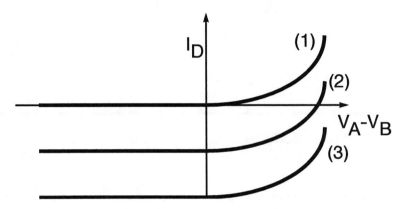

Figure 2.97: The Current Versus Voltage Diagram for a Typical Photodiode. Amount of Light Hitting the Diode Increases From 1 Toward 3.

## 2.10.2 PHOTODIODES.

**Objectives**
*When you complete this section, you should be able to*

1. *Explain how a photodiode differs from other forms.*

2. *Explain the effect of light on a photodiode.*

3. *Draw an "equivalent circuit" for a photodiode.*

4. *Explain what reverse biasing a photodiode is and why it has advantage over other types of diode biasing.*

5. *Solve for voltage and current in circuits containing photodiodes by approximating the diode with an idealized conduction.*

A photodiode is a semiconductor device which has the ability to convert light striking it into current. Figure 2.96 shows a schematic diagram of a photodiode. A photodiode responds to voltages like a regular diode. It has a forward bias and reverse breakdown. When light strikes a photodiode, an induced current flows from terminal B to A, opposite the normal conduction of a diode. Figure 2.97 shows a typical current versus bias voltage for a photodiode. Each of the three curves in figure 2.97 are similar in shape to the conductance of a regular diode (see figures 2.92 and 2.93). Curve (1) of figure 2.97 corresponds to no light hitting the diode, curve (2) shows the response for some light, and curve (3) shows the effect of even more light. Notice that the effect of light is to pull the curves vertically downward but their shape remains roughly the same. Understand that this is ideal behavior. (The actual response is slightly more complex). Figure 2.97 is a plot of the photodiode's conductance – the output (flow) $I_D$ versus input (effort) $V_A - V_B$. Assuming the description of

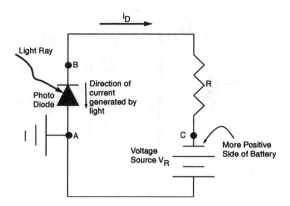

Figure 2.98: A Circuit Showing a Photodiode

the photodiode is correct, describe the energy and charge content of light. Are energy and charge conserved in the photodiode?

## Example 2.25

## A Photodiode as a Light Intensity Sensor

**PROBLEM:**

Figure 2.98 shows a circuit that provides a voltage that is a linear function of the intensity of light striking the circuit. The circuit contains a photodiode, a resistor, and a battery. Notice that the battery has the photodiode reverse biased. Therefore, there will be little battery current flowing through the diode therefore the battery will last a long time. This is one advantage of reverse bias. Determine the behavior of the circuit by determining a mathematical formulation for the current $i_D$.

**FORMULATION:**

The system is zero [NIA] since there are no accumulations, and it has one [NIF]. Therefore we expect to be able to write a single algebraic equation for the variable $i_D$. First, we write the impedance relation for the resistor (note that $V_C = V_R$).

$$V_B - V_R = i_D R \qquad (2.77)$$

Since there are two unknowns in this one equation, we cannot solve for the current directly, we need another equation. We anticipate an algebraic equation.

Let's consider the photodiode. What we will be looking for is an equation (or relationship) between our two unknowns, $i_D$ and $V_B$. Accounting for electrical energy in the photodiode gives

**Electrical Energy Accounting**

system[**photodiode fig.2.98**]      time period [**differential**]

*input − output + generation − consumption = accumulation*

| input/output: | $V_A i_D - V_B i_D$ |
|---|---|
| **consumption (thermal):** | $E_{consumed}$ |
| **generation:** | |
| power input from light | $E_{light}$ |
| **accumulation:** | |

$$\boxed{V_A i_D + E_{light} - V_B i_D} \ - \ \boxed{E_{consumed}} \ = \ \boxed{0}$$

*input/output*                    *consumption*          *change in accum.*

$$V_A i_D - V_B i_D + E_{light} - E_{consumed} = 0$$

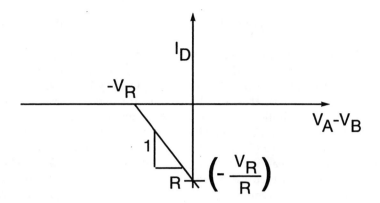

Figure 2.99: Impedance Relationship for the Resistor in the Photodiode Circuit

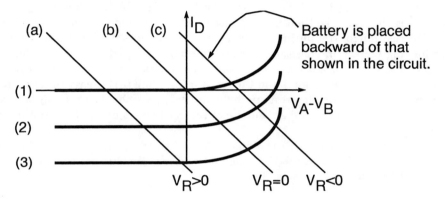

Figure 2.100: Operating Points Corresponding to Two Values of $V_R$

Unfortunately, we have not discussed the inner workings of the diode enough to be able to quantify the energy introduced ($E_{\text{light}}$), nor that consumed ($E_{\text{consumed}}$) so this equation is not of much use to us. What we want is a relation between $i_D$ and $V_B$ and figure 2.97 is exactly the relationship we need.

To determine the current, we could determine an equation for the curve in figure 2.97, then solve this equation simultaneously with equation 2.77. This is rather difficult. Therefore, what we will do instead is solve for the current graphically. We did this earlier when we discussed operating points. We can rewrite equation 2.77 in terms of current as (we are trying to get $V_A - V_B$ in the equation so it looks like the horizontal axis of the diode's conduction diagram)

$$i_D = \frac{V_B - V_R}{R} = -\frac{-V_B}{R} - \frac{V_R}{R}$$

Since $V_A = 0$, we can write

$$i_D = -\frac{(V_A - V_B)}{R} - \frac{V_R}{R} \tag{2.78}$$

The reason we want $V_A$ in the equation is to make it look like what appears in figure 2.97. This is an equation of a straight line and it appears in figure 2.99. To satisfy equation 2.78 the current and voltage must lie somewhere on the line of figure 2.99. Also, according to figure 2.97, the current and voltage must lie somewhere on a funny shaped curve, (defined once we know the light intensity) or else it does not satisfy the diode relationship.

If the current and voltage must lie on both curves simultaneously, it must lie at the intersection of the two. Figure 2.100 shows the operating point (intersection) formed for three values of $V_R$ and three different values of light intensity. Line (b) represents zero battery voltage ($V_R = 0$ or battery missing). Line (a) represents a positive $V_R$ just as in the original circuit. Line (c) is for a battery installed backward to what is shown in the figure and corresponds to a photodiode that is forward biased. An operating point occurs where each of the three lines (a, b, and c) intersect a diode curve (1, 2 or 3).

You understand that we show three lines to show the effect of three different designs you could only have one of the three. For example if you have a battery reverse biasing the diode (as shown in figure 2.98) then you have only line (a). If you had no battery in the circuit (not shown in any figure), you would have only line (b). If you had a battery installed backward you would have only line (c). We show 3 diode curves indicating what happens under 3 lighting conditions. Only one of the curves would be valid at a time. If no light hits the diode, curve (1) is valid. If a huge amount of light hits, curve (3) is valid. Curve (2) has a "modest" amount of light.

Each operating point determines the current $i_D$ and voltage $V_B$. No matter what line (a, b, or c) we are on, the circuit is capable of measuring light intensity because as intensity increases (moving from diode curve 1 toward 3) the value of $V_B$ increases. By measuring $V_B$ we can determine the light intensity. What we have is a light meter.

The only thing remaining is to determine which line (a, b, or c) we want to be on. This is equivalent to determining the desired battery voltage $V_R$. Notice that if $V_R$ is less than or equal to zero (lines (b) and (c)), then the increase in $V_B$ for an increase in intensity is a nonlinear relationship. The relation for computing the intersection would be very difficult to determine because we would be looking for the intersection between the funny shaped curve and a line. If however, we choose a positive value for $V_R$ (reverse biasing the diode), the intersection between the diode curve (approximately a horizontal line) and the resistor line (line (a)) is easy to determine. If we are looking for a simple method for determining light intensity, we should probably choose to reverse bias the photodiode. This is a second advantage of reverse bias, it makes the solution linear.

**SOLUTION:**

Suppose we do an experiment with the diode. We will pick up the diode all by itself, connect terminal A to B, hit it with various light levels and measure the current $i_D$. We would expect to see current from B to A and suppose we find:

$$i_D = -kL$$

where $k$ is some constant and $L$ is the magnitude of the light. With the photodiode reverse biased, we have equation 2.77 defining the voltage at B in terms of $i_D$. Plugging in the relationship for current we find:

$$V_B - V_R = -kLR \rightarrow L = \frac{V_R - V_B}{kR}$$

we know $V_R$, $k$, and $R$ so if we measure $V_B$ we can find the light intensity very easily. What are the practical limits on values of $V_R$ and $R$? Why would you want them to be large or small?
**End 2.25**

## 2.10.3   TRANSISTORS

**Objectives**
*When you complete this section, you should be able to*

1. *Describe two types of transistors.*

2. *Explain how transistors are used as amplifiers and as switches.*

3. *Sketch conductance diagrams for FET and BJT devices.*

4. *Configure FET and BJT devices as switches in a circuit.*

5. *Describe what the input impedance is for the gate of an FET.*

6. *Describe what the input impedance is for the base of a BJT.*

Transistors are semiconductors with three terminals. A small signal at one terminal controls a much larger signal at the other two. Transistors are typically used to either switch a current on and

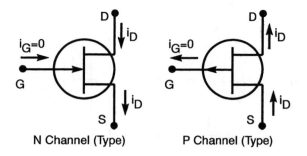

Figure 2.101: A Junction FET (Field Effect Transistor)

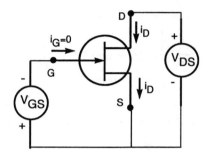

Figure 2.102: Circuit for Measuring the Impedance of an n type JFET

off or to amplify. There are several different kinds of transistors. This text will discuss only a few of them.

## THE JUNCTION FIELD EFFECT TRANSISTOR

Figure 2.101 shows the schematic diagram for a Junction Field Effect Transistor (JFET). The three terminals are the drain, D, the gate G, and the source S. The currents $i_D$ and $i_G$ represent the flow of conventional current through the device. Voltages present on the gate control the conductance (the resistance) between the source and drain.

There are two types of JFETs, a p and an n. Type n JFETs will be discussed in this text. Type p JFETs are similar to type n but the voltage polarities must be reversed. After designing an n type JFET application, we can reverse the polarities and insert the same type number p type JFET.

One of the important characteristics of JFETs is that the gate current is essentially zero provided the gate voltage has the proper polarity. For an n type JFET the gate voltage should be below the source voltage. For pJFETs the gate voltage should always be above the source. The reason for this is that the G to S interface is a diode junction. As long as the gate polarity is proper, the diode from G to S is reverse biased and no flow occurs. If G has incorrect polarity, the G to S diode is forward biased and current will flow in the gate terminal. When polarities are correct, $i_G \approx 10^{-8}$ [amps] for a typical JFET.

A JFET's operation is best understood by considering its conductance relationship between drain and source (D to S). Consider an n type JFET connected the way figure 2.102 shows. If one measures the current $i_D$ as a function of $V_D - V_S = V_{DS}$ for various values of gate voltage $V_G - V_S = V_{GS}$ a family of curves similar to those in figure 2.103 appear. Notice that for small $V_{DS}$ the relation is curved. In the middle section, there is a linear (horizontal) conductance. The starting point of the linear region (when $V_{GS} = 0$) occurs at a voltage of $V_p$, the pinch-off voltage. The linear region (when $V_{GS} = 0$) ends at a voltage of $BV_{DSS}$, the breakdown voltage, source shorted. The various type number n JFETs have their own $V_p$ and $BV_{DSS}$ but the shape of their conductance diagrams are the same. Notice that the effect of negative $V_{GS}$ is to cause the linear region to begin and end sooner, and lower its current.

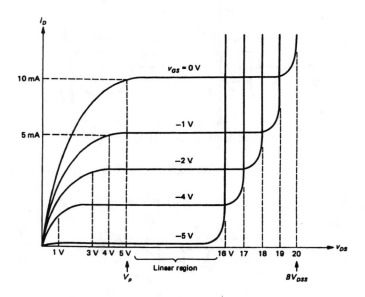

Figure 2.103: Typical Conductance of a JFET

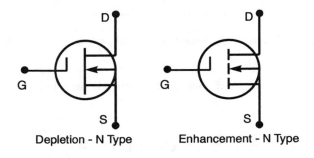

Depletion - N Type          Enhancement - N Type

Figure 2.104: Common Symbols for MOSFETs

## THE MOSFET TRANSISTORS

A second kind of FET is the MOSFET. There are four types of MOSFETs, n type depletion, p type depletion, n type enhancement, and p type enhancement. Like JFETs the n and p types differ in polarities so this text will discuss only the n types. Figure 2.104 shows common symbols used for MOSFETs. All MOSFETs have the characteristic that the gate is electrically insulated from the rest of the transistor. As a result, the resistance between the gate and other terminals is typically $10^{12}$ [$\Omega$]. Unfortunately, the gate does have some capacitance, which depends on the size of the device. Because the gate is insulated, we can apply positive and negative voltages to the gate.

The depletion type MOSFETs have characteristics similar to the JFET. Figure 2.105 shows the typical drain to source response of a depletion n type MOSFET. Figure 2.106 shows the typical response on an enhancement type MOSFET. Note that in the enhancement type, no current flows until the gate is excited.

## Concept 2.3 ───────────────────────────────

# A MOSFET Switch.

We can use an enhancement n type MOSFET as an electrical switch as figure 2.107 shows. The current flowing through the switch is negligible, whereas the current from D to S can be significant. A circuit similar to the one in figure 2.107 is of greatest value if the switch were actually another electronic device, like a computer output or a photodiode, that cannot source or sink enough current to drive the light bulb directly. For example, if the controlling device has voltage but little current

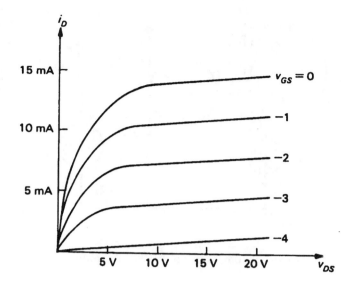

Figure 2.105: Conductance Relation of a Typical Depletion Type MOSFET

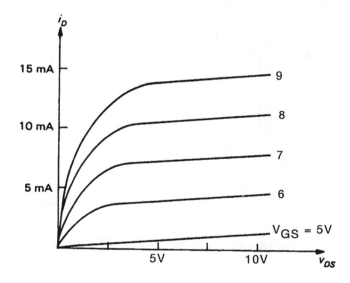

Figure 2.106: Conductance Relation of a Typical Enhancement Type MOSFET

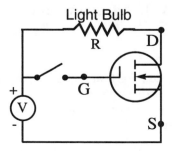

Figure 2.107: A Power (Large) Enhancement MOSFET Used as a Switch

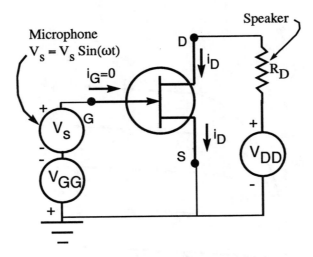

Figure 2.108: A Simple JFET Amplifier

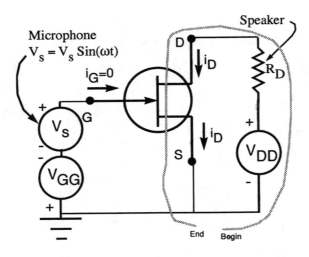

Figure 2.109: A Loop for Applying Kirchhoff's Law

(small power), the MOSFET amplifies the controller's power. Another use of a switching circuit is to isolate a human operator (the person throwing the switch) from potentially large currents in the main circuit. We might do this for safety reasons.

**End 2.3** ————————————————————————————————————————————————————

**Example 2.26** ————————————————————————————————————————————————

## A JFET Amplifier.

**PROBLEM:**

Figure 2.108 shows a simple Junction FET amplifier. The circuit is expected to amplify the small voltages $V_s \sin(\omega t)$ coming from a microphone. The purpose of the source $V_{GG}$ is to prevent the gate voltage from becoming positive causing a forward biased S-G junction and large current exiting the gate. Determine the currents and voltages of the JFET amplifier.

**FORMULATION:**

To determine the currents and voltages in the two [NIF] zero [NIA] amplifier, we will apply Kirchhoff's law. Figure 2.109 shows the first path we will use.

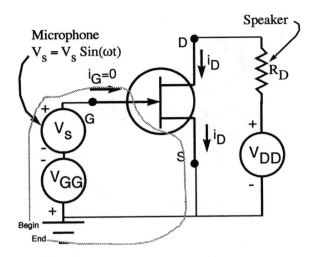

Figure 2.110: A Second Loop for Applying Kirchhoff's Law

---

**Kirchhoff's Voltage Law (KVL)**

system[ **loop fig.2.109**]     time period [**differential**]

$\sum^{\text{loop}} V = 0$

| | |
|---|---|
| **rise:** | $V_{DD}$ |
| **fall:** | $I_D R_D + V_{DS}$ |

$$\boxed{V_{DD}} \ - \ \boxed{I_D R_D + V_{DS}} \ = \ \boxed{0}$$

*rise*          *fall*

---

To determine whether the voltage rises or falls over the transistor junction (from D to S), we assume that $V_D$ is above $V_S$. When this is the case, the voltage falls as we cross the junction. $V_{DS} = V_D - V_S$.

$$V_{DD} - I_D R_D - V_{DS} = 0 \qquad (2.79)$$

Figure 2.110 shows a second loop to walk around.

---

**Kirchhoff's Voltage Law (KVL)**

system[ **loop fig.2.110**]     time period [**differential**]

$\sum^{\text{loop}} V = 0$

| | |
|---|---|
| **rise:** | $V_s \sin(\omega t)$ |
| **fall:** | $V_{GG} + V_{GS}$ |

$$\boxed{V_s \sin(\omega t)} \ - \ \boxed{V_{GG} + V_{GS}} \ = \ \boxed{0}$$

*rise*          *fall*

---

To determine whether the voltage rises or falls over the transistor junction (from G to S), we assume that $V_G$ is above $V_S$ ($V_{GS} > 0$). When this is the case, the voltage falls as we cross the junction. $V_G - V_S = V_{GS}$.

$$V_{GS} = -V_{GG} + V_s \sin(\omega t) - V_{GS} = 0$$

This means that

$$V_{GS} = -V_{GG} + V_s \sin(\omega t) \qquad (2.80)$$

**SOLUTION:**

Equations 2.79 and 2.80 contain three unknowns ($V_{GS}$, $V_{DS}$ and $I_D$). To solve for these we need an additional relationship. This extra relation comes from the conductance diagram for the JFET.

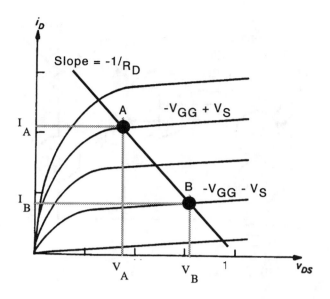

Figure 2.111: Operating Point for the Simple JFET Amplifier

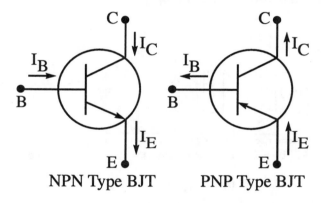

Figure 2.112: Schematic Diagrams of the npn and pnp Type BJT

Since both the conductance diagram (figure 2.103) and equation 2.79 relates $I_D$ to $V_{DS}$, if we plot the equation on top of the conductance, their intersection will define the operating point (values $V_{DS}$ and $I_D$). Figure 2.111 shows the superposition of the two graphs. As the microphone operates, the gate voltage ($V_{GS}$) varies between points A and B (see equation 2.80). Because the gate voltage varies, the current $I_D$ varies between $I_A$ and $I_B$ and the voltage $V_{DS}$ varies between $V_A$ and $V_B$ (see figure 2.111). As a result, the speaker has a voltage and current across/through it which varies with the same frequency as the microphone but has considerable magnitude.
**End 2.26**

## BIPOLAR JUNCTION TRANSISTORS

The last type of transistor this text will discuss is the Bipolar Junction Transistor or BJT. Figure 2.112 shows the symbols used for the two types of BJTs, the npn and the pnp. The pnp is similar to the npn except the currents and voltage polarities are reversed therefore this text will discuss only the npn.

The arrow between the base (B) and emitter (E) leads implies the presence of a diode junction. In the case of the npn, when the base voltage rises slightly above the emitter voltage ($V_B - V_E = V_{BE} > 0$), this diode junction is forward biased and current flows from base to emitter. Provided the collector (C) voltage is approximately one [volt] above the emitter voltage ($V_C - V_E = V_{CE} > 1$),

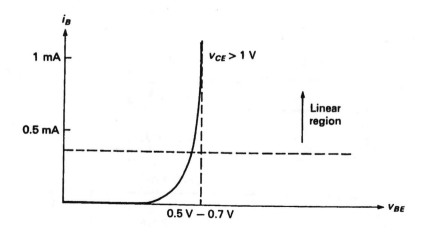

Figure 2.113: Conductance of the BE Junction of a BJT

the conductance of the base-emitter junction is independent of the collector. Figure 2.113 shows the conductance for the base-emitter (BE) junction. Notice it looks like a diode (which it is).

The junction between collector and base is also a diode. The C-B diode is reversed biased when the collector voltage is above the base voltage. This is the normal operating condition. There is normally just a little current flowing from C to B. $I_{CBO}$ denotes the current from the collector to base with the emitter open circuited. The current $I_{CBO}$ is strongly dependent on temperature. As temperature increases, so does $I_{CBO}$. If we want to build an amplifier, this hurts. If we want to build a temperature sensor, it helps.

Figure 2.114 shows the conductance of the collector terminal for various values of base current. The voltage $BV_{CEO}$ is the breakdown voltage with emitter open circuited. An approximate relation between the base and collector currents for values in the linear region is

$$I_B \approx \frac{I_C}{\beta} \tag{2.81}$$

where a typical value of $\beta$ is 100.

**Concept 2.4** _____

### A BJT Switch

We can use a BJT transistor as a switch by connecting it as figure 2.115 shows. When the switch is open, $V_B \approx 0$ therefore there is zero $i_B$ and (according to figure 2.114) very little $I_C$; the transistor is off. When the switch closes, $V_B$ increases (hopefully above 0.7 [volts]) forward biasing the BE junction causing $I_B$ and allowing $I_C$; the transistor is on. Naturally, the real power of a switch like this is when the base B is connected to another electrical device not a manual switch!

**End 2.4** _____

**Example 2.27** _____

### A Simple BJT Amplifier.

**PROBLEM:**

Figure 2.116 shows a simple, almost working, amplifier. The amplifier almost works because the base voltage (that the microphone provides) is most of the time below what the BE junction requires for forward biasing (0.7 [volts]). As a result, the transistor rarely conducts current through the collector. What the circuit needs is to boost the base voltage above the forward bias level. Figure 2.117 shows how we can accomplish this. Analyze the amplifying characteristics of this circuit by determining its voltages and currents.

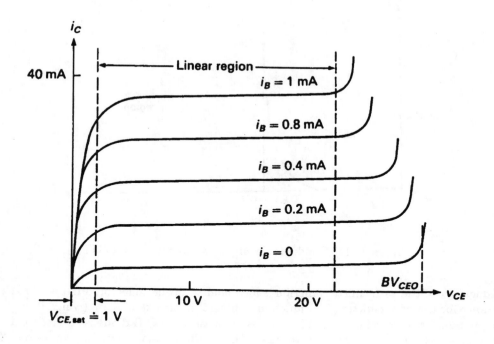

Figure 2.114: Conductance of the Collector for Various Values of Base Current

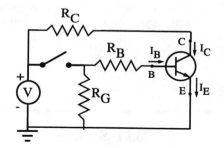

Figure 2.115: An npn BJT Used As a Switch

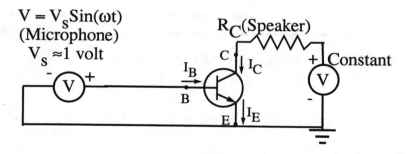

Figure 2.116: An Almost Working BJT Amplifier

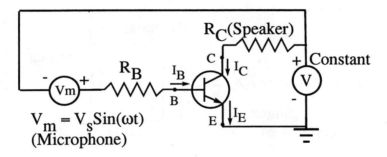

Figure 2.117: A Simple BJT Amplifier

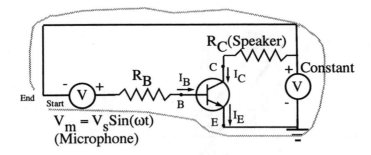

Figure 2.118: A Loop for Kirchhoff's Law

**FORMULATION:**

The first step in solving this problem is to determine the [NIF] and [NIA] of the circuit. Doing this, we find that the circuit has zero [NIA] and two [NIF].

Taking Kirchhoff's law around BE (see figure 2.118), we find that(let $V_{BE} = V_B - V_E$)

**Kirchhoff's Voltage Law (KVL)**

system[ **BE fig.2.118**]      time period [**differential**]

$\sum^{\text{loop}} V = 0$

| rise: | $V_o \sin(\omega t) + V$ |
|-------|--------------------------|
| fall: | $R_B I_B + V_{BE}$ |

$$\boxed{V_o \sin(\omega t) + V} \quad - \quad \boxed{R_B I_B + V_{BE}} \quad = \quad \boxed{0}$$
$\;\;\;\;\;rise \qquad\qquad\qquad\quad fall$

To determine the rise or fall across the BE junction, we assume that $V_B$ is positive. If it is, then as we step over the junction the voltage falls.

$$V_o \sin(\omega t) - R_B I_B - V_{BE} + V = 0 \qquad (2.82)$$

Now consider the second loop in figure 2.119(let $V_{CE} = V_C - V_E$) .

**Kirchhoff's Voltage Law (KVL)**

system[ **loop fig.2.119**]      time period [**differential**]

$\sum^{\text{loop}} V = 0$

| rise: | $V$ |
|-------|-----|
| fall: | $R_C I_C + V_{CE}$ |

$$\boxed{V} \quad - \quad \boxed{R_C I_C + V_{CE}} \quad = \quad \boxed{0}$$
$\;rise \qquad\qquad fall$

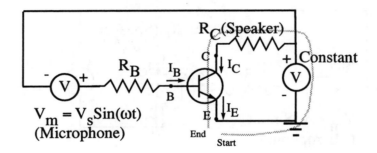

Figure 2.119: Second Loop for Kirchhoff's Law

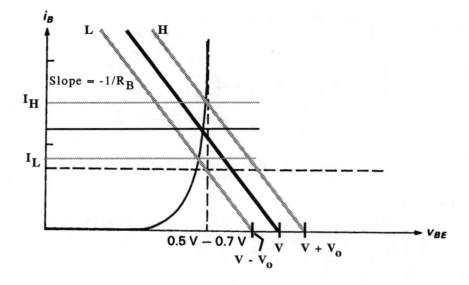

Figure 2.120: The Operating Point for the Base-Emitter Junction

$$V - R_C I_C - V_{CE} = 0 \tag{2.83}$$

**SOLUTION:**

Equations 2.82 and 2.83 contain four unknowns ($I_B$, $V_{BE}$, $I_C$ and $V_{CE}$) so we need more relationships. In the case of transistors, the relations are easiest to obtain by looking at the conductance diagrams.

We know the BE junction must behave as figure 2.113 indicates and it must also satisfy equation 2.82. We can plot equation 2.82 on top of figure 2.113 to determine the operating point. Figure 2.120 shows the operating point. Notice that as the microphone changes, the line representing equation 2.82 shifts. The L and H lines in figure 2.120 show the minimum and maximum microphone voltages. Since the microphone voltage changes sinusoidally, the base current shifts sinusoidally between the values of $I_H$ and $I_L$ (as the figure shows). Note that the slope of the line in figure 2.120 is $-\frac{1}{R_B}$. Therefore, the steeper the slope, the greater the change in base current. If the line gets too steep, we run the danger of having the operating point enter the nonlinear region. Figure 2.120 shows one line slightly in the nonlinear region already. The problem with the nonlinearity is that the amplifier will begin to distort the sounds if it is driven nonlinear. If we approximate the intersections in figure 2.120 as occurring at $V_{BE} \approx 0.5$, then we can solve equation 2.82 for $I_B$.

$$I_B \approx \frac{V + V_o \sin(\omega t) - 0.5}{R_B}$$

Now consider the C-E junction conductance. If we plot equation 2.83 on top of figure 2.114 we obtain figure 2.121. Notice that as the base current fluctuates between $I_H$ and $I_L$, the collector

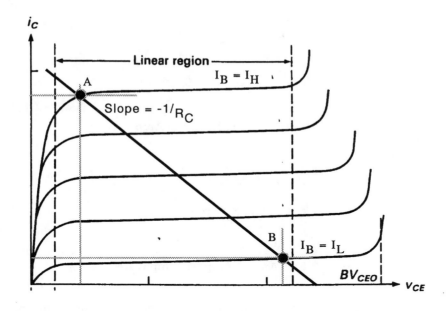

Figure 2.121: The Operating Point for the Collector-Emitter Junction

voltage and current fluctuate between points A and B. According to equation 2.81 typical fluctuations in collector current would be approximately $100(I_H - I_L)$. Notice also the effect of decreasing $R_C$ is to increase the slope of the line in figure 2.121 causing a smaller range in voltage $V_{CE}$.

**End 2.27**

## 2.10.4  OPERATIONAL AMPLIFIERS

**Objectives**
*When you complete this section, you should be able to*

1. *Explain why Op-Amps are important and list some applications of them.*

2. *List the two typical assumptions made about Op-Amps that allows one to design with them.*

3. *Calculate the output voltage relation for an Op-Amp using the simplifying assumptions used in the design process.*

4. *Explain what the CMRR is and why it is important.*

5. *Derive the output voltage relation for an Op-Amp when the CMRR is given.*

An operational amplifier is a semiconductor device that we can use to isolate electrical circuits and to perform low power voltage amplification. It isolates circuits because it has a high input impedance (relative to typical electrical circuits) and a low output impedance. As a result, it draws little power from a circuit connected to its input (it loads the input very little) thus having little impact on the input circuit. Its low output impedance means it has the ability to supply large currents[36] at constant voltages. Therefore, it allows us to provide the output circuit with a given voltage without regard to the required current. The amp acts like a voltage source. It should be noted however that with respect to most mechanical loads such as motors and relays, the Op

---

[36] Large is a relative term. When connected to other low power electrical circuits, the output current is large. If we connect it to a motor (or some other current demanding device), the op amp will have a small output current.

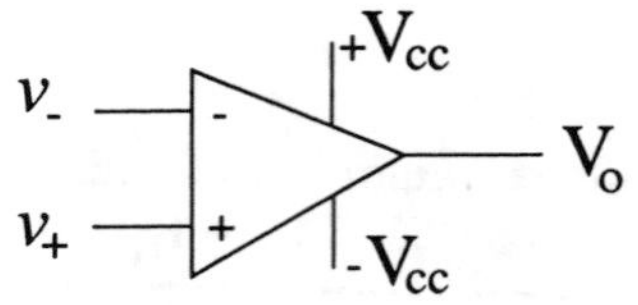

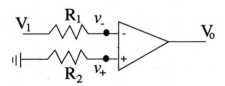

Figure 2.122: An Op Amp

Figure 2.123: An OP Amp with Input Resistances

Amp has high output impedance which means the motor typically demands more current at a given voltage than the Amp can supply. For this reason, Op Amps typically never drive motors directly. If we must drive a large current device, we typically place a power transistor between the Op Amp and the load. The Op Amp acts like the "manual switch" in the transistor switching circuits and the transistor is connected to the motor or other load. The high current demand comes from switching the transistor on.

Figure 2.122 shows a typical schematic for an Op Amp. Note that there are 5 connections. Two of them $v_-$ and $v_+$ are inputs and they typically have high input impedance which means very little current flows into them (perhaps $10^{-6}$ amps). One terminal $V_o$ is an output connection and the other two are connected to a power supply (a battery for example). Often the power is connected to + and - voltages (two batteries with their - ends common) but they can be connected to + and ground (one battery). Often, no one cares about the power connections (as long as they are connected nothing interesting happens with them) and so most diagrams do not show the power connections. Even if the power connections are not shown they are still there. The Amp also has a very large voltage amplification $(G)$.[37] This means the output voltage is a large number times the difference in input voltages. The constant of proportionality is typically $10^6$. Typically the voltage amplification gain $(G)$ is constant for harmonic inputs with frequencies less than 10 [Mhz]. The gain does tend to vary with temperature.

If we connect the Op Amp to its inputs through resistors as figure 2.123 shows, the high input impedance implies very little current flows into the Amp. Therefore, we know the following:

$$v_- \approx V_1$$

$$v_+ \approx 0$$

Furthermore, the large gain implies that

$$V_o = G(v_+ - v_-) \tag{2.84}$$

Combining these relations, we find that

$$V_o = -GV_1$$

For very big $G$ (typical of Op Amps) the usable range of $V_1$ shrinks. As figure 2.124 shows, the maximum output voltage is less than or equal to the supply voltage $V_{cc}$ used to power the Amp. For very small values of input voltage, the output amplifies the input. As the output approaches $V_{cc}$, the response levels off and is no longer amplifying the input. This leveling off is called saturation.

---

[37]Engineers often call voltage amplification gain.

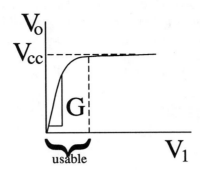

Figure 2.124: Open Loop Op Amp Response

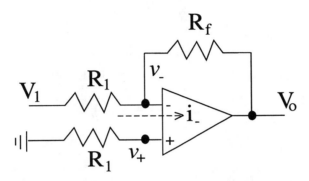

Figure 2.125: Resistance Feedback on an OP Amp

When the gain is large, saturation occurs at small values of input. As configured the amp is next to useless.

Another problem with the amp configuration in figure 2.123 is that small percent shifts in $G$ (because $G$ is large) can cause significant changes in the output. This can be an exceptionally serious problem because Op Amps are manufactured in a batch process much like we bake cookies. Because of variations from batch to batch, the properties of Op Amps will vary by several percent.

To overcome these deficiencies, the Op Amp is used with some form of feedback. Figure 2.125 shows a typical configuration using a resistor as feedback. The resistor feeds the output back to the input (hence the term feedback). Applying the conservation of charge at the node labeled negative gives

**Conservation of Charge**

system[− **terminal fig.2.125**]    time period [**differential**]

*input − output = accumulation*

| **input/output:** the - node **accumulation:** | $\frac{V_1 - v_-}{R_1} + \frac{V_o - v_-}{R_f} - i_-$ |
|---|---|

| $\frac{V_1 - v_-}{R_1} + \frac{V_o - v_-}{R_f} - i_-$ | $=$ | $0$ |
|---|---|---|
| *input/output* | | *change in accum.* |

In equation form, we have

$$\frac{V_1 - v_-}{R_1} - i_- + \frac{V_o - v_-}{R_f} = 0 \tag{2.85}$$

If you're thinking "I thought $i_-$ was small", you're right. We are carrying it along to make a point. At the very end, we'll drop it out. Applying the conservation of charge to the positive terminal, we find that

**Conservation of Charge**

system[**+ terminal fig.2.125**]      time period [**differential**]

$input - output = accumulation$

| input/output: the + node | $\frac{-v_+}{R_1} - i_+$ |
|---|---|
| accumulation: | |

| $\frac{-v_+}{R_1} - i_+$ | $=$ | $0$ |
|---|---|---|
| *input/output* | | *change in accum.* |

In equation form, we have

$$\frac{-v_+}{R_1} - i_+ = 0 \tag{2.86}$$

Solving equation 2.86 for $v_+$ gives

$$v_+ = -i_+ R_1$$

Using this equation and equation 2.84 to solve for $v_-$, we find that

$$v_- = -\left(\frac{V_o}{G} + R_1 i_+\right) \tag{2.87}$$

Using equation 2.87 in 2.85, we get that

$$0 = V_1 + \left(R_1 + \frac{R_1^2}{R_f}\right) i_+ - R_1 i_- + V_o \left(\frac{R_1 + R_f}{R_f G} + \frac{R_1}{R_f}\right)$$

Solving for the output voltage, we find that

$$V_o = \frac{-V_1 - \left(R_1 + \frac{R_1^2}{R_f}\right) i_+ + R_1 i_-}{\frac{R_1 + R_f}{R_f G} + \frac{R_1}{R_f}}$$

This last expression tells you something about what would happen if $i_-$ and $i_+$ were not small and if $G$ were not big. We will not say much about it at this point. Using the approximations that $i_+ \approx 0$, $i_- \approx 0$ and $G \approx 10^6$, we find that the output is approximately

$$V_o \approx \frac{-V_1}{\frac{R_1}{R_f}} = -\frac{R_f V_1}{R_1}$$

Notice that since the output is not a function of $G$, it is no longer sensitive to changes in the gain $G$. This is one typical advantage of feedback, it makes the system insensitive to changes in some parameters. Also note that the output amplification is adjustable by properly modifying the resistances $R_f$ and $R_1$. In this feedback configuration, the amp is much more useful because we can make it amplify voltages. Keep in mind it will still saturate so we cannot expect miracles from our device.

A more direct and simplified method for deriving the output equation is to assume from the start (based on large gain and no saturation) that

$$v_- \approx v_+$$

If the gain $G$ is not huge, this assumption is not valid and you must derive the operation of the amp the long way. If you are willing to assume large input impedance of the amp (small currents $i_-$ and $i_+$) then

$$i_- = i_+ = 0$$

Combing these assumptions with the conservation of charge for the + terminal, we find for the same amp as before

**Conservation of Charge**

system[**+ terminal fig.2.125**]        time period [**differential**]

*input − output = accumulation*

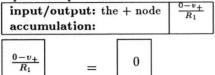

| **input/output:** the + node **accumulation:** | $\frac{0-v_+}{R_1}$ |

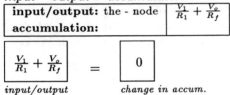

| $\frac{0-v_+}{R_1}$ | $=$ | $0$ |

*input/output*            *change in accum.*

In equation form, we have

$$\frac{-v_+}{R_1} = 0$$

which states $v_+ = 0$. Since $v_- \approx v_+$ (large gain $G$) we have $v_- \approx 0$. Applying conservation of charge to the − terminal, we find that

**Conservation of Charge**

system[**− terminal fig.2.125**]        time period [**differential**]

*input − output = accumulation*

| **input/output:** the - node **accumulation:** | $\frac{V_1}{R_1} + \frac{V_o}{R_f}$ |

| $\frac{V_1}{R_1} + \frac{V_o}{R_f}$ | $=$ | $0$ |

*input/output*            *change in accum.*

In equation form, we have

$$\frac{V_1}{R_1} + \frac{V_o}{R_f} = 0$$

Solving this equation for $V_o$, we get that

$$V_o = -\frac{R_f V_1}{R_1}$$

By using different electrical components in the feedback and input path, we can obtain output responses other than a voltage gain.

Notice that the simplified equations do not contain $R_2$ hence it says that the resistor is not needed. There are other more sophisticated aspects of the amp that we will not go into that will tell you the magnitude of $R_2$ ought to be the same order of magnitude of $R_1$. Also, the equation does not say anything about the values of the individual resistors $R_1$ and $R_f$ only their ratio. What are some reasons why you would not want the resistors too small or too large? If they were too small, there may need to be significant current coming from going to $V_1$, this current would not flow into the amp, it flows around it through the resistors. This large current may be a bad thing since we have seen some examples of how circuits do not perform as we want when we pull significant current out of them. The voltage divider circuit is just one example (see figure 2.60). If the resisters are too large, the amount of current coming in through $R_1$ will be very small. If the current coming in gets so small it becomes approximately equal to $i_-$ and when that happens our assumption that $i_-$ is so small it can be ignored is no longer true. What happens with large $R_1$? The amp does not amplify exactly as we expect.

**Example 2.28**

## Design of an Integrator

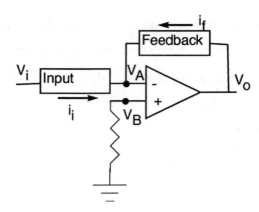

Figure 2.126: An Op Amp with Unknown Input and Feedback Blocks

**QUESTION:**
Design an Op Amp that provides an output voltage that is the integral the amp's input voltage.
**ANSWER:**
When faced with designing a system response, it is often advantageous to find simple explanations for the essential operation of the system. Consider the Op Amp with unknown input and feedback elements in figure 2.126. Because the Op Amp has a very high gain, the voltage at terminals A and B must be (essentially) identical. Since terminal B is ground, its voltage is zero. Therefore, the voltage at terminal A must also be nearly zero. Since the Op Amp also has relatively high input impedance, very little current actually flows into the Amp. As a result, current $i_i + i_f = 0$. We can express this basic Op Amp function as follows:

$$i_i = -i_f \qquad (2.88)$$

Equation 2.88 is the design rule we will use.

Notice that since $V_A$ is zero, $i_i$ is a function of only $V_i$ and the element placed in the input. Likewise, $i_f$ is a function of only $V_o$ and the element placed in the feedback block.

The response we want from our system is

$$\int V_i dt = V_o$$

or equivalently

$$V_i = \dot{V}_o \qquad (2.89)$$

If we compare equations 2.89 and 2.88 we find that if we allow

$$i_i = V_i \qquad (2.90)$$

and

$$-i_f = \dot{V}_o \qquad (2.91)$$

then we would have our desired response.

The two remaining questions are the following:

1. what element can we put in the forward path so that the current $i_i$ equals $V_i$ as equation 2.90 requires?

2. what element can we put in the feedback path so that the current $i_f$ equals $\dot{V}_o$ as equation 2.91 requires?

Clearly a resistor of 1 [$\Omega$] works in the input path, and a capacitor works for the feedback. Once we know what type of elements to use, we can work out an analysis to determine the sizes of each one.
**End 2.28**

## 2.11   Questions and Problems

An * after a problem number indicates a solution is available.

**Resistor Problems**

1. In figure 2.127, $i_1$ and $i_3$ are not independent. Explain why. Are $i_1$ and $i_4$ independent?

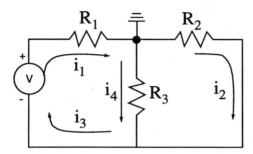

Figure 2.127:

2. For the resistor network in figure 2.128 (figure A), determine the value of $R_{eq}$ in (figure B) so the current in both figures are the same. The value of each resistance is 1000 times its number.

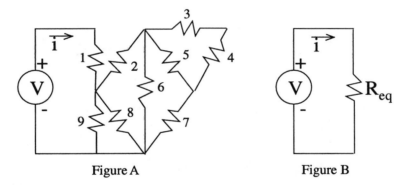

Figure 2.128:

3. What is the current flowing through the resistor marked with an 8 (figure 2.128). The voltage applied to the circuit is 100 [volts].

4. * Figure 2.129 shows resistors $R_1$, $R_2$, $R_3$ and $R_4$ configured in what is called a bridge circuit. A bridge is used to measure changes in resistance of one of the resistors in the bridge. For example, a strain gage is a resistor that changes its resistance when it is stretched. Anyway, the objective is to measure the voltage across various places on the bridge and equate this voltage back to the change in resistance. Most bridges are excited by a voltage source rather than a current source but we use the current source to make the problem more challenging. Suppose we want to know the voltage across $R_2$ so we place a meter (shown as the dashed object) across $R_2$. The problem is all meters have an input impedance (shown as a simple resistor $R_m$, usually the impedance is more complex). The input impedance causes the measured voltage to differ from what is present without the meter. You cannot measure something without disturbing it slightly. What does the resistance $R_m$ have to be for the voltmeter to read the voltage drop across $R_2$ to within 1% of true value.

5. * Determine all the node voltages and all the currents of the circuit in figure 2.130 through figure 2.132.

6. Formulate a mathematical model that we can use to determine all the node voltages or all the currents of the circuit in figure 2.133. How many initial conditions are needed to solve the equations?

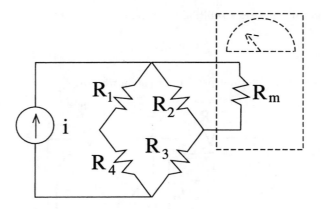

Figure 2.129:

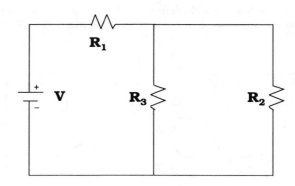

Figure 2.130:

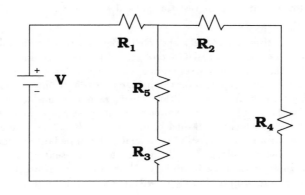

Figure 2.131:

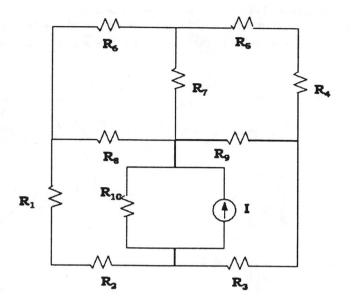

Figure 2.132:

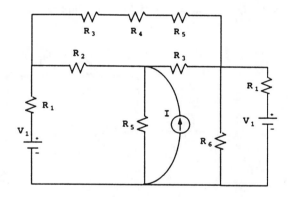

Figure 2.133:

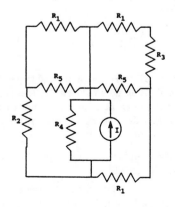

Figure 2.134:

7. Formulate a mathematical model that we can use to determine all the node voltages or all the currents of the circuit in figure 2.134. How many initial conditions are needed to solve the equations?

8. * In figure 2.135, let $R_1 = 1$ $R_2 = 2$ and $R_3 = 3$, $k = 4$ and $\alpha = 5$. Find all the node voltages and element currents in the circuit.

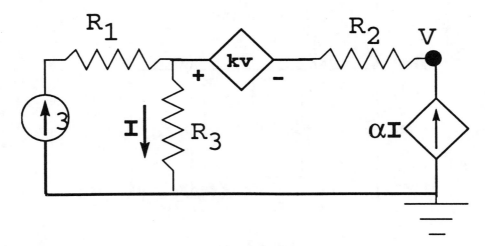

Figure 2.135:

9. * In figure 2.136 let $R_L = 1$, $R_d = 2$, $g_m = 3$ and $V_1 = 4$. Find all the node voltages and element currents in the circuit.

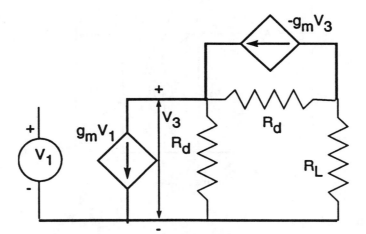

Figure 2.136:

10. * The circled components in figure 2.137 represent a special type of photosensor. There are three wires protruding from the device. When light strikes the sensor, a current is generated (as indicated by the current source). The magnitude of the current is proportional to the light intensity. The resistance $R$ changes from 0 to $R_p$ ($R_p = 100$ ohms) depending on where the light strikes the sensor. The resistors $R_1$ are connected to the sensor as shown.

    (a) Demonstrate that the quantity $\frac{V_a - V_b}{V_a + V_b}$ is a linear equation for the value of $R$ which includes only the values of the resistors $R_1$, $R_p$ and $R$.

    (b) Determine a "good" value for $R_1$ if the maximum current generated by light striking the sensor is 0.001 amps and neither voltage $V_a$ nor $V_b$ is to exceed 0.7 volts relative to ground.

11. * For the circuit shown in figure 2.138, resistor $R_3$ varies between 100 and 300 ohms. Determine resistors $R_1$ and $R_2$ so the voltage across $R_3$ is always between 6 and 7 volts.

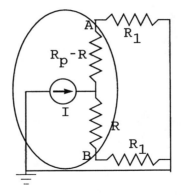

Figure 2.137:

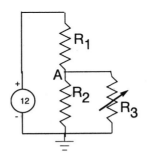

Figure 2.138:

**Capacitor Problems**

12. Explain in what ways does a capacitor act like a voltage source when we attempt to change the voltage across it.

13. Determine all the node voltages and all the currents as a function of time for the circuit in figure 2.139. Let all initial conditions be zero. How many initial conditions are needed?

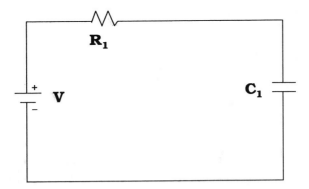

Figure 2.139:

14. Determine all the node voltages and all the currents as a function of time for the circuit in figure 2.140. Let all initial conditions be zero. How many initial conditions are needed?

15. Formulate a mathematical model that we can use to determine all the node voltages or all the currents of the circuit in figure 2.141. How many initial conditions are needed to solve the equations?

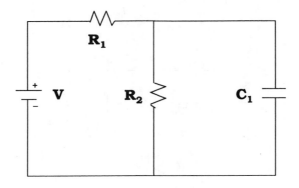

Figure 2.140:

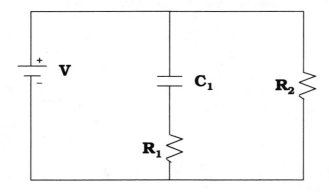

Figure 2.141:

16. Formulate a mathematical model that we can use to determine all the node voltages or all the currents of the circuit in figure 2.142. How many initial conditions are needed to solve the equations?

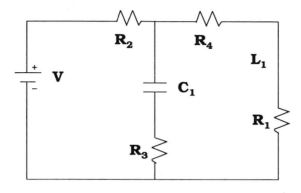

Figure 2.142:

17. Formulate a mathematical model that we can use to determine all the node voltages or all the currents of the circuit in figure 2.143. How many initial conditions are needed to solve the equations?

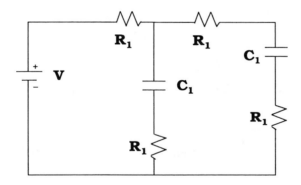

Figure 2.143:

18. Formulate a mathematical model that we can use to determine all the node voltages or all the currents of the circuit in figure 2.144. How many initial conditions are needed to solve the equations?

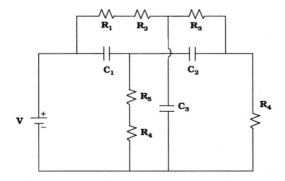

Figure 2.144:

19. Formulate a mathematical model that we can use to determine all the node voltages or all the currents of the circuit in figure 2.145. How many initial conditions are needed to solve the equations?

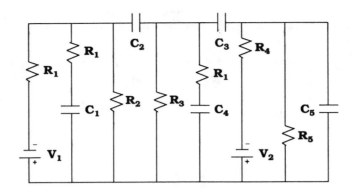

Figure 2.145:

20. * In figure 2.146, the current source and $R_s$ represent a light intensity sensor. The current source is equal to $2 + 2\sin(\omega t)$ where $\omega$ is 120 cycles / second. Think of the constant coming from sunlight and the sinusoid as coming from fluorescent lights. Fluorescent lights turn off 120 times a second when running on 60 hertz power. Both $R_m$ and $R_s = 100$ ohms. Determine the value of C that makes the time varying component of V less than 10% of the constant value.

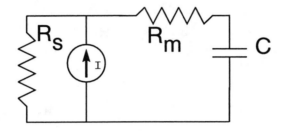

Figure 2.146:

**Inductance**

21. It was planned at one time to perform a power generation experiment using the Space Shuttle. The Shuttle was supposed to drag a satellite through orbit while being connected to it with an electrical cable. The Earth's magnetic field was expected to induce electrical currents in the cable which would be measured by instruments on the Shuttle. Since the electrical cable has resistance, there would be power dissipation when the current flows. Explain:

   (a) where the power being dissipated comes from,

   (b) how the electrical circuit (required for flowing current) can be completed,

   (c) the effect of this experiment on the Earth, and

   (d) the effect of the experiment on the Shuttle.

22. Explain why a mercury switch would not be a good choice for use in the spark plug circuit of example 2.14?

23. List and explain some ways that we can use the force induced by the interaction of current and magnetic fields.

24. If we place a conductive loop in a changing magnetic field, then a voltage will be induced in the loop that is related to the rate of change of the magnetic flux through the area enclosed by the loop. If the magnetic flux changes quickly, then the voltage induced will be large. List some possible ways that we can use this phenomenon.

25. In what ways does an inductor act like a current source when we attempt to change current through it?

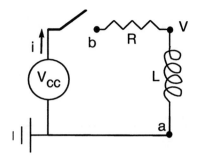

Figure 2.147:

26. Explain what would happen if we placed a wire loop made of a superconducting material within a changing magnetic field.

27. For the circuit in figure 2.147, if the voltage at b becomes low enough, a spark will jump across the switch. Since the resistance of the air gap separating the ends of the switch is large, it would not take much spark current to dissipate a large amount of energy (because $E_{\text{dissipated}} = i_{\text{spark}}^2 R_{\text{gap}}$). Describe what would happen to the switch if a spark did occur. Which side of the switch would be damaged the most by a spark jumping across the gap? Explain. How could we protect the switch from the dangers of sparking? Explain how we could use this phenomenon productively.

28. A high speed train is to be supported magnetically above an aluminum track. The train makes no contact with the track and the magnetic field used for support is generated with a superconducting permanent magnet. The train has a propulsion system which provides it with a forward thrust. Describe qualitatively how the levitation works. Can the system operate if the train is stationary? What are the sources of resistance the thrust must overcome? If you needed to carry extra weight on the train, what could be modified to provide the lift needed?

29. Consider again the inductor in figure 2.36. Each element of the coil is in the presence of magnetic fields of all other elements. Is there a net force acting on an element of the coil? If so, in what direction do the forces act? Do they try to hold the coil together, or push it apart?

30. Give several examples of how eddy currents can be beneficial and detrimental to something. Think about levitation, vibration damping and energy conversion.

31. Figure 2.148 shows a circuit tuned to a frequency of $\sqrt{\frac{1}{LC}}$. Solve for the voltage at B assuming that $V_A = \sin\left(\frac{k}{LC}\right)$. Determine the solution for several values of $k \neq 1$. Based on the results, what do you think the solution is when $k = 1$?

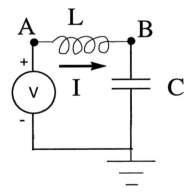

Figure 2.148:

32. Modify Dr. Zap's circuit so that he can obtain the peak current in figure 2.19 for a sustained period of time.

33. Determine all the node voltages and all the currents as a function of time for the circuit in figure 2.149. Let all initial conditions be zero. How many initial conditions are needed?

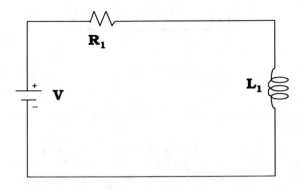

Figure 2.149:

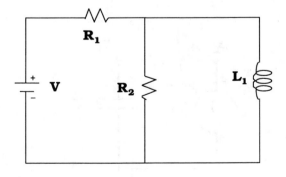

Figure 2.150:

34. Determine all the node voltages and all the currents as a function of time for the circuits in figure 2.150. Let all initial conditions be zero. How many initial conditions are needed?

35. Formulate a mathematical model that we can use to determine all the node voltages or all the currents of the circuit in figure 2.151. How many initial conditions are needed to solve the equations?

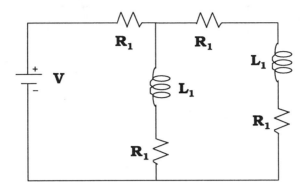

Figure 2.151:

36. Formulate a mathematical model that we can use to determine all the node voltages or all the currents of the circuit in figure 2.152. How many initial conditions are needed to solve the equations?

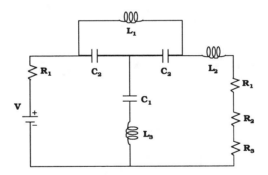

Figure 2.152:

37. Formulate a mathematical model that we can use to determine all the node voltages or all the currents of the circuit in figure 2.153. How many initial conditions are needed to solve the equations?

38. Formulate a mathematical model that we can use to determine all the node voltages or all the currents of the circuit in figure 2.154. How many initial conditions are needed to solve the equations?

39. Formulate a mathematical model that we can use to determine all the node voltages or all the currents of the circuit in figure 2.155. How many initial conditions are needed to solve the equations?

40. Formulate a mathematical model that we can use to determine all the node voltages or all the currents of the circuit in figure 2.156. How many initial conditions are needed to solve the equations?

41. Formulate a mathematical model that we can use to determine all the node voltages or all the currents of the circuit in figure 2.157. How many initial conditions are needed to solve the equations?

42. Formulate a mathematical model that we can use to determine all the node voltages or all the currents of the circuit in figure 2.158. How many initial conditions are needed to solve the equations?

43. Formulate a mathematical model that we can use to determine all the node voltages or all the currents of the circuit in figure 2.159. How many initial conditions are needed to solve the equations?

44. Formulate a mathematical model that we can use to determine all the node voltages or all the currents of the circuit in figure 2.160. How many initial conditions are needed to solve the equations?

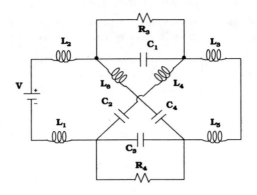

Figure 2.153:

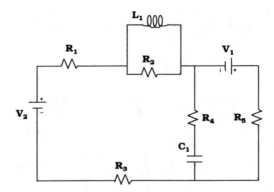

Figure 2.154:

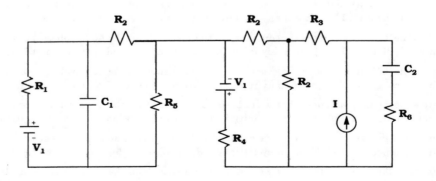

Figure 2.155:

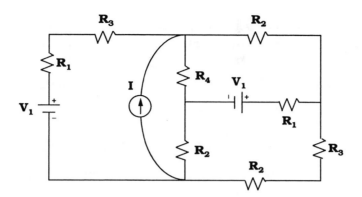

Figure 2.156:

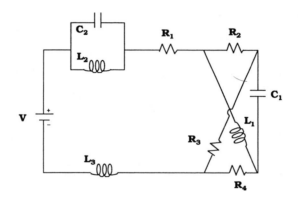

Figure 2.157:

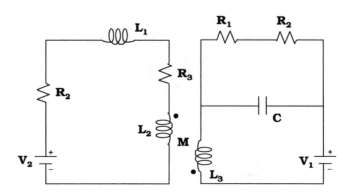

Figure 2.158:

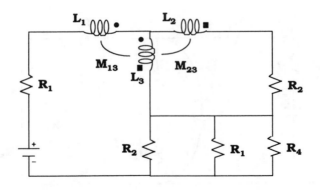

Figure 2.159:

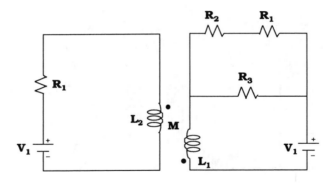

Figure 2.160:

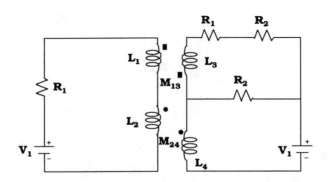

Figure 2.161:

45. Formulate a mathematical model that we can use to determine all the node voltages or all the currents of the circuit in figure 2.161. How many initial conditions are needed to solve the equations?

**Switched Problems**

46. For the circuit in figure 2.162 do the following. The switch has been open for a long time. At time 0, the switch closes. All sources are constant. Determine the equations that will describe all node voltages for times greater than $0^+$. How many initial conditions are needed to solve your equations? Find the necessary initial conditions.

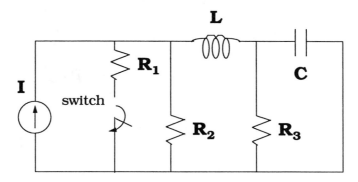

Figure 2.162:

47. Repeat problem 46 except the switch was originally closed and it is opened at time 0.

48. For the circuit in figure 2.163 do the following. The switch has been open for a long time. At time 0, the switch closes. All sources are constant. Determine the equations that will describe all node voltages for times greater than $0^+$. How many initial conditions are needed to solve your equations? Find the necessary initial conditions.

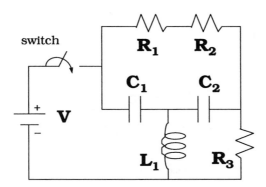

Figure 2.163:

49. Repeat problem 48 except the switch was originally closed and it is opened at time 0.

50. For the circuit in figure 2.164 do the following. The switch has been open for a long time. At time 0, the switch closes. All sources are constant. Determine the equations that will describe all node voltages for times greater than $0^+$. How many initial conditions are needed to solve your equations? Find the necessary initial conditions.

51. Repeat problem 50 except the switch was originally closed and it is opened at time 0.

52. For the circuit in figure 2.165 do the following. The switch has been open for a long time. At time 0, the switch closes. All sources are constant. Determine the equations that will describe all node voltages for times greater than $0^+$. How many initial conditions are needed to solve your equations? Find the necessary initial conditions.

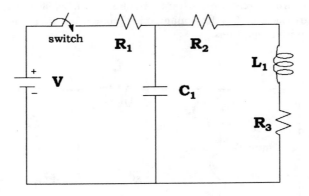

Figure 2.164:

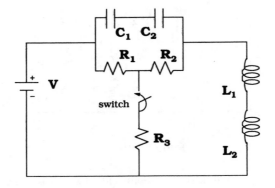

Figure 2.165:

53. Repeat problem 52 except the switch was originally closed and it is opened at time 0.

54. For the circuit shown in figure 2.166 the switch has been open a long time and is suddenly closed. Determine the voltage at A as a function of time. Let $R = 1$, $C = 1$ (unrealistic but makes the numbers nicer), $L_1 = 1$, $L_2 = 2$, $M = 0.5$ and $V = 10$.

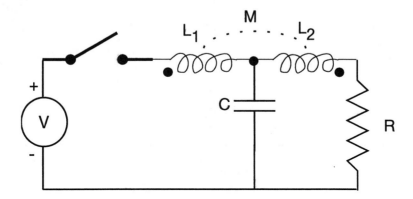

Figure 2.166:

55. Repeat problem 54 except let the switch be closed a long time then opened.

56. For the circuit in figure 2.167 do the following. The switch has been open for a long time. At time 0, the switch closes. All sources are constant. Determine the equations that will describe all node voltages for times greater than $0^+$. How many initial conditions are needed to solve your equations? Find the necessary initial conditions.

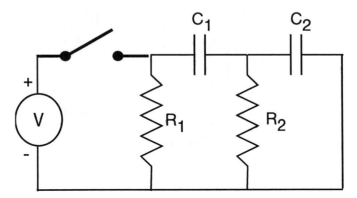

Figure 2.167:

57. Repeat problem 56 except the switch was originally closed and it is opened at time 0.

58. For the circuit in figure 2.168 do the following. The switch has been open for a long time. At time 0, the switch closes. All sources are constant. Determine the equations that will describe all node voltages for times greater than $0^+$. How many initial conditions are needed to solve your equations? Find the necessary initial conditions.

59. Repeat problem 58 except the switch was originally closed and it is opened at time 0.

60. For the circuit in figure 2.169 do the following. The switch has been open for a long time. At time 0, the switch closes. All sources are constant. Determine the equations that will describe all node voltages for times greater than $0^+$. How many initial conditions are needed to solve your equations? Find the necessary initial conditions.

61. Repeat problem 60 except the switch was originally closed and it is opened at time 0.

62. For the circuit in figure 2.170 do the following. The switch has been open for a long time. At time 0, the switch closes. All sources are constant. Determine the equations that will describe all node voltages for times greater than $0^+$. How many initial conditions are needed to solve your equations? Find the necessary initial conditions.

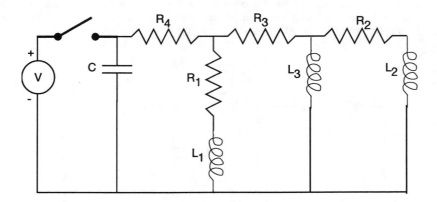

Figure 2.168:

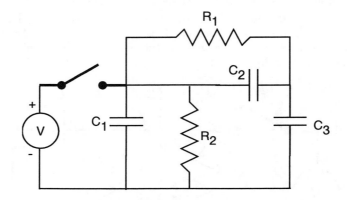

Figure 2.169:

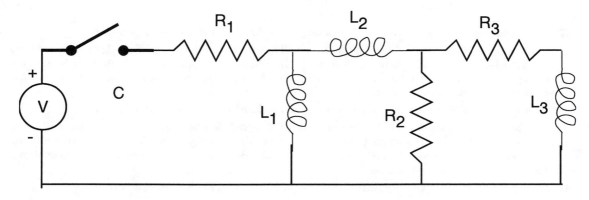

Figure 2.170:

63. Repeat problem 62 except the switch was originally closed and it is opened at time 0.

64. For the circuit in figure 2.171, the switch has been closed for a long time. Then, at t=0, we open it. Determine

    (a) the node voltages at t=0$^+$ and t=$\infty$

    (b) the first derivative ($\frac{d}{dt}$ ()) of the node voltages at t=0$^+$ and t=$\infty$

    (c) the second derivative ($\frac{d^2}{dt^2}$ ()) of the node voltages at t=0$^+$ and t=$\infty$

    (d) the currents at t=0$^+$ and t=$\infty$

    (e) the first derivative ($\frac{d}{dt}$ ()) of the currents at t=0$^+$ and t=$\infty$

    (f) the second derivative ($\frac{d^2}{dt^2}$ ()) of the currents at t=0$^+$ and t=$\infty$

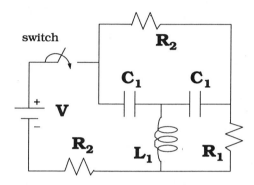

Figure 2.171:

65. For the circuit in figure 2.172, the switch has been closed for a long time. Then, at t=0, we open it. Determine

    (a) the node voltages at t=0$^+$ and t=$\infty$

    (b) the first derivative ($\frac{d}{dt}$ ()) of the node voltages at t=0$^+$ and t=$\infty$

    (c) the second derivative ($\frac{d^2}{dt^2}$ ()) of the node voltages at t=0$^+$ and t=$\infty$

    (d) the currents at t=0$^+$ and t=$\infty$

    (e) the first derivative ($\frac{d}{dt}$ ()) of the currents at t=0$^+$ and t=$\infty$

    (f) the second derivative ($\frac{d^2}{dt^2}$ ()) of the currents at t=0$^+$ and t=$\infty$

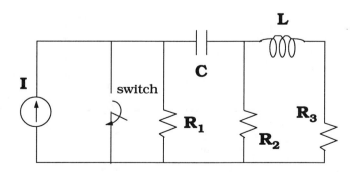

Figure 2.172:

66. For the circuit in figure 2.173, the switch has been closed for a long time. Then, at t=0, we open it. Determine

(a) the node voltages at $t=0^+$ and $t=\infty$

(b) the first derivative ($\frac{d}{dt}()$) of the node voltages at $t=0^+$ and $t=\infty$

(c) the second derivative ($\frac{d^2}{dt^2}()$) of the node voltages at $t=0^+$ and $t=\infty$

(d) the currents at $t=0^+$ and $t=\infty$

(e) the first derivative ($\frac{d}{dt}()$) of the currents at $t=0^+$ and $t=\infty$

(f) the second derivative ($\frac{d^2}{dt^2}()$) of the currents at $t=0^+$ and $t=\infty$

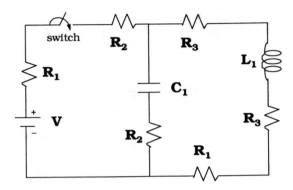

Figure 2.173:

67. For the circuit in figure 2.174, the switch has been closed for a long time. Then, at $t=0$, we open it. Determine

(a) the node voltages at $t=0^+$ and $t=\infty$

(b) the first derivative ($\frac{d}{dt}()$) of the node voltages at $t=0^+$ and $t=\infty$

(c) the second derivative ($\frac{d^2}{dt^2}()$) of the node voltages at $t=0^+$ and $t=\infty$

(d) the currents at $t=0^+$ and $t=\infty$

(e) the first derivative ($\frac{d}{dt}()$) of the currents at $t=0^+$ and $t=\infty$

(f) the second derivative ($\frac{d^2}{dt^2}()$) of the currents at $t=0^+$ and $t=\infty$

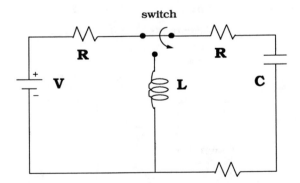

Figure 2.174:

68. For the circuit in figure 2.175, the switch has been closed for a long time. Then, at $t=0$, we open it. Determine

(a) the node voltages at $t=0^+$ and $t=\infty$

(b) the first derivative ($\frac{d}{dt}()$) of the node voltages at $t=0^+$ and $t=\infty$

(c) the second derivative ($\frac{d^2}{dt^2}()$) of the node voltages at $t=0^+$ and $t=\infty$

(d) the currents at $t=0^+$ and $t=\infty$

(e) the first derivative ($\frac{d}{dt}$ ()) of the currents at $t=0^+$ and $t=\infty$

(f) the second derivative ($\frac{d^2}{dt^2}$ ()) of the currents at $t=0^+$ and $t=\infty$

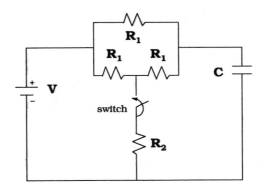

Figure 2.175:

**Electronics**

69. Why do transistors have a maximum current rating? What would happen if we drove more current than recommended?

70. Under what conditions would it be desirable to eliminate the bias in a light intensity sensor similar to the one in figure 2.98? Show a schematic diagram of a circuit that has no bias on the photodiode.

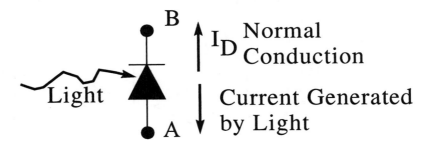

Figure 2.176:

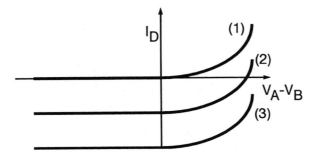

Figure 2.177:

71. We want to use the photodiode in figure 2.176 to build a sensor. Figure 2.177 shows the effort versus flow relationship for the photodiode for various intensities of light. Curve (1) corresponds to no light striking the diode. Curve (3) corresponds to intense light striking the diode. Describe the energy and charge content of light. Are energy and charge conserved in the photodiode? Explain.

72. Under what conditions (values of $V_R$ and light intensity) does a photodiode, configured as figure 2.98 shows, add electrical energy to the circuit? Where does this energy come from? Where does it go once inside the circuit? When is it possible that the energy added by the photodiode is greater than that consumed by the resistor?

73. Explain how we could protect the switch in example 2.10?

74. Explain what would happen to the spark circuit of example 2.14 if we installed a clamping diode in the primary circuit.

75. When using an Op Amp amplifier, the input and feedback resistors are never really small nor really big, why not?

76. If the current generated by a photodiode is proportional to the light intensity striking it, determine the magnitude of reverse bias required in the circuit in figure 2.98 such that voltage at terminal B is always a linear function of intensity. Assume intensity ranges from zero to some maximum.

77. We have a volt meter with a 3 digit display that is capable of measuring 5 [volts] maximum. We want to measure light intensity which ranges from very dark to very bright. The lowest light level produces practically no current in a photodiode and the highest light level produces 0.1 [amp] at zero bias voltage. Design a circuit that will make the most accurate light intensity measurements possible.

78. Design a BJT switch that turns on when the base voltage is pulled low.

79. Determine the maximum power dissipation in the diode of example problem 2.10 and explain why this is important to compute.

80. When catalogs characterize Op Amps, two gains are listed. One is the differential gain ($G \approx 10^6$) and the other is the common mode gain ($C \approx 0$). Often these gains are specified with the gain number $G$ and the common mode rejection ratio number $CMRR$ (defined as CMRR $= \frac{G}{C}$). The relation of these gains to the open loop (no feedback) output ($V$) is as follows (see figure 2.122 for definitions):

$$V = G\left(V_+ - V_-\right) + \frac{1}{2}C\left(V_+ + V_-\right)$$

Derive the output voltage relationship for the device in terms of the resistances, $G$ and the CMRR.

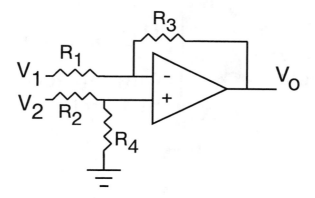

Figure 2.178:

81. Call the expression determined in problem 80 the "real" solution. Take the limit of the expression as $G \to \infty$ and $CMRR \to \infty$. Call this expression the "ideal" solution. Compute error as (real - ideal)/ideal. Let the $CMRR = \infty$ and determine the range of $G$ that makes the error magnitude less than 0.1 %.

82. Replace the grounded potential of figure 2.125 with a voltage labeled $V_2$. Derive the relationship between the output voltage $V_o$ and the input voltages $V_1$ and $V_2$.

83. The current flow through a capacitor is

$$i_c = C\frac{dV_c}{dt},$$

where $V_c$ is the voltage across the capacitor, and $C$ is a constant (capacitance). The voltage across an inductor is

$$V_l = L\frac{di_l}{dt},$$

where $i_l$ is the current flowing in the inductor and $L$ is a constant (inductance). Using the figure 2.125 as a guide, design two circuits that will provide an output voltage that is proportional to the integral of the input voltage.

84. Design two circuits that will produce an output that is proportional to the derivative of the input.

85. Design a circuit that will produce an output voltage that is proportional to the difference of two input voltages.

86. Design a circuit (a differential integrator) that will produce an output voltage proportional to the integral of the difference of two input voltages.

87. Design a circuit that will sum two input voltages and produce an output voltage that is proportional to the sum.

88. For the Op Amp in figure 2.179, determine the output voltage ($v_o$) in terms of the input voltages ($v_1$ and $v_2$).

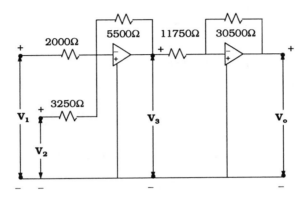

Figure 2.179:

89. For the Op Amp in figure 2.180, determine the output voltage ($v_o$) in terms of the input voltages ($v_1$ and $v_2$).

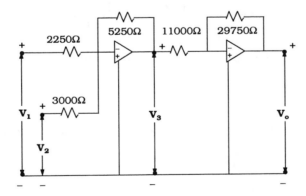

Figure 2.180:

90. * For the circuit shown in Figure 2.181, derive an approximate expression for the current $I$. Hint: Don't focus on $L$ or $R_2$. Draw a single idealized element that approximates all the elements except $L$ and $R_2$.

## 2.12 MAPLE PROGRAMS

This section includes the Maple files referenced in the text. Plots generated by Maple are displayed in the text, not in this section.

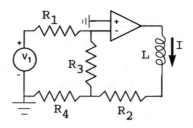

Figure 2.181:

## 2.12.1   File: ex7p1.ms - Maple Program for Example 2.1.

```
> restart;
> e1 := (vb + v) * i1 - vb * i3 - (vb + v)^2/r1 - vb^2/r3 - vb^2/r2 = 0:
> e2 := (vb + v)*i1 - i1^2 * r1 = 0:
> e3 := i1 - i3 = 0:
> ans := solve({e1,e2,e3},{i1,i3,vb});
```

$$ans := \left\{ i1 = 0,\ i3 = 0,\ vb = \text{RootOf}(\%1\_Z^2 + r3\ r2 + 2\ r3\ r2\ \_Z)\,v \right\},$$

$$\left\{ i3 = \frac{v}{r1},\ vb = 0,\ i1 = \frac{v}{r1} \right\},\ \left\{ vb = -\frac{r3\ r2\ v}{\%1},\ i3 = \frac{(r2 + r3)\,v}{\%1},\ i1 = \frac{(r2 + r3)\,v}{\%1} \right\}$$

$$\%1 := r3\ r2 + r1\ r2 + r1\ r3$$

Note that there are 3 answers, which one is correct? Well we know vb is not 0. so divide equation1 by vb and solve again.

```
> ee1 := e1/vb;
> aans := solve({ee1,e2,e3},{i1,i3,vb});
```

$$ee1 := \frac{(vb + v)\,i1 - vb\,i3 - \dfrac{(vb + v)^2}{r1} - \dfrac{vb^2}{r3} - \dfrac{vb^2}{r2}}{vb} = 0$$

$$aans := \left\{ i1 = 0,\ i3 = 0,\ vb = \text{RootOf}(\%1\_Z^2 + r3\ r2 + 2\ r3\ r2\ \_Z)\,v \right\},$$

$$\left\{ vb = -\frac{r3\ r2\ v}{\%1},\ i3 = \frac{(r2 + r3)\,v}{\%1},\ i1 = \frac{(r2 + r3)\,v}{\%1} \right\}$$

$$\%1 := r3\ r2 + r1\ r2 + r1\ r3$$

It eliminates one answer, which is correct? Use some numbers to make it look better.

```
> r1 := 10:   r2 := 20:   r3 := 30:   v:=10:
> ans;
> aans;
```

$$\left\{ i1 = 0,\ i3 = 0,\ vb = 10\,\text{RootOf}(11\_Z^2 + 6 + 12\_Z) \right\},\ \left\{ vb = 0,\ i3 = 1,\ i1 = 1 \right\},$$

$$\left\{ i1 = \frac{5}{11},\ vb = \frac{-60}{11},\ i3 = \frac{5}{11} \right\}$$

$$\left\{ i1 = 0,\ i3 = 0,\ vb = 10\,\text{RootOf}(11\_Z^2 + 6 + 12\_Z) \right\},\ \left\{ i1 = \frac{5}{11},\ vb = \frac{-60}{11},\ i3 = \frac{5}{11} \right\}$$

Ok the second solution in aans is the right one.

## 2.12.2   File: maplezap.ms - Maple Program for Example 2.7.

```
> restart;
> e1 := (V - Va(t))/R - C* diff(Va(t),t) = 0;
```

$$e1 := \frac{V - \text{Va}(t)}{R} - C\left(\frac{\partial}{\partial t}\,\text{Va}(t)\right) = 0$$

```
> e2 := (V - Va(t))/R - C*diff(Va(t),t) - Va(t)/Rl = 0;
```

$$e2 := \frac{V - \text{Va}(t)}{R} - C\left(\frac{\partial}{\partial t}\,\text{Va}(t)\right) - \frac{\text{Va}(t)}{Rl} = 0$$

The solution when closed.

```
>   closed := dsolve({e2,Va(0)=V},Va(t));
```

$$closed := \mathrm{Va}(t) = \dfrac{Rl\,V + \dfrac{V\,R\,e^{(-\frac{t\,(Rl+R)}{c\,R\,Rl})}\,Rl}{Rl+R} + \dfrac{V\,R^2\,e^{(-\frac{t\,(Rl+R)}{c\,R\,Rl})}}{Rl+R}}{Rl+R}$$

```
>   closesol := simplify(closed);
```

$$closesol := \mathrm{Va}(t) = \dfrac{V\,(Rl + e^{(-\frac{t\,(Rl+R)}{c\,R\,Rl})}\,R)}{Rl+R}$$

The initial condition when it opens is what it was closed at the time 0.02

```
>   openinit := subs(t=0.02,closesol);
```

$$openinit := \mathrm{Va}(.02) = \dfrac{V\,(Rl + e^{(-.02\,\frac{Rl+R}{c\,R\,Rl})}\,R)}{Rl+R}$$

The solution when switch is open.

```
>   opend := dsolve({e1,openinit},Va(t));
```

$$opend :=$$
$$\mathrm{Va}(t) = V - 1.\dfrac{e^{(-\frac{t}{c\,R})}\,V\,R\,(e^{(.02000000000\,\frac{Rl+R}{c\,R\,Rl})}-1.)\,e^{(\frac{.02000000000}{c\,R}-.02000000000\,\frac{Rl+R}{c\,R\,Rl})}}{Rl+R}$$

```
>   opensol := simplify(opend);
```

$$opensol :=$$
$$\mathrm{Va}(t) = -1.\dfrac{(-1.\,Rl - 1.\,R + R\,e^{(-.02000000000\,\frac{50.\,t-1}{C\,R})} - 1.\,R\,e^{(-.02000000000\,\frac{50.\,t\,Rl+R}{c\,R\,Rl})})\,V}{Rl+R}$$

Define a procedure that says the solution is closed < t=0.02 and open otherwise

```
>   sol := tau -> if tau<0.02 then
>   evalf(subs(t=tau,rhs(closesol)))
>   else evalf(subs(t=tau,rhs(opensol))) fi;
```

$$sol := \mathbf{proc}(\tau)$$
$$\quad \mathbf{option}\,operator,\ arrow;$$
$$\quad\quad \mathbf{if}\ \tau < .02\ \mathbf{then}\ \mathrm{evalf}(\mathrm{subs}(t = \tau, \mathrm{rhs}(closesol)))\ \mathbf{else}\ \mathrm{evalf}(\mathrm{subs}(t = \tau, \mathrm{rhs}(opensol)))\ \mathbf{fi}$$
$$\quad \mathbf{end}$$

Let's look at the solution for some specific values.

```
>   Rl := 10:R := 100:V := 5:C := .0001:
>   plot(sol,0..0.06);
```

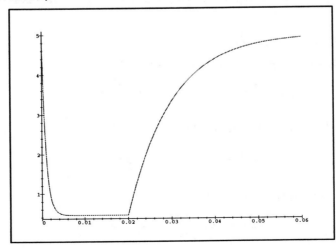

Determine the currents.

```
>  i1 := tau -> (V - sol(tau) )/R;
>  i2 := tau -> if tau < .02 then (sol(tau)/Rl) else 0 fi;;
>  i3 := tau -> i1(tau) - i2(tau);
```

$$i1 := \tau \to \frac{V - \text{sol}(\tau)}{R}$$

$$i2 := \mathbf{proc}(\tau)\,\mathbf{option}\,operator,\ arrow;\ \mathbf{if}\ \tau < .02\ \mathbf{then}\ \text{sol}(\tau)/Rl\ \mathbf{else}\ 0\ \mathbf{fi}\ \mathbf{end}$$

$$i3 := \tau \to \text{i1}(\tau) - \text{i2}(\tau)$$

PLot all the currents on a single axis.

```
>  with(plots):
>  pi1 := plot(i1,0..0.06):
>  pi2 := plot(i2,0..0.06):
>  pi3 := plot(i3,0..0.06):
>  display({pi1,pi2,pi3});
```

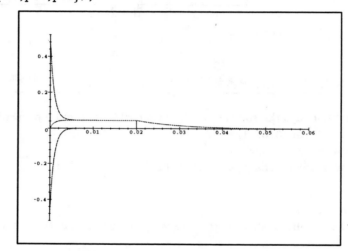

## 2.12.3   File: ex7p3.ms - Maple Program for Example 2.10.

```
>  restart;
```

The (t) means it is a function of t (time)

```
>  e1 := Vcc - i(t) * R = L * diff(i(t),t);
```

$$e1 := Vcc - \text{i}(t)\,R = L\left(\frac{\partial}{\partial t}\,\text{i}(t)\right)$$

```
>  closed := dsolve({e1,i(0)=0},i(t));
```

$$closed := \text{i}(t) = \frac{Vcc - e^{(-\frac{Rt}{L})}\,Vcc}{R}$$

```
>  closesol := rhs(simplify(closed));
```

$$closesol := -\frac{Vcc\,(-1 + e^{(-\frac{Rt}{L})})}{R}$$

```
>  R := 10:L := 1:Vcc := 5:
>  plot(closesol,t=0..0.6);
```

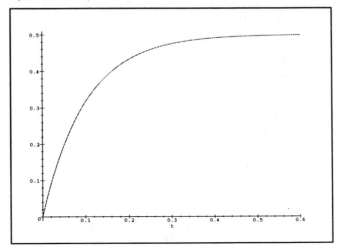

### 2.12.4   File: clampms.ms - Maple Program for Example 2.24.

```
>  e1 := -vf - r*i(t) = L * diff(i(t),t);
```

$$e1 := -vf - r\,\mathrm{i}(t) = L\left(\frac{\partial}{\partial t}\,\mathrm{i}(t)\right)$$

```
>  sol := dsolve({e1,i(0)=vcc/r},i(t));
```

$$sol := \mathrm{i}(t) = \frac{-vf + e^{(-\frac{r\,t}{L})}\,(vf + vcc)}{r}$$

```
>  nsol := subs(r=10,L=1,vf=7/10,vcc=5,rhs(sol));
```

$$nsol := -\frac{7}{100} + \frac{57}{100}\,e^{(-10\,t)}$$

```
>  plot(nsol,t=0..1.5);
```

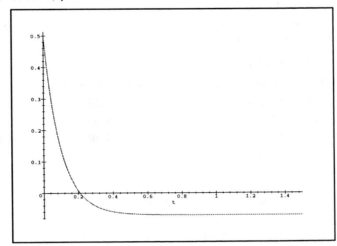

### 2.12.5   File: 2caps.ms - Maple Program for Example 2.9.

```
>  restart;with(plots):
```

Set some values up front so the algebra looks better.

```
>  R1 := 1:  R2 := 2:  C1 := 1/1000000:  C2 := 2/1000000:
```

The following says E is ground.

```
>  VE:=0;
```

$$VE := 0$$

Equations 1 and 2 are KVL, 3 and 4 are the "cheater" equations, 5 6 and 7 are nodes.
Notice that the (t) are put on so Maple knows they change with time.

```
>   e1 := V - R1 *I1(t) - (VB(t)-VE) = 0;
>   e2 := (VB(t)-VE) - R2 *I2(t) - (VD(t)-VE ) = 0;
>   e3 := I3(t) = C1 *diff(VB(t) - VE,t);
>   e4 := I4(t) = C2 * diff(VD(t) - VE,t);
>   e5 := I5(t) - I1(t) = 0;
>   e6 := I1(t) - I3(t) - I2(t) = 0;
>   e7 := I2(t) - I4(t) = 0;
```

$$e1 := V - \text{I1}(t) - \text{VB}(t) = 0$$

$$e2 := \text{VB}(t) - 2\,\text{I2}(t) - \text{VD}(t) = 0$$

$$e3 := \text{I3}(t) = \frac{1}{1000000}\,(\frac{\partial}{\partial t}\,\text{VB}(t))$$

$$e4 := \text{I4}(t) = \frac{1}{500000}\,(\frac{\partial}{\partial t}\,\text{VD}(t))$$

$$e5 := \text{I5}(t) - \text{I1}(t) = 0$$

$$e6 := \text{I1}(t) - \text{I3}(t) - \text{I2}(t) = 0$$

$$e7 := \text{I2}(t) - \text{I4}(t) = 0$$

The following statement was "built up" piece by piece.
First we substitute V=10 into the equations.
Second we added the initial conditions VB(0) = 0 etc.
Third we ask Maple to solve the differential equations.
Fourth we ask for it as a number so it will "clean up" the algebra a little.
Last we set the answer equal to ans.

```
>   ans := evalf(dsolve(subs(V=10, {e1,e2,e3,e4,e5,e6,e7,VB(0)=0,VD(0)=0}), {I1(t),
I2(t), I3(t), I4(t), I5(t), VD(t),VB(t)}));
```

$ans := \{\text{I5}(t) = 4.12961173\,\%2 + 5.870388276\,\%1,$
$\quad \text{VD}(t) = -11.09271796\,\%2 + 1.092717958\,\%1 + 10.00000001,$
$\quad \text{I2}(t) = 3.48155312\,\%2 - 3.481553118\,\%1,\ \text{I1}(t) = 4.12961173\,\%2 + 5.870388276\,\%1,$
$\quad \text{VB}(t) = -4.12961173\,\%2 - 5.870388276\,\%1 + 10.00000001,$
$\quad \text{I4}(t) = 3.48155312\,\%2 - 3.481553118\,\%1,\ \text{I3}(t) = .6480586023\,\%2 + 9.351941398\,\%1\}$
$\quad \%1 := e^{(-.1593070331\,10^7\,t)}$
$\quad \%2 := e^{(-156929.6691\,t)}$

You can see the currents and voltages above but this simplifies it a little.

```
>   subs(ans,I5(t));
```

$$4.12961173\,e^{(-156929.6691\,t)} + 5.870388276\,e^{(-.1593070331\,10^7\,t)}$$

Okay the next commands look ugly. I played around with it until I got the plot to look the way
I wanted. A simple plot command would work but wouldn't look as pretty! :)

```
>   pvb := plot(subs(t=tau/100000, subs(ans,VB(t))), tau=0..3,
>   title='VB Solid, VD Dashed in Volts', axesfont=[TIMES,ROMAN,10], font=[TIMES,ROMAN,14],
labels =['Time X 100,000 Sec','']  , labelfont=[TIMES,ROMAN,10]):
>   pvd := plot(subs(t=tau/100000,subs(ans,VD(t))), tau=0..3, linestyle=2):
>   display(pvb,pvd);
```

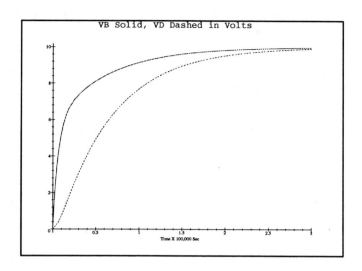

Now do the same for the sinusoidal source. I wasn't sure what "frequency" I wanted so I made it a variable T so I could change it easily. Note that the solution statement is practically the same as the last one.

```
>  T := 1000000:
>  ans := evalf(dsolve(subs(V=10*sin(T*t), {e1,e2,e3,e4,e5,e6,e7,VB(0)=0,VD(0)=0}),
{I1(t), I2(t), I3(t), I4(t), I5(t), VD(t),VB(t)})):
```

The plots are about the same as before. A simple plot command would work.

```
>  pvb := plot(subs(t=tau/T, subs(ans,VB(t))), tau=0..8,
>  title='VB Solid, VD Dashed in Volts', axesfont=[TIMES,ROMAN,10], font=[TIMES,ROMAN,14]
labels =['Time X 100,000 Sec',''], labelfont=[TIMES,ROMAN,10]):
>  pvd := plot(subs(t=tau/T,subs(ans,VD(t))), tau=0..8, linestyle=2):
>  display(pvb,pvd);
```

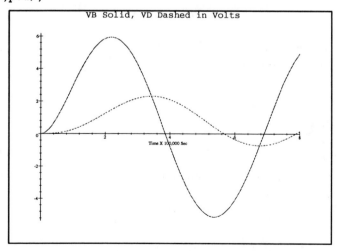

This is the current through the sinusoidal source.

```
>  simplify(subs(ans,I5(t)));
```

$$-.6324825060\, e^{(-156929.6691\, t)} - 2.643379565\, e^{(-.1593070331\, 10^7\, t)}$$
$$+ 3.275862068 \cos(.1000000\, 10^7\, t) + 5.689655173 \sin(.1000000\, 10^7\, t)$$

## 2.12.6   File: transformer.ms - Maple Program for Example 2.17.

This is file Transformer.mws

```
>  restart;
>  e1 := v(t) - R1*i1(t) - L1*diff(i1(t),t) - M*diff(i2(t),t) = 0;
>  e2 := -L2*diff(i2(t),t) - M*diff(i1(t),t) - R2*i2(t) = 0;
>  e3 := V(t) = R2*i2(t);
```

$$e1 := v(t) - R1\,i1(t) - L1\,(\frac{\partial}{\partial t}\,i1(t)) - M\,(\frac{\partial}{\partial t}\,i2(t)) = 0$$

$$e2 := -L2\,(\frac{\partial}{\partial t}\,i2(t)) - M\,(\frac{\partial}{\partial t}\,i1(t)) - R2\,i2(t) = 0$$

$$e3 := V(t) = R2\,i2(t)$$

Make some definitions, define the inductances in terms of the number of windings.

```
>  v(t) := sin(omega*t);L2 := N2^2*L; L1 := N1^2*L; M := N1*N2*L;
```

$$v(t) := \sin(\omega t)$$

$$L2 := N2^2\,L$$

$$L1 := N1^2\,L$$

$$M := N1\,N2\,L$$

Solve the equations, we are only concerned with the steady state. We don't care about the initial conditions so leave them off. Its ugly so we don't print it.

```
>  ans := dsolve({e1,e2},{i1(t),i2(t)}):
```

All we really want is the voltage so plug the solution into equation 3. Its ugly.

```
>  a := solve(simplify(subs(ans,e3)),V(t)):
```

Maple uses _C1 and _C2 since it was not given initial conditions. Set these to zero to get rid of the initial condition garbage.

```
>  temp := simplify(subs({_C1=0,_C2=0},a));
```

$$temp := -\frac{N1\,N2\,L\,\omega\,(R2\,R1\cos(\omega t) + \omega\,L\sin(\omega t)\,N2^2\,R1 + \omega\,L\sin(\omega t)\,R2\,N1^2)\,R2}{R2^2\,R1^2 + \omega^2\,L^2\,N2^4\,R1^2 + 2\,\omega^2\,L^2\,N2^2\,R1\,R2\,N1^2 + \omega^2\,L^2\,R2^2\,N1^4}$$

Notice that if R1 is zero, then it simplifies to the simple equations.

```
>  subs({R1=0},temp);
```

$$-\frac{N2\sin(\omega t)}{N1}$$

Pull off the numerator and expand it. We are going to find the magnitude of the voltage.

```
>  n := expand(numer(temp));
```

$$n :=$$
$$-L\,N1\,N2\,\omega\,R2^2\,R1\cos(\omega t) - R2\,L^2\,N1\,N2^3\,\omega^2\sin(\omega t)\,R1 - L^2\,N1^3\,N2\,\omega^2\sin(\omega t)\,R2^2$$

Pull off the denominator.

```
>  d := denom(temp);
```

$$d := R2^2\,R1^2 + \omega^2\,L^2\,N2^4\,R1^2 + 2\,\omega^2\,L^2\,N2^2\,R1\,R2\,N1^2 + \omega^2\,L^2\,R2^2\,N1^4$$

Get the cosine and sine terms from the numerator.

```
>  cn := simplify(subs(t=0,n));
>  sn := simplify(subs(t=Pi/(2*omega),n));
```

$$cn := -L\,N1\,N2\,\omega\,R2^2\,R1$$

$$sn := -R2\,L^2\,N1\,N2^3\,\omega^2\,R1 - L^2\,N1^3\,N2\,\omega^2\,R2^2$$

The magnitude is the square root of the cosines squared plus sines squared.

```
>  mag := simplify(sqrt(cn^2+sn^2)/d);
```

$$mag :=$$
$$(L^2\,N1^2\,N2^2\,\omega^2\,R2^2\,(R2^2\,R1^2 + \omega^2\,L^2\,N2^4\,R1^2 + 2\,\omega^2\,L^2\,N2^2\,R1\,R2\,N1^2 + \omega^2\,L^2\,R2^2\,N1^4))$$
$$^{1/2}\Big/(R2^2\,R1^2 + \omega^2\,L^2\,N2^4\,R1^2 + 2\,\omega^2\,L^2\,N2^2\,R1\,R2\,N1^2 + \omega^2\,L^2\,R2^2\,N1^4)$$

Put in some numbers to make it easier to look at.

```
>   mag2 := subs({N1=1,N2=100}, subs({L=1/N1^2,omega=60*2*Pi},mag));
>   evalf(limit(subs(R1=1,mag2),R2=infinity));
```

$$mag2 := \sqrt{14400}\,\sqrt{10000}$$
$$\frac{}{\sqrt{\pi^2\,R2^2\,(R2^2\,R1^2 + 1440000000000\,\pi^2\,R1^2 + 288000000\,\pi^2\,R1\,R2 + 14400\,\pi^2\,R2^2)}} \Big/ \Big($$
$$R2^2\,R1^2 + 1440000000000\,\pi^2\,R1^2 + 288000000\,\pi^2\,R1\,R2 + 14400\,\pi^2\,R2^2\Big)$$

$$99.99964817$$

Observe the effect of 1 ohm on R1.

```
>   plot(subs(R1=1,mag2),R2=0..100000,
>   title='VA Versus R2 With R1 = 1');
```

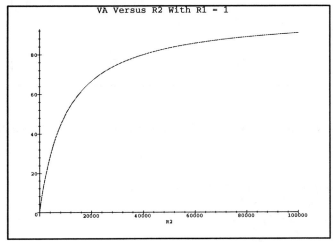

Increasing R1 just makes it worse.

```
>   plot(subs(R1=10,mag2),R2=0..1000000,
>   title='VA Versus R2 With R1 = 10');
```

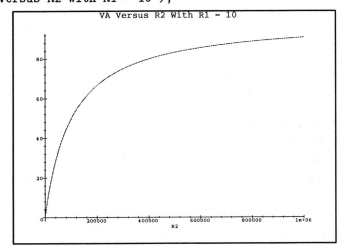

```
>   power := mag2^2/R2;
```

$$power :=$$
$$144000000\,\frac{\pi^2\,R2}{R2^2\,R1^2 + 1440000000000\,\pi^2\,R1^2 + 288000000\,\pi^2\,R1\,R2 + 14400\,\pi^2\,R2^2}$$

Observe the effect of R2 on power.
```
>    plot(subs(R1=1,power),R2=0..20000,color='BLAC K',
>    title='Power Dissipation Versus Load');
```

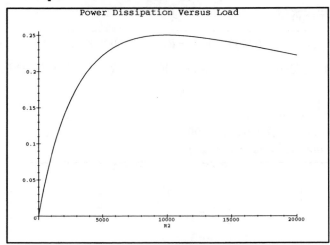

### 2.12.7    File: lowpass.ms - Maple Program for a LowPass filter.

This is low pass.
```
>    restart;
```
You could use KVL but this is the impedance relations.
```
>    e1 := v*sin(omega*t) - va(t) = r*ir;
>    e2 := ic = c*diff(va(t),t);
>    e3 := va(t) = i*rl;
```
This is conservation of charge.
```
>    e4 := ir-ic-i = 0;
```

$$e1 := v\sin(\omega\,t) - \mathrm{va}(t) = r\,ir$$

$$e2 := ic = c\,(\frac{\partial}{\partial t}\,\mathrm{va}(t))$$

$$e3 := \mathrm{va}(t) = i\,rl$$

$$e4 := ir - ic - i = 0$$

Solve the algebra equations first.
```
>    ans := solve({e1,e3,e4},{ir,ic,i});
```

$$ans := \{i = \frac{\mathrm{va}(t)}{rl},\ ir = -\frac{-v\sin(\omega\,t) + \mathrm{va}(t)}{r},\ ic = \frac{rl\,v\sin(\omega\,t) - rl\,\mathrm{va}(t) - \mathrm{va}(t)\,r}{r\,rl}\}$$

Now solve the differential equation.
```
>    difeq := subs(ans,e2);
>    dans := simplify(dsolve(difeq,va(t)));
```

$$difeq := \frac{rl\,v\sin(\omega\,t) - rl\,\mathrm{va}(t) - \mathrm{va}(t)\,r}{r\,rl} = c\,(\frac{\partial}{\partial t}\,\mathrm{va}(t))$$

$$dans := \mathrm{va}(t) = \left(-c\,r\,v\,\omega\,rl^2\cos(\omega\,t)\,e^{(\frac{t\,(rl+r)}{c\,r\,rl})} + v\,rl^2\sin(\omega\,t)\,e^{(\frac{t\,(rl+r)}{c\,r\,rl})}\right.$$

$$+ r\,v\,rl\sin(\omega\,t)\,e^{(\frac{t\,(rl+r)}{c\,r\,rl})} + \_C1\,rl^2 + 2\,\_C1\,r\,rl + \_C1\,r^2 + \_C1\,\omega^2\,c^2\,r^2\,rl^2)e^{(-\frac{t\,(rl+r)}{c\,r\,rl})}$$

$$\left./\,(rl^2 + 2\,r\,rl + r^2 + \omega^2\,c^2\,r^2\,rl^2)\right.$$

We only want the steady state so throw away the initial conditions.
```
>    temp := simplify(rhs(subs({_C1=0},dans)));
```

$$temp := \frac{rl\,v\,(-c\,r\,\omega\,rl\cos(\omega\,t) + rl\sin(\omega\,t) + r\sin(\omega\,t))}{rl^2 + 2\,r\,rl + r^2 + \omega^2\,c^2\,r^2\,rl^2}$$

Find the magnitude of the answer.
```
>  n := numer(temp); d := denom(temp);
>  cn := simplify(subs(t=0,n));
>  sn := simplify(subs(t=Pi/(2*omega),n));
>  mag := simplify(sqrt(cn^2+sn^2)/d);
```
$$n := -rl\,v\,(cr\,\omega\,rl\cos(\omega\,t) - rl\sin(\omega\,t) - r\sin(\omega\,t))$$

$$d := rl^2 + 2\,r\,rl + r^2 + \omega^2\,c^2\,r^2\,rl^2$$

$$cn := -rl^2\,v\,c\,r\,\omega$$

$$sn := rl\,v\,(rl + r)$$

$$mag := \frac{\sqrt{rl^2\,v^2\,(rl^2 + 2\,r\,rl + r^2 + \omega^2\,c^2\,r^2\,rl^2)}}{rl^2 + 2\,r\,rl + r^2 + \omega^2\,c^2\,r^2\,rl^2}$$

Simplify so it is easier to look at by letting Rl be infinite.
```
>  temp := limit(mag,rl=infinity);
```
$$temp := \frac{\sqrt{v^2 + v^2\,\omega^2\,c^2\,r^2}}{1 + \omega^2\,c^2\,r^2}$$

Notice that what is important is omega*c*r.
```
>  subs({v=1,omega=Omega/(r*c)},temp);
```
$$\frac{1}{\sqrt{1 + \Omega^2}}$$

```
>  plot(",Omega=0..100);
```

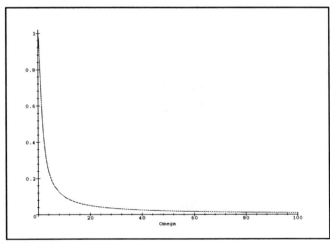

Notice that if omega*c*r > 20 there is less than 10% of the voltage V appearing at node a. Also note that if omega = 0, 100% of the voltage shows up at node a

### 2.12.8    File: highpas.ms - Maple Program for a High Pass filter.

This is high pass.
```
>  restart;
```
You could use KVL but this is the impedance relations.
```
>  thev := v*sin(omega*t);
>  e1 := thev = r*ir;
>  e2 := ic = c*diff(thev-va(t),t);
>  e3 := va(t) = ic*rl;
```
$$thev := v\sin(\omega\,t)$$

$$e1 := v\sin(\omega\,t) = r\,ir$$

$$e2 := ic = c\left(v\cos(\omega t)\,\omega - \left(\frac{\partial}{\partial t}\,\text{va}(t)\right)\right)$$

$$e3 := \text{va}(t) = ic\,rl$$

Solve the algebra equations first.

```
> ans := solve({e1,e3},{ir,ic});
```

$$ans := \{ic = \frac{\text{va}(t)}{rl},\ ir = \frac{v\sin(\omega t)}{r}\}$$

Now solve the differential equation.

```
> difeq := subs(ans,e2);
```

$$difeq := \frac{\text{va}(t)}{rl} = c\left(v\cos(\omega t)\,\omega - \left(\frac{\partial}{\partial t}\,\text{va}(t)\right)\right)$$

```
> dans := simplify(dsolve(difeq,va(t)));
```

$$dans :=$$
$$\text{va}(t) = \frac{v\,\omega\,c\,rl\cos(\omega t) + v\,\omega^2\,c^2\,rl^2\sin(\omega t) + \_C1\,e^{(-\frac{t}{crl})} + \_C1\,\omega^2\,c^2\,rl^2\,e^{(-\frac{t}{crl})}}{1 + \omega^2\,c^2\,rl^2}$$

We only want the steady state so throw away the initial conditions.

```
> temp := simplify(rhs(subs({_C1=0},dans)));
```

$$temp := \frac{v\,\omega\,c\,rl\,(\cos(\omega t) + \omega\,c\,rl\sin(\omega t))}{1 + \omega^2\,c^2\,rl^2}$$

Find the magnitude of the answer.

```
> n := numer(temp); d := denom(temp);
> cn := simplify(subs(t=0,n));
> sn := simplify(subs(t=Pi/(2*omega),n));
> mag := simplify(sqrt(cn^2+sn^2)/d);
```

$$n := v\,\omega\,c\,rl\,(\cos(\omega t) + \omega\,c\,rl\sin(\omega t))$$
$$d := 1 + \omega^2\,c^2\,rl^2$$
$$cn := v\,\omega\,c\,rl$$
$$sn := v\,\omega^2\,c^2\,rl^2$$
$$mag := \frac{\sqrt{v^2\,\omega^2\,c^2\,rl^2\,(1 + \omega^2\,c^2\,rl^2)}}{1 + \omega^2\,c^2\,rl^2}$$

Well I'll be, the R plays no role. We could leave it out.
Notice that what is important is omega*c*rl.

```
> subs({v=1,omega=Omega/(rl*c)},mag);
```

$$\frac{\sqrt{\Omega^2\,(1 + \Omega^2)}}{1 + \Omega^2}$$

```
> plot(",Omega=0..10);
```

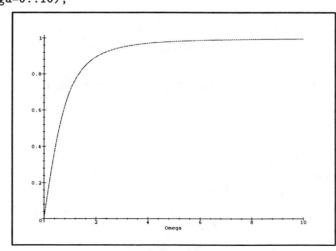

Notice that if omega*c*rl > 2 there is more than 90% of the voltage V appearing at node a. Also note that if omega = 0, 0% of the voltage shows up at node a.

### 2.12.9   File: voltagedivider.ms - Maple Program for a Voltage Divider.

This is voltage divider.

```
>  restart;
```

You could use KVL, we use the resistor's impedance relationships.

```
>  e1 := V-Va = R1*i1;
```
$$e1 := V - Va = R1\ i1$$

```
>  e2 := Va = R2*i2;
```
$$e2 := Va = R2\ i2$$

This is the conservation of charge at a.

```
>  e3 := i1 - ia - i2 = 0;
```
$$e3 := i1 - ia - i2 = 0$$

```
>  ans := solve({e1,e2,e3},{i1,i2,Va});
```
$$ans := \{i2 = \frac{V - R1\ ia}{R2 + R1},\ Va = \frac{R2\ (V - R1\ ia)}{R2 + R1},\ i1 = \frac{ia\ R2 + V}{R2 + R1}\}$$

```
>  va := subs(ans,Va);
```
$$va := \frac{R2\ (V - R1\ ia)}{R2 + R1}$$

Pick some convenient values.

```
>  temp := subs({V=2,R2 = R1},va);
```
$$temp := 1 - \frac{1}{2}\ R1\ ia$$

Note from above that the "designed" voltage at a is 1 volt and that ia causes it to drop.

```
>  with(plots):
>  r1 := plot(subs(R1=1,temp),ia = 0..1,color='BLACK'):
>  r2 := plot(subs(R1=2,temp), ia = 0..1,color='BLACK', linestyle=2):
```

Note from the plot that larger R1 makes it worse.

```
>  display({r1,r2}, font=['TIMES','ROMAN',12]);
```

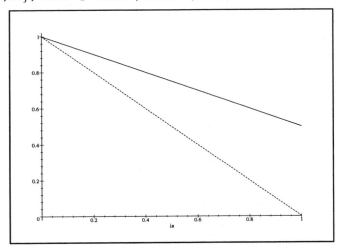

### 2.12.10   File: resonant.ms - Maple Program for a Resonant Circuit.

This is resonant.

```
>  restart;
```

You could use KVL but this is the impedance relations. This is a 2 NIA system.

```
>  e1 := v*sin(omega*t) - vb(t) = L*diff(i(t),t);
>  e2 := i(t) = c*diff(vb(t),t);
```

$$e1 := v\sin(\omega t) - \text{vb}(t) = L\left(\frac{\partial}{\partial t}\,\text{i}(t)\right)$$

$$e2 := \text{i}(t) = c\left(\frac{\partial}{\partial t}\,\text{vb}(t)\right)$$

Now solve the differential equation.

```
>  dans := dsolve({e1,e2},{i(t),vb(t)});
```

$$\begin{aligned}
\textit{dans} := \{\,\text{i}(t) = &\Big(-\tfrac{1}{2}\_C1\sqrt{-Lc}\,e^{(\frac{\sqrt{-Lc}\,t}{Lc})} + \tfrac{1}{2}\_C1\sqrt{-Lc}\,e^{(\frac{\sqrt{-Lc}\,t}{Lc})}\,\omega^2 Lc \\
&-\tfrac{1}{2}\_C1\sqrt{-Lc}\,\%1 + \tfrac{1}{2}\_C1\sqrt{-Lc}\,\%1\,\omega^2 Lc - \tfrac{1}{2}c\_C2\,\%1 + \tfrac{1}{2}c^2\_C2\,\%1\,\omega^2 L \\
&+\tfrac{1}{2}c\_C2\,e^{(\frac{\sqrt{-Lc}\,t}{Lc})} - \tfrac{1}{2}c^2\_C2\,e^{(\frac{\sqrt{-Lc}\,t}{Lc})}\,\omega^2 L - cv\omega\cos(\omega t)\sqrt{-Lc}\Big)\Big/\big(\sqrt{-Lc} \\
&(-1+\omega^2 Lc)\big),\ \text{vb}(t) = \Big(\tfrac{1}{2}L\_C1\,\%1 - \tfrac{1}{2}L^2\_C1\,\%1\,\omega^2 c - \tfrac{1}{2}L\_C1\,e^{(\frac{\sqrt{-Lc}\,t}{Lc})} \\
&+\tfrac{1}{2}L^2\_C1\,e^{(\frac{\sqrt{-Lc}\,t}{Lc})}\,\omega^2 c - \tfrac{1}{2}\_C2\sqrt{-Lc}\,e^{(\frac{\sqrt{-Lc}\,t}{Lc})} + \tfrac{1}{2}\_C2\sqrt{-Lc}\,e^{(\frac{\sqrt{-Lc}\,t}{Lc})}\,\omega^2 Lc \\
&-\tfrac{1}{2}\_C2\sqrt{-Lc}\,\%1 + \tfrac{1}{2}\_C2\sqrt{-Lc}\,\%1\,\omega^2 Lc - v\sin(\omega t)\sqrt{-Lc}\Big)\Big/\big(\sqrt{-Lc}\,(-1+\omega^2 Lc)\big) \\
&\}
\end{aligned}$$

$$\%1 := e^{(-\frac{\sqrt{-Lc}\,t}{Lc})}$$

We only want the steady state so throw away the initial conditions.

```
>  temp := simplify(subs({_C1=0,_C2=0},dans));
```

$$temp := \Big\{\,\text{vb}(t) = -\frac{v\sin(\omega t)}{-1+\omega^2 Lc},\ \text{i}(t) = -\frac{cv\omega\cos(\omega t)}{-1+\omega^2 Lc}\Big\}$$

Find the magnitude of the answer.

```
>  mag := simplify(subs(t=Pi/(2*omega), subs(temp,vb(t))));
```

$$mag := -\frac{v}{-1+\omega^2 Lc}$$

Notice that what is important is omega^2*c*L.

```
>  ans := abs(subs({v=1,L=1,c=1},mag));
```

$$ans := \frac{1}{|-1+\omega^2|}$$

Notice that if omega^2 = CL the magnitude is infinite.

```
>  with(plots):
>  small := plot(ans,omega=0..0.9):
>  big := plot(ans,omega=1.1..3):
>  display({small,big});
```

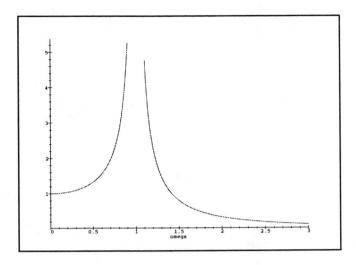

# Chapter 3

# MECHANICS

For this chapter, we assume you are somewhat familiar with the concept of linear and angular momentum and how to write basic conservation equations for these properties. The purpose of this chapter is to develop a set of heuristics which you can follow to analyze more complex problems.

We will use a six step heuristic for attacking the problems in this chapter. The heuristic is given in table 3.1. We will build on each step throughout the chapter. You might want to list these steps in your notes or put a bookmark here.

Occasionally you will face problems in which you do not have enough equations to solve for the unknowns. When this happens the problem may be ill posed or ill conditioned which has been discussed in the book's introduction or it may be that you need to "find other equations". The following table lists some typical ways of finding these equations. Although some of the steps may be foreign to you now, we will cover each of them as we move through the chapter. Again put these steps in your notes or use another bookmark.

## 3.1 FORCES AND MOMENTS

**Objectives**
*When you complete this section, you should be able to*

1. *Explain why forces and moments can only occur between two objects.*

2. *Determine all points on a body where a force or moment might exist.*

3. *Calculate the moment caused by a force.*

4. *Explain why two objects applying a force or moment to each other apply them equally and opposite to each other.*

---

There are only two ways momentum can enter or leave a system: (1) mass flow, or (2) applied forces/torques. This section discusses the origins of forces and moments and may be a bit of a review for you.

The key principles of most mechanic's problems are momentum conservation (linear and angular) and mechanical energy accounting. Forces and moments play an important role in these principles. For this reason, it is important to understand how forces and moments originate. There are several excellent physics books that can be consulted for a more philosophical meaning of force [13]. For our work, we will consider a force to represent a momentum exchange. The units of force are momentum per time, in other words, a rate of momentum.

For the extent of our work in this text, forces can arise in only two ways. They can be applied through contact between bodies or through action at a distance phenomena (*e.g.* gravity, magnetics, and electrostatics). If we view forces as momentum in transit, all forces (even action at a distance forces) must occur between bodies because momentum is a conserved property (ignoring relativistic effects). In other words, all forces on a body are caused by another body. They never just happen.

The six basic steps we will follow in this chapter are the following:

1. Organize Your Thoughts.

    (a) Develop a plan of attack. What is given, what is required.
    (b) Identify what motions you will worry about and what you will ignore.
    (c) What assumptions need to be made.
    (d) **Use the [NIF] and [NIA] to assist with this.**

2. Choose Coordinates.

    (a) Define enough coordinates for the greatest [NIF].
    (b) Draw a picture showing the positive directions of the coordinates.

3. Define the System.

    (a) Choose a body or bodies.
    (b) Determine the inputs/outputs.
    (c) Choose a time period.

4. Apply a Conservation Equation if Needed.

5. Find "Extra" Equations.

    (a) If the number of flows exceeds the [NIF], write kinematic relations.
    (b) Look for other systems to choose.

6. Solve and Interpret.

    (a) Verify assumptions.
    (b) Check that solution matches your intuition. If it doesn't determine if your intuition needs to be updated or the solution is incorrect.
        i. Attempt to solve the problem another way.
        ii. Solve a simplified version of the problem to gain insight to what is happening.

Table 3.1: The Problem Solving Process.

## Steps for Finding Additional Equations

1. Look for an additional conservation equation for the current system. If you have written linear momentum for a block, think about the angular momentum or mass or...

2. If the equations have more flow variables than [NIF], then there IS a kinematic relation between the extra flows. There is a complete section devoted to writing these relations. If our equations have $n$ flow variables and $m$ [NIF], then we can eliminate $n-m$ flow variables using kinematics. As another example suppose there are 4 [NIF], 7 flow variables in the equations and 2 of the 7 flow variables are known. What this means is there NEED ONLY BE 4 (the [NIF]) flow variables in the equations. There are actually 7 in use so there are 7-4 = 3 "too many" flow variables. Which further means that, provided the problem is well defined, there MUST BE a way to relate these 3 "extras" to the other 4.

3. Choose another system and express conservation equations for it. A reasonable heuristic is to choose one system for each mass or object.

4. Look for information that may have been overlooked in the problem statement.

Table 3.2: Steps for Finding Additional Equations.

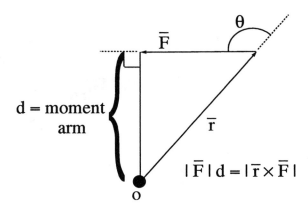

Figure 3.1: Vectors Used in Calculating the Moment of Force $\vec{F}$ About Point O

Furthermore, interacting bodies exert equal and opposite forces on each other. This concept is traditionally thought of as Newton's third law. If we view force as an exchange of momentum then the forces must be equal and opposite because whatever momentum is given up by one body is received by the other because momentum is conserved.

Contact forces can only be transmitted through contact between bodies. Contrary to what an illusionist wants us to believe, we cannot apply a force to an object merely by waving our hand over it. Even action at a distance forces occur between two objects. For example, the Earth's gravity are forces (think of invisible hands) that reach up from every particle of the Earth and grasp every particle of an object.[1] Likewise, we can visualize the other action at a distance forces in the same way.

All forces produce moments (sometimes called torques). A moment is a tendency to twist or rotate. The amount of twist a force produces depends on two things, the force (size and direction), and the location of the force relative to the point about which you're finding the moment. We can calculate the moment produced by a force using $\vec{r} \times \vec{F}$ where $\vec{F}$ is a vector expression for the force, and $\vec{r}$ is a vector whose tail touches the point that we are calculating the moment about and whose tip touches the force (see figure 3.1). This mathematical definition of moments is convenient for three dimensional calculations. However, it is sometimes easier to calculate two dimensional moments geometrically. By the definition of cross product, the size of the moment of a force is $|\vec{r}||\vec{F}|\sin\theta$. From the figure, we see that $|\vec{r}|\sin\theta = d$ where $d$ is the moment arm – the perpendicular distance from the point. Thus, we can express the size of the moment as $|\vec{F}|d$. Notice that $d$ is the **perpendicular** distance from the point the moment is about to the **line of action of the force**. Most students use the wrong value of $d$.

The direction of a moment vector is always perpendicular to the plane containing both the force and the position vector. If you use the cross product definition of the moment, the mathematical result will provide you with the correct direction. To find the direction of the moment when you use the force and moment arm calculation first find the plane containing both the force and the position vector. In two dimensional problems this plane is the paper on which the force and position vectors are drawn. The moment points perpendicular to this plane. In two dimensions, the moment is either out of the paper (counterclockwise) or into the paper (clockwise). To determine if the moment is into or out of this plane, we use the right hand rule. To do this, draw the force and position on paper. Next extend your right hand placing your right wrist at point O (the point you want the moment about) and curl your fingers in the general direction of the force. Your right thumb naturally points in the vector direction of the moment. The direction of your curved fingers is the actual twist that the moment generates.

A moment is an exchange of angular momentum between two systems. It represents angular momentum flow rate. Accordingly, a moment has the units of angular momentum per time.

$$[\text{mass}][\text{length}]^2/[\text{time}]^2.$$

---

[1]Although we often approximate gravitational forces as occurring between the centers of mass of two objects, in reality each particle of one object pulls each particle of the other object.

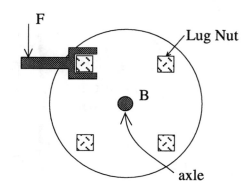

**Analysis Assumptions**

(1) B is the point of contact between the axle and the wheel.

(2) The lugs are right handed.

(3) $F$ is the force that we apply to the wrench.

Figure 3.2: (autowheel): An Automobile Wheel

## Example 3.1

### Changing Flat Tires

**QUESTION:**

We want to remove the wheel in figure 3.2 from our car. To do this, we have to unscrew the four lug nuts. Not having enough strength with our bare hands, we place a wrench on one of the lugnuts as the figure shows. By pushing on the wrench with a force of $F$, we generate a moment (a desire to twist or rotate) about the lug nut. The wrench allows us to generate a large moment by giving us a larger moment arm (than our bare hands). The question is will the lug nut loosen.

**ANSWER:**

To answer this, we must determine the magnitude and direction of the twist applied by the wrench. As the figure shows, the wrench would tend to rotate the lug counterclockwise (practice using the right hand rule), since the twist is counterclockwise, the moment vector points out of the page. Assuming the lug is right handed, the twist tendency is in the correct direction.

Next, we have to guarantee that the magnitude of the twist can overcome the friction force moment holding the lug nut in place. The wrench is not the only thing in contact with the lug nut. The lug nut also touches the wheel. Thus, the friction between the wheel and nut can also produce a moment that resists the moment of the wrench. If the wrench moment exceeds the friction moment, the nut will loosen.

If the nut does not loosen, we must increase the wrench moment until it overcomes the friction moment. We can do this either by increasing the force we apply to the wrench or by increasing the length of moment arm. The latter is the reason why a wrench with a long handle feels easier to use than a wrench with a short handle. It requires less force to produce the same moment. The longer handle gives a larger perpendicular distance (a larger moment arm) to an applied force. Thus, the longer handle produces a larger moment given a known force.

The moment generated by the wrench may affect more than just the nut touching the wrench. If the lug nuts and wheel remain attached, the moment will affect the wheel and its attachments as well. For example, the wrench will generate a twist or moment about the wheel's axle that may cause the wheel to turn. If we jack up the car, there will be little friction in the wheel's bearing. Thus, if we try to remove the lug while the wheel is elevated, the moment we apply will most likely cause the wheel to spin. The lug will move with the wheel and will not loosen if there is more friction between the lug and the wheel than between the wheel and axle. For this reason, it is common practice to crack the lug nuts loose while the car is still on the ground and the force from the ground holds the tire in place.

**OTHER CONSIDERATIONS:**

What would happen if we mounted the wheel in figure 3.2 on a very smooth bearing, jacked up the car, and tried to remove the lug nut. If we apply our wrench as figure 3.2 shows, we will not only

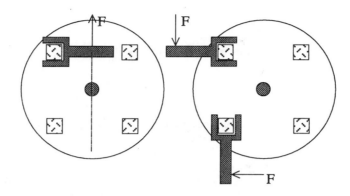

Figure 3.3: Wrench Orientation for No Wheel Rotation

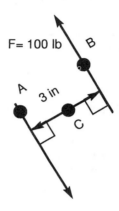

Figure 3.4: A Couple is Made of Two Equal and Opposite Forces Which Are Separated.

produce a moment about the nut, but also about the axle. In fact, the moment produced about the axle would be greater than that about the nut because its moment arm is larger. To get the nut off, we have to generate enough moment about the nut to force nut rotation without producing a moment about the axle. We show two ways this could be accomplished in figure 3.3. In both cases, the torque about the axle is zero implying the wheel has no tendency to rotate about its axle. What is a disadvantage of the configuration in the right part of figure 3.3?

**End 3.1** _____

Quite often an object is acted on by two forces that are equal and opposite to each other. Such a load condition is called a couple or "pure" torque (torque for short). Unfortunately the terms are used a bit loosely. For example the terms moment, torque and couple are often used interchangeably to refer to two equal and opposite forces. The terms moment and torque are also used at times to refer to the tendency to twist caused by a single force. Alas, the only way to know for sure is to see how the word is used in context.

Consider a steering wheel of an automobile. Although power steering makes it pretty easy to turn corners, imagine what happens if the power dies. It can get pretty hard to turn the wheel. If you have to turn without power, you might find yourself pulling on one side of the wheel while pushing on the opposite side. If you pull nearly as hard as you push you are applying a couple. Perhaps you have used a wrench or other tool with two handles. A common tool with two handles is the device that holds dies used for cutting screw threads on shafts. The purpose of the two handles is to allow you to apply a couple to the tool. Some electric motors also produce couples because for every force on one side of the rotating shaft there is an equal (nearly equal) and opposite force on the opposing side.

Figure 3.4 shows a diagram of a couple. So what is so special about a couple? Well if the couple is applied to a single rigid body, you can forget about the forces in the conservation of linear momentum. Why? Because the effect of one force is exactly counteracted by the equal and opposite

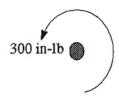

Figure 3.5: The Symbol Used to Indicate a Couple or Torque.

one. They cancel each other out! Can you ignore the individual forces if the body is rubber? Explain. What can you say about the moment produced by a couple? As an exercise, determine the moment of the couple in figure 3.4 about point A. The left force generates 0 in-lbs (convince yourself). The right force causes 300 in-lbs into the page. Add them to get 300 in-lbs into the page. Now determine how much moment they produce about point B. Its 300 in-lbs into the page. What is the moment about point C? Figure it out. You'll find it to be 300 in-lbs into the page. What do you think the moment of the couple is about the moon? The funny thing about couples or pure torques is that they do not contribute to the linear momentum but they do contribute to the angular momentum, more on this later.

To be more intuitive, engineers often draw couples as circular arrows. Figure 3.5 shows the typical symbol used to represent a couple. The circular arrow indicates a tendency to twist and the arrowhead shows what direction the tendency has. For example using the right hand rule the couple shown in figure 3.5 is out of the page.

What does it matter where a couple is applied to a body? It does not matter if the body is rigid. Since the moment produced by a couple is the same everywhere, the effect of a 3 in-lb out of the page couple applied anywhere on a rigid body is to produce a 3 in-lb out of the page moment about every point on the body. It does not matter where the couple is located.

## 3.2   FRICTION

**Objectives**
*When you complete this section, you should be able to*

1. *Determine when dry friction occurs.*

2. *Calculate the magnitude of a dry frictional force.*

3. *Sketch the impedance diagram for dry friction.*

4. *Explain the difference between kinetic and static friction.*

5. *Determine the direction that a dry friction force points.*

6. *Solve for dry friction when it is unknown.*

7. *Explain the concept of impending motion.*

8. *Identify when impending motion is present and know how to handle such problems.*

This section discusses dry friction. When two surfaces come in contact there are two types of forces which can be present. One type of force is perpendicular to the contact plane and will be called the Normal force. You can think of normal force as the pressure between the bodies times contact area. Of course if the bodies are glued together, the normal of one body on the other can pull, but it is still perpendicular to the contact. A second force which can act between bodies is parallel to the contact and can be called a shear force. One type of shear force is called friction. Friction is a complex phenomenon and considerable research has been devoted to developing a good mathematical representation for the forces. Typically, we deal with only two types of friction forces

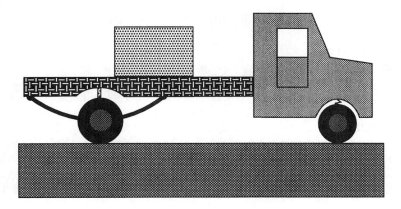

Figure 3.6: A Package Placed in the Bed of a Truck

– dry and viscous.

Dry friction occurs between two nonlubricated objects. It consists of two forms – static friction and kinetic friction. Static friction occurs when two contacting bodies do not slide with respect to each other. Static friction is always **less than or equal to** $\mu_s N$ where $\mu_s$ is a constant, and $N$ is the normal force between the bodies. Of course this is a simplification, but it is commonly used. When static friction equals $\mu_s N$, the bodies are in impending motion. Kinetic friction occurs only when two contacting bodies slide with respect to each other. Kinetic friction is always equal to $\mu_k N$ where $\mu_k$ is a constant that is usually less than $\mu_s$. We can feel this phenomenon when we push a heavy object. Once the object begins to slide, it is easier to push. This is a result of the fact that $\mu_k$ is less than $\mu_s$.

Viscous friction occurs between two wet lubricated surfaces. Viscous friction occurs in shock absorbers, automatic door closers, and in some aerodynamic situations. We call viscous friction devices such as shock absorbers dashpots. The magnitude of viscous friction is proportional to the sliding velocity. We can write this as $F = Cv$ where $v$ is the sliding (or slip) velocity.

Friction always opposes the tendency to slip. A common misconception is that friction opposes motion. This is not correct. Friction always opposes slip or the tendency to slip, but not necessarily motion. In most cases, the slip direction is obvious. However, in more complex situations, it is not obvious. We have to be careful. The following methodology can be used to determine the direction of a friction force when we encounter these cases.

Like all contact forces, two objects generate friction through contact. To determine the direction of a friction force on body 1, we imagine standing on body 2. While standing on body 2, we observe the direction that body 1 appears to slide. The friction produced by body 2 opposes the apparent slip of body 1.

## Example 3.2

### Friction in a Truck Bed

**QUESTION:**

A package is in the back of a truck traveling to the right as figure 3.6 shows. If the truck slams on its brakes, the package may slide. If the package does slide, it will most certainly slide toward the cab of the truck. Even if the package does not slide, it will still have a tendency to slide toward the cab. Determine the direction of the friction force on the package.

**ANSWER:**

Since the truck bed causes the friction on the package, we will imagine that we are standing on the truck (actually we can imagine that we are sitting behind the wheel and observing the package). It takes little imagination to realize what we will observe while sitting in the truck. The package will appear to approach the cab. Friction on the package opposes this slip and therefore points away from the cab. In this case, the friction force on the package points to the left. The package's actual motion is to the right (at least until the truck comes to a stop). Therefore, for this example, the friction does oppose the package's motion (because the package slips and moves in the same direction).

For the sake of this example, let's also determine the direction of the friction force on the truck. Since the forces generated between bodies are always equal and opposite to each other,[2] we know the direction must be to the right. Just for practice, however, we will use our imagination to determine the direction. Since the package causes the friction on the truck, we will imagine that we are standing on the package. From this perspective, we see that when the brakes are applied, the cab of the truck appears to approach the package (the truck appears to slip to the left). The friction on the truck opposes this apparent slip and points to the right. The truck's actual motion (although decelerating) is to the right and the friction on the truck is to the right. They point in the same direction! To avoid any confusion, remember that friction always opposes slip but does not always oppose motion. Now, for practice, determine the direction of friction on the truck and the package if the driver hits the accelerator pedal.

**End 3.2** _____

If two objects are in contact and are sliding relative to each other, we do not know the relative motion of the bodies. In this case, however, we know the magnitude of the friction force between the bodies (friction = $\mu_k N$). Notice that the flow is unknown, but the effort is known. This happens often. If we know the magnitude of friction (or any force) and want to use it as a known quantity, we must know its correct direction. If we use the wrong direction for the force, we will calculate the flow incorrectly.

## 3.3   [NIF] AND [NIA]

_____

**Objectives**

*When you complete this section, you should be able to*

1. *Calculate [NIF] and [NIA] of mechanical systems.*

2. *Recognize when assumptions about motion are being made.*

3. *Make assumptions concerning motion and explain how to test them.*

4. *Be able to determine whether or not a conservation equation is required for the solution of a problem.*

_____

As in most disciplines, there is a formal mathematical definition of the number of independent flows. There is also a formal definition for [NIF] of a mechanical system. However, trying to apply the formal definition generally provides little *a-priori* insight into a problem. Therefore, in this text, we will use an informal method for calculating the [NIF]. In an earlier lesson, we introduced a process for finding the [NIF] of mechanical systems. We will continue to use this process but we will discuss an important complication. In many mechanical systems, we cannot be sure what motion actually occurs, hence we may not be sure what the [NIF] is before solving the problem. When this occurs, the best approach is to construct a table of possibilities. The following example explains what we are talking about.

**Example 3.3** _____

### What's the [NIF]?

**QUESTION:**

Figure 3.7 shows two masses. The masses are tied with a rope and are held in place by someone's hands (not shown). Suppose the person suddenly lets go of the masses, determine the [NIF] of the two mass system. Let there be dry friction between the incline and the blocks.

**SOLUTION:**

_____

[2]A consequence of Newton's third law – every action causes an equal and opposite reaction.

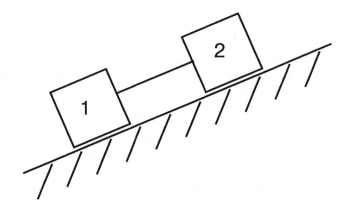

Figure 3.7: Two Masses Tied With a Rope On an Incline

| Assumption | [NIF] |
|:---:|:---|
| Both Blocks Stick | 0 |
| Block 1 Sticks, 2 Slides | 1 |
| Block 1 Can Slide, 2 Sticks, Rope Firm | 0 |
| Block 1 Can Slide, 2 Sticks, Rope Breaks | 1 |
| Both Blocks Slide, Rope Firm | 1 |
| Both Blocks Slide, Rope Breaks | 2 |

Table 3.3: Possible [NIF] of the System.

The method for finding the [NIF] is to ask yourself if something can move. In this case, the answer is that it depends.

1. If the friction forces beneath the blocks are large, it may be that neither block can move and there is no motion. The [NIF] is zero.

2. If the friction beneath block 1 prevents it from moving, but the friction beneath block 2 does not prevent it from moving, then there is one motion. The [NIF] is one.

3. If the friction beneath block 1 is small and the friction beneath block 2 is large, and the rope does not break, then block 2 holds block 1 and no motion occurs. The [NIF] is zero.

4. If the conditions are the same as in 3 except that the rope breaks, then block 1 can move. The [NIF] is one.

5. If there is very little friction beneath either block and the rope remains taut without breaking, then both blocks move together with the same motion. The [NIF] is one.

6. If the conditions are the same as in 5 except that the rope breaks or goes limp, then the blocks may have different motions. The [NIF] is two.

To organize all of the possibilities, you should construct a table as we have done in Table 3.3. In all of these cases, there is an implicit assumption that both blocks remain in contact with the incline. This may seem obvious; however, if a situation arises in which the blocks could lift off of the incline, then we should include that possibility in our analysis too.

**End 3.3** ─────────────────────────────────────────────

Determining the [NIF] of a mechanical system is the first thing to do after reading the problem. It is a time for you to gather your thoughts about what motion you are going to consider reasonable. It helps you organize your thoughts. Most importantly, it also plays a key role in deciding how to approach the solution. We will discuss this in more detail later. For now, get in the habit of determining the [NIF]. The [NIA] is not as critical so we will not discuss it as often.

## 3.3.1   MAKING ASSUMPTIONS

When there are multiple possibilities for the [NIF] of a system, we generally have to make an assumption about the motion. We will approach the problem by assuming that one of the possibilities is true, then solve the problem and test the assumption. If the assumption holds, we are finished. If it does not, we assume that another possibility is true, and check it. We continue this approach until we find an assumption that holds.

Since you will need to check assumptions, it is important to know how to test them. If you assume to know something about a flow, then you will determine a corresponding effort and verify its magnitude or direction. If you assume to know an effort, then you will determine a flow and verify it. The following example, discusses this for the system in example 3.3.

## Example 3.4

### Testing Assumptions.

**Question:**

Explain how to test each of the possibilities described in example 3.3 (table 3.3).

**SOLUTION:**

Suppose we *assume* that the first option is true, both blocks stick, [NIF] is 0. In this case, we *assume* the linear momentum of the blocks are zero. We would then calculate the friction forces while pretending to *know* all motion is zero. If both friction forces are less than the maximum possible friction (less than the respective $\mu_s N$) then our assumption is correct. Notice that we assumed something about two flows (zero momentum), we calculated two efforts (friction forces) and tested the efforts to verify our assumption.

By the way, we implicitly assumed the blocks remained on the incline (assumed flows). We should verify this by computing the normal forces between the incline and blocks (compute efforts) and demonstrating that the efforts are acceptable (both of them are positive). In the remainder, these implicit assumptions are present but will not be discussed.

If we *assume* that the second option is true, block 1 sticks and 2 slides. We *know* the momentum of 1 (zero), but we do not *know* the momentum of 2. Since we *know* that block 2 is slipping, we *know* the dry friction force beneath it is $\mu_k N_2$. A momentum that we knew last time has become unknown this time, and a force that was unknown last time has become known. We will still have the same number of unknowns, but they are different than before. In this case, we will calculate: (1) the friction beneath 1 (an effort) and, (2) the momentum of 2 (a flow). To test the assumptions, we make sure (1) the calculated friction (the effort) is less than or equal to the maximum possible friction ($\mu_s N_1$) and (2) the calculated momentum is possible. For example, a momentum directed up the incline would not be possible. Note that for the assumed flow, we tested an effort. For the assumed effort, we tested a flow. This is the pattern.

If we *assume* that the third option is true, block 1 slides, block 2 sticks, and the rope does not break. The friction beneath block 1 is maximum and both momentums are zero. We calculate the friction beneath block 2 and the tension in the rope. To test our assumption, we will verify if the friction is possible and the tension is less than what it takes to break the rope. We *assume* two flows (both zero), and test two efforts.

If we *assume* that the fourth option is true, block 1 slides, block 2 sticks, and the rope breaks. The friction beneath block 1 is maximum ($\mu_s N_1$), the rope tension is zero (it broke), and the momentum of block 2 is zero. We calculate the friction beneath block 2 and the momentum of block 1. We test if the friction beneath 2 is possible (is its magnitude less than or equal to $\mu_k N_2$) and the momentum of 1 is possible to verify our assumptions.

If we *assume* that the fifth option is true, both blocks slide and the rope remains firm. We assume that block 2 has the same momentum as block 1. The friction forces beneath both blocks is the maximum ($\mu_k$ times the respective normal), momentum of block 1 is unknown and computed and the rope tension is computed. To test this assumption, we have to verify if (1) the tension in the rope is positive but less than the maximum tension for breaking and (2) the momentum of block 1 is reasonable.

Finally, if the last option is true, both blocks slide and the rope breaks. Two momentums are unknown, the friction forces are maximum ($\mu_k$ times the respective normal) and the tension in the

rope is zero. To test this assumption, we have to verify that both the momentums are reasonable.
**End 3.4** ─────────────────────────────────

In general, it is very difficult to test whether or not a momentum is reasonable. This is because few people have good feel for motion. As a result, you should begin a problem by making as many flow assumptions as possible. For example, in the case of friction, if we do not know better, we always:

1. Assume no slip.

2. Solve for the unknown friction force.

3. Verify that the magnitude of friction is $\leq \mu_s N$. If it is, we are finished.

4. If magnitude of friction is $> \mu_s N$, then rework the problem with the friction known as $\mu_k N$ and the motion unknown. **This is critical!** When you substitute $\mu_k N$ for the friction, you must draw and use its correct direction because you will not be solving for friction and will not have a +- sign to help you. If you use the incorrect direction for $\mu_k N$, you WILL obtain an incorrect motion and (since motion is hard to think about) you may not realize you made a mistake.

In summary, this means to begin with the option with the least [NIF]. If more than one option has the same minimum [NIF], choose one. If the initial assumptions do not check, then eliminate that possibility, and choose the possibility with the next fewest [NIF]. In the previous example, we would begin by assuming that option 1 or 3 is correct. Suppose you start with 1, if it fails to verify, try option 3. If 3 fails to verify, then one by one choose option two, four, or five. If all these fail to verify, then choose option six. Once an assumption is found to be true, we do not need to check the other options.

Similar methods for testing assumptions apply to other domains. For example, in the case of electrical systems a common assumption that may need to be checked is whether or not a fuse blows. A fuse is a safety device that stops current flow if it ever exceeds a predetermined value. For example, if 9 amps flow through a 10 amp fuse, there will be a negligible voltage drop (the fuse is very much like a perfect wire). If however 11 amps flows through a 10 amp fuse, the fuse will "blow" and prevent flow. The voltage across a blown fuse is unknown but the current through it is known to be zero. If you are analyzing a circuit with a fuse, the [NIF] will depend on whether or not the fuse is blown. The easy test to see if a fuse blows is to compare the flow (current) to some value. As a consequence, you should assume you know the effort across the fuse (zero volts for an unblown fuse), compute the flow and compare the computed value to the rating. If the computed flow is less than the fuse rating, you are done, the fuse is not blown. If the computed flow (the current) is greater than the rating, the fuse is blown and you start over knowing the flow (zero amps) and solve for the effort (the voltage across the fuse).

The point here is that when you make an assumption, you either assume an effort or a flow. You then compute the opposite of what you assumed. If you assume an effort, compute a flow. If you assume a flow, compute an effort. You then test your assumption by comparing what you computed against what you know to be reasonable. In the case of a fuse, you know a maximum current so you want to solve for current hence assume the effort (zero for unblown). In the case of mechanics, most of the time we know what reasonable forces (efforts) are hence we should assume we know the flows.

### 3.3.2 Deciding Whether or Not to Write a Conservation Equation.

Another importance use of [NIF] is to help you decide whether or not to write a conservation equation. For example consider the truck and crate shown in figure 3.6. Suppose we know the truck is moving at a constant 30 mph. Furthermore suppose we *know* that the crate is not slipping and that the objective is to determine the motion of the crate. Well what a simple question! If the crate is stuck to the truck it must be moving the same as the truck! Who cares what the conservation laws say!

Now what we need is a general way to make similar observations when the problem is not so obvious. This is the procedure to use.

1. Determine the [NIF]. The [NIF] is equal to the number of independent flows. If you know [NIF] flows, you can compute all other flows without any conservation/accounting equations. It may be difficult to do, but it can be done.

2. If you are given [NIF] flows and asked to find some other flow(s) then you do not need conservation/accounting equations.

3. If you MUST find an effort, you MUST use a conservation/accounting equation.

Apply this procedure to the problem in figure 3.6. If we know the crate sticks to the truck, there is one [NIF]. If we are given the motion of the truck (given a flow) and asked for the motion (flow) of the crate then there is no need for conservation/accounting equations. If we want to determine the friction force between the crate and truck, perhaps to test a no slip assumption then we need conservation/accounting.

## 3.4   FREEBODY DIAGRAMS

**Objectives**
*When you complete this section, you should be able to*

1. *Explain what freebody diagrams are and why they are important.*

2. *Isolate a system and draw correct freebodies of them.*

3. *Apply conservation of linear momentum to a rigid body.*

4. *Apply conservation of angular momentum to a stationary body.*

5. *Draw proper free body diagrams when the body has a known acceleration.*

---

To help keep track of the forces acting on a system, we use a freebody diagram. As the name implies, a freebody diagram is a picture of a body completely removed from its surroundings. It is free of external bodies and connections. The purpose of a freebody is to graphically show the forces acting on a system. An experienced engineer can write accounting and conservation equations without using a freebody diagram. However, a freebody diagram reduces confusion and makes the accounting laws easier to apply. Not using a freebody is like balancing a checkbook without a calculator. It can be done, but it is easier to get the job done with fewer mistakes with the right tool.

To draw a correct freebody diagram, the free body must be loaded so that the system responds exactly as if connected to its surroundings. To do this, it is necessary to put forces and moments in appropriate places on the freebody. A simple way to do this is to imagine cutting the body out of its environment and isolating it in a plastic bag. For example, suppose we choose the mass resting on the table in figure 3.8 as a system. To draw a proper freebody diagram, imagine cutting the mass out of its environment, we imagine cutting the air away from the mass and removing the mass from the table. Figure 3.9 shows the mass inside the imaginary bag. We illustrate the bag with the dotted curve.

To place the forces and torques on this isolated body, consider only the locations where the body **used** to touch another object. Remember, forces can only occur through contact.[3] We can determine the existence of all forces acting on our freebody by looking for contact between something outside the bag and the body in the bag. The freebody must include only external forces; that is force between the body in the system and something touching it from outside the system.[4]

Since we cut the air away from the body, the air is outside the bag. Since the air was in contact with the body, it exerted forces on the body. Even if the air was motionless, it still exerted a

---

[3]Think of action at a distance forces as invisible hands grabbing the body.

[4]This is a result of the fact that forces are momentum in transit. If the force occurs between something outside and something inside, it represents momentum crossing the system boundary which is what we want in the conservation/accounting equation.

Figure 3.8: A Mass Resting on a Horizontal Table

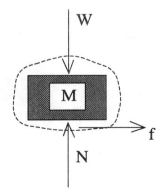

**Analysis Assumptions**

(1)  $N$ force from the table

(2)  $f$ friction force from the table

(3)  $W$ gravitational force acting at the center
of mass

Figure 3.9: (fbdmass): The Freebody Diagram of a Mass on a Table

hydrostatic force (a pressure) on the body. This hydrostatic force would be constant (if the height of the mass is small) and would act in all directions. When we sum the forces on the freebody, the net force that results because of this hydrostatic pressure is nearly zero. This force is so small, we will always ignore it in this text. Making simplifying assumptions like this about the forces on a freebody diagram can make our analysis of a system much easier. However, keep in mind that if we ignore a force because of a mistaken assumption, we will analyze the system's response incorrectly. Be careful. For example, bizarre as it may seem, suppose we take two objects (let's say they are metal plates) and suppose we smooth them off so they are quite flat. If we put these objects together and slowly slide them one over the other we can "squeeze" all the air from between them. If this happens, the air pressure will be exerting on all sides of the blocks **except** for the two mating surfaces. The air pressure will actually hold the parts together! This makes a great parlor trick. Unless otherwise stated, we will ignore these effects in this chapter.

Continuing with the freebody, the table is a body outside the system. Since the table was in contact with the body, it exerted a normal and possibly even a friction force on the body. We represent these forces as $N$ and $f$ respectively. Since the mass is at rest on the table (no slip), the friction force's magnitude must be less than or equal to $\mu_s N$. The only other contact is the invisible hand of gravity between the center of the Earth and the center of mass of the body ($W$ in the figure).

If a body has a known motion, do not confuse that motion with a force. Motion (acceleration) is not a force and it does not belong in a freebody. Acceleration is the effect of a force. For example, when a jet takes off down a runway, you feel as if something is forcing you backward. In reality, your body wants to remain motionless while the seat (attached to the plane) wants to move forward due to the thrust from the jet engine. As a result, the seat pushes your back because it wants to occupy the space your body is taking up. The force in on your back because that is where the seat touches your body. Your back responds to this force moving forward trying to occupy the space your chest has. Your body moves your chest out of the way by passing force along to your chest via your muscles and skeleton. The forces inside your body give you the sensation that your body is being squashed. It is. From the back forward, not the front backward.

As another example, though you most likely have not experienced this, when astronauts are launched they lie on their backs. The reason for this is that the acceleration of the launch vehicle is so large that if they were standing, their skeletons would push their body tissues upward to get the body out of the way of the vehicle. Although the skeleton may be able to handle this, the forces acting on the blood are viscous and pressure created by the heart. These forces are too small to make the blood move at the same rate as the body so the body moves up faster than the blood. Hence the blood drains from the head slightly and collects in the feet. The lack of blood in the astronaut's brain makes him/her blackout. Not a good thing! Similar problems occur with jet fighter pilots when they accelerate too much.

The same sensation can be felt while riding an elevator. When an elevator suddenly accelerates upward, you feel a sensation that you are being pushed downward. In reality, you feel your weight beneath your shoes. When the elevator suddenly moves upward, it causes an increased pressure on the bottom of your feet. Your skeleton supports this increased force. As a result, your spine compresses. The sensation you feel is your body becoming shorter from the bottom up, not the top down.

The sensation can also be felt when riding in an car. If you turn hard to the left, then you feels like you are being thrown right. In reality, your body is trying to maintain constant linear momentum. To make your body change momentum, you have to force it. When the car moves left, friction between you and the seat pulls you left. If friction is not enough, the right side door hits your body on the right and pressures you to the left. The sensation of being thrown right is caused by the forces pulling and pushing you from the left, not from the right, as you would expect.

In summary,

1. Your sensations generally tell you the opposite direction that you are actually being forced.

2. Applied forces cause the acceleration of a body.

3. Acceleration is not a force and does not belong in a freebody diagram.

Some textbooks teach you to draw freebody diagrams by putting the negative of mass times acceleration on the body then summing forces (including the mass times acceleration) to zero. This is known as D'Alembert's principle. Although the principle is valid and useful in some situations this text will not use it for the following reasons:

1. The principle does not provide any advantages for many problems. If there is not an advantage, why bother?

2. By drawing mass times acceleration on the freebody diagram students often get confused and handle it incorrectly. Especially in conservation of energy formulations.

3. Most of the time, you do not know what the acceleration is so you cannot draw it anyway.

A good heuristic to follow when you draw freebody diagrams is to always draw the body in some arbitrary position in which the distances are positive. Drawing a picture in some "general" position helps you avoid making the mistake of looking at a picture and thinking some line is always horizontal or some force is always pointed up when it isn't. By drawing the system in the positive direction it often cuts down on the number of negative signs in the formulation and the fewer the negative signs the fewer chances of silly mistakes.

## 3.5 APPLYING CONSERVATION OF LINEAR MOMEN-TUM

**Objectives**
*When you complete this section, you should be able to*

*1. Explain how to wisely choose coordinate directions for conserving linear momentum.*

Since linear and angular momentum are vector quantities, you must choose a coordinate system and application directions when writing conservation equations for these quantities. When doing this, think about the motion. In general, expressing the motion of an object will be the hardest thing to do. Resolving forces into various directions, although it can be difficult, is usually trivial compared to expressing the motion of an object. Therefore, you should always choose coordinate directions that make expressing motion (acceleration) easy.

If you are working on a problem for which your are making assumptions you should define enough position variables to describe all the motions that are possible. For example, even though you may be working a problem where you are assuming there are 0 [NIF], define enough variables to describe the most complex motion you are considering.

The following example is used to demonstrate these concepts and to make a couple of important points that will be used throughout the text. These important concepts are highlighted so watch for them.

**Example 3.5**  ───────────────────────────────

### Stopping a Truck

**PROBLEM:**
Determine if the package on the truck's bed in figure 3.10 will slip when the truck driver slams on the brakes and decelerates at a rate of $d_r$. Note that the term deceleration means the speed is decreasing. If the object is moving in the positive direction, then deceleration is a negative change in momentum. If the object is moving in the negative direction, then deceleration is a positive change in momentum. In this problem, let's assume the truck is moving forward (to the right) so a positive number $d_r$ means the driver is slowing down and momentum is changing to the left.

**FORMULATION:**
First let's choose the crate and truck as a system. Later we will choose only the crate and discuss the differences. Next we count the [NIF]. This is where we think about the possible motions. At this point, we have no idea whether the crate slips or not, therefore we construct the possibilities. Table 3.4 lists the motion possibilities. Notice that in both possibilities we do not worry about

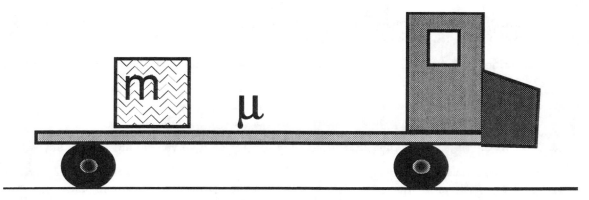

Figure 3.10: A Truck with a Massive Package

| Assumption | [NIF] |
|---|---|
| The Mass Sticks | 1, the mass and truck move together |
| The Mass Slides | 2, the mass and truck move separately |

Table 3.4: Motion Possibilities for the Mass and Truck.

motion vertically. What we are saying is we are willing to bet that neither object moves vertically. This is a pretty good bet, don't you think.

Since we do not know if the mass slips, we start with the minimum [NIF] assumption. We assume that the block does not slip. With this assumption, there is one [NIF], one motion is given (the acceleration of the truck is $d_r$ left) therefore we need not write a single conservation/accounting equation. If our assumption is correct, the block has the same motion as the truck, so the acceleration of the crate is $d_r$ to the left. If our assumption is incorrect however, this conclusion is wrong.

We need to test our assumption. Since we assumed a flow (no slip), we must compute an effort (the friction force) and verify its feasibility (that the friction force under the block is $\leq \mu_s N$). Since we need to compute a force, we must draw a freebody diagram and apply a conservation/accounting equation.

Choose the crate as a system and draw the freebody. **Take a note of this point.** We chose the crate as the system because we want the friction force beneath it to be external. If we chose the truck and crate together, the friction force is internal. This would do us no good. If we chose the truck as the system, we would have to worry about the forces under the truck tires. So it seems the easiest thing to do is choose the crate. The proper freebody diagram of the crate appears in figure 3.11. Note the following items:

1. Notice the direction of $f$ is incorrect.

2. $N$ is the normal force from the truck.

3. We assume the block does not tip over. If we wanted to worry about this, it should have been included in the [NIF] table. Angular momentum is needed to deal with rotation or tipping so let's forget about it for the moment.

We drew the friction in the wrong direction to emphasize a point. Since we plan to calculate the unknown friction force, its solution sign ($+$ or $-$) will tell us which direction it should be. For example, if you write your equations consistent with the arrows drawn in the freebody then a positive solution always tells us the direction drawn on the freebody is correct. A negative solution tells us the opposite direction is correct. Because we drew friction incorrectly, we expect a negative magnitude for $f$. Notice that the motion (acceleration) of the mass does not appear anywhere in the freebody diagram. The freebody only shows the forces and moments caused by an external object touching an internal object. If you want, you can draw lines to represent angles and other information that will help you keep things straight.

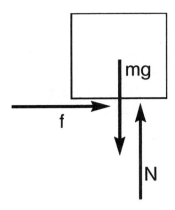

Figure 3.11: Freebody Diagram of the Block on the Truck

Next we need to determine the directions for applying the conservation equations. To do this define two positions. Let $X_c$ be the horizontal distance measured from a fixed point on the road (say a grease spot behind the truck) to the center of mass of the crate. Let $X_t$ be the horizontal distance measured from the same grease spot to the center of mass of the truck. These two positions will give us the motion of both objects which defines the maximum number of flows in the problem (two [NIF]). **Note this point.** Notice that the positions we chose were measured from a fixed point. Always try to measure positions from fixed points so that when you use them to define momentum you will have the momentum measured from an inertial reference. There will be some examples later where we measure from a moving object, but we are careful to convert the velocity to an inertial component.

Based on the assumption that the crate sticks to the truck, the rate of change of the crate's momentum is the crate mass times the rate of change of the truck's velocity ($d$ to the left). Applying the conservation of linear momentum in the horizontal and vertical directions, we find the following:

**Conservation of Linear Momentum (Vector)**
system[**box fig.3.11**]    time period [**differential**]

$input - output = accumulation$

| input/output: | mass flow | |
|---|---|---|
| | external forces | $f\,\vec{i} + N\,\vec{j}$ |
| | gravity force | $-mg\,\vec{j}$ |
| **accumulation:** | | $m(-d_r)\,\vec{i}$ |

$x$:  $\boxed{f}$  $=$  $\boxed{m(-d_r)}$
     *external*      *accum.*

$y$:  $\boxed{N}$  $-$  $\boxed{mg}$  $=$  $\boxed{0}$
     *external*     *gravity*     *accum.*

This gives us two equations (the linear momentum equations) and two unknowns ($N,f$). Therefore, we have a solvable formulation. We use $-d_r$ in the horizontal equations because the momentum is to the right but is slowing. Note that the magnitude of the change in momentum of the box is equal to $d_r$ not zero. Some students incorrectly think that since the crate is not moving on the truck the momentum is zero. When you write conservation of linear (and angular) momentum, you must write inertial or Newtonian momentum. This means that you must express momentums relative to an inertial (Newtonian) reference frame. Basically, this means that momentum must be measured from a point that is either fixed or moving on a line at constant speed. If we measure the box's position from the bed of the truck, we would conclude that its position is constant. After all, the box sticks to the truck. If the box's position is constant, then its change in position (speed) is zero

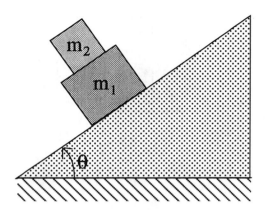

Figure 3.12: Two Blocks on an Incline

and constant (with respect to the truck).  A constant speed indicates zero change in momentum. Does this mean that the change in momentum for the box is zero?  No.  To determine this, we have to measure the box's position from a stationary point.  In this case, the box's position clearly changes.  It always has the speed of the truck (because it sticks to the truck).  Since the truck changes momentum, so does the box.  They have the same motion.

**SOLUTION:**

Using the $x$ equation, we determine the unknown friction force $f = -md_r$.  The negative sign means the direction drawn for $f$ is wrong (we knew that).  Friction should point to the left.  **Pay attention to how to handle a quantity when it is negative.**  Now that we know that $f$ is in the wrong direction, we do not attempt to change the diagram.  We leave everything alone and continue to generate equations as if the diagram is correct.  When it is time to plug a number in for $f$, we simply plug in a negative quantity.  **Here is another heuristic.**  If we start changing things in the middle of a problem, it is very easy to make a mistake.

Using the $y$ equation, we determine $N = mg$.  The positive sign means we drew it in the correct direction (obviously, because the truck cannot pull the block downward).  Now if $md_t \leq \mu mg$ or rather if $d_t \leq \mu g$ then our assumption is correct.  If it is incorrect, we have to begin again.

If we did begin again, we would know that the block slips and that the friction force magnitude equals $\mu_k N$.  We would not know the motion of the block.  We would only know that it does slip.  If we substitute the known value for the friction force into our conservation equation, we would have to be sure that the direction of the force in the freebody diagram is correct.  We could not use the solution of the conservation equation to verify the direction of the force because this time we are solving for the motion of the box, not the friction force.  Often a little reasoning and the solution obtained for $f$ when we assumed no slip will help you find the proper direction for the friction force when it does slip.  Just remember friction always opposes the slip but it doesn't always oppose the actual motion.

**End 3.5**  ───────────────────────────────────────────────

**Example 3.6**  ───────────────────────────────────────────────

## Two Sliding (?) Blocks

**PROBLEM:**

Determine the motion of the two blocks in figure 3.12.  Let $g = 9.8$ [m/s$^2$], $m_1 = 5$ [kg], $m_2 = 2$ [kg], $\theta = 15^o$, $\mu_{s1} = 0.1$, $\mu_{s2} = 0.3$, and $\mu_k = \mu_s$.  For the time being, we will assume none of the blocks tip.  Tipping is best handled using angular momentum which we don't want to discuss right now.

**FORMULATION: [I]**

To determine what the blocks do, we begin by determining the [NIF] of the system.  At this point, we do not know what happens.  If the top block slides off of the bottom block, the motion gets very complex.  For simplicity, we will restrict our analysis to the time period that the top block

| Assumption | [NIF] |
|---|---|
| Both Stick | 0 |
| Top Slides, Bottom Sticks | 1 |
| Bottom slides, Top Sticks to Bottom | 1 |
| Both slide independently | 2 |

Table 3.5: Possible Motions for the Two Blocks.

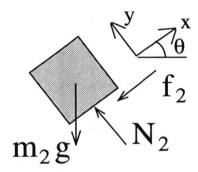

Figure 3.13: Free Body Diagram of the Top Mass

remains on top of the bottom block. We still are not certain of the [NIF] of the system. Table 3.5 shows the possibilities.

Since we do not know which is correct, we will begin by assuming that the option with the fewest [NIF] is true. Thus, we will begin by assuming that both block's stick (0 [NIF]).

At this point, we must decide what to choose as a system. In the next section, we will explore this in detail, but for now, we will simply choose each mass, one at a time. To start, we arbitrarily choose the top mass. The freebody diagram of the top mass appears in figure 3.13. Note the following about this freebody:

- $\theta$ is the angle of the incline.

- $N_2$ is the normal force on block 2 due to block 1.

- $f_2$ is the friction force on block 2 due to block 1.

Since the bottom block is outside of this system, the forces of interaction between the two blocks, $N_2$ and $f_2$, appear in this freebody. Gravity acts through an invisible hand from the center of the Earth (something outside the system) to the center of the top mass (something inside the system) so we include it, as well. The gravitational force between the center the Earth and the bottom block does not appear in our freebody diagram because it does not directly touch the isolated body. It is an interaction between the Earth (something outside the system) and the lower block (also something outside the system).

Now for the coordinate directions. Since the maximum [NIF] in the table of possibilities is two, we will choose two coordinates that define the positions. Measure $x_1$ along the incline starting from the bottom of the hill. Measure $x_2$ also parallel to the incline from the bottom to the center of mass 2. Always choose the coordinate system that makes expressing the motion of the object easy. Generally, finding motions is much harder to do than finding forces. The reason for this is that the momentum and energy equations (achieved from the conservation laws) usually require us to differentiate motion. We rarely differentiate forces. Therefore, we choose coordinates that make the hard part (differentiation) as easy as possible. Now since we are assuming no motion, we could have chosen any direction and the motion would have been easy to express (its zero), but we chose what we did for practice. The momentum of the bottom block is $m_1 \dot{x}_1 \vec{i}$ and the momentum of the top block is $m_2\dot{x}_2$. Note that if $\dot{x}_1$ is positive, the block is sliding up the incline. Obviously, if the block slides it will come down so if we ever solve for $\dot{x}_1$ we will expect a negative number. The same is true for the top block.

Applying the conservation of linear momentum to the system in figure 3.13 gives:

**Conservation of Linear Momentum (Vector)**
system[**top block fig.3.13**]      time period [**differential**]

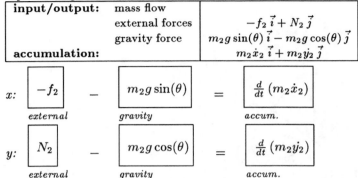

*input* − *output* = *accumulation*

| input/output: | mass flow |  |
|---|---|---|
|  | external forces | $-f_2\,\vec{i} + N_2\,\vec{j}$ |
|  | gravity force | $m_2 g \sin(\theta)\,\vec{i} - m_2 g \cos(\theta)\,\vec{j}$ |
| accumulation: |  | $m_2\dot{x}_2\,\vec{i} + m_2\dot{y}_2\,\vec{j}$ |

$x$:   $\boxed{-f_2}$  −  $\boxed{m_2 g \sin(\theta)}$  =  $\boxed{\dfrac{d}{dt}(m_2\dot{x}_2)}$

     *external*      *gravity*        *accum.*

$y$:   $\boxed{N_2}$  −  $\boxed{m_2 g \cos(\theta)}$  =  $\boxed{\dfrac{d}{dt}(m_2\dot{y}_2)}$

     *external*      *gravity*        *accum.*

Notice that we put the forces in as either positive or negative depending on their orientation in the freebody diagram. As usual, if we solve for one of these forces, a positive answer tells us that the direction in the freebody is correct. A negative tells us that the opposite direction is correct. We must correctly draw all forces that we are not calculating. Also note that if we get lazy and leave the unit vectors off, we would let positive $x$ direction to be up the incline (the positive direction that $x_2$ is positive.

Notice that even though we are assuming 0 [NIF], we substituted general expressions for motion into our equations. In this way, if our assumption is incorrect, we can begin from this point and not lose too much time. In the next paragraph, we will discuss how we can simplify these equations using our assumptions. Because we substituted the velocities as positive $\dot{x}$ and $\dot{y}$, positive solutions mean the top mass is moving up the incline (the positive direction). Intuition tells us the top mass cannot move up the incline. Thus, we will be very suspicious of a positive answer for $\dot{x}$. It may indicate that we made a mistake. Intuition can be a great asset for verifying solutions, but keep an open mind because it is not hard to fool intuition.

Now let's consider each term in the linear momentum equations to determine which (if any) we know. Since we are assuming 0 [NIF], we should have 0 unknown flows (motions). In this case, we know all positions are constant and all velocities are zero. With all the motion terms set equal to zero, we could solve for $N_2$ and $f_2$, and then verify part of our assumption. Is $f_2 \le \mu_{s2} N_2$? If it is then part of our no slip assumption could be valid. We can only say "could be valid" because if any one of the assumptions are found invalid, all results become unreliable. Only if all assumptions check out can we say with confidence that we have the solution.

To check another assumption ask, is $N_2 > 0$? Which ensures that the top block remains in contact with the bottom block because if $N_2 < 0$ it says the bottom block is pulling the top block toward the incline. Come on how in the world could the bottom block pull the top one closer to the incline? Maybe if there were gum or glue stuck between them and some really weird motion was happening with the incline, but not in the present case. Most of the time in this text, we don't bother to test these "obvious" assumptions, but you should just to make sure you haven't done something really dumb like overlooking a negative sign somewhere.

Notice we expressed only conservation of linear momentum for the system. We should always express all of the independent conservation equations for the systems that we select. In this case, application of angular momentum would require knowledge of the dimensions of the top mass.[5] Because the dimensions are unknown, it is not possible to express angular momentum. We expect that our solution satisfies angular momentum but have no way to check it. We should also express conservation of energy but we will forgo that for this simple example. Besides, in the presence of

---
[5]Angular momentum would tell us if the masses tip over.

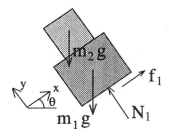

Figure 3.14: Free Body Diagram of the Top and Bottom Mass

only mechanical energy, the momentum and energy equations are dependent. Conservation of mass and charge are trivial (typical of mechanical systems). Therefore, we will not discuss them further.

Now to verify that the bottom block sits on the incline without slipping. To do this we need to determine the friction force under the bottom block. We cannot find this force with any of our existing equations. Therefore, we need more equations. Table 3.2 gives us some hints about what to do. Since we have no flow variables in our equations it is pretty clear that unless we overlooked some information in the problem statement, it is time to choose another system. Since there are two masses in the problem, we should have expected as much.

Just for kicks, choose the top and bottom masses together as a system. Figure 3.14 shows a correct freebody diagram.[6] We still make the following assumptions:

- The blocks stick to the incline.

- $f_1$ and $N_1$ are forces caused by the interaction between the incline and the two block system.

- The top block sticks to the bottom block.

Notice that this time the weights of both blocks appear in the diagram because both blocks are in the system. The Earth (outside the system) touches both blocks with the invisible hand of gravity causing the weights. The interaction between the upper and lower blocks does not appear because these forces are internal. They are caused by something inside the system touching something else inside the system. They do not represent momentum crossing the system boundary. The incline plane (outside) touches the bottom mass (inside) and therefore potentially exerts the two forces $N_1$ and $f_1$. Our new system still has zero [NIF] and zero [NIA] (assuming the top mass sticks to the bottom). Expressing the conservation of linear momentum for the two block system gives the following:

**Conservation of Linear Momentum (Vector)**
system[**both blocks fig.3.14**]      time period [**differential**]

*input − output = accumulation*

| input/output: | mass flow external forces gravity force | $f_1\,\vec{i} + N_1\,\vec{j}$ $-m_2 g \sin(\theta) - m_1 g \sin(\theta)\,\vec{i} - m_2 g \cos\theta - m_1 g \cos(\theta)\,\vec{j}$ |
|---|---|---|
| **accumulation:** | | $(m_1 \dot{x}_1 + m_2 \dot{x}_2)\,\vec{i}$ |

$x$:  $\boxed{f_1}$ − $\boxed{m_2 g \sin(\theta) + m_1 g \sin(\theta)}$ = $\boxed{\frac{d}{dt}\left((m_1 \dot{x}_1 + m_2 \dot{x}_2)\right)}$

   *external*   *gravity*   *accum.*

$y$:  $\boxed{N_1}$ − $\boxed{m_2 g \cos\theta + m_1 g \cos(\theta)}$ = $\boxed{0}$

   *external*   *gravity*   *accum.*

---

[6] We did not have to choose this system. We could have chosen the bottom mass alone. In fact, as we continue with the problem, we will consider the bottom block separately.

Again, we used general expressions for the force and motion just in case our assumptions are incorrect and we have to start over.

Since we assumed 0 [NIF], there should be no flow. This allows us to set all the flows in the problem ($\dot{x}_1$ and $\dot{x}_2$) equal to zero and solve for all the efforts ($N_1$ and $f_1$). To verify the last assumption, we have to check that the magnitude of $f_1 \leq \mu_{s1}N_1$ and $N_1 > 0$ where the latter tests ensures that the block does not fly off the incline. Although the test for $f_1$ really says something about the magnitude of the force, if we were to get a negative value for $f_1$ we would be really suspicious of a wrong answer.

**SOLUTION: [I]**

To solve our model, we need the parameters of the problem. $g = 9.8$ [m/s$^2$], $m_1 = 5$ [kg], $m_2 = 2$ [kg], $\theta = 15^\circ$, $\mu_{s1} = 0.1$, $\mu_{s2} = 0.3$.

Using these parameters in the model, we can solve for the motion of the two blocks. We used Maple (see Maple output 3.15.1) to find the following: $f_1 = 17.77$, $f_2 = -5.078$, $N_1 = 66.33$ and $N_2 = 18.95$. The maximum friction beneath mass 1 is $f_{1\max} = \mu_{s1}N_1 = 5.685$. The maximum possible under mass 2 is $f_{2\max} = \mu_{s2}N_2 = 6.633$.

**VERIFICATION: [I]**

Notice that the friction force under block 2 is $-5.08$ [N]. This magnitude is less than the maximum possible $\mu_{s2}N_2 = 5.68$ [N]. This means our no slip assumption for the top is valid. The negative sign means that the force acts opposite the direction drawn in the freebody. This result does not mean that we have to rework the problem. However, we should verify that this result is what we expect. If the top block were to slip, it would slip down the incline. The friction force would oppose this downward slip and point upward – opposite the direction that appears in the freebody diagram. We therefore should expect to get a negative force. Thus, we have confidence that our calculated friction force is correct.

Unfortunately, the magnitude of friction required under the bottom block to prevent slip is 17.77 [N]. This is far greater than the maximum allowable value of $\mu_{s1}N_1 = 6.633$ [N]. Thus, our no slip assumption here is invalid. We must rework the problem. This time we will assume 1 [NIF]. Since the bottom block slid last time, we make another guess and assume the bottom block slides, but the top block sticks. Of course this is just a guess, an educated guess maybe, but still just a guess. By the way, what is the [NIA]? It is two, 1 [NIA] comes from the linear momentum (knowing the LM in one block gives us the LM in the other since they are stuck together, we assume) and the second independent accumulation is the gravity potential energy.

**FORMULATION: [II]**

To do this, we must treat the rate of change of $\dot{x}_1$ as an unknown, but the friction force $f_1$ is known to be $\mu_k N_1$. Thus, we switch one unknown for another, the number of unknowns does not change. Thus, we can still solve the problem. One important point here, since we are not solving for the bottom friction force, we must be certain the arrowhead is drawn correctly. Why? Because we cannot solve for $f_l$ and check the + - signs to determine whether it points in the correct direction, hence we better draw it correctly. Since the bottom block can only slide down the incline and since friction always opposes the slip, the friction must point up which is what the freebody (figure 3.14) shows.

Let's review our conservation equations in light of our new assumption. We will first repeat the previous equations. For the top block (2), we wrote the equations:

$$-f_2 - m_2 g \sin(\theta) = \frac{d}{dt}(m_2 \dot{x}_2)$$

$$-m_2 g \cos(\theta) + N_2 = 0$$

For the bottom block (1), we had:

$$-m_2 g \sin(\theta) - m_1 g \sin(\theta) + f_1 = \frac{d}{dt}((m_1 \dot{x}_1 + m_2 \dot{x}_2))$$

$$-m_2 g \cos\theta - m_1 g \cos(\theta) + N_1 = 0$$

Now based on our new assumption, we have the following new constraints:

$$\dot{x}_2 = \dot{x}_1$$

$$f_1 = \mu_{k1} N_1$$

These two equations say the blocks stick together while sliding, and that the friction beneath the bottom block is kinetic.

**SOLUTION: [II]**

The Maple transcript in 3.15.2 gives the modified solution for the motion of the two blocks. Notice that in this case, we know the effort variable (the friction under the bottom block) but do not know the flow variable (the velocity of the bottom block). Additionally, $\mu_k = 0.9\mu_s$ as given in the problem statement. The solutions to this formulation are $N_1 = 66.33$, $N_2 = 18.95$, $f_2 = -1.706$, and $\ddot{x}_1 = \ddot{x}_2 = -1.686$.

**VERIFICATION: [II]**

This time notice the friction force under block 2 is $-1.706$ [N]. This magnitude is far less than the maximum allowable $\mu_{s2} N_2 = 5.68$ [N]. This implies that our stick assumption for the top block is valid. Can you explain why its magnitude dropped below what is was before? The acceleration of the top and bottom blocks is $-1.686[\text{m/s}^2]$. The negative sign implies that our assumed direction is incorrect. We guessed that it had an acceleration up the incline when we plugged in a $+m\dot{x}$ (+ is up the incline). Based on the negative answer it accelerates down the incline. This is obviously correct.

**End 3.6** ─────────────────────────────

## 3.6  Springs and Dampers

─────────────────────────────

**Objectives**

*When you complete this section, you should be able to*

1. *Derive the impedance relationship of a linear spring.*

2. *Calculate forces in a spring.*

3. *Draw the freebody diagram of bodies with springs.*

4. *Explain what a damper is, how it can be constructed and how to draw a schematic of one.*

5. *Derive the impedance relationship of a damper and be able to calculate forces in a system with dampers.*

6. *Draw the freebody diagram of bodies with dampers.*

7. *Derive equations predicting the motion of masses connected with springs and dampers.*

─────────────────────────────

Any device that deforms under a force is a spring. Using this definition, almost everything acts like a spring (*e.g.* a diving board and a rubber band).

The force that causes a spring to deform can be a complicated function of the deformation and will be studied in other classes [27, 11]. We will deal mainly with simple linear springs in which the force is directly proportional to the deformation. This is a very good approximation to most springs, besides, the processes we will demonstrate are applicable to other impedance relationships with very minor modifications so you should be able to adapt what you learn to more complex situations easily. Unless otherwise stated, when we use the term spring, we mean a linear spring. Figure 3.15 shows the schematic diagram of a linear spring. The distance $l_o$ is the spring free length, the length when the spring feels no applied forces. Figure 3.16 shows a linear spring compressed an amount $x$. The magnitude of force acting on a linear spring is given by the equation:

$$|F_{\text{spring}}| = k\,|l - l_o| = k\,|x| \tag{3.1}$$

The constant $k$ is called the stiffness, spring constant or spring rate.

In addition to coiled springs that push and pull, there are torsional springs. Figure 3.17 shows a torsional spring. It sounds funny but in a linear torsional spring, a moment is generated proportional

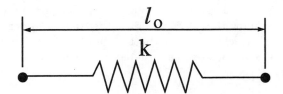

Figure 3.15: A Schematic Diagram of Linear Spring

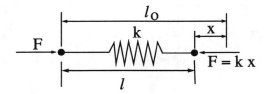

Figure 3.16: A Deformed Linear Spring

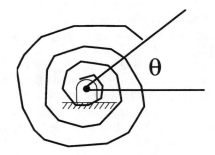

Figure 3.17: Schematic of a Torsional Spring

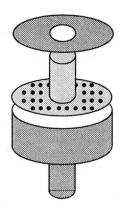

Figure 3.18: Three Dimensional View of a Simple Damper

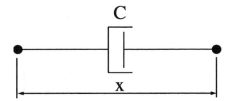

Figure 3.19: A Schematic Diagram of a Damper.

to the angle of deformation. For example the torque ($T$) generated by a linear torsional spring is:

$$T = G(\theta - \theta_o) \tag{3.2}$$

. The term $G$ is the stiffness, spring constant or spring rate. Angle $\theta_o$ is the free angle which corresponds to the freelength of a linear spring. A linear torsional spring is called linear because equation 3.2 is a linear equation. Usually the spring is called a torsional spring for short. Just as in "regular" springs, there are nonlinear torsional springs in which the torque/angle relationship is more complex. The process we discuss here will work for these complex relationships with minor modification.

Idealized springs are massless. This is often a reasonable assumption because for many man made springs, the magnitude of the forces they exert are much greater than their weights. If you pick up a spring used on an automobile, it feels heavy to you but just think of the force it exerts on the car! If you need to include the mass of the spring, it can get complicated. You'll have to study other texts to find ways to handle the situation [11].

Figure 3.18 shows a simple damper (dashpot). A damper could consist of a plate with small holes drilled in it immersed in a sealed can of fluid. When the plate is moved in the can, fluid must flow through the holes and around the edges of the plate. It takes effort to make the fluid flow. The composition of the fluid (*e.g.* air, water, molasses, oil), the size of the holes, and the velocity of the plate all effect the amount of force it takes to move the plate in the cup. Figure 3.19 shows the schematic diagram of a damper. An idealized damper is massless. The force required to pull a linear, idealized damper apart is directly proportional to the velocity of the damper. The impedance relationship is:

$$F_{\text{damper}} = C\dot{x} \tag{3.3}$$

See figure 3.20, the constant $C$ is called the damping constant or damping factor. The device is called a linear damper because equation 3.3 is a line.

Just as in springs, it is possible to construct a linear rotational damper. A rotational damper provides a resisting torque proportional to angular speed of motion.

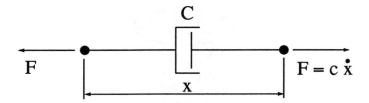

Figure 3.20: The Force Required to Move a Damper

---

When isolating an external spring or damper, do the following:

1. Draw a line where the force is applied.

2. Determine the magnitude of the spring or damper force in terms of the flow variables of the problem.

$$F_{\text{spring}} = k(\text{deflection})$$

$$F_{\text{damper}} = C(\text{rate of deformation})$$

3. Label the line draw in step 1 with the magnitude of the force.

4. Pretend that the flows are such that the magnitude of the force is a positive number. Look at the figure and determine if these flows would make the spring or damper stretched or compressed.

5. Determine the direction of the force and draw an arrowhead on the line representing the force.

---

Table 3.6: A Process for Drawing FBD of Springs and Dampers.

### 3.6.1  Freebody Diagrams of Springs and Dampers

In complex systems, it can be difficult to draw proper freebody diagrams when springs and dampers are involved. Students often get confused trying to determine whether springs are pulling or pushing. You may use the procedure outlined in Table 3.6 to assist you. The following example shows how to apply these steps in a simple system. If you are handling nonlinear springs simply substitute the proper impedance relationship for the linear ones shown in step 2 of Table 3.6.

**Example 3.7** ━━━━━━━━━━━━━━━━━━━━━━━━━━━━━━━━━━━━━━━━

### Equations of Motion for a Mass, Spring, and Damper

**PROBLEM:**

Figure 3.21 shows a mass hanging from a roof. The roof is 25 [m] high, the mass is 1 [kg], the

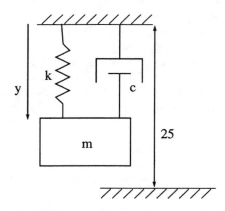

Figure 3.21: A Mass With a Spring and Damper

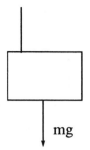

Figure 3.22: Step One for Drawing the Freebody Diagram for the Spring Force

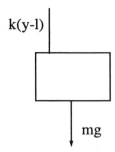

Figure 3.23: Steps Two and Three for the Spring Force

spring constant is 1 [N/m], and the spring's freelength is 2 [m]. Determine the motion of the mass if it is dropped from rest from the ceiling.

**FORMULATION:**

First, as usual, determine the [NIF] and [NIA] of the system. If we assume that the mass moves up and down only and there is no rotation of the mass, then the [NIF] is one. The way the mass is drawn, it looks like it could rotate; however, assume that it does not. Hey, let's keep it simple while we do this one. Since there is one [NIF] if we knew one flow, we would not need any conservation equations. Unfortunately, we do not know any flows, so we definitely need conservation equations.

To determine the system's [NIA], count the number of independent accumulations in the system. (1) The mass can accumulate linear momentum in the vertical direction. The vertical velocity of the mass will quantify this. The same variable will quantify the kinetic energy so only one is independent. (2) The spring can accumulate potential energy. The spring's length quantifies this energy. What about gravity potential? Why is it dependent on the spring potential energy?

The spring's elongation is independent of the mass's velocity. Therefore, the spring's potential energy accumulation is independent of the mass's kinetic energy accumulation. The gravitational potential energy of the mass can also be quantified by the spring's elongation. Therefore, this accumulation is not independent of the spring potential energy accumulation. If we ignore the damper's thermal energy accumulation, then there are no other independent accumulations. The [NIA] of the system is 2.

We will use the variables $y$ and $\dot{y}$ to quantify the spring's elongation and the mass's velocity. (See figure 3.21). These variable choices are convenient because they are measured from a fixed point. Note that positive is downward.

Next choose the mass as the system and draw a freebody diagram to help us keep track of the important forces. We use the steps in Table 3.6. We begin by determining the force caused by the spring. Figure 3.22 shows the first step for drawing this force, draw a line. Figure 3.23 shows the second and third steps (label the magnitude of force). If the spring is nonlinear, put in the proper impedance relationship on the line. Note that $l$ is the spring's free length. Now assume that $y$ is greater than $l$ so the value put on the diagram ($k(y-l)$) is positive, don't worry just pretend. This does not mean that $y$ is always greater than $l$, we are simply making sure our diagram is consistent for one possibility. If it is, you find it consistent with all possibilities. If you don't trust this, simply check other possibilities AFTER the diagram is finished. If $y$ is greater than $l$, then the spring will

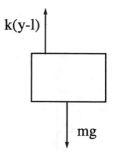

Figure 3.24: Fourth and Fifth Steps for the Spring Force

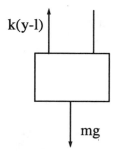

Figure 3.25: First Step for the Damper Force

be stretched, which means we (the mass) is pulling the spring to stretch it therefore the spring is pulling us. See figure 3.21. If the spring pulls the mass, then the spring's force points upward. Figure 3.24 shows the fourth and fifth steps for drawing the spring's freebody diagram.

We now determine the force caused by the damper. Figure 3.25 shows the first step for drawing this force. Figure 3.26 shows the second and third steps. We assume that $\dot{y}$ is greater than zero to make the label on the line positive. If $\dot{y}$ is positive, then the damper is being pulled apart. Again we are not saying $\dot{y}$ is always positive, we only want to make sure the diagram is correct if it were. You will find that once the diagram is correct, it works for negative $\dot{y}$ too. If the damper is being pulled apart, then the mass is pulling the damper so the damper is pulling the mass. If the damper pulls the mass, then the force on the mass is upward. Figure 3.27 shows the fourth and fifth steps for drawing the damper's freebody diagram. Note that **IF** the damper were connected beneath the mass **AND** the damper were pulling **THEN** the damper force would be downward. Pulling means exactly what its dictionary definition says.

Now that we have a freebody diagram for the system, we apply the conservation of linear momentum in the vertical direction. Since $y$ is positive downward, we choose the downward direction as positive. As a habit always choose positive direction is the direction of positive motion which is in this case downward.

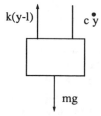

Figure 3.26: Second and Third Steps for the Damper Force

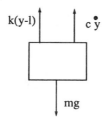

Figure 3.27: Fourth and Fifth Steps for the Damper Force

---

**Conservation of Linear Momentum (Vector)**
system[**mass fig.3.27**]      time period [ **differential**]

*input − output = accumulation*

| **input/output:** | mass flow | |
| | external forces | $-k\,(y-l)-C\dot{y}+mg$ |
| | gravity force | |
| **accumulation:** | | $m\dot{y}$ |

$y$:  $\boxed{-k(y-l)-C\dot{y}+mg}$ $=$ $\boxed{\frac{d}{dt}\left(m\dot{y}\right)}$

   *external*                    *accumulation*

---

$$-k\,(y-l)-C\dot{y}+mg = m\ddot{y} \qquad (3.4)$$

This is a second order differential equation that contains a single flow variable ($y$). This result is exactly what we expected based on the system's [NIF] and [NIA].

**SOLUTION:**

Before we proceed with solving these equations, let's just take a look at them a bit. Imagine the system in static equilibrium. That is, look for a solution to equation 3.4 where all derivatives of position ($y$) are zero. In other words, solve the following:

$$-k\,(y_s - l) + mg = 0$$

Note that we labeled $y$ as $y_s$ so you will remember the value is the point of static equilibrium. The value of $y_s$ is $y_s = \frac{mg}{k} + l$. Okay, now for the fun. Imagine drawing a line a distance $\frac{mg}{k} + l$ below the ceiling. Measure the position of the mass from this line. This is possible because the line is fixed so the measurement is from a fixed point. Let this position be called $z$. We can relate $z$ to $y$ as:

$$y = z + \frac{mg}{k} + l$$

Differentiate to find:

$$\dot{y} = \dot{z}$$

and

$$\ddot{y} = \ddot{z}$$

substitute all these values into equation 3.4 to find:

$$-k\left[\left(z + \frac{mg}{k} + l\right) - l\right] - C\dot{z} + mg = m\ddot{z}$$

simplify to find:

$$-kz - C\dot{z} = m\ddot{z}$$

Notice that if you measure the mass position from the static equilibrium point, the freelength of the spring disappears and so does the weight. This occurs so often, many textbooks do it all the time.

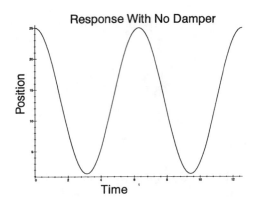

Figure 3.28: Response of the System When $C = 0$

Usually however they simply ignore the weight and freelength and not say anything about it which really causes students confusion. In this text, we will always leave in the freelength and weight. If they drop out, they drop out, who cares?

Maple program 3.15.3 shows the solution to the formulation in equation 3.4. As you look over this solution you may feel like it is overwhelming but keep in mind that the solution is computerized. How to find solutions to equations is a well understood process that continues to be more and more automated. This will become less and less a problem for Engineers and will be relegated to computer tools. What is important is an ability to "know what to expect." For example, there are some equations which cause problems for the automated solution methods. It is important for an engineer to have a feel for the answer so its validity can be determined. The solution of equation 3.4 will be solved with four values of $C$ to demonstrate some of the "expected" behaviors.

Notice that when the damping factor $C$ is zero, the solution is

$$y = -11.81\cos(t) + 11.81$$

Figure 3.28 shows the position of the mass measured from the floor $(25-y)$. Notice that the response oscillates and does not stop because there is nothing in the system that reduces its mechanical energy. There is no damping in this system (because $C = 0$). We say the system is undamped.

If we add a small damper with $C = \frac{1}{2}$, the solution becomes more complex. From the results of the maple program, we see that in this case the solution becomes

$$y = \Big($$
$$8\,e^{\left(1/4\left(-1+I\sqrt{15}\right)t\right)} + 7\,e^{\left(-1/4\left(1+I\sqrt{15}\right)t\right)} + I\,e^{\left(-1/4\left(1+I\sqrt{15}\right)t\right)}\sqrt{15} - I\sqrt{15} - 15\Big)$$
$$\left(-\frac{1181}{1600} + \frac{1181}{24000}I\sqrt{15}\right)$$

Note that the complex number $I = \sqrt{-1}$ is merely a convenient method of representing sines and cosines [30]. Figure 3.29 shows the response $(25 - y)$ versus time. Note that the system still oscillates but quickly comes to rest. The damper has converted all the system's mechanical energy into other forms. This system is said to be underdamped.

If the damping is increased to $C = 2$, the exponentials in the solution are "raised to the same power." This system is called critically damped. Now, the solution becomes

$$y = \frac{1181}{100} - \frac{1181}{100}e^{(-t)} - \frac{1181}{100}e^{(-t)}t$$

Figure 3.30 shows the response of this system. Note that the oscillation now disappears.

Finally, if $C$ is increased to 3, the exponentials in the solution are raised to two distinct real powers. This system is called overdamped. Its response is

$$y = -\frac{1181}{1000}\Big(-2\sqrt{5} + e^{\left(1/2\left(-3+\sqrt{5}\right)t\right)}\sqrt{5} + 3\,e^{\left(1/2\left(-3+\sqrt{5}\right)t\right)}$$
$$-3\,e^{\left(-1/2\left(3+\sqrt{5}\right)t\right)} + e^{\left(-1/2\left(3+\sqrt{5}\right)t\right)}\sqrt{5}\Big)\sqrt{5}$$

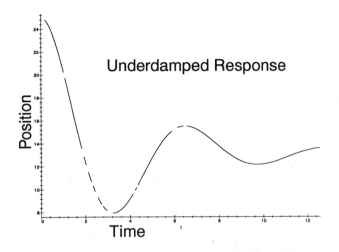

Figure 3.29: Response of the System With Little Damping

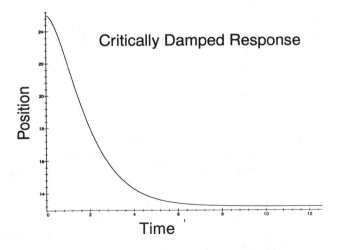

Figure 3.30: Response of the System With Critical Damping

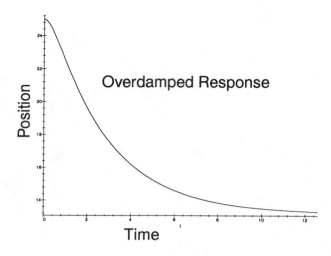

Figure 3.31:  System Response When Overdamped

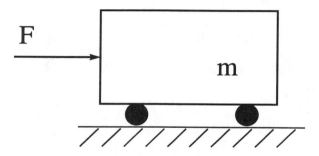

Figure 3.32:  A Force Trying to Store Energy in a Mass

Figure 3.31 shows the response of this system. Note that it is difficult to distinguish this response from that of a first order system.

**End 3.7** ────────────────────────────────────────────

The following example is intended to show two things.  First the variables used to express the length of the spring might be considered "weird" and second, it shows a phenomenon called resonance.

## Example 3.8 ────────────────────────────────────────

### A Paddle-Ball Game.

**Problem:**

A paddle ball game consists of a rubber ball connected to a wooden paddle with a rubber band. The object of the game is to continuously hit the ball with the paddle. Let's represent the ball as a mass of 1 [kg],[7] let the rubber band be a spring with stiffness of 1 [N/m] and the freelength of the band be 2 [m].

Since the point of this example is not to figure out all the weird ways the ball can move when you play this game but to take another look at some specific responses, let's keep it simple. Let the ball move up and down only and let's forget about the collision between ball and paddle. Figure 3.33 shows a simplified schematic of the game. The mass represents the ball and the spring represents the rubber band. The paddle, which is not shown, moves up and down and is represented by the known distance $z$. Formulate a mathematical model of the motion of the ball ($m$ in the figure) and solve the formulation.

**Formulation:**

─────────────────────────────

[7]Do you think our choice of ball mass is realistic?  We chose it just to make the numbers we get in the solution look nice.  You can repeat the example with more reasonable numbers if you want.

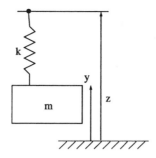

Figure 3.33: A Paddle-Ball Game

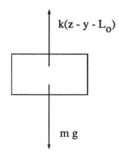

Figure 3.34: Freebody Diagram of the Paddle-Ball

First, count the [NIF] and [NIA] of the system. If we assume that the ball only moves vertically, there is two [NIF](the ball and paddle can move independently). One of the flows (the top position of the spring) is given. The independent accumulations are the ball's kinetic energy and the spring's potential energy. Therefore, there are at least two [NIA]. The mass's gravity potential energy is related to the spring's potential energy because both accumulations can be quantified using the single unknown variable $y$. Since there are two [NIF] and only one flow is given, we must apply a conservation equation.

Next, choose the mass as the system and draw a freebody of it. Figure 3.34 shows a correct freebody diagram. We follow the procedure in table 3.6 to draw the spring force correctly in the diagram. Here was the first point, even though we used some "weird" distances, you should be able to express the stretch of the spring in terms of the variables.

Now, apply the conservation of linear momentum in the vertical direction.

**Conservation of Linear Momentum (Vector)**
system[**paddle-ball  fig.3.34**]      time period [ **differential**]

*input − output = accumulation*

| input/output: | mass flow external forces gravity force | $-mg + k\,(z - y - L)$ |
|---|---|---|
| accumulation: | | $m\dot{y}$ |

| | | | |
|---|---|---|---|
| $y$: | $\boxed{-mg + k\,(z - y - L)}$ | $=$ | $\boxed{\frac{d}{dt}\,(m\dot{y})}$ |
| | *external* | | *accumulation* |

$$-mg + k\,(z - y - L) = m\ddot{y} \tag{3.5}$$

Do you think $mg$ will cancel provided the variables are measured from the static equilibrium point of the spring? Yes. Give it a try and see what "unexpected terms" show up.
**Solution:**

To solve this formulation, we have to choose some initial conditions. Since there are 2 [NIA], we need 2 initial conditions. Let the ball start from rest ($\dot{y} = 0$) when the spring force balances the ball's weight (at static equilibrium). This says that at $t = 0$, all derivatives of $y$ are zero and solving equation 3.5 for $y$ at $t = 0$ gives:

$$y = z - L - \frac{m}{k}g$$

In summary then the initial conditions we'll use are $y = z - L - \frac{mg}{k}$ and $\dot{y} = 0$. Maple program 3.15.4 shows the solution to the formulation when $z = \sin \omega t$ (where $\omega$ is constant).

The general solution is

$$y = -\frac{1}{2}\left(-2k^2 L \sqrt{-mk} + 2L\omega^2 mk\sqrt{-mk} + 2k^2 \sin(\omega t)\sqrt{-mk}\right.$$

$$\left. -2mgk\sqrt{-mk} + 2m^2 g\omega^2\sqrt{-mk} - k^2\omega me^{\left(\frac{\sqrt{-mk}\,t}{m}\right)} + k^2\omega me^{\left(-\frac{\sqrt{-mk}\,t}{m}\right)}\right)$$

$$\Big/\left((-k + \omega^2 m)\,k\sqrt{-mk}\right)$$

To simplify this, substitute some numbers. Let $m = 1$ [kg], $g = 9.81$ [m/s$^2$], $k = 1$ [N/m], and $L = 2$ [m]. Doing this, the solution reduces to

$$y = -\frac{1}{100}\,\frac{-100\,\omega\sin(t) - 1181 + 1181\,\omega^2 + 100\sin(\omega t)}{-1 + \omega^2}$$

Figure 3.35 shows $y$ as a function of $\omega$ and $t$. Although the graph is coarse, we can see that there is little activity except near $\omega = 1$. Taking the limit of $y$ as $\omega \to 1$, we find that

$$y = \frac{1}{2}\sin(t) - \frac{1181}{100} - \frac{1}{2}\cos(t)\,t$$

Now the sine term is bland, nothing of real interest there, but the real interesting term is the cosine. Note that its amplitude builds continuously. Take the derivative of $y$ and look at the magnitudes of the terms. You will find that there is a term with a continuously increasing magnitude. This means the mass will have a higher and higher speed as time goes on. This means the ball is storing more and more energy. Where do you think this energy is coming from? Yes, from whoever is moving the upper end of the rubber band.

Figure 3.36 shows $y$ along with the input $z$. Note that the two responses are out of phase with each other. For example, when $z$ is maximum, $y$ is close to minimum. It is this out of phase phenomenon and the large amplitude of the game that allows the paddle-ball game to work. The large amplitude says that the ball moves a significant amount. The out of phase phenomenon means that the ball rises up to hit the paddle as the paddle comes down to meet the ball. Do you think it would work if the ball was rising at the same time as the paddle?

If $\omega$ is smaller than 1, the amplitude of the ball does not increase indefinitely, hence it will not absorb or store very much energy. The solution with $\omega = 0.01$ is

$$y = -\frac{100}{9999}\sin(t) - \frac{1181}{100} + \frac{10000}{9999}\sin\left(\frac{1}{100}t\right)$$

Notice how the sinusoid with the largest influence (the one with the 10000/9999 in front) has the same phase as $z$ (the same angle). We say the output is "in phase" with the input. Basically, the spring is not deforming much and the ball is merely riding along with $z$.

For large $\omega$, there is a similar response. For example, for $\omega = 2$, 3, and 10, the solutions to the formulation are respectively

$$y = \frac{2}{3}\sin(t) - \frac{1181}{100} - \frac{1}{3}\sin(2t)$$

$$y = \frac{3}{8}\sin(t) - \frac{1181}{100} - \frac{1}{8}\sin(3t)$$

$$y = \frac{10}{99}\sin(t) - \frac{1181}{100} - \frac{1}{99}\sin(10t)$$

The magnitudes of the sine terms diminish as $\omega$ increases. This means that the spring is being pulled too quickly for the mass to respond.

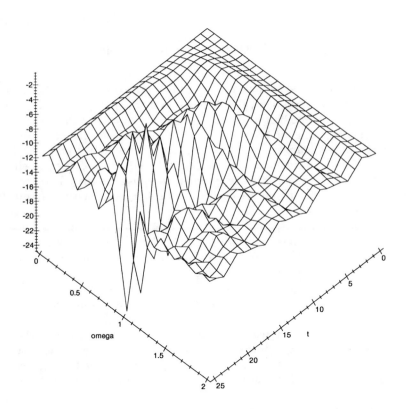

Figure 3.35: Plot of Response Versus Time and $\omega$

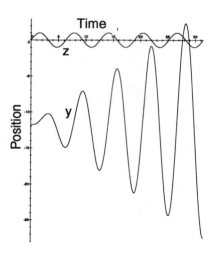

Figure 3.36: Response $y$ Shown With Input $z$

The point of this solution is to show how energy can be stored in a mass. Once started, a mass on a spring will oscillate at a frequency called the natural frequency. If there is a damper, this motion will die away. In the paddle ball, the natural frequency was 1 [radian/second]. The natural frequency can be determined from the transient solution of the system's response. You will study this in detail in a mathematics course. Considerable energy is stored in the mass when it is excited (bumped or pulled) at the same frequency as the natural frequency of the system. This is what happened in this example. The same thing occurs when you push a swing. If you push and wait for the swing to return, then push again you are exciting the swing at its natural frequency and before long the swing will be really moving.

**End 3.8** ────────────────────────────────────────────────────

Why is it important to know how to store energy? Sometimes, we want to put mechanical energy into a mass. For example, if we want to shake a can of paint to mix its colors, we have to put energy into the paint. When we swing, we want to put energy into our body so that the swing will move. Other times, we want to keep energy out of a system. For example, if a building or bridge is subjected to an earthquake or high winds, we want to keep the energy out of them so that they do not sway.

Energy can be kept out of a system using a damper. For example, consider what would happen to a spring-mass system when a damper is installed. In general, damping prevents the response from becoming infinite when energy is input at the natural frequency. As the forcing frequency increases above the natural, the damper tends to lock up causing relatively high forces to be transmitted to or from the mass.

The following example shows how to draw a freebody diagram of a system with several masses, dampers, and springs.

**Example 3.9** ────────────────────────────────────────────────

### Equations of Motion

**PROBLEM:**

Formulate a mathematical model predicting the motion of the two masses in figure 3.37. Assume that we know $F$ and want to determine the motions of the two masses.

**FORMULATION:**

We begin by counting the [NIF] and [NIA]. Since there are two possible motions, there are two [NIF]. Both masses can accumulate linear momentum and kinetic energy. However, since linear

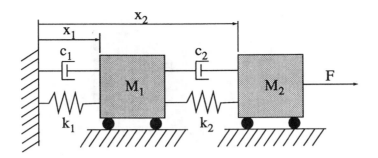

Figure 3.37: A Two Mass System

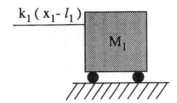

Figure 3.38: Spring 1 Force Magnitude Acting on Mass 1

momentum and kinetic energy are dependent the masses contribute one [NIA] each. Continuing our count, we note that both springs can accumulate energy independent of the energy accumulated in the masses. The energy accumulated in the springs depends on the deflection in the spring while the energy accumulated in the masses depends on the velocity of the masses. The two spring energies are independent of each other because knowing one spring's deflection does not determine the other's. They each contribute an additional [NIA]. Ignoring thermal energy effects, there are no other independent accumulations. Thus, the system's [NIA] is four. Since there are two masses in the problem, we anticipate having to choose two systems before the problem is over.

Using our [NIF] information, we define two motion variables. Arbitrarily, we will choose $x_1$ and $x_2$ both measured from the wall. These variables could be measured from any convenient nonaccelerating point. Since there is two [NIF] and no flows are given, we must write conservation equations.

Since there are two masses, Table 3.2 suggests expecting to choose two systems. If we choose both masses together as one system, it will be a little confusing when we try to write the linear momentum. We will Choose the two masses separately and begin by drawing freebody diagrams. Since only horizontal motion is of interest, we ignore the vertical forces. Remember the heuristic for drawing freebody diagrams of bodies with an unknown motion (as we are doing here), it is helpful to assume that the body moves in the positive direction.

We begin by drawing the freebody for mass one. If we choose spring one to be outside the system, it exerts a force on the mass. The horizontal line in figure 3.38 represents this force (we assume the mass moves horizontally). The magnitude of the spring force is $k_1 (x_1 - l_1)$ (where $l_1$ is the freelength of spring 1). Now, we will reason out the correct direction of the force. According to Table 3.6, we assume $x_1$ is greater than $l_1$ (*i.e.* if we assume the body has moved to the right). If this is true, the mass stretches (pulls) the spring so the spring pulls the mass. The force points to the left as figure 3.39 shows.

If we choose damper one to be outside the system, it exerts a force whose magnitude is $C_1 \dot{x}_1$. We now assume that $\dot{x}_1$ is positive. If this is true, the mass pulls the damper apart so the damper pulls the mass, therefore the force points to the left. Figure 3.40 shows the forces exerted by spring and damper one.

If damper two is outside the system, it exerts a force of magnitude

$$C_2(\dot{x}_1 - \dot{x}_2)$$

It is important to realize the rate of deformation is the difference in velocities, it is not important to put $\dot{x}_1$ first. Assuming $\dot{x}_1$ is greater than $\dot{x}_2$ (positive magnitude), then the damper is compressing,

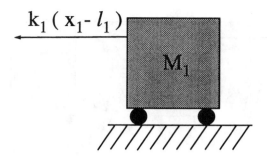

Figure 3.39: Spring 1 Force Magnitude and Direction Acting on Mass 1

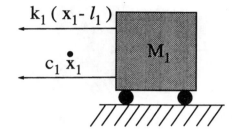

Figure 3.40: The Effect of Spring and Damper 1 on Mass 1

masses 1 and 2 push the damper compressing it, so the damper pushes the masses. Mass 1 is pushed to the left. As a check, draw the force from damper 2 if $\dot{x}_1$ is subtracted from $\dot{x}_2$. Explain how it will affect the solution.

If spring two is outside the system, it exerts a force of magnitude

$$k_2 \left( x_2 - x_1 - l_2 - w_1 \right)$$

Here it is important to realize the deformation is the position difference, it does not matter whether $x_2$ or $x_1$ goes first. Also note that we are subtracting the freelength and the width of mass 1. If we lump both constants into one ($L_2$), we can write

$$k_2 \left( x_2 - x_1 - L_2 \right)$$

this just makes it less cluttered. If $x_2 - x_1$ is greater than $L_2$ the masses stretch (pull) the spring, so the spring pulls the masses. Figure 3.41 shows the complete freebody excluding the vertical forces for mass 1.

The complete freebody for mass 2 appears in figure 3.42. We can determine the forces and directions in the same manner as for mass 1 or by simply making the forces shared by masses 1 and 2 equal and opposite.

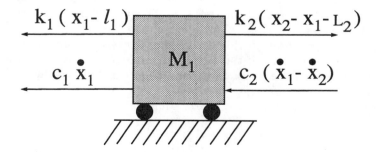

Figure 3.41: The Complete Freebody Diagram for Mass 1 (Excluding Vertical Forces)

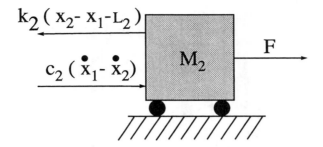

Figure 3.42: The Complete Freebody Diagram for Mass 2 (Excluding Vertical Forces)

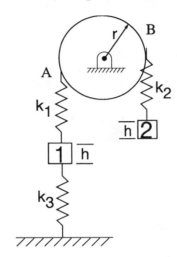

Figure 3.43: A Massless Pulley With a Spring and Two Masses.

Applying conservation of linear momentum for masses 1 and 2 gives:

$$-k_1 \left( x_1 - l_1 \right) - C_1 \dot{x}_1 + k_2 \left( x_2 - x_1 - L_2 \right) - C_2 \left( \dot{x}_1 - \dot{x}_2 \right) = \frac{d}{dt} \left( M_1 \dot{x}_1 \right)$$

$$-k_1 \left( x_2 - x_1 - L_2 \right) + C_2 \left( \dot{x}_1 - \dot{x}_2 \right) + F = \frac{d}{dt} \left( M_2 \dot{x}_2 \right)$$

These equations contain two flows (matching the [NIF]) and are each second order (matching the [NIA]). We are finished.

What do you think the equations will be like if we measured the distances from static equilibrium rather than from the wall? The freelengths and widths of the masses would cancel.

**End 3.9** ─────────────────────────────────────────────

**Example 3.10** ─────────────────────────────────────────────

## Blocks Pulleys and Springs

**PROBLEM:**

Now for a more challenging example. Formulate a mathematical model predicting the motion of the two masses in figure 3.43. As usual, let's assume we know all the parameters (masses, stiffnesses, diameters and freelengths). Before we get started let's discuss some simplifications. Since we have not covered rotation yet let's assume the motions are small enough that we can "assume" points a and b move small amounts. What this means is the motion of a is equal and opposite to the motion of b. In other words if a goes up, b goes down. Also let's assume the springs are attached at a and b, that there is no slip between the springs and the pulley. Also, since rotation will be discussed later, let's assume the pulley is very lightweight (massless) so we can ignore its angular momentum.

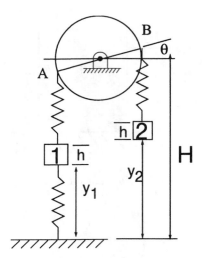

Figure 3.44: Three Coordinates for Pulley and Masses.

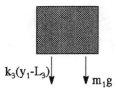

Figure 3.45: Partial Freebody Diagram of Mass 1, Leaving Out the Force Due to Spring 1.

**FORMULATION:**

We begin by counting the [NIF] and [NIA]. If we ignore all motions side to side, then there are 3 [NIF]. The two masses can move up and down, hold them and the pulley can rock back and forth. The two masses can accumulate linear momentum so they contribute 2 [NIA]. Mass 2 can accumulate gravity potential so add one making 3 [NIA]. Mass 1 can accumulate gravity potential so we're up to 4 [NIA]. Spring $k_3$ is dependent on others. Why? Now for springs 1 and 2. They can accumulate energy, are they dependent or independent? Suppose we know all the accumulations listed thus far. This means we know where the masses are located (we also know their velocities but that is irrelevant for springs). If we know the positions of the masses, do we know the energy stored in springs 1 and 2? That energy depends not only on the positions of the masses but also the location of the pulley. Given the mass positions and the pulley position, it seems possible to determine all the accumulations hence springs 1 and 2 could only add one [NIA] if any. We have either 4 or 5 [NIA]. Which do you think it is? We'll come back to it later.

Since there are 3 [NIF] and there are no flows given, we must apply a conservation equation. Since there are two masses in the diagram, we expect two systems, let's take the masses individually.

Figure 3.44 shows three coordinates which we will use for describing the motions. We chose three coordinates because there are three [NIF]. We measure from fixed positions to make it easy to write the total momentum.

Now for the hard part, the freebody diagrams. Start with mass 1. Figure 3.45 shows a partial freebody showing everything except the force from spring 1. Now follow the steps to determine the force caused by spring 1. The magnitude of the force is approximately (the arc length $r\theta$ is not exactly correct):

$$k_1 \left( H - y_1 - h - r\theta - L_1 \right)$$

Now imagine that $H - y_1 - h - r\theta$ is greater than $L_1$ then the spring is in tension which means the mass pulls the spring and the spring pulls the mass. Hence the spring force is up. Figure 3.46 shows the complete freebody diagram for mass 1.

Now for mass 2. Figure 3.47 shows a partial freebody showing everything except the force from

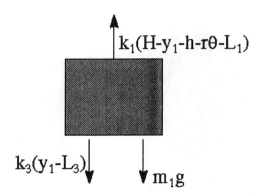

Figure 3.46: Freebody Diagram of Mass 1.

Figure 3.47: Partial Freebody Diagram of Mass 2, Leaving Out the Force Due to Spring 2.

spring 2. Follow the steps to determine the force caused by spring 2. First the magnitude of the force is approximately:

$$k_2 \left( H - y_2 - h + r\theta - L_2 \right)$$

If we pretend $H - y_2 - h + r\theta$ is greater than $L_2$ then the spring is stretched which means the mass is pulling the spring so the spring is pulling the mass and the force is upward. The complete freebody is shown in figure 3.48.

Now all that's left is to apply the conservation of linear momentum to each system. For mass 1 we have:

$$k_1 \left( H - y_1 - h - r\theta - L_1 \right) - k_3 \left( y_1 - L_3 \right) - m_1 g = \frac{d}{dt} \left( m_1 \dot{y}_1 \right) \tag{3.6}$$

For mass 2 we have:

$$k_2 \left( H - y_2 - h + r\theta - L_2 \right) - m_2 g = \frac{d}{dt} \left( m_2 \dot{y}_2 \right) \tag{3.7}$$

Now these equations are each second order so if the [NIA] is 4, we have all the differential equations we need. Notice however that there are three unknown flow variables ($y_1$, $y_2$ and $\theta$) exactly matching

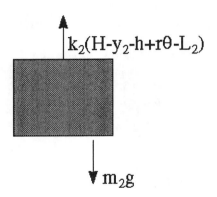

Figure 3.48: Freebody Diagram of Mass 2.

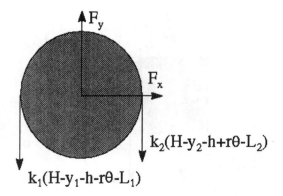

Figure 3.49: A Freebody Diagram of the Pulley.

the [NIF]. Unfortunately, since there are only two equations, we cannot solve for all three unknowns. We need another equation. What to do. Since we expect three flow variables and that is exactly the number we have, there is no reason to attempt eliminating any of them from the equations. The process of eliminating flow variables will be covered later, for now we realize (using the [NIF]) that elimination is not an option.

What is needed is the realization that there really are three "bodies" in the problem not two. The pulley is a body even though it has very little (no) mass. Draw a freebody diagram of the pulley as figure 3.49 shows. Now if we apply conservation of linear momentum, we could determine the forces $F_x$ and $F_y$ but who cares? Apply instead conservation of angular momentum. Choose the center of the pulley and compute moments about the center. Since the pulley has no mass, there is no angular momentum no matter what motion the pulley possesses. Hence the conservation of angular momentum is:

$$k_1 \left( H - y_1 - h - r\theta - L_1 \right) r - k_2 \left( H - y_2 - h + r\theta - L_2 \right) r = \frac{d}{dt}(0) = 0 \qquad (3.8)$$

Now this makes a third equation for the three unknowns hence we are finished. Notice also that it is algebraic not differential. What then can you say about the [NIA] of the mechanism? It obviously has a [NIA] of 4.

Now that we know there is a [NIA] of 4, how can we justify this? Well look back at the discussion on [NIA]. Given the positions of the masses, why is it that we automatically "know" the energy accumulated in springs 1 and 2? Well by fixing the positions of the masses, we fix the bottom ends of the springs. Since there is no mass on the pulley, the pulley will automatically and immediately move to where the moments balance hence there is only one angle $\theta$ that can exist. How would this differ if the pulley HAD mass? If the pulley had mass, it could not respond immediately to the moments and hence it is not possible to know exactly how much energy is in the springs 1 and 2 and they would be independent increasing the [NIA].

Do you think a conservation of energy equation would be useful? Yes provided the pulley is used in the formulation.

What would happen if the distances $y$ were measured from the static equilibrium point? Well, first determine the values of $y_1$, $y_2$ and $\theta$ at static equilibrium. Find this by setting all derivatives to zero (everything is at rest) and solve equations 3.6 through 3.8 for the variables at static equilibrium. There is considerable algebra, we used Maple to find at static equilibrium (see Maple program 3.15.7):

$$y_1 = \frac{k_3 L_3 - m_1 g + m_2 g}{k_3}$$

$$y_2 = -\frac{\left( \begin{array}{c} -2k_2 H k_1 k_3 + 2k_2 h k_1 k_3 + k_2 k_1 k_3 L_3 \\ -k_2 k_1 m_1 g + k_2 k_1 m_2 g + k_2 k_3 m_2 g + k_2 k_3 k_1 L_1 \\ +k_2 L_2 k_1 k_3 + m_2 g k_1 k_3 \end{array} \right)}{k_2 k_1 k_3}$$

$$\theta = -\frac{k_1 k_3 L_3 - k_1 m_1 g + k_1 m_2 g + k_3 m_2 g - k_3 k_1 H + k_3 k_1 h + k_3 k_1 L_1}{k_3 r k_1}$$

Wow! Now define new variables as (note that we are merely subtracting the static solution) $Y_1 = y_1 - \frac{k_3 L_3 - m_1 g + m_2 g}{k_3}$, $Y_2 = y_2 +$Stuff from above and $\Theta = \theta + \frac{k_1 k_3 L_3 - k_1 m_1 g + k_1 m_2 g + k_3 m_2 g - k_3 k_1 H + k_3 k_1 h + k_3 k_1 L_1}{k_3 r k_1}$. Notice that these new variables are zero if the system is at static equilibrium point. Finally differentiate the new variables twice to find: $\ddot{Y}_1 = \ddot{y}_1$, $\ddot{Y}_2 = \ddot{y}_2$ and $\ddot{\Theta} = \ddot{\theta}$. Substitute these new variables into the original set to find (again we used Maple):

$$-Y_1 k_1 - \Theta r k_1 - Y_1 k_3 = m_1 \ddot{Y}_1$$

$$k_2 \left(-Y_2 + \Theta r\right) = m_2 \ddot{Y}_2$$

$$Y_1 k_1 + \Theta r k_1 - Y_2 k_2 + k_2 \Theta r = 0$$

Now aren't these much nicer to use? How did we know all the cancellation would occur? Well it usually does when you measure distances from the static equilibrium. In many classes considerable time is spent showing how to derive the final set of equations from scratch. The big problem is how do you justify leaving out the gravity and freelengths? Its a hard sell and often novices make mistakes. If you use Maple or similar tool, you can stick to the fundamentals and use the tool to subtract off all the constants. Things have changed. In the "old days" it was important to cut through the algebra, now there are tools to handle these things. Use the tools.

Solving the equations is not too bad. Use the last equation to solve for $\Theta$ then eliminate $\Theta$ from the previous two leaving two second order differential equations for two unknowns $Y_1$ and $Y_2$. We would use Maple to solve these.

**End 3.10**

---

## 3.7 Choosing Systems

---

**Objectives**
*When you complete this section, you should be able to*

1. *Estimate how many independent linear and angular momentum equations a system will generate.*

2. *Estimate the number of unknowns a particular system will contain.*

3. *Explain how idealized supports (attachment points) can simplify a mechanics problem.*

4. *Choose systems in mechanics problems wisely.*

5. *Identify two-force bodies and explain how this simplification helps in mechanics problems.*

6. *Explain what would happen if one did not recognize two-force bodies.*

---

In the previous examples, the isolated bodies were easy to cut loose because they merely sat on top of each other. When isolating some systems, however, it may be necessary to cut into a body. When this is necessary, it is important to know what kinds of forces and moments are present at the cuts. In addition, some systems are composed of several bodies, links or masses. It is important to know how many systems to choose and how to choose them wisely.

Since linear momentum is a vector quantity, we can generally write three conservation of linear momentum equations for every system. Often, systems are assumed to be two dimensional. If it is, then linear momentum in the third direction produces a useless equation such as $0 = 0$. The equation is valid, but it does not yield any results. In addition, since angular momentum is a vector, a single body will have three angular momentum equations. Likewise, if the body is considered to be two dimensional, then rotation in two directions will be trivial. In summary, a three dimensional system will have a total of six momentum equations. A two dimensional system will have a total of three. In the absence of non-mechanical energies, the momentum and energy equations are identical.

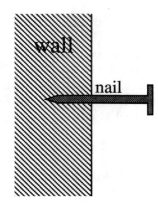

Figure 3.50: A Nail in a Garage Wall

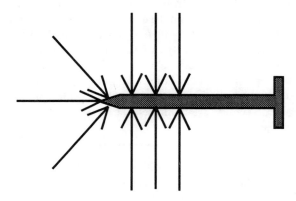

Figure 3.51: Freebody Diagram of the Entire Nail

Therefore, we cannot write momentum and energy and expect them to be independent unless we have non-mechanical energies.

When isolating systems for analysis, avoid choosing systems that contain bodies with different motions because it can be difficult to calculate the time rate of change of momentum when the system has multiple momentums. This is not impossible, but generally is more difficult than choosing separate systems. A classic exception to this rule is when the system of multiple bodies has zero external forces or moments. When this happens, we can often gain insight by choosing a system that includes all the bodies together.

When there are several choices for systems. Consider choosing a system that has a minimal number of unknowns. In particular, try finding a system in which the number of unknowns is less than or equal to the expected number of equations for the system.

Occasionally, it is necessary to cut into a body to isolate it from the surroundings. Consider for example, the forces acting on a nail in a wall (see figure 3.50). To draw a freebody diagram of this nail, we must first isolate it. There is more than one way to do this, therefore we have to decide which way best suits our purpose. We can isolate the nail from the wall in the following ways:

1. We can cut between the nail and the wall so that our system consists of the entire nail. The freebody diagram for this system appears in figure 3.51. Since contact between the nail and the wall occurs over an area, the wall can exert a distributed load on the nail. The effect of this distributed load is unknown, but we can generalize it as a vertical force, a horizontal force, and a moment (*i.e.* twist). As long as the nail is rigid, the distributed load in figure 3.51 reduces to the equivalent set of loads in figure 3.52. It is possible for all these simplified loads to be positive, negative, or zero depending on the situation.

2. We can isolate only the part of the nail that sticks out of the wall. By cutting the nail where it intersects the wall, we get the freebody diagram in figure 3.53. Note the following regarding

| Analysis Assumptions |
|---|
| (1) These simplified loads can be positive, negative, or zero depending on the situation. |

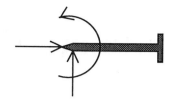

Figure 3.52: (simplewholenail): A Simplified Freebody Diagram

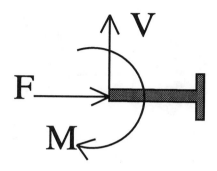

Figure 3.53: Freebody Diagram of Part of the Nail

this piece of nail:

- $V$ denotes a possible vertical force. Since $V$ acts perpendicular to the long axis of the nail, it would be called a shear.

- $F$ denotes a possible compressive horizontal force. Since $F$ is parallel to the long axis, it is called an axial force.

- $M$ denotes a possible moment.

If you do not believe that all three loads (axial, shear and moment) are possible, imagine that your hand is the wall holding the nail (ouch!). Since your hand could push the nail vertically, there is a possible vertical shear force ($V$) in the freebody. Likewise, since you could push the nail horizontally, there is a possible horizontal axial force ($F$) in the freebody. Finally, since you could rotate the nail, there is a possible moment ($M$) in the freebody.

In a general case, when we cut through a body to isolate it (as we did here) we will end up having to draw several forces and moments at the cut. This unfortunately introduces several unknowns into the problem.

Many bodies are connected with idealized joints that require fewer forces and moments in their freebody when cut. For example, we can idealize the model of a door hinge as a frictionless pin joint. The door hinge prevents the wall from resisting the opening and closing of the door. Figure 3.54 shows how an ideal pin joint is drawn. The ideal pin operates such that body 1 can push and pull body 2 in all directions, but body 1 cannot twist body 2 (in the plane of the paper). If we want to isolate body 2, we have to cut it away from body 1. It is best to do this by cutting through the pin. If we do this, the freebody diagram looks like figure 3.55. The forces can be either positive or negative, but there can be no moment at the pin because body 1 cannot twist body 2 (in the plane of the paper).

There are several other idealized connections. Table 3.7 shows the types of loads they can apply. In general, we try to cut a body loose such that we introduce a minimum number of unknowns. If a body has ideal joints, the cut point that introduces the fewest unknowns will almost always be an idealized connection.

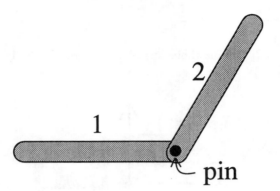

Figure 3.54: An Ideal Frictionless Pin (Hinge) Joint

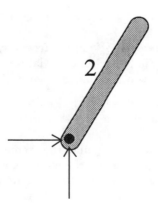

Figure 3.55: The Freebody Diagram of a Pin Joint

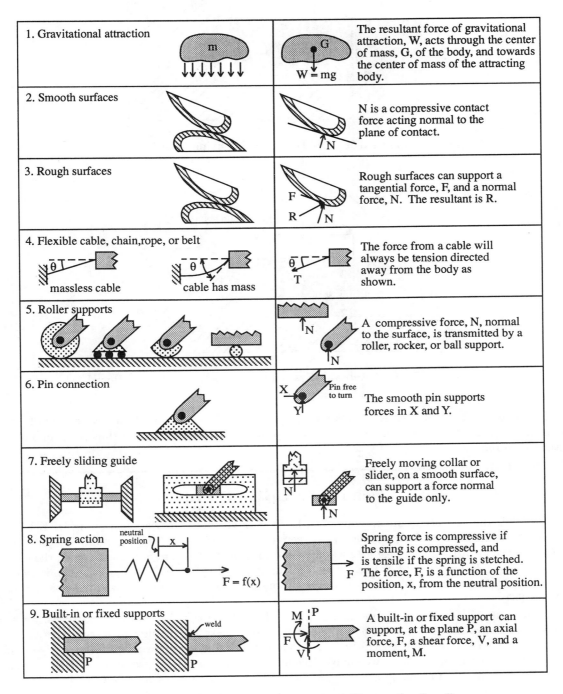

| | | |
|---|---|---|
| 1. Gravitational attraction | | The resultant force of gravitational attraction, W, acts through the center of mass, G, of the body, and towards the center of mass of the attracting body. |
| 2. Smooth surfaces | | N is a compressive contact force acting normal to the plane of contact. |
| 3. Rough surfaces | | Rough surfaces can support a tangential force, F, and a normal force, N. The resultant is R. |
| 4. Flexible cable, chain, rope, or belt | massless cable      cable has mass | The force from a cable will always be tension directed away from the body as shown. |
| 5. Roller supports | | A compressive force, N, normal to the surface, is transmitted by a roller, rocker, or ball support. |
| 6. Pin connection | | The smooth pin supports forces in X and Y. |
| 7. Freely sliding guide | | Freely moving collar or slider, on a smooth surface, can support a force normal to the guide only. |
| 8. Spring action | $F = f(x)$ | Spring force is compressive if the sring is compressed, and is tensile if the spring is stetched. The force, F, is a function of the position, x, from the neutral position. |
| 9. Built-in or fixed supports | weld | A built-in or fixed support can support, at the plane P, an axial force, F, a shear force, V, and a moment, M. |

Table 3.7: Idealized Connections and Their Respective Loadings

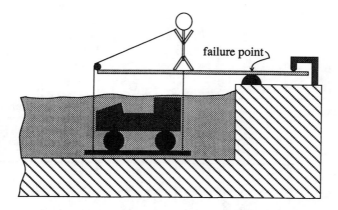

Figure 3.56: The Defective(?) Diving Board

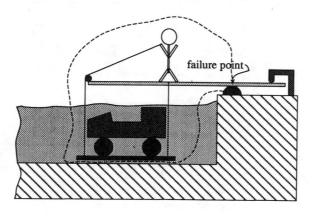

Figure 3.57: An Acceptable System for the Defective Diving Board

The following example demonstrates how a system can be chosen so that several complicated forces can be ignored. It also demonstrates how to properly choose a system so that forces of interest enter the formulation.

## Example 3.11

### Failure of a Diving Board

DEEP END Incorporated builds diving boards.  A pool cleaner (Mr. Poole) is suing DEEP END because he claims one of their boards was defective.  The pool cleaner was standing on the board attempting to lift a 100 [lbs] (weight in water) golf cart from the bottom of the pool when the board snapped.  The CEO of DEEP END has requested our assistance.  Figure 3.56 shows the configuration that the pool cleaner used.  Determine the system we should choose to analyze why the board broke.

**ANSWER:**

To help the CEO, we have to determine the loading in the board.  To accomplish this, we will choose a section of the diving board and model it.  The hard part is deciding which section.  Realizing that internal forces (stresses) caused the board to fail and that internal forces in a rigid body do not appear in the conservation laws, we choose the section in which the failure point is at the system boundary and in which as many other forces as possible are internal.  This allows us to determine the forces present at the failure point and ignore other superfluous factors.

An acceptable system appears inside the dashed curve in figure 3.57.  The freebody diagram of this system appears in figure 3.58.  This system makes the problem of dealing with the forces caused by the rope and the interaction of Mr. Poole's tennis shoes with the board immaterial because they are internal forces.  The only external forces are Mr. Poole's weight, the golf cart's weight, the board's weight, and the forces and torques present in the board at the failure point.  Note that since none of the bodies are moving, it is easy to express the change in momentum of all the bodies in the

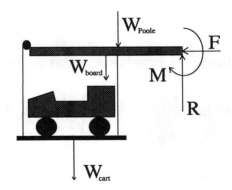

Figure 3.58: A Freebody Diagram for the Defective Diving Board System

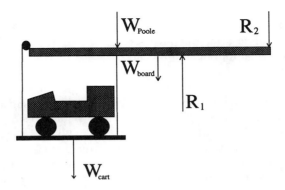

Figure 3.59: Freebody Diagram of Diving Board Including Mr. Poole and the Golf Cart

system, its zero. Since it is easy to express the change in momentum, it is okay to choose a system that has multiple bodies in it.

The reason we cut through the diving board was to bring the forces present in the board into the analysis. Since the magnitudes of these forces are related to the failure of the board, we wanted to determine them. Some other systems are shown here along with some comments about their usefulness.

If we are interested in the supports of the board, we might cut the system loose at its support locations. If we include the vehicle and Mr. Poole in the system, the freebody diagram for this system appears in figure 3.59. If we are interested in the interaction of Mr. Poole and the diving board, we could choose a system without Mr. Poole. The freebody diagram of this system appears in figure 3.60 where $T$ is the tension in the rope; $N_r$ and $N_l$ are the normal forces caused by Mr. Poole's tennis shoes pressing down on the board; and $f_r$ and $f_l$ are the friction forces caused by Mr. Poole's right and left tennis shoes respectively. All of these forces are present in the freebody because they are external. Notice that Mr. Poole's weight does not enter directly into the freebody diagram of figure 3.60 because both he and the Earth are outside the system. Why did we not include the buoyant forces (those tending to float the golf cart) in the free body diagrams in figures 3.58, 3.59, and 3.60?

**End 3.11**

There are many engineering systems that consist of multiple bodies. Bridges, building supports, and machines are all common examples. The loads that these systems carry are generally far greater than the weight of the individual members of the systems. Because of this, engineers often ignore the weight of the individual members.

Engineers often distinguish between two types of links or elements they use to construct buildings, bridges and machines. These links or elements are called truss and frame.. Some truss elements are designed to resist tension. A tension or tensile load is one that pulls along the long axis of

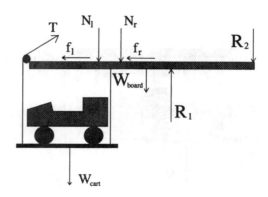

Figure 3.60: Freebody Diagram of Diving Board without Mr. Poole

the element.  A string can support a tension.  A truss element may also be designed to support
a compression.  A compression is a load applied along the long axis which would tend to shorten
the element.  String does not support compression.  A stack of pennies can support considerable
compression but no tension.  A concrete pillar can support considerable compression but relatively
little tension.  Sure maybe concrete can resist any attempts you make to pull it apart with your
bare hands, but what you can do is nothing compared to what happens inside a building! Relatively
speaking, ordinary concrete is not expected to resist tension.  What truss elements are never expected
to do is resist bending.  To understand what bending is, think about what you would do if we handed
you a long metal rod and asked you to bend it.  What you might do is try to bend it over your knee.
Think about what forces you are applying to do this.  Two forces down from your hands on each
end of the rod and one force up in the middle from your knee.  If you are strong enough, the rod will
bend or break.  Another attempt to bend the element might consist of clamping it in a vise which
produces a "fixed support" as shown in the bottom row of table 3.7.  Once it is clamped, you might
twist the end or push it perpendicular to the long axis.  Think about the uniqueness of these types
of loads compared to the simple tension and compression.  Truss elements are not expected to carry
(resist) a bending load, frame elements are.  A stack of pennies could not support a bending load.  A
string cannot support a bending load.  A diving board is expected to support some bending.  Could
you use an element designed to be a truss as a frame element?  Sure but don't expect a lot from it
because it probably isn't shaped to handle much bending.  Why would anyone want a truss element?
Well basically it is easier for materials to support tensions/compressions than bending (you will
learn why in follow on classes) therefore IF you CAN design a support (a bridge for example) so
that the loading on the members is only tension or compression then you can build the support
(the bridge) using only truss elements.n general a truss design will be stronger for a given element
weight.  In other words a truss bridge will generally be lighter and carry more load than one with
frame elements.  Usually truss structures are large.  Many bridges, roofs and many other structures
are designed as trusses.  A truss bridge is a bridge constructed predominately with truss elements.

If you can recognize certain loading conditions, the conservation equations can be simplified
making your design and/or analysis job easier.  Table 3.8 summarizes the simplifications.  In this
text the only simplification used is the so-called two force body, a body where forces are applied
at only two locations.    All of the simplifications listed in Table 3.8 can be derived by applying
conservation of linear and angular moment to an element.  These "proofs" are left as an exercise.

An important question is how can you tell the difference between truss and frame elements?
Basically you look at the loading applied to the element, Table 3.8 shows that truss elements are
straight and are loaded only by two forces.  By the way, it is possible to take a piece of metal which
was designed to operate like a truss element and load it like a frame.  It may bend or break easily
however.  If a member is supporting a couple applied to it, or it supports at least three forces, it is
a frame element.  Stationary members loaded by forces at only two positions (usually the ends) are
sometimes called two force bodies.

Consider the tripod in figure 3.61 (the third leg is not shown, imagine that it disappears into the
page).  This tripod consists of three legs tied together at the top and gently placed on the surface

| Simplification for **STATIONARY** Elements | | |
|---|---|---|
| Loading | Shape | Simplifications |
| Two Locations Where Forces Are Applied, No Couples | Any Shape Straight | Forces Must be Collinear The Member is a Truss. |
| Three Locations Where Forces Are Applied, No Couples | Any Shape | Net Forces Must Intersect at a Point. |
| Anything Else | | No General Simplifications |

Table 3.8: Some Simplifications for Stationary Elements.

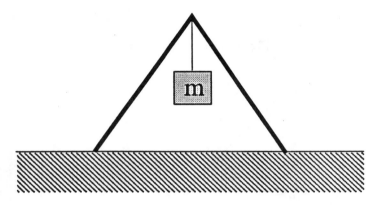

Figure 3.61: A Simple Tripod

of the ground. The tripod can support a relatively heavy object by hanging it from the legs. The tripod stands up because of friction at its feet.

Ignoring the weight of each member, forces are exerted at only two places (at the top and at the feet). Because only two forces are present, the force components must be collinear. Because the members are two force bodies, they are truss members. Straight truss members do not have a tendency to bend. A structure constructed exclusively of two force bodies is called a truss. A two force body (a truss element) is said to be in tension if the ends are being pulled as if being stretched. The member is in compression if the ends are being pushed as if being shortened. The legs of the tripod in figure 3.61 are all in compression. If the legs are real thin, they might "bow" slightly. You have probably made a thin metal ruler or thin wire bow by compressing it. This bowing is actually called buckle and is actually very distinct from bending in its characteristics. This text will not get into the differences.

As an alternative example, figure 3.62 shows a different type of tripod. This tripod also consists

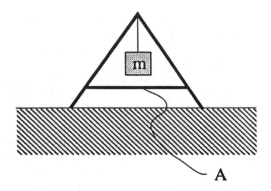

Figure 3.62: A Simple Tripod That Does Not Rely on Friction

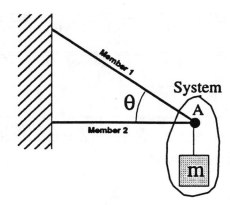

Figure 3.63: A Simple Truss

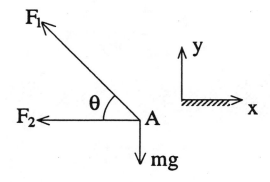

Figure 3.64: The Freebody Diagram of Point A of the Truss

of three members (third leg not shown) tied together at the top and connected with braces slightly above the bottom of their legs (member A in the figure is a brace). Again, ignoring all member weights, member A has forces applied only at two locations. It is a truss member. The legs have forces applied at three locations (top, bottom, and at A). They are frame elements and may have a tendency to bend.

## Example 3.12

## A Simple Truss

**PROBLEM:**

Formulate a mathematical model of the truss in figure 3.63. The structure supports a mass $m$ as the figure shows. If the members are really truss members, the connections at the wall can be pin connections because all the load will act along the axis of the member. If this were a real system, the members might be fixed into the wall and if they were, it could be possible for them to carry a little bending load. By taking into consideration the material of each one, we could determine how much they bend. Deformation in bodies is too advanced for this text so we'll leave it alone. To make it simple, we'll assume the bodies are rigid and they are pinned to the wall.

**FORMULATION:**

To determine the forces present in each member of the truss, we begin by drawing the freebody diagram of point A. See figure 3.64. Gravity pulls downward on the mass at point A with an external force of $mg$, therefore, we include it in the freebody diagram. The truss members also exert external forces at A, therefore, we include these forces in the freebody as well. The forces from the members must be collinear with the members because they are two force bodies, therefore we arbitrarily draw the forces as if they act in tension. If we get a positive solution to the forces, then the direction in our freebody is correct and the members are in tension. If we get a negative solution, then the direction drawn is incorrect and that member is in compression. Note that if member 1 is in tension, then point A pulls member 1 so member 1 pulls point A which is what we show. Likewise, if member 2 is in tension, then point A pulls member 2 and member 2 pulls point A. Since we are drawing a freebody diagram of point A, we show the members do to A not what A does to the members. That is why the forces are shown pulling.

We know our system is a wise choice because the number of unknowns (two) is less than the maximum number of equations we expect for the system.

Applying conservation of linear momentum for the resulting freebody (figure 3.64), we get the following:

**Conservation of Linear Momentum (Vector)**
system[A fig.3.64]        time period [**differential**]

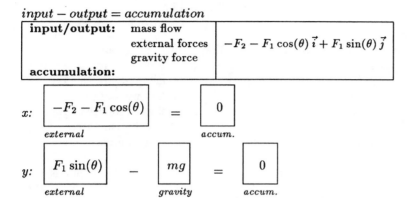

$input - output = accumulation$

| input/output: | mass flow external forces gravity force | $-F_2 - F_1 \cos(\theta)\,\vec{i} + F_1 \sin(\theta)\,\vec{j}$ |
|---|---|---|
| **accumulation:** | | |

$$x: \quad \boxed{-F_2 - F_1 \cos(\theta)} \atop \text{external} \quad = \quad \boxed{0} \atop \text{accum.}$$

$$y: \quad \boxed{F_1 \sin(\theta)} \atop \text{external} \quad - \quad \boxed{mg} \atop \text{gravity} \quad = \quad \boxed{0} \atop \text{accum.}$$

**SOLUTION:**
Solving these for the forces gives

$$F_1 = \frac{mg}{\sin(\theta)} \qquad F_2 = -F_1 \cos(\theta) = -\frac{mg}{\tan(\theta)}$$

Because $F_1$ is positive, the direction in our freebody is correct. Member 1 pulls point A and point A pulls member 1 (tension). Because $F_2$ is negative, the direction in our freebody is incorrect. Member 2 pushes point A and point A pushes member 2 (compression). What would happen to the forces at point A if $\theta$ approached zero? Is this what we should expect?

**VERIFICATION:**
Notice that because we assumed both members were truss elements, we ignored any transverse (non-axial) forces in the members. Do you think this assumption is accurate? Can you determine whether the assumption is valid based solely on the information given in the problem? Under what conditions could member 2 carry a transverse force? The answers to these questions depends on the stiffnesses of the two members. Suppose for example, that member 2 is rigid (*e.g.* a nail) and member 1 is flexible (*e.g.* a rubber band). In this case, member 2 (the nail) would support (if its strong enough) nearly all the vertical weight. However, if member 1 was rigid and member 2 was flexible in the transverse (non-axial) direction (*e.g.* a stack of pennies dipped in oil), then member 1 would bend rather than support the vertical load. In this case, member 1 would support most of the vertical weight. What this shows is that the load a member carries is dependent on its flexibility. This text will not consider flexibility, but other courses will.

**End 3.12**  ━━━━━━━━━━━━━━━━━━━━━━━━━━━━━━━

Consider the frame figure 3.65 shows. The horizontal arm carries significantly more bending than axial load. It has to carry bending loads. Why? Well how else could it support the load at the right end? So what are the loads in the horizontal member? Are there moments applied to it? Yes, there must be or else the load will fall. Where do the moments come from? They come from the vertical part of the frame. Do this experiment. Pick up a pencil between your thumb and forefinger of one hand then push perpendicular to the pencil axis with you other hand. What do you feel between your finger and thumb? Do you feel the moment your fingers are applying to the pencil? Can you

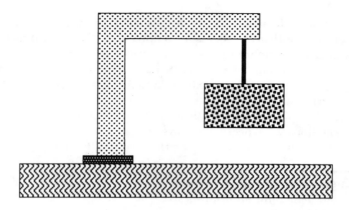

Figure 3.65: A Frame with Transverse Load Capability

see or feel which way your finger and thumb are pulling and pushing? You are supporting the load on the pencil by applying a moment. The vertical member in figure 3.65 is doing the same thing. Since the horizontal member has a couple applied to it, it is not a truss element. Is the vertical member a truss element? No. Can the support at the floor be a pin? No. When we analyze a system like this, we follow the same steps that we did when analyzing the simple truss. However, we also have to apply conservation of angular momentum because we cannot assume collinear forces. They really are not difficult, they just have a couple extra conservation equations, the angular momentum ones.

If the supported load in figure 3.65 increases significantly, the frame will break. What is the most probable location for the break? The answer depends on the material. For example if the vertical member is cardboard and the horizontal is steel it may break in a different place than if it were all steel. It also depends on the size and shape of the bodies. If one were thinner than the other for example. It also depends on where the "greatest" loads are. It is very complex and beyond the scope of this class but basically we would be looking for places that have the greatest forces and/or moments. All things being equal (size, shape etc.) I would guess it would break either at the junction between vertical and horizontal, or from the base. What do you think? Why?

## Example 3.13

### A Failed Roof

**PROBLEM:**

The owner of RAINOFF Incorporated (a manufacture of roofs) came to us with the following problem: "A few days ago, after I installed the roof design in figure 3.66 on a lumber yard building, section AB snapped and the roof collapsed. All the All the members in the roof were I-beams with a support capacity of 25 [tons] in tension and 18 [tons] in compression. The only load was a 10 [ton] pallet of bricks that hung from joint A. I cannot believer the design was bad; can you tell me what happened?"

**FORMULATION: [I]**

We decided to look for evidence that the collapse was due to a poor design, a bad I-beam, or sabotage. First do you think the roof is designed as a truss? Why? If you do, what does it tell you about the joints between the members? I think it is a truss because the members are two force and only tension and compression loads were quoted. If it is a truss, then I will model the joints as pins because they were not designed to carry very much moment loads hence they act like pins.

Since member AB snapped, we will determine the force present in that member. To determine this force, we recognize that it must appear in an equation. Realizing that only external forces appear in the momentum equations, we select a system that cuts through member AB. This way the force present in AB is external. We chose the system which cuts through the pin support on the left, cuts through members AC, AB and DB. This system was chosen because it has only a few unknown forces and AB is one of them. The freebody diagram of the system appears in figure 3.67.

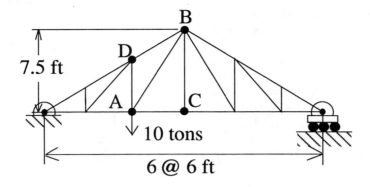

Figure 3.66: A Failed Roof

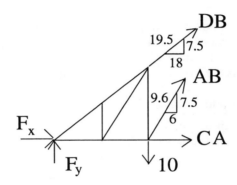

Figure 3.67: Freebody Diagram of the Failed Roof

Incidentally, there is a system which has one less unknown and still has $AB$ external, can you find it? If you can, then you may be developing the ability to choose systems wisely. Generally you want to find systems that have the unknowns you want and as few others as possible.

Since the cut members are trusses (loaded only at their ends), we know the external forces are collinear with the members in either tension (as shown) or compression. Note that

- AB, CA, and DB are the forces in members AB, CA, and DB respectively and are assumed to be in tension.

- 10 is the load of bricks.

- $F_x$, $F_y$ are the supporting forces from the roof.

At this point, we should not expect to get a solution using only this system. Why? Because we should expect a maximum of three conservation of momentum equations. Two linear and one angular. Since there are five unknowns, we should not expect to find a solution. What would we do ordinarily? Choose another system. In fact, if we chose a system as the entire truss structure, we would have three equations maximum for the entire roof and three unknowns ($F_x$, $F_y$ and $R$ the force under the roller on the right). We could solve the equations for the entire roof for $F_x$ and $F_y$ then come back the current problem and find the forces in members AC, AB, and DB. Rather than do that, we will demonstrate a trick that works for some problems.

Expressing conservation of angular momentum about the left end point of the system in figure 3.67 gives

**Conservation of Angular Momentum (Vector) [left end]**
system[**roof fig.3.67**]      time period [**differential**]

$input - output = accumulation$

| input/output: | mass flow | |
|---|---|---|
| | external moment | $-10(12) + AB\frac{7.5}{9.6}(12)\,\vec{k}$ |
| | gravity moment | |
| **accumulation:** | | |

$z:$  $\boxed{-10(12) + AB\frac{7.5}{9.6}(12)}$  $=$  $\boxed{0}$
        *external*                              *accum.*

Notice that there is one unknown in this equation. This happened because all the other unknown forces pass through the point we are applying the angular momentum equation. Do not expect this to happen often.

**SOLUTION: [I]**

Solving for the force in member AB, we get $AB = 12.8$. The positive sign says we drew the arrowhead correctly so it is in tension. This is clearly below the 25 [tons] tension that the manufacturer claims the member should support. Perhaps, the owner hung more than 10 [tons] from the roof or perhaps the I-Beam was defective. We decide to evaluate some of the other members nearby.

**FORMULATION: [II]**

Keeping the same freebody and expressing linear momentum, we get

**Conservation of Linear Momentum (Vector)**
system[**roof fig.3.67**]      time period [**differential**]

$input - output = accumulation$

| input/output: | mass flow | |
|---|---|---|
| | external forces | $\left(CA + AB\frac{6}{9.6} + DB\frac{18}{19.5}\right)\vec{i} + \left(-10 + F_y + AB\frac{7.5}{9.6} + DB\frac{7.5}{19.5}\right)\vec{j}$ |
| | gravity force | |
| **accumulation:** | | |

$x:$  $\boxed{F_x + CA + AB\frac{6}{9.6} + DB\frac{18}{19.5}}$  $=$  $\boxed{0}$
        *external*                                          *accum.*

$y:$  $\boxed{-10 + F_y + AB\frac{7.5}{9.6} + DB\frac{7.5}{19.5}}$  $=$  $\boxed{0}$
        *external*                                          *accum.*

Recognizing that these two equations have four unknowns ($F_x$, $F_y$, CA, and DB), we choose another system. We look for a system that has as few new unknowns as possible. We choose the complete roof and draw the freebody diagram in figure 3.68. Note that it has only one new unknown, $F_r$. Applying conservation of angular momentum about the right end gives:

**Conservation of Angular Momentum (Vector) [right end]**
system[**entire roof fig.3.68**]      time period [**differential**]

$input - output = accumulation$

| input/output: | mass flow | |
|---|---|---|
| | external moment | $24(10) - F_y(36)\,\vec{k}$ |
| | gravity moment | |
| **accumulation:** | | |

$z:$  $\boxed{24(10) - F_y(36)}$  $=$  $\boxed{0}$
        *external*                      *accum.*

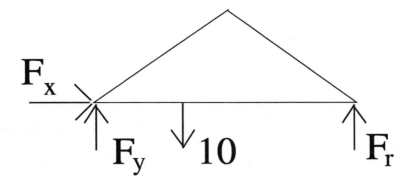

Figure 3.68: Freebody Diagram of the Entire Roof

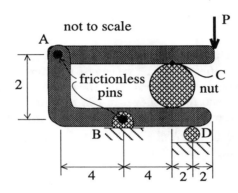

Figure 3.69: A Nutcracker

**SOLUTION: [II]**

Solving for the vertical force at the left end, we get $F_y = 6\frac{2}{3}$. Since $F_x$ is the only horizontal force, conservation of linear momentum for the entire roof would indicate that $F_x = 0$ (write the equation yourself and see). Using this with the previous equations allows us to solve for the rest of the forces in the system. They are $CA = 8$ and $DB = -17.3$. Ha! Member DB is much closer to the 18 [ton] compression limit than member AB is to the tensile limit. So if the owner had overloaded the roof, member DB would probably have broken before $AB$. Since this did not happen, we conclude that the owner did not overload the roof. This leads us to believe that member AB was either defective or sabotaged.

**End 3.13** ━━━━━━━━━━━━━━━━━━━━━━━━━━━━━━━━━━━━━━━━━━━━━━━

**Example 3.14** ━━━━━━━━━━━━━━━━━━━━━━━━━━━━━━━━━━━━━━━━━━━━━━━

## A Modern Nutcracker

**PROBLEM:**

An inventor has a new nut cracker design that he wants to sell (see figure 3.69). Before they buy the product, NUTSAREUS Incorporated wants us to determine how strong the members have to be to prevent breaking. Not knowing much about the strength of materials, we call a materials specialist. Burdened with her own problems, she does not offer much help; however, she does say that the required size of the member has a lot to do with the maximum shear force and bending moment in the members. We decide to determine these for each member of the nut cracker.

**ANSWER:**

To determine the shear forces and bending moment in the members, we will analyze systems that cut through the members. This allows us to determine the forces inside a member by making them external to the system.

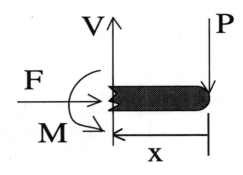

**Analysis Assumptions**

(1)  $M$ is the bending moment in the member.

(2)  $F$ is the axial force in the member.

(3)  $V$ is the shear force in the member.

Figure 3.70: (upperright): Freebody Diagram of the Upper Right Side of the Nutcracker

**FORMULATION: [I] The Upper Member $0 \leq x \leq 4$**

Analyzing member AP (the upper member), we choose a system whose freebody diagram appears in figure 3.70. The freebody diagram is only valid for the sections to the right of point C. The force $V$ is the shear force, $F$ is the axial force (shown in compression), and $M$ is the bending moment in the member. These loads are present because we cut the solid body at a non-idealized joint. Applying conservation of linear and angular momentum gives

**Conservation of Linear Momentum (Vector)**
system[**nutcracker fig.3.70**]      time period [**differential**]

*input − output = accumulation*

| **input/output:** | mass flow external forces gravity force | $F\,\vec{i} + (V - P)\,\vec{j}$ |
|---|---|---|
| **accumulation:** | | |

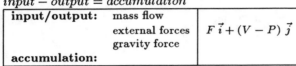

$x$:   $\boxed{F}$  $=$  $\boxed{0}$
       *external*      *accum.*

$y$:   $\boxed{V - P}$  $=$  $\boxed{0}$
       *external*        *accum.*

**Conservation of Angular Momentum (Vector) [cut point]**
system[**nutcracker fig.3.70**]      time period [**differential**]

*input − output = accumulation*

| **input/output:** | mass flow external moment gravity moment | $(M - Px)\,\vec{k}$ |
|---|---|---|
| **accumulation:** | | |

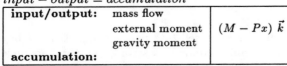

$z$:   $\boxed{M - Px}$  $=$  $\boxed{0}$
       *external*          *accum.*

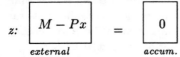

**SOLUTION: [I] The Upper Member $0 \leq x \leq 4$**

$$F = 0$$

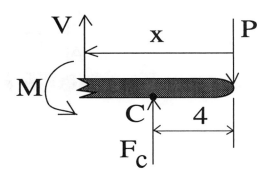

**Analysis Assumptions**

(1) $M$ is the bending moment in the member.

(2) $F_C$ is the force at C due to the nut.

(3) $V$ is the shear force in the member.

(4) Axial force in the member is zero.

Figure 3.71: (fbdwithc): Freebody Diagram of the Upper Member Including Point C

$$V = P$$

$$M = Px \quad \overset{\text{reduces}}{\rightsquigarrow} \quad M_{\max} = P(4) \text{ at } x = 4$$

Thus, the shear in the upper right part of the member is constant and the moment increases as we move to the left. The maximum moment (so far) occurs at point C and equals $M_{\max} = 4P$.

**FORMULATION: [II] The Upper Member $4 \leq x \leq 12$**

Continuing with our analysis of the upper member, we take a system whose freebody diagram appears in figure 3.71. This time point C is part of the system and therefore the freebody diagram has $F_C$ (the contribution from the nut at point C). Note that we have applied the conservation of linear momentum in our head and realized that there cannot be any axial force in the member. Applying conservation of linear and angular momentum gives

**Conservation of Linear Momentum (Vector)**
system[**nutcracker fig.3.71**]    time period [**differential**]

*input − output = accumulation*

| **input/output:** | mass flow | |
| | external forces | $(V + F_C - P)\,\vec{j}$ |
| | gravity force | |
| **accumulation:** | | |

$y$:  $\boxed{V + F_C - P}$ $=$ $\boxed{0}$
     *external*              *accum.*

**Conservation of Angular Momentum (Vector) [cut end]**
system[**nutcracker fig.3.71**]    time period [**differential**]

*input − output = accumulation*

| **input/output:** | mass flow | |
| | external moment | $M - Px + F_C(x - 4)$ |
| | gravity moment | |
| **accumulation:** | | |

$z$:  $\boxed{M - Px + F_C(x - 4)}$ $=$ $\boxed{0}$
     *external*                        *accum.*

**SOLUTION: [II] The Upper Member $4 \leq x \leq 12$**

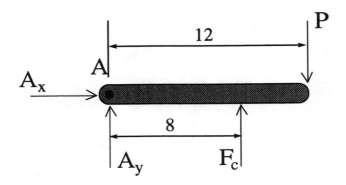

Figure 3.72: The Freebody Diagram of the Top Member

$$V + F_C - P = 0 \qquad\qquad\qquad (3.9)$$

$$M - Px + F_C(x - 4) = 0 \qquad\qquad (3.10)$$

These equations are valid only for $4 \leq x \leq 12$. Recognizing that there are three unknowns ($M$, $F_C$, and $V$) and only two useful equations, we look for another equation. To determine what to do, we follow the steps laid down in Table 3.2. Since it is possible to write at most three momentum equations for each two dimensional freebody diagram (two linear and one angular),[8] there are no more applicable conservation equations for our system. Note that one of the linear momentum equations, the one in the $x$ (axial) is trivial $0 = 0$ so why bother with it? Since the number of flow variables (zero) equals the [NIF] (zero) we have the correct number of flows. Since there is not any information given in the problem statement we have not used, we must choose a new system and analyze it.

**FORMULATION: [III] The Upper Member** $0 \leq x \leq 12$

Choosing the entire top member by cutting it loose at the idealized joint gives the freebody in figure 3.72. Taking conservation of angular momentum about point A, we get

**Conservation of Angular Momentum (Vector) [A]**

system[**nutcracker fig.3.72**]      time period [**differential**]

*input − output = accumulation*

| **input/output:** | mass flow | |
|---|---|---|
| | external moment | $(F_C(8) - P(12))\ \vec{k}$ |
| | gravity moment | |
| **accumulation:** | | |

$z$: $\boxed{F_C(8) - P(12)}$ = $\boxed{0}$

     *external*         *accum.*

**SOLUTION: [III] The Upper Member** $0 \leq x \leq 12$

Solving for the force at C gives

$$F_C = \frac{3}{2}P.$$

Now, we return to equations 3.9 and 3.10. Substituting $F_C = \frac{3P}{2}$ into these equations gives

$$V + \frac{3}{2}P - P = 0$$

---

[8]If only mechanical forms of energy are present, the energy equation reduces to the momentum. Therefore, we cannot expect the energy equation to provide additional information. If we write it, it will be dependent on those we already have.

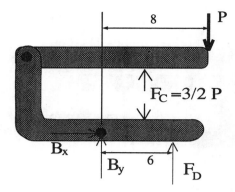

Figure 3.73: The Complete Nutcracker Assembly

$$M - Px + \frac{3}{2}P(x-4) = 0$$

Solving for the shear and moment at the cut gives $V = -\frac{1}{2}P$ and $M = 6P - \frac{1}{2}Px$. The maximum moment occurs at C ($x = 4$) and is $M_{\max} = 4P$. The shear is constant at $-\frac{1}{2}P$.

Therefore, the maximum moment on the upper section is $M_{\max} = 4P$ and occurs at point C. The maximum shear is $V_{\max} = P$ and occurs between the C and the applied force $P$.[9]

**FORMULATION: [IV] The Entire Nutcracker (Excluding the Nut)**

Shifting our interest to the lower member, we determine the reactions at the supports first. This is a common practice, in fact we did this for the upper member but not until after we analyzed it. Most engineers do this step first. Choosing the entire nutcracker (excluding the nut) as our system (figure 3.73) and applying the conservation of angular momentum about B, we get that

**Conservation of Angular Momentum (Vector) [B]**
system[**nutcracker fig.3.73**]     time period [**differential**]

*input − output = accumulation*

| **input/output:** | mass flow | |
| | external moment | $(-8P + 6F_D)\ \vec{k}$ |
| | gravity moment | |
| **accumulation:** | | |

$z$:  | $-8P + 6F_D$ | $=$ | $0$ |

       *external*         *accum.*

Applying conservation of linear momentum to the same system, we get that

**Conservation of Linear Momentum (Vector)**
system[**nutcracker fig.3.73**]     time period [**differential**]

*input − output = accumulation*

| **input/output:** | mass flow | |
| | external forces | $\left(B_y + \frac{4}{3}P - P\right)\ \vec{j}$ |
| | gravity force | |
| **accumulation:** | | |

$y$:  | $B_y + \frac{4}{3}P - P$ | $=$ | $0$ |

       *external*         *accum.*

[9]With this in mind, explain why we can sometimes break a stick with our knee that we cannot break using only our hands.

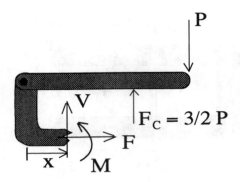

Figure 3.74: Cutting the Lower Member to the Left of point B

**SOLUTION: [IV] The Entire Nutcracker (Excluding the Nut)**

$$F_D = \frac{4}{3}P.$$

$$B_x = 0 \quad \text{and} \quad B_y = -\frac{1}{3}P.$$

**FORMULATION: [V] The Lower Member $0 \le x \le 4$**

Now let's cut the bottom member open to determine the forces inside it. Cutting the lower member $x$ from the left gives the freebody diagram in figure 3.74. Applying linear momentum and angular momentum (about the cut point) for this system gives

**Conservation of Linear Momentum (Vector)**
system[**nutcracker fig.3.74**]      time period [ **differential**]

*input − output = accumulation*

| **input/output:** | mass flow | |
| | external forces | $F\,\vec{i} + \left(V + \frac{3}{2}P - P\right)\vec{j}$ |
| | gravity force | |
| **accumulation:** | | |

$x$:  | $F$ | = | $0$ |
     *external*      *accum.*

$y$:  | $V + \frac{3}{2}P - P$ | = | $0$ |
     *external*      *accum.*

**Conservation of Angular Momentum (Vector) [cut end]**
system[**nutcracker fig.3.74**]      time period [ **differential**]

*input − output = accumulation*

| **input/output:** | mass flow | |
| | external moment | $\left(M + \frac{3}{2}P(8 - x) - P(12 - x)\right)\vec{k}$ |
| | gravity moment | |
| **accumulation:** | | |

$z$:  | $M + \frac{3}{2}P(8 - x) - P(12 - x)$ | = | $0$ |
     *external*        *accum.*

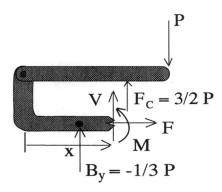

**Analysis Assumptions**

(1) $M$ is the bending moment in the member.

(2) $F$ is the axial force in the member.

(3) $V$ is the shear force in the member.

Figure 3.75: (lowermid): A Freebody Diagram for the Middle Section of the Lower Member

**SOLUTION: [V] The Lower Member** $0 \leq x \leq 4$

$$F_x = 0$$

$$V = -\frac{1}{2}P$$

$$M + 12P - \frac{3}{2}Px - 12P + Px = 0 \quad \overset{\text{reduces}}{\rightsquigarrow} \quad M = \frac{1}{2}Px$$

$$M_{\max} = 2P \text{ at } x = 4$$

These equations are valid for $0 \leq x \leq 4$. The shear is a constant $V = -\frac{1}{2}P$. The maximum moment is $M_{\max} = 2P$ and occurs at $x = 4$. Notice that the system in figure 3.74 consists of two bodies connected by a pin. Some students develop the idea that we cannot apply the momentum equations to collections of bodies. However, this idea is not correct. We can apply the momentum equation to whatever system we choose.

**FORMULATION: [VI] The Lower Member** $4 \leq x \leq 8$

Choosing a system that cuts the lower member between B and the nut yields the freebody diagram in figure 3.75. Applying linear momentum and angular momentum (about the cut point) for this system gives

**Conservation of Linear Momentum (Vector)**

system[**nutcracker fig.3.75**]    time period [ **differential**]

*input − output = accumulation*

| **input/output:** | mass flow external forces gravity force | $F\,\vec{i} + \left[ -P + \frac{3}{2}P + V + \left( -\frac{1}{3}P \right) \right]\,\vec{j}$ |
|---|---|---|
| **accumulation:** | | |

$x$:    $\boxed{F}$    =    $\boxed{0}$
     *external*         *accum.*

$y$:    $\boxed{-P + \frac{3}{2}P + V + \left(-\frac{1}{3}P\right)}$    =    $\boxed{0}$
     *external*                                              *accum.*

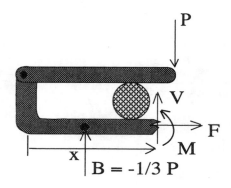

**Analysis Assumptions**

(1) $M$ is the bending moment in the member.

(2) $F$ is the axial force in the member.

(3) $V$ is the shear force in the member.

Figure 3.76: (lowermidright): The Freebody Diagram for a Section Between the Nut and D

---

**Conservation of Angular Momentum (Vector) [B]**
system[**nutcracker fig.3.75**]      time period [ **differential**]

*input − output = accumulation*

| input/output: | mass flow | |
| :--- | :--- | :--- |
| | external moment | $\left[-\left(-\frac{1}{3}P(x-4)\right) + M + \frac{3}{2}P(8-x) - P(12-x)\right]\,\vec{k}$ |
| | gravity moment | |
| **accumulation:** | | |

$z$:  $\boxed{-\left(-\frac{1}{3}P(x-4)\right) + M + \frac{3}{2}P(8-x) - P(12-x)}$  $=$  $\boxed{0}$

      *external*                                         *accum.*

---

**SOLUTION: [VI] The Lower Member** $4 \leq x \leq 8$

$$F = 0$$

$$V = -\frac{1}{2}P + \frac{1}{3}P$$

$$\frac{1}{3}Px + Px - \frac{3}{2}Px - \frac{4}{3}P + 12P - 12P + M = 0$$

$$M = \frac{4}{3}P + \frac{1}{6}Px$$

$$M_{\max} = \frac{8}{3}P \text{ at } x = 8$$

The shear is a constant $V = -\frac{1}{6}P$ over the middle section. Since the equations are only valid for $4 \leq x \leq 8$, the maximum moment for this section is $M_{\max} = \frac{8}{3}P$ and occurs at the nut (at $x = 8$).

**FORMULATION: [VII] The Lower Member** $8 \leq x \leq 10$

The freebody diagram for the section of the nutcracker between the nut and D appears in figure 3.76. Applying the conservation of linear and angular momentum (about the cut point) for this system gives

## Conservation of Linear Momentum (Vector)
system[nutcracker fig.3.76]      time period [ **differential**]

*input − output = accumulation*

| input/output: | mass flow | |
|---|---|---|
| | external forces | $F\,\vec{i} + \left[ -P + \left( -\frac{1}{3}P \right) + V \right] \vec{j}$ |
| | gravity force | |
| **accumulation:** | | |

$x:$ 
$$\boxed{F} = \boxed{0}$$
*external*     *accum.*

$y:$ 
$$\boxed{-P + \left( -\frac{1}{3}P \right) + V} = \boxed{0}$$
*external*     *accum.*

## Conservation of Angular Momentum (Vector) [cut end]
system[nutcracker fig.3.76]      time period [ **differential**]

*input − output = accumulation*

| input/output: | mass flow | |
|---|---|---|
| | external moment | $\left[ M - \left( -\frac{1}{3}P(x-4) \right) - P(12-x) \right] \vec{k}$ |
| | gravity moment | |
| **accumulation:** | | |

$z:$ 
$$\boxed{M - \left( -\frac{1}{3}P(x-4) \right) - P(12-x)} = \boxed{0}$$
*external*     *accum.*

**SOLUTION: [VII] The Lower Member** $8 \leq x \leq 10$

$$F = 0$$

$$V = \frac{4}{3}P$$

$$M + \frac{1}{3}Px - \frac{4}{3}P - 12P + Px = 0 \quad \overset{\text{reduces}}{\rightsquigarrow} \quad M = \frac{40}{3}P - \frac{4}{3}Px$$

$$M_{\max} = \frac{8}{3}P \text{ at } x = 8$$

Since these are valid for $8 \leq x \leq 10$, the maximum moment is $M = \frac{8}{3}P$ and occurs at the nut (at $x = 8$). The shear is a constant $V = \frac{4}{3}P$.

Therefore, the maximum moment on the lower section is $M_{\max} = \frac{8}{3}P$ and occurs at the nut. The maximum shear is $V_{\max} = \frac{4}{3}P$ and occurs between the nut and point D.

Where does the maximum shear occur in the nutcracker? Where does the maximum moment occur? Did you expect this?

**End 3.14** ▬▬▬▬▬▬▬▬▬▬▬▬▬▬

So far, when we speak of choosing systems wisely we have referred to choosing systems so the unknowns we want to find are in the system and the number of unknowns is as few as possible (hopefully no more than the number of equations the system will generate). When there are multiple objects or bodies in a problem and when these bodies move we usually choose systems made up of the individual bodies taken one at a time. In general, we are not finished with a problem until we have used all the bodies in a system. For example, consider the two blocks shown in figure 3.12. If we know there is two [NIF], the two blocks will have different motions. Usually, we would analyze two systems, one made of the top block the second made of the bottom block. We would do this because we want to keep the expression for the change in momentum simple. Since there are two blocks, we would expect to use two systems before we "know everything" about the blocks and their motion. Of course there are exceptions to this. For example, in many truss problems, by clever choice of a system, we can get the information we want without using every body in a system.

There are some occasions when we want to choose a system with multiple moving bodies. These occasions occur when we can choose a system such that there are no external forces present and no mass entering or leaving. Suppose we had or could choose a system where there were no external forces and no mass flow in or out. For the conservation of linear momentum, there would be no input/output hence the accumulation is zero. We could write $0 = \frac{d}{dt}(L)$ where $L$ is some potentially ugly expression for the linear momentum. Now even though the equation is differential, it is pretty easy to solve. What the equation says is that there is no change in linear momentum. That $L$ is constant because $\frac{d}{dt}()$ of a constant is zero. As a result the linear momentum equation can be written as $L = L_o$ where $L_o$ is a constant. Sometimes is can be difficult to solve differential equations so if you can choose a system such that there are no external forces and no mass flow in or out, the differential equations will be trivial. The following example shows how this can work in our favor.

**Example 3.15** ───────────────────────────────

### You and a Canoe.

Have you ever been in a canoe on a calm lake? If so, did you notice that if you move quickly from one end to the other the canoe moves too? Why does this happen and how far can you make the canoe move?

**Formulation:**

Let's assume we move in one direction only so there is two [NIF]. Let's say that we know where we are walking so we are specifying one flow. With two [NIF] and one specified flow, we must apply a conservation law.

Think about choosing a system. What should we choose? You alone? The canoe alone? If you did either, you would have the problem of determining the forces between you and the canoe. Do you know these forces? Do you care about them? No in both cases. What if you took both the canoe and you? Now think about the freebody diagram. You have vertical forces (weight and force from the lake holding the canoe afloat). What about horizontal forces? If there is friction drag between the canoe and water there is a horizontal force. How large do you think the friction is? Not very large at all so let's ignore it. If you apply conservation of linear momentum in the horizontal direction what do you get? You get the following:

$$0 = \frac{d}{dt}((m_c v_{cx} + m_u v_{ux})) \tag{3.11}$$

where $m_c$ is the mass of the canoe, $m_u$ is the mass of you, $v_{cx}$ is the horizontal speed of the canoe and $v_{ux}$ is your horizontal speed. Now notice that the advantage of this formulation is the fact that there are no input/outputs hence the 0 on the left. The ugly part of it is the fact that the linear momentum term is made up a term for each body. In this case its not hard to write the terms but in general the more bodies you take in your system, the more ugly the momentum term becomes.

**Solution:**

What's really good about this formulation is that the zero on the left makes it really easy to solve the differential equation 3.11. The solution is:

$$0 = (m_c v_{cx} + m_u v_{ux})_{\text{final}} - (m_c v_{cx} + m_u v_{ux})_{\text{initial}} \tag{3.12}$$

It says the momentum does not change whatever you have at the beginning, you have at the end because there is no momentum flowing into the canoe and you. You are merely exchanging it between you and the canoe.

Let's solve for positions. Let's assume we know the initial velocities so we can compute the initial values in equation 3.12. When you do, you now have a number for the initial stuff, let's call that number $L_o$. Rather than fixing the final time, let's let it be a variable time so we can write equation 3.12 as:

$$0 = \left[m_c \frac{d}{dt}((x_{cx})) + m_u \frac{d}{dt}((x_{ux}))\right] - L_o$$

where $x_{cx}$ is the canoe's horizontal position measured from some fixed point and $x_{ux}$ is your position. Since the masses are constant we can "factor out" the derivative:

$$0 = \frac{d}{dt}\left([m_c x_{cx} + m_u x_{ux}]\right) - L_o$$

This one is not too hard to solve either, separating variables and integrating gives:

$$0 = [m_c x_{cx} + m_u x_{ux}]_{\text{final}} - [m_c x_{cx} + m_u x_{ux}]_{\text{initial}} - L_o t$$

Do you recognize the terms which look like $[m_c x_{cx} + m_u x_{ux}]$? They are expressions for the combined mass times the center of mass of the combined system. Let's divide everything by the combined mass $m_c + m_u$ to obtain:

$$0 = \frac{[m_c x_{cx} + m_u x_{ux}]}{m_c + m_u}\bigg|_{\text{final}} - \frac{[m_c x_{cx} + m_u x_{ux}]}{m_c + m_u}\bigg|_{\text{initial}} - \frac{L_o}{m_c + m_u}t \qquad (3.13)$$

Okay so what does this mean? Well the first two terms on the right are the positions of the mass center at the end minus the position of the mass center at the beginning. The last term, if you look carefully, is the initial velocity of the mass center times time.

So what does this tell us? First let's do the easy one. Let $L_o = 0$ in other words we start out at rest on the lake. What the equation says is the initial position of the center of mass equals the final position of the center of mass. How far can you move yourself by walking? Not far. Now what if $L_o \neq 0$, you are initially coasting. Now the equation says the final mass center position equals the initial mass center position plus the distance you coast in time $t$.

Surely you have seen a small child in a shopping cart in a grocery store. Have you ever seen such a child wiggle their body to make the cart scoot along the floor until he can reach a box of cookies on a shelf? The kid is too small to reach the floor so how does this happen? What is different about the child in the shopping cart? What is present on the cart that is not present on the canoe?

**End 3.15**

Another reason to be clever with the way you choose systems is to avoid dealing with ugly complex forces acting on a body. In the previous example, if we chose the canoe alone as the system, we would need to consider the forces between you and the canoe. Do you know what these forces are? Can you imagine the difficulty in finding them? By choosing the system wisely these forces don't matter. The negative side of choosing several objects in the system is that you have more "things" in the accumulation terms. In some problems, the accumulation terms are so hard to find, you might have difficulty.

Another "clever" tool used by experts is choosing the time period wisely. We imagine that you, at some point in your life, have been hit by a ball. Perhaps you noticed that the harder the ball, the more the impact hurts. Why? Consider the ball. Before it makes contact it has some momentum. After the contact, it has some other momentum. This change in momentum can only occur because of external forces, the force of your body hitting the ball and gravity, air resistance and whatever. Now think about how long the ball is in contact with your body. Not very long at all. All of the ball's momentum change must occur over a very small time period. What does this say about the rate of change of momentum? It must be very large, you have a finite reasonable amount of momentum change divided by a very small time change giving a very large rate of change. What does this say about the forces acting on the ball? Think about the conservation of linear momentum. If the rate of change of momentum is large, there must be large forces. Which ones are large? Weight? Come on that cannot be more than a pound or two or you would not be playing with it. Air resistance? Nope. The force of your body on the ball? Yep, the force of contact, sometimes called the force of impact, is very large. It has to be and that is why it hurts. What can you do to make it hurt less? You could pad yourself or the ball. With a padded ball, as contact is made, the padding deforms doing two things: (1) increasing the time of contact making the rate of change of momentum smaller, hence less force and (2) spreading the contact over a larger area making it hurt less. Forces that are very large and act over very short periods of time are called impulsive.

So how does this information help in an analysis? Suppose you have a situation in which one or more forces acting on an object are impulsive and a handful of other forces are not impulsive. Call the impulsive forces $\vec{I}$ and the nonimpulsive forces $\vec{N}$. Choose the time period that the impulsive forces are acting and express the linear momentum equation as:

$$\vec{I} + \vec{N} = \frac{d}{dt}\left(\vec{L}\right)$$

where $\vec{L}$ is the linear momentum. Manipulate this to express integrate it as:

$$\int \vec{I}dt + \int \vec{N}dt = \int d\left(\vec{L}\right) = \vec{L}_{\text{final}} - \vec{L}_{\text{initial}}$$

Now consider the left hand side terms. The value $dt$ is very small therefore unless the force is large, the integral is going to be negligible. So a reasonable approximation is:

$$\int \vec{I}dt = \vec{L}_{\text{final}} - \vec{L}_{\text{initial}}$$

In other words, when a system has an impulsive force acting on it, choose a time period over which the impulsive force acts and ignore all nonimpulsive forces. The term $\int \vec{I}dt$ is called the impulse of the force $\vec{I}$. All it is the time integral of the force. In an engineering situation, it is quite often possible to measure a force as a function of time. If you plot the force versus time you can then determine the area under the plot and have the impulse of the force.

## Example 3.16 ───────────────────────────────────

### Abra Cadabra

Surely you have seen the common feat where a magician pulls a table cloth out from under a pile of dishes. How does he do that?

Take the dishes as a system and think about the freebody diagram. There will be a weight, a normal and friction from the cloth. Now think about the cloth as a system. The forces are what he applies on the cloth, pulling it, the dish normal and friction, and a normal and friction from the table. Now suppose the magician pulls really hard on the cloth so the pull is much greater than the frictions. Is this possible? Yes because the frictions have a maximum value so if the magician has worked out recently he should be able to manhandle the cloth no? So, suppose he really pulls hard, what is the cloth's motion? It will have a large acceleration because the cloth is lightweight. Well if the cloth accelerates a large amount then it will be out from under the dishes in no time. Now look back at the dishes. They have had a friction force applied to them for a very short length of time, how much will their momentum change? Not much, basically their change in momentum is the integral of force (friction) over time. Imagine what the plot of friction force versus time looks like. What's the area (the integral) under the force? Not much so the change in momentum of the dishes is small. If the dishes have mass then how much does their velocity change? Even less. So in summary, if the magician will just have faith and really pull hard, the dishes will not have time to begin moving hence the cloth comes out and the dishes remain standing.

A word of warning: don't try this for the first time at Thanksgiving dinner!

## End 3.16 ───────────────────────────────────

So what good is the concept described in the previous example? Suppose you want to move screws up a ramp. How could you "vibrate" the ramp and make the screws slide up? Think about it.

## 3.8   KINEMATICS

### Objectives
*When you complete this section, you should be able to*

*1. Explain what a kinematic relationship is.*

Kinematics is the study of motion. The relationship between position, velocity and acceleration is kinematics. If two bodies are connected by a rope that is not limp, then when one body moves away the other follows. This is a kinematic relation. Kinematics is important in the analysis of mechanics problems because we need to express linear and angular momentum.

## 3.8.1 Position, Velocity, and Acceleration

**Objectives**
*When you complete this section, you should be able to*

1. *Calculate the velocity and acceleration given position.*

2. *Determine velocity given acceleration and proper initial conditions.*

3. *Determine position given velocity or acceleration and proper initial conditions.*

4. *Explain why velocity is tangent to the path of motion.*

5. *Express velocity and acceleration in path variables.*

The position of a point is a vector quantity that tells where the point is relative to some other point. When we specify the position of a point, we must always express it in reference to some other point. If you say a point is at the 2 inches left and 1 inch up from the corner of a table, it is assumed you know where the corner of the table is located. If we want to specify the position of point P measured from O, we write it as the vector $_O\vec{r}_P$ whose tail is on the left subscript and whose tip is on the right subscript. Alternatively the expression $_O\vec{r}_P$ (the position of P relative to O) can be interpreted as a "road map" telling how to go from O to P. Although position is always relative to some known point, we sometimes get lazy and leave off the left subscript. When the left subscript is missing the vector means that the position is measured from some arbitrary stationary point.

The absolute velocity of a point P is the total time derivative of the position of P measured from any stationary point. It is true that when we discuss the time derivative, it is usually important to indicate where we are standing when we calculate the derivative. In this course, all time derivatives will be taken while we are standing on the ground (or other suitable stationary place). Therefore, if we desire the absolute velocity of a point, we must take the derivative of a position measured from a stationary point along unit vector directions that are constant. Furthermore, absolute acceleration is the total time derivative of the absolute velocity.

We write the velocity of a point P measured from point O as $_O\vec{v}_P$. The acceleration of P measured from O is $_O\vec{a}_P$. If point O is not fixed, we say that we measure the velocity or acceleration from O. If O is fixed, we say the expressions are the absolute velocity and acceleration of P. Note that we often write absolute velocities and acceleration by leaving off the left subscript. Also it is common practice to drop the word absolute when saying velocity or acceleration. Lazily, we say the velocity and the acceleration to imply the left subscript is a fixed point. We will only deal with absolute velocity and accelerations in this text.

We have already computed velocities and accelerations of masses moving on straight lines. In this section, we will begin computing velocities and accelerations of points with more complex motion.

Two of the prerequisites for understanding the material in this section is a good knowledge of the chain rule for differentiation and the concept of finding derivatives at specific locations. To help review consider the following problem. Suppose $y = x^2$ and $x = t^3$ where $x$, $y$ and $t$ are variables. Suppose we want to compute $\frac{dy}{dx}$. No problem, its simply $2x$. Suppose we want to compute $\frac{dy}{dt}$, now there is a slight complication. One way to proceed is to substitute $x = t^3$ into the equation for $y$ to obtain $y = t^6$ then take a derivative to find $\frac{dy}{dt} = 6t^5$. An alternative is to use the chain rule and say:

$$\frac{dy}{dt} = \frac{dy}{dx}\frac{dx}{dt} = (2x)\left(3t^2\right) = 6t^3 t^2 = 6t^5$$

Note that this is exactly the same as before. Why use the chain rule? Because what if we knew $x$ was a function of $t$ but did not know exactly what the function was? In this case we could write

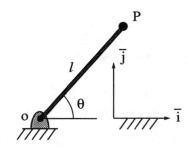

---

**Analysis Assumptions**

(1) The length of the rod is 2 [m] (*i.e.* $l = 2$).

(2) The rod is free to rotate in the plane of the paper.

(3) $\theta = 30^\circ$, $\dot{\theta} = 2$ [rad/s], and $\ddot{\theta} = 5$ [rad/s$^2$].

---

Figure 3.77: (singlerod): A Single Rigid Rod with Endpoint P

$\frac{dy}{dt} = \frac{dy}{dx}\frac{dx}{dt} = 2x\dot{x}$ and leave it at that. Now for the second point. Suppose we want to determine $\frac{dy}{dt}$ when $t = 1$. What we MUST do is first take the derivative, then substitute $t = 1$ into the result. The correct answer is:

$$\left.\frac{dy}{dt}\right|_{t=1} = 6t^5\big|_{t=1} = 6$$

Never substitute in a special value and then differentiate. If we do the incorrect order we would find at $t = 1$ that $x = 1^3 = 1$ and $y = 1^2 = 1$ the derivative then would be the derivative of 1 which is 0. WRONG!

Before we do an example, we want to remind you of three points. One, that velocity and acceleration refer to points not bodies. The only time it makes sense to discuss the velocity or acceleration of a body is when all of the points in the body have the same motion. For example when your car is traveling down a straight road we can talk about the velocity of the front seat. Second, when we specify one flow it means we provide not only a variable, but derivatives as well. This is consistent with the definition of flow variables. Third, although it is not obvious from most examples, it is possible to specify a flow by providing the position of one point, the velocity of a second and the acceleration of a third.

## Example 3.17

### Velocity and Acceleration of a Point on a Rod

**PROBLEM:**

Determine the velocity and acceleration of point P at the end of the rigid rod in figure 3.77.

Let $l = 2$ [m], $\theta = 30^\circ$, $\dot{\theta} = 2$ [rad/s], and $\ddot{\theta} = 5$ [rad/s$^2$].

**FORMULATION:**

The system consists of an infinite number of points but has only one [NIF]. This means that once we have one independent flow we can determine all the other flows. Also note that since $\theta$ and its derivatives are given, we have one flow. We can therefore determine the flow of point P without any conservation equation.

We can easily write the position of point P as $_O\vec{r}_P$ which is a vector whose tail is on point O and tip is on point P. Since velocity is the time derivative of position, we have

$$_O\vec{v}_P = \frac{d}{dt}\left(_O\vec{r}_P\right)$$

Expressing the position of the rod in a general location (remember differentiate then evaluate at $\theta = 30^\circ$), we get

$$_O\vec{r}_P = l\cos(\theta)\vec{i} + l\sin(\theta)\vec{j}$$

Differentiating to obtain velocity, we get

$$_O\vec{v}_P = -l\sin(\theta)\dot{\theta}\vec{i} + l\cos(\theta)\dot{\theta}\vec{j}$$

We should take note of a few things at this point.

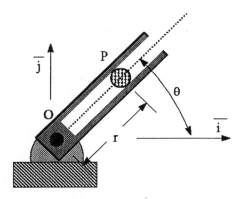

Figure 3.78: A Peg Moving in a Slot

1. Until we have completed differentiating, we do not substitute specific values for variables (*e.g.* we do not substitute $\theta = 30°$ because $\theta$ is not always 30). We can however substitute values for constant variables (*e.g.* we can substitute $l = 2$ because the length of the rod is constant).

2. The vectors $\vec{i}$ and $\vec{j}$ are constants because they have constant length and direction (they do not rotate).

3. Do not forget the calculus concepts concerning implicit differentiation and the chain rule. For example,
$$\frac{d}{dt}(\cos\theta) = \frac{d(\cos\theta)}{d\theta}\frac{d\theta}{dt} = -\sin(\theta)\dot{\theta}$$

To determine the acceleration, we differentiate the velocity

$$_O\vec{a}_P = \frac{d}{dt}\left(_O\vec{v}_P\right) = \frac{d}{dt}\left(-l\sin(\theta)\dot{\theta}\vec{i} + l\cos(\theta)\dot{\theta}\vec{j}\right)$$

$$_O\vec{a}_P = \left(-l\cos(\theta)\dot{\theta}^2 - l\sin(\theta)\ddot{\theta}\right)\vec{i} + \left(-l\sin(\theta)\dot{\theta}^2 + l\cos(\theta)\ddot{\theta}\right)\vec{j}$$

**SOLUTION:**
Substituting the values given for $l$, $\theta$, $\dot{\theta}$, and $\ddot{\theta}$; we get that

$$_O\vec{v}_P = -2\vec{i} + 3.46\vec{j}$$

$$_O\vec{a}_P = -11.9\vec{i} + 4.66\vec{j}$$

Our previous notes also apply to acceleration. We will be differentiating a large number of equations. Therefore, become comfortable with this process.
**End 3.17** ──────────────────────────────

In the next example we drop the subscripts because the points we are discussing should be obvious.

## Example 3.18 ──────────────────────────────

### Velocity of a Peg in a Slot

**PROBLEM:**
Determine the velocity of the peg in figure 3.104 in terms of the angle $\theta$ and distance $r$. Also find the time rate of change of momentum for the peg.
**FORMULATION:**
Though the wording may sound funny when a problem statement says to do something in terms of, it means those items are known. In this case, we assume the $r$ and $\theta$ and all their derivatives are known and derive the velocity and acceleration. In many cases these things are not known but we want them to show up in the conservation equations because we want to solve for them.

The point P has two [NIF], that is why its velocity must be expressed in terms of two variables. We find the velocity by writing a position vector from ANY fixed point to the point of interest then taking the derivative with respect to time. We start by writing the position of the peg from point O at all time as:

$$\vec{r} = r\cos(\theta)\vec{i} + r\sin(\theta)\vec{j}$$

The derivative of this vector is (in this example $r$ is variable)

$$\frac{d\vec{r}}{dt} = \dot{r}\cos(\theta)\vec{i} - r\sin(\theta)\dot{\theta}\vec{i} + \dot{r}\sin(\theta)\vec{j} + r\cos(\theta)\dot{\theta}\vec{j}$$

Note that we used the product formula to find this expression. Also carefully notice how the derivative was taken for the sine and cosine terms. It isn't magical, just carefully go through it yourself and get fast at it. One quick check you can apply is to verify that each addend in the velocity expression has one and only single dot term. This does not guarantee the expression is correct, but the lack of it proves it wrong. The velocity of the peg is:

$$\vec{v} = \left(-\sin(\theta)\dot{\theta}r + \dot{r}\cos(\theta)\right)\vec{i} + \left(\cos(\theta)\dot{\theta}r + \dot{r}\sin(\theta)\right)\vec{j}$$

hence the momentum is simply the mass of the peg times velocity:

$$m\vec{v} = m\left(-\sin(\theta)\dot{\theta}r + \dot{r}\cos(\theta)\right)\vec{i} + m\left(\cos(\theta)\dot{\theta}r + \dot{r}\sin(\theta)\right)\vec{j} \tag{3.14}$$

To find the time rate of change of equation 3.14, simply take the derivative. If mass is constant it factors out and we find ourselves determining simply mass times acceleration. Constant mass problems are so common, many texts focus on teaching you to compute acceleration quickly expecting to simply multiply acceleration by mass in the end. Be careful because this will not work for variable mass problems. Formally you should always find momentum then differentiate and if mass is constant, factor it out.

Anyway, let's get to the heart of the matter. Take the derivative of equation 3.14. This gets ugly so be careful. There are hundreds of ways to do this, you try it and compare to our result. We'll do it one piece at a time in case you need help.

$$\frac{d}{dt}(m\vec{v}) = \frac{d}{dt}\left(m\left(-\sin(\theta)\dot{\theta}r + \dot{r}\cos(\theta)\right)\vec{i} + m\left(\cos(\theta)\dot{\theta}r + \dot{r}\sin(\theta)\right)\vec{j}\right)$$

$$= m\frac{d}{dt}\left(\left(-\sin(\theta)\dot{\theta}r + \dot{r}\cos(\theta)\right)\vec{i}\right) + m\frac{d}{dt}\left(\left(\cos(\theta)\dot{\theta}r + \dot{r}\sin(\theta)\right)\vec{j}\right)$$

Note that we factored the constant $m$ and distributed the derivative over the sum. Now factor out the constants $\vec{i}$ and $\vec{j}$:

$$\frac{d}{dt}(m\vec{v}) = m\frac{d}{dt}\left(\left(-\sin(\theta)\dot{\theta}r + \dot{r}\cos(\theta)\right)\right)\vec{i} + m\frac{d}{dt}\left(\left(\cos(\theta)\dot{\theta}r + \dot{r}\sin(\theta)\right)\right)\vec{j}$$

$$= m\left(\frac{d}{dt}\left(-\sin(\theta)\dot{\theta}r\right) + \frac{d}{dt}\left(\dot{r}\cos(\theta)\right)\right)\vec{i} + m\left(\frac{d}{dt}\left(\cos(\theta)\dot{\theta}r\right) + \frac{d}{dt}\left(\dot{r}\sin(\theta)\right)\right)\vec{j}$$

Now notice that each $\frac{d}{dt}()$ term is going to give several terms (use the product differentiation formula).

$$\frac{d}{dt}(m\vec{v}) = m\left(-\cos(\theta)\dot{\theta}\dot{\theta}r - \sin(\theta)\ddot{\theta}r - \sin(\theta)\dot{\theta}\dot{r} + \ddot{r}\cos(\theta) - \dot{r}\sin(\theta)\dot{\theta}\right)\vec{i}$$

$$+ m\left(-\sin(\theta)\dot{\theta}\dot{\theta}r + \cos(\theta)\ddot{\theta}r + \cos(\theta)\dot{\theta}\dot{r} + \ddot{r}\sin(\theta) + \dot{r}\cos(\theta)\dot{\theta}\right)\vec{j}$$

Wow! Notice there are a couple of terms which can be added to obtain:

$$\frac{d}{dt}(m\vec{v}) = m\left(-r\cos(\theta)\dot{\theta}^2 - r\sin(\theta)\ddot{\theta} - 2\dot{r}\sin(\theta)\dot{\theta} + \ddot{r}\cos(\theta)\right)\vec{i}$$

$$+ m\left(-r\sin(\theta)\dot{\theta}^2 + r\cos(\theta)\ddot{\theta} + 2\dot{r}\cos(\theta)\dot{\theta} + \ddot{r}\sin(\theta)\right)\vec{j} \tag{3.15}$$

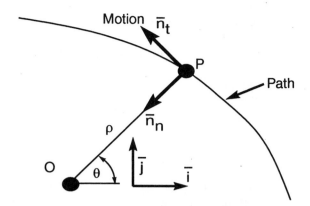

Figure 3.79: Point P Moving on a Known Path

Is this correct? What did you get? Hint: Yes it is correct. These calculations can be simplified tremendously by using the correct tools. A simple Maple program does the same thing. See Maple program 3.15.8.

**End 3.18** ────────────────────────────────────────────

This example shows how we can use kinematics to determine the velocity and acceleration of an object. We begin by defining the object's position from a fixed point and differentiate. Later we will come back to this problem and define the position in terms of moving vectors. We'll get the same result with less ugly algebra. Sometimes our results are easier to understand if we use moving vectors.

In a manner uncharacteristic of this text, we will digress slightly and derive a couple of general equations. There are many experts who claim the following ideas are critically important and this seems like a natural way to discuss them. When a body moves on a known path like a roller coaster, it is sometimes convenient to express its velocity and acceleration in terms related to the path. What we will do now is derive a couple of equations which are convenient to use when the point's path is known.

Consider a point P moving on a known path as figure 3.79 shows. Point O is the center of curvature of the path[10] at the position of P. The distance from O to P is $\rho$, the radius of curvature.

We can write the position of P as

$$_O\vec{r}_P = -\rho\vec{n}_n$$

with the unit vector $\vec{n}_n$ pointing from P toward the center of curvature O

$$\vec{n}_n = -\cos(\theta)\vec{i} - \sin(\theta)\vec{j}$$

Differentiating the position, we get the velocity of P

$$_O\vec{v}_P = -\dot{\rho}\vec{n}_n - \rho\dot{\vec{n}}_n$$

Since the path appears to be a circle (constant radius about the center of curvature),[11] $\dot{\rho} = 0$, and velocity is $_O\vec{v}_P = -\rho\dot{\vec{n}}_n$. The derivative of the unit vector is

$$\dot{\vec{n}}_n = \left(\sin(\theta)\vec{i} - \cos(\theta)\vec{j}\right)\dot{\theta} \tag{3.16}$$

Substituting this into the velocity relation, we get

$$_O\vec{v}_P = -\rho\left(\sin(\theta)\vec{i} - \cos(\theta)\vec{j}\right)\dot{\theta}$$

$$_O\vec{v}_P = \rho\left(-\sin(\theta)\vec{i} + \cos(\theta)\vec{j}\right)\dot{\theta}$$

────────────────────────────────

[10] An infinitesimal section of the path appears to be a circle whose center is at the center of curvature.
[11] This is the definition of center of curvature.

Defining an additional unit vector ($\vec{n}_t$) perpendicular to $\vec{n}_n$ that points in the direction of motion, we get

$$\vec{n}_t = -\sin(\theta)\vec{i} + \cos(\theta)\vec{j} \tag{3.17}$$

By inspection, the velocity simplifies to

$$_O\vec{v}_P = \rho\dot{\theta}\vec{n}_t = v\vec{n}_t \tag{3.18}$$

where

$$v = \rho\dot{\theta} \tag{3.19}$$

Equation 3.18 shows that if P moves, it always moves tangentially along its path with a speed of $v$. In fact, all points move so their velocity is ALWAYS tangent to the path they are on. ALWAYS.

Differentiating velocity gives acceleration.

$$_O\vec{a}_P = \dot{v}\vec{n}_t + v\dot{\vec{n}}_t \tag{3.20}$$

Using equation 3.17, $\dot{\vec{n}}_t$ (the derivative of $\vec{n}_t$) simplifies to

$$\dot{\vec{n}}_t = \left(-\cos(\theta)\vec{i} - \sin(\theta)\vec{j}\right)\dot{\theta} = \dot{\theta}\vec{n}_n = \frac{v}{\rho}\vec{n}_n \tag{3.21}$$

Thus, using equations 3.20 and 3.21, the acceleration simplifies to

$$_O\vec{a}_P = \dot{v}\vec{n}_t + \frac{v^2}{\rho}\vec{n}_n \tag{3.22}$$

This final equation indicates that the acceleration of P has a tangential component (provided $v$ changes) and a radial (normal) component (provided P moves on a curve). This is a general result, valid for any point. A easier way to think about this result is imagine yourself driving an automobile. Suppose you are traveling around a left corner at a constant speed. Since you are on a curve, you have a finite radius of curvature $\rho$. Since you have a speed $v$, you will have an acceleration equal to $\frac{v^2}{\rho}$ and it points toward the center of the turn, left. You are not speeding up or slowing down but you are accelerating. Now if you hit the gas during the turn then you have a positive $\dot{v}$ term also. If you hit the brake, then $\dot{v}$ is negative. If you move on a curve, you HAVE and acceleration.

Just remember the last set of equations are valid all the time but are useful only when you know with certainty the path and hence can find $\rho$ or you know the acceleration and want to know the path or $\rho$. Otherwise leave them alone.

## 3.8.2   Relating Flow Variables

**Objectives**
*When you complete this section, you should be able to*

  *1. Determine when kinematic relationships are needed.*

When applying the conservation equations to mechanical devices, you will often encounter times when there are more unknown variables than there are equations. When this happens you need to look for "extra" equations. Table 3.2 highlights some of the common things to think about when you look for these extra equations.

When you find that there are more than [NIF] flow variables in your equations, some of the flow variables are related to each other. Under these conditions you will need to find the kinematic relationships between the variables before you can proceed.

In this text, we have three types of kinematic relationships. The first of these are relationships in rope and pulley systems.

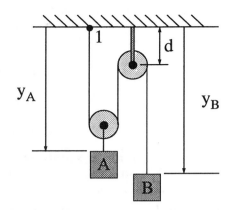

**Analysis Assumptions**

(1) Point 1 is an attachment point of the rope.

(2) The pulleys have the same circumference ($2C$).

(3) $d$ is a constant length.

(4) $y_A$ and $y_B$ are the positions of the masses.

Figure 3.80: (abonrope): Two Masses Tied Together with an Inextensible Rope

## Ropes and Pulleys

**Objectives**

*When you complete this section, you should be able to*

*1. Determine the number of relations present in a rope and pulley system.*

*2. Derive the kinematic relationships for a rope and pulley system.*

This section discusses the kinematic relations present in a rope and pulley system. When we analyze a rope attached to several masses and pulleys, we can find kinematic relations for the system by writing equations for the length of the rope. You will obtain one kinematic relationship for every rope in the system.

To express the kinematic relationships, define the length of rope in terms of the positions of the objects. After expressing the rope length, differentiate once or twice to relate terms with single or double dotted terms respectively.

**Example 3.19** ════════════════════════════════════════════════

### Balancing Masses

**PROBLEM:**

Formulate a mathematical model of the motion of the two masses in figure 3.80 using kinematics.

**FORMULATION:**

The system has one [NIF]. Therefore, we can express all the motions in terms of a single flow variable. To accomplish this, we will express the length of rope ($L$) and differentiate twice.

When we express the length of the rope, we keep two important factors about this type of problem in mind.

1. All our numbers must convey positive rope length because the rope cannot have a negative length.

2. We do not worry about constants because when we differentiate all the constants will become zero.

For figure 3.80, we can express the length as

$$L = y_A - h + C + (y_A - d) + C + (y_B - d)$$

where $L$ is the constant length of the rope, $C$ is half the circumference of the pulleys, $h$ is the constant distance from the left pulley to mass A, $d$ is the constant distance in the figure, $y_A$ is the position of mass A, and $y_B$ is the position of mass B. Differentiating this positional relation gives

$$0 = v_A + v_A + v_B = 2v_A + v_B$$

where $v_A$ and $v_B$ are the velocities of A and B, respectively. Differentiating this velocity relation gives

$$0 = a_A + a_A + a_B = 2a_A + a_B$$

where $a_A$ and $a_B$ are the accelerations of A and B respectively. These equations tell us that $v_B = -2v_A$ (*i.e.* B moves twice as fast as A except in the opposite direction).
**End 3.19** ────────────────────────────────────────────
## Example 3.20 ──────────────────────────────────────────

### Unbalanced Masses

**PROBLEM:**

Suppose the masses in figure 3.80 are released from rest. If the mass of A is 5 [kg] and B is 10 [kg], determine how long it takes body B to move 2 meters.

**FORMULATION:**

The strategy we plan to use is to use the conservation of momentum to solve for the time rate of change of momentum of block B, then integrate to determine the position of B with respect to time. This device has one [NIF] provided we ignore any swinging back and forth but because we do not know any flows we must apply a conservation equation. Let's consider linear momentum in the vertical direction applied at a differential time period. Since there are two masses and each one has different motions, we will choose the masses separately so the momentum will be easy to express. First for mass B, we have the following (take downward positive since $y_B$ is measured positive down, remember take directions to make motion easy to express):

$$-T + m_B g = \frac{d}{dt}\left(m_B \dot{y}_B\right)$$

where $T$ is the rope tension, it is not equal to the weight of the block! Since there are two unknowns in this one equation ($T$ and $\dot{y}_B$) we need another equation. Not to worry, we expected this since we expect to need systems with both masses. Apply linear momentum to block A with its attached pulley, again down is positive:

$$-2T + m_A g = \frac{d}{dt}\left(m_A \dot{y}_A\right)$$

Note that there are two places where the external rope touches the internal pulley attached to A hence the $2T$. Also note that the tension in this expression is the same as the tension applied to mass B. Why? Apply angular momentum to each of the pulleys one at a time. What you will find is if the pulleys are lightweight and if the pins on the pulleys are smooth, then there cannot be any unbalanced moment about the center of the pulley, hence the rope tensions on either side are identical. This isn't true for all pulleys in all problems, you should consider each one individually.

Now with the two momentum equations there are three unknowns $\dot{y}_A$, $\dot{y}_B$ and $T$. What to do? Well, it is a one [NIF] system but there are two flows in the equations. There should only be one, the flows are related. We found the relationship in example 3.19. The relationship is: $2\dot{y}_A + \dot{y}_B = 0$. With this kinematic relation we can find all the unknowns to be:

- $\ddot{y}_A = -3.27$ [m/s] the negative means our assumed direction for acceleration (downward is positive) is wrong it accelerates up.

- $\ddot{y}_B = 6.54$ [m/s], the positive means it is accelerating in the positive direction (downward).

- $T = 32.7$ [N].

Now that we have the acceleration we can integrate backward to find what an expression for the position. Start with:

$$\ddot{y}_B = \frac{d}{dt}\left(\dot{y}_B\right) = 6.54$$

Separating the variables and integrating (it isn't always this easy):

$$\dot{y}_B = 6.54t + C$$

The value $C$ is an arbitrary constant. Since B is at rest at $t = 0$ the value of $C = 0$, hence:

$$\dot{y}_B = \frac{d}{dt}\left(y_B\right) = 6.54t$$

Again separating the variables and integrating:

$$y_B = 3.27t^2 + C_1$$

If we mark a spot on the wall where B was released and measure $y_B$ from the spot, then nothing would have changed but the value of $C_1 = 0$ so:

$$y_B = 3.27t^2$$

Now what time is it when B falls 2 meters?

$$2 = 3.27t^2$$

$t = 0.782$ [sec].

**End 3.20** ─────────────────────────────────

In the last example we were able to find the time derivative of a length and were able to integrate to find the position as a function of time. Basically what we did was solve a simple differential equation. We cannot always solve for the derivatives and then simply integrate. In the general case, we'll apply conservation of momentum and end up with a set of differential equations which we must then solve. You'll see what we mean a bit later.

**Example 3.21** ─────────────────────────────────

## Handling Changing Rope Lengths

**PROBLEM:**
Consider the two masses connected with a rope which figure 3.81. Suppose pulley P is connected to a motor which pulls the rope in at a constant rate of 3 [ft/sec] and if block B is pulled downward at a constant speed of 4 [ft/sec], determine the speed of block A.

**FORMULATION:**
First determine that the [NIF] of the device is two. Also note that two flows are given (the speed of the rope and the motion of B). Since we only desire to determine a flow, not an effort, we need not write any conservation equations.

Next define enough positions to describe the positions of the masses. Although only two variables are really required (two [NIF]) it is often easier to define several variables and sort them out later. Define $y_A$ as the distance from the ceiling to the center of mass A (down is positive). Also define $y_B$ as the distance from the ceiling to the center of mass B. Let $L$ be the length of the rope. If there were other quantities that change, we could define other variables.

Next define the length of the rope. We will ignore constants because we plan to differentiate the expression once we have it. The rope length $L$ is (all $C$ quantities are constants):

$$L = (y_A - C) + (y_A - C - C) + (y_B - C - C) + (y_B - C - C)$$

Now differentiate to find:

$$\frac{d}{dt}(L) = \dot{y}_A + \dot{y}_A + \dot{y}_B + \dot{y}_B = 2\dot{y}_A + 2\dot{y}_B$$

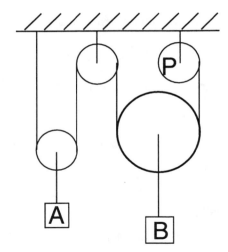

Figure 3.81: Two Masses Connected Via a Rope and Pulleys Where the Rope is Pulled in by a Wench.

Differentiate again to find:

$$\ddot{L} = 2\ddot{y}_A + 2\ddot{y}_B$$

If the rope were a constant length then $\dot{L}$ and $\ddot{L}$ would be 0. Since the rope is being pulled in at a constant rate, the rope is getting shorter so $\dot{L} = -3$ [ft/sec]. The negative says $L$ is getting shorter. For the second derivative $\ddot{L} = 0$ because it is being pulled in at a constant rate. The terms $\dot{y}_B$ and $\ddot{y}_B$ are $+4$ [ft/sec] and 0. The positive value of $\dot{y}_B$ is positive because the value of $y_B$ is getting bigger (mass B is pulled downward). Therefore solving for the motion of A gives $\ddot{y}_A = 0$ and $\dot{y}_A = -5.5$ [ft/sec]. The value is negative meaning $y_A$ is getting smaller, it is moving up.

**End 3.21**

## Vector Loops

**Objectives**
*When you complete this section, you should be able to*

1. *Be able to describe when vector loops are used to relate motions.*

2. *Be able to write valid vector loop equations.*

3. *Be able to solve vector loop equations to relate motions.*

The basis of all kinematics is relating the motions of several points together. It is pretty simple to relate the positions. For example to get from O to P we could go from O to A then from A to P see Figure 3.82. The following equation expresses this position relationship.

$$_O\vec{r}_P = {_O\vec{r}_A} + {_A\vec{r}_P}$$

To find the velocity of P, we differentiate the position.

$$_O\vec{v}_P = {_O\vec{v}_A} + {_A\vec{v}_P}$$

Likewise to find the accelerations of P, we differentiate the velocity.

$$_O\vec{a}_P = {_O\vec{a}_A} + {_A\vec{a}_P}$$

When searching for a kinematic relation between a set of points when there are no ropes present, the first approach to try is the vector loop. We apply a vector loop by starting at any point and then

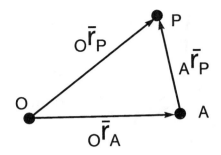

Figure 3.82: Relating the Positions of Three Points

summing position vectors as we walk around a loop returning to the starting point. The general idea is to add vectors as you go around a loop, when you get back to the starting point you will have a vector sum that adds up to zero.

Generally, when we write vector loops, we only use special points in the loop. A special point is a point that

1. we know something about

2. we want to know something about

3. touches two different bodies (*i.e. a contact point*).

There are many correct ways to write vector loops for most mechanisms. There are some rules of thumb that make it a little easier to find correct, worthwhile loops. When looking for vector loops try to apply the following rules of thumb:[12]

1. Start at any special point you want.

2. Move from special point to special point until you return to the start.

3. Do not backtrack. If you go from point B to C do not move from C back to B. If you do backtrack, you are just wasting your time.

4. As you move from point to point try to move to a special point which introduces the minimum number of unknowns. This is often accomplished by moving down links, along slots and the like. If possible walk in either a known direction or a known distance.

## Example 3.22

### A Simple Machine

**PROBLEM:**

Figure 3.83 shows a simple mechanism. The mechanism has two links connected by a pin at A. The link AC is pinned to a block at B and the block is trapped in a vertical slot so it can move up and down only. Suppose the hydraulic cylinder moves point A so it has a known horizontal position, speed and acceleration. Determine the velocity and acceleration of point C.

**FORMULATION:**

The mechanism has one [NIF] and since one flow is given and only a flow is desired, we do not need any conservation equations. To find the velocity and acceleration of any point, we start at any fixed point and write the position of the point we want information about, then differentiate.

To find the position of point C, start at the hydraulic cylinder (which is fixed) and walk to point C. When you do this, define whatever variables make the job easy and worry about them later. An obvious thing to do (this is not the only thing) is walk from the cylinder to A, then along the link to C. This can be written as:

$$_D\vec{r}_C = {}_D\vec{r}_A + {}_A\vec{r}_C = X_A\vec{i} - 2L\cos\theta\vec{i} + 2L\sin\theta\vec{j} \tag{3.23}$$

---

[12]A rule of thumb is a rule that works most of the time. Keep your head about you and use the rules until you get some experience.

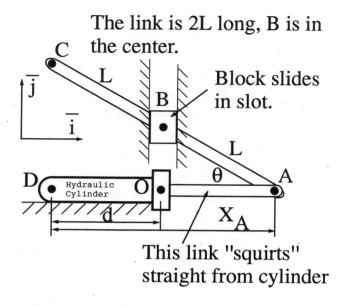

The link is 2L long, B is in the center.

Block slides in slot.

This link "squirts" straight from cylinder

Figure 3.83: A Hydraulic Cylinder Moves a Link.

We can differentiate this to find the velocity and acceleration as:

$$_D\vec{v}_C = \dot{X}_A\vec{i} + 2L\sin\theta\dot{\theta}\vec{i} + 2L\cos\theta\dot{\theta}\vec{j} \tag{3.24}$$

$$_D\vec{a}_C = \ddot{X}_A\vec{i} + 2L\sin\theta\ddot{\theta}\vec{i} + 2L\cos\theta\ddot{\theta}\vec{j} + 2L\cos\theta\dot{\theta}^2\vec{i} + 2L\sin\theta\dot{\theta}^2\vec{j} \tag{3.25}$$

Now note the following. In equation 3.23 there are three unknowns (two on the left hand side plus $\theta$), let's assume the constants $L$, $d$ are known. There are only two scalar equations in 3.23 so there are too many unknowns to solve anything. The same problem occurs in the velocity and acceleration equations.

What is missing is that there is a relationship between the variables. The way we find the relationship is to write a vector loop equation. Remember all loops do is find relationships between flow variables and that is exactly what we want to do at this moment. The obvious special points in the problem are D, A, B, and C. We will also use O, a fixed point below the vertical slot, although it is not immediately obvious why we want O, hang on and you'll see why in a minute. There are several loops which you can form, we like to start at fixed points although we could start anywhere. To form a loop start at D move to A (this is a variable but known distance and known direction). Traveling from D to A introduces no unknowns. From A go to B which is a known distance but unknown variable direction. From B go to O, this is a known direction but unknown distance (call the distance $Y$). From O go back to D, this is a known distance and known direction. We are back at the start and should have no more than 2 unknown variables in it. Mathematically, the vector loop is written as:

$$X_A\vec{i} - L\cos\theta\vec{i} + L\sin\theta\vec{j} - Y\vec{j} - d\vec{i} = 0 \tag{3.26}$$

Note that if the constants are known (if this were a real object you could measure the constants) then there are three variables $X_A$, $\theta$ and $Y$. Since $X_A$ is known there are two unknown variables and there are two scalar equations so we can solve for both unknowns. This is done as follows. First pull out the $\vec{i}$ terms:

$$X_A - L\cos\theta - d = 0$$

now the $\vec{j}$ terms:

$$L\sin\theta - Y = 0$$

Solve these two for $\theta$ and $Y$

$$\theta = \cos^{-1}(\frac{X_A - d}{L}) \tag{3.27}$$

$$Y = L \sin \theta = L \frac{\pm \sqrt{L^2 - (X_A - d)^2}}{X_A - d} \tag{3.28}$$

The last substitution in equation 3.28 comes from finding the $\sin(\theta)$ using the identity $\cos^2(\theta) + \sin^2(\theta) = 1$. The $\pm$ comes from the fact that it is "theoretically" possible for $\theta$ to be negative. The real machine would probably always have positive $\theta$ and we would take the + term rather than the -.

Back to the problem, plug the constants and $X_A$ into equation 3.27 to find the angle $\theta$. Then use $\theta$ in equation 3.23 to find the position of $C$ as:

$$_D\vec{r}_C = X_A \vec{i} - 2L \frac{X_A - d}{L} \vec{i} + 2L \frac{\pm \sqrt{L^2 - (X_A - d)^2}}{X_A - d} \vec{j}$$

If we put $\theta$ into equation 3.24 we find that $\dot{\theta}$ is still unknown. So we need to find $\dot{\theta}$.

To find $\dot{\theta}$ we can proceed two ways. We will perform both methods here as a demonstration but you need to do one or the other. First, we can differentiate equation 3.27 to find $\dot{\theta}$:

$$\dot{\theta} = -\frac{\dot{X}_A}{\sqrt{L^2 - (X_A - d)^2}} \tag{3.29}$$

Now we could plug in numbers for the constants, for $X_A$ and $\dot{X}_A$ to find $\dot{\theta}$. We would then put $\dot{\theta}$ and the value for $\theta$ from before in equation 3.24 to find the velocity.

A second technique to find $\dot{\theta}$ is to differentiate the vector loop equation 3.26 as:

$$\dot{X}_A \vec{i} + L \sin \theta \dot{\theta} \vec{i} + L \cos \theta \dot{\theta} \vec{j} - \dot{Y} \vec{j} = 0 \tag{3.30}$$

Now pull off the $\vec{i}$ and $\vec{j}$ terms:

$$\dot{X}_A + L \sin \theta \dot{\theta} = 0$$

$$L \cos \theta \dot{\theta} - \dot{Y} = 0$$

Now solve for $\dot{\theta}$ and what the heck, get $\dot{Y}$ while we are at it:

$$\dot{\theta} = -\frac{\dot{X}_A}{L \sin \theta} \tag{3.31}$$

$$\dot{Y} = L \cos \theta \dot{\theta} = -L \cos \theta \left( \frac{\dot{X}_A}{L \sin \theta} \right)$$

Now having $\theta$ from before, put it into equation 3.31 to get $\dot{\theta}$ then take both $\theta$ and $\dot{\theta}$ into equation 3.24 to find the velocity of $C$.

So which of the two methods for finding $\dot{\theta}$ is best? Well it depends really. The first method is very good when you are trying to solve the problem analytically. The second technique is very good when you are solving the problem numerically. For example, if we were solving the problem numerically, we could plug numbers into equation 3.27 and with one subtraction, one division and an inverse cosine button push on the calculator, we have a number for $\theta$. Take the number for $\theta$ and compute its sine, multiply by $L$ and pow, $Y$ is known. Take these numbers and turn to equation 3.31 and with a couple of key strokes, we have $\dot{\theta}$. Finally, take the numbers for $\theta$ and $\dot{\theta}$ and it trivial to calculate each of the terms in the velocity equation 3.24.

So how do we get the acceleration? Notice that the only "number" we need that we do not already have is $\ddot{\theta}$. You guessed it, there are two ways to get $\ddot{\theta}$, (a) differentiate equation 3.29, (b) differentiate equation 3.30. We will do the later this time. The derivative of equation 3.30 is:

$$\ddot{X}_A \vec{i} + L \cos \theta \dot{\theta}^2 \vec{i} + L \sin \theta \ddot{\theta} \vec{i} - L \sin \theta \dot{\theta}^2 \vec{j} + L \cos \theta \ddot{\theta} \vec{j} - \ddot{Y} \vec{j} = 0$$

Pulling off the $\vec{i}$ and $\vec{j}$ gives:

$$\ddot{X}_A + L \cos \theta \dot{\theta}^2 + L \sin \theta \ddot{\theta} = 0$$

$$-L\sin\theta\dot\theta^2 + L\cos\theta\ddot\theta - \ddot Y = 0$$

Solve the first of these for $\ddot\theta$:

$$\ddot\theta = -\frac{\ddot X_A + L\cos\theta\dot\theta^2}{L\sin\theta}$$

Now if we are solving numerically, we have numbers for $\theta$ and $\dot\theta$ so with a handful of button pushing on the calculator, we have $\ddot\theta$.

Before we leave this example note the following points. Note that the position equations and loops are used to solve for the undotted variables. The velocity equation and the derivative of the loop are used to solve for the single dot terms. The acceleration equation and the second derivative of the loop are used to solve for the double dotted terms. This pattern occurs in all loop problems.

Let's discuss what would happen if we defined a different loop. Consider a loop from D to A (zero unknown variables), from A to C ($\theta$ is unknown), from C back to D (two unknown variables). This loop has a total of three unknown variables compared to ours. Why? What rule was broken? The step from C to D had neither a known direction nor known length. You see when you make a step in which there is an unknown direction and unknown magnitude, you have just introduced two unknowns. Since you only get 2 equations for each two dimensional vector loop, you have in the single step, introduced all the unknowns your loop can determine. Take steps that reduce the number of unknown variables introduced in the problem.

What would happen if we went from D to A, A to B then B to D? Nothing different really. The only real change was the jump from B to D. Note that in that jump there is only one variable. What was the difference in this jump and the one from C to D? One variable. Why? Because we can think of the jump from B to D as a variable jump down the slot and a constant jump horizontally. You cannot say this about the jump from C to D.

What can you do to see how to write the loop we used. First find the special points, you MUST realize that point B is special. If you don't realize this, you are dead in the water. How did we know to introduce point O? Because when writing the loops, we attempt to "walk along" members or slots and the like. Once we arrived at B, we thought about "walking down" the vertical slot and it just made sense to put O in the middle of the slot even with the starting point D.

**End 3.22** ━━━━━━━━━━━━━━━━━━━━━━━━━━━━━━━━━━━━━━━━━━━━━━━━━━━━━━━━━━

Before you lose sight of what is going on. Remember the process for solving problems. Compute the [NIF]. Based on the [NIF] and what is given and required determine whether you need to write a conservation equation. If you need a conservation equation write it. To determine the momentum of an object determine the velocity of the center of mass. The ONLY way to find velocity is to write a position from a fixed point to the center of mass and differentiate for velocity and acceleration. **Note we did not say write a loop, we said write a position. Loops only relate one motion to another, they do not find velocity or acceleration.** After finding the velocity or momentum count the flow variables, if there are too many (more than the [NIF]) then they must be related. A vector loop is one way to find these relationships. The rope length equations is another way to find relationships. Now for another example.

**Example 3.23** ━━━━━━━━━━━━━━━━━━━━━━━━━━━━━━━━━━━━━━━━━━━━━━━━━━━━━━━━

### A Circular Slider

**PROBLEM:**

The circular disk in figure 3.84 has a circular slot with a radius of 200 [mm]. The disk rotates about O with a constant $\dot\beta = 15$ [rad/s]. Determine the acceleration of the slider (A), at the instant when it passes through the center of the disk, if at that moment, $\dot\theta = 12$ [rad/sec] and $\ddot\theta = 0$ [rad/s$^2$].

**FORMULATION:**

Notice that the system has two [NIF]. The disk can spin and if we stop it, the slider can still move. Because the system has two [NIF], we specify two motions, $\beta$, $\theta$ and their derivatives. We will accomplish our objective by writing the position of A from a fixed point, then differentiate twice. **Note that we do not start with a loop. Since the problem asks for acceleration, we start by writing a position and see where it leads.**

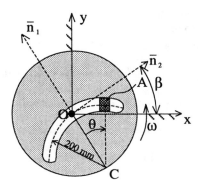

Figure 3.84: A Rotating Disk with a Circular Slot

Begin by writing a position from O (a fixed point) to C and then to A. We will choose these points because they are special. O is fixed, A is what we want, and C is known to be at the center of the slot. The reason for bringing in point C is because we know how to write the position of something on a circle by starting from the center, and we haven't ever (in this book anyway) attempted to write the position any other way.

Expressing the position of slider A kinematically using a vector position, we get

$$_O\vec{r}_A = {_O}\vec{r}_C + {_C}\vec{r}_A$$

We can easily write these positions in terms of vectors $\vec{n}_1$ and $\vec{n}_2$ painted on the disk. By painted on, we mean the vectors are literally painted on. They move with the disk. The result follows:

$$_O\vec{r}_A = -0.2\vec{n}_1 + 0.2\left(\cos(\theta)\vec{n}_1 + \sin(\theta)\vec{n}_2\right)$$

The unit vectors are (look at the figure and resolve the vectors just like you would do to find the x and y components of a force)

$$\vec{n}_1 = -\sin(\beta)\vec{i} + \cos(\beta)\vec{j}$$
$$\vec{n}_2 = \cos(\beta)\vec{i} + \sin(\beta)\vec{j}$$

We do not really need to use these painted on unit vectors, but there would be considerable algebra involved. If we did not use the vectors $\vec{n}_1$ and $\vec{n}_2$ then the position is:

$$_O\vec{r}_A = -0.2\left(-\sin(\beta)\vec{i} + \cos(\beta)\vec{j}\right) + 0.2\left[\cos(\theta)\left(-\sin(\beta)\vec{i} + \cos(\beta)\vec{j}\right) + \sin(\theta)\left(\cos(\beta)\vec{i} + \sin(\beta)\vec{j}\right)\right]$$

If you want to, you can follow along by taking derivatives of this expressions but we plan to use the shorter one with $\vec{n}_1$ and $\vec{n}_2$.

Differentiating the position relations, we get the velocity

$$\frac{d}{dt}\left(_O\vec{r}_A\right) = {_O}\vec{v}_A = -0.2\dot{\vec{n}}_1 + 0.2\left(-\sin(\theta)\vec{n}_1 + \cos(\theta)\vec{n}_2\right)\dot{\theta} + 0.2\left(\cos(\theta)\dot{\vec{n}}_1 + \sin(\theta)\dot{\vec{n}}_2\right)$$

Notice that $\vec{n}_1$ and $\vec{n}_2$ are variables, they change hence we used the product formula to find the correct derivatives. Basically we treated the vectors just like any other variable, no big deal. We can also differentiate the expressions for the vectors:

$$\dot{\vec{n}}_1 = \left(-\cos(\beta)\vec{i} - \sin(\beta)\vec{j}\right)\dot{\beta} = -\vec{n}_2\dot{\beta}$$

$$\dot{\vec{n}}_2 = \left(-\sin(\beta)\vec{i} + \cos(\beta)\vec{j}\right)\dot{\beta} = \vec{n}_1\dot{\beta}$$

If we did not see that $\dot{\vec{n}}_1$ was $-\vec{n}_2\dot{\beta}$, we could proceed, but there would be considerably more algebra. If you work several problems using moving vectors you will find a similar pattern. The derivative of one vector is most often a simple expression of another so keep your eyes open and expect some simplifications. In fact, a later section gives a formal technique of finding derivatives of unit vectors. Back to the example, the velocity is (plug in the expressions for the derivatives of the unit vectors)

$$_O\vec{v}_A = .2\vec{n}_2\dot{\beta} + .2\left(-\sin(\theta)\vec{n}_1 + \cos(\theta)\vec{n}_2\right)\dot{\theta} +$$

$$.2\left(-\cos(\theta)\vec{n}_2\dot{\beta} + \sin(\theta)\vec{n}_1\dot{\beta}\right)$$

Differentiating this velocity relation, we get the acceleration

$$_O\vec{a}_A = .2\dot{\vec{n}}_2\dot{\beta} + .2\vec{n}_2\ddot{\beta} +$$

$$.2\left(-\cos(\theta)\vec{n}_1\dot{\theta} - \sin(\theta)\dot{\vec{n}}_1 - \sin(\theta)\vec{n}_2\dot{\theta} + \cos(\theta)\dot{\vec{n}}_2\right)\dot{\theta} +$$

$$.2\left(-\sin(\theta)\vec{n}_1 + \cos(\theta)\vec{n}_2\right)\ddot{\theta}$$

$$+.2\left(\sin(\theta)\vec{n}_2\dot{\beta}\dot{\theta} - \cos(\theta)\dot{\vec{n}}_2\dot{\beta} - \cos(\theta)\vec{n}_2\ddot{\beta}\right.$$

$$\left. + \cos(\theta)\vec{n}_1\dot{\beta}\dot{\theta} + \sin(\theta)\dot{\vec{n}}_1\dot{\beta} + \sin(\theta)\vec{n}_1\ddot{\beta}\right)$$

Wow! If you think this is bad, try it without using $\vec{n}_1$ and $\vec{n}_2$. This reduces to (substitute in the derivatives of the vectors and algebra, algebra, algebra, skip ahead if you want)

$$_O\vec{a}_A = .2\vec{n}_1\dot{\beta}\dot{\beta} + .2\vec{n}_2\ddot{\beta} +$$

$$.2\left(-\cos(\theta)\vec{n}_1\dot{\theta} + \sin(\theta)\vec{n}_2\dot{\beta} - \sin(\theta)\vec{n}_2\dot{\theta} + \cos(\theta)\vec{n}_1\dot{\beta}\right)\dot{\theta} +$$

$$.2\left(-\sin(\theta)\vec{n}_1 + \cos(\theta)\vec{n}_2\right)\ddot{\theta} +$$

$$.2\left(\sin(\theta)\vec{n}_2\dot{\beta}\dot{\theta} - \cos(\theta)\vec{n}_1\dot{\beta}\dot{\theta} - \cos(\theta)\vec{n}_2\ddot{\beta}\right.$$

$$\left. + \cos(\theta)\vec{n}_1\dot{\beta}\dot{\theta} - \sin(\theta)\vec{n}_2\dot{\beta}\dot{\beta} + \sin(\theta)\vec{n}_1\ddot{\beta}\right)$$

Now collect up similar terms:

$$_O\vec{a}_A = .2\left(\dot{\beta}\dot{\beta} - \cos(\theta)\dot{\beta}\dot{\beta} - \cos(\theta)\dot{\theta}\dot{\theta} + \cos(\theta)\dot{\beta}\dot{\theta} +\right.$$

$$\left.\cos(\theta)\dot{\beta}\dot{\theta} - \sin(\theta)\ddot{\theta} + \sin(\theta)\ddot{\beta}\right)\vec{n}_1 +$$

$$.2\left(\ddot{\beta} - \cos(\theta)\ddot{\beta} + \sin(\theta)\dot{\beta}\dot{\theta} + \sin(\theta)\dot{\beta}\dot{\theta} - \sin(\theta)\dot{\theta}\dot{\theta} + \cos(\theta)\ddot{\theta} - \sin(\theta)\dot{\beta}\dot{\beta}\right)\vec{n}_2$$

**SOLUTION:**

Now that the differentiation is finished, we can substitute the specific values of $\theta = 0$ (the slider is passing through point O at the moment of interest), $\dot{\beta} = 15$ radians/sec (given in the problem statement), $\ddot{\beta} = 0$ (given), $\dot{\theta} = 12$ radians/sec (given), and $\ddot{\theta} = 0$ (given) into the expression above and solve for $_O\vec{a}_A$.

$$_O\vec{a}_A = .2\left(-144 + 2(15)(12)\right)\vec{n}_1 = 43.2\vec{n}_1$$

**OTHER CONSIDERATIONS:**

Notice that the most difficult part of this example is the mathematics. Later in this chapter we will discuss easier ways of differentiating vectors that will make this part of the example easier. Notice also that we did not use a vector loop. The only purpose of a loop is to relate one flow variable to another. If we are not relating variables, we do not need the loop. Don't start a problem by writing a loop, think about it first and determine what it is you really want.

**End 3.23** ──────────────────────────────────────────────

In the next and final example, we show that you may sometimes need more than one vector loop. The way you discover this is by counting the [NIF] and the number of unknown variables in each loop. If there are too many unknowns (and you haven't made a mistake) then there is most likely another vector loop somewhere. This will happen rarely enough that you will likely panic when you experience it. It happens frequently enough however that you **will** most likely experience it. So, be warned, you will probably have a problem sometime in your work which looks like the loops do not work but don't panic, try another loop.

Most students find it easiest to write loops by thinking about walking from point to point. This is a good way to do it. For others, they like more formal ways to doing things. For these people we will use a vector formulation in the next example. If this confuses you, simply write the loop by thinking about walking from point to point.

For some people, drawing a picture of the loop often makes it easier to write the loop equation than thinking about walking. Once you draw the loop, you can express it in terms of subscripted vectors. You can also check your loop equations by observing the subscripts on the position vectors. For example, all valid loop equations can be rewritten in a standard form such that

1. all the adjacent subscripts are identical

2. the left most subscript is the same as the rightmost, and

3. all the terms are added, none are subtracted.

If you are trying to get a loop equation into the standard form (to check its validity) you can use the following identity:

$$_O\vec{r}_A = -_A\vec{r}_O$$

As an example, the following loop equation is valid:

$$_O\vec{r}_A + _A\vec{r}_B + _B\vec{r}_C + _C\vec{r}_O = 0$$

Whereas, these loop equations are not.

$$_O\vec{r}_A + _A\vec{r}_B + _B\vec{r}_C + _C\vec{r}_P = 0$$

$$_O\vec{r}_A - _A\vec{r}_B + _B\vec{r}_C + _C\vec{r}_O = 0$$

$$_O\vec{r}_A + _A\vec{r}_B + _E\vec{r}_C + _C\vec{r}_O = 0$$

## Example 3.24

### Kinematic Analysis of a Bell-Crank

**PROBLEM:**

The flanges of the collar B in figure 3.85 slide with a constant velocity $v_B$ to the right along the fixed shaft. The flanges guide the pin A in the bell-crank $AOD$ of the figure. The radial slot in the bell-crank ($AOD$) positions the plunger's upper end.

In other words, B slides horizontally. When it does, it pushes and/or pulls A. A moves horizontally when B moves and A also moves vertically a little because it is rotating about O. AOD is a solid object so when A moves, the body AOD rocks back and forth. Since C is caught in the slot of AOD, the rocking motion of AOD makes C move vertically.

Determine the velocity and acceleration of the plunger $CE$ (find the velocity and acceleration of point C on the plunger) when $\theta = 30, 45,$ and $60°$. Let $v_B$ be constant at $v_B = 3$ [ft/sec].

**FORMULATION:**

Notice that the system has one [NIF] and one flow is given which means we do not need any conservation equations to find the flow we want. Since the problem asks for the acceleration of point C, we write a position vector from any fixed point to C, then differentiate twice. Label a point F at the intersection of the horizontal rod B moves on and the vertical line C moves on. The position of C is:

$$_F\vec{r}_C = -y\vec{j}$$

where $y$ is the variable distance from F to C. The velocity and acceleration are:

$$_F\vec{v}_C = -\dot{y}\vec{j} \tag{3.32}$$

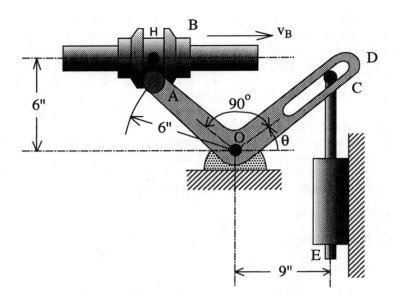

Figure 3.85: A Bell-Crank Mechanism

$$_F\vec{a}_C = -\ddot{y}\vec{j} \tag{3.33}$$

What we really need is to find $\ddot{y}$ given $\theta$ and the velocity of B. In other words we need to relate $\ddot{y}$ to $\theta$ and $v_B$. To determine a relationship, we need a vector loop.

Since we only write loops using special points, we need to determine which points are special. In our case, points A and C are special because they are contacts between bodies. Points O and H are special because we know their motions (O is fixed). Since we are given the motion of H and asked for the motion of C, we will define two fixed points that make expressing their motions easy. Figure 3.86 shows them as G and F. The positions of H and C are $_G\vec{r}_H$ and $_F\vec{r}_C$.

One loop is shown in figure 3.87. Note a couple of things. First we start the loop from G to H so we can conveniently bring in the motion of point H which is what was given. Second, we drew a vector from O to A because we think it is easier to write than one from A to O. We think A to O makes the vector confusing to write in terms of $\vec{i}$, $\vec{j}$ and $\theta$. True its a minor point but this is just an example. Third, note that all the vectors except from O to G walk along the mechanism. The vector between O and G is constant so it doesn't really matter that it does not follow the mechanism because it does not introduce any variables. If you had to, you could get a ruler and measure the vector between O and G, its known. Mathematically, the loop is

$$_G\vec{r}_H + {}_H\vec{r}_A - {}_O\vec{r}_A - {}_G\vec{r}_O = 0$$

Note that this is a valid vector loop, the negatives are due to the vectors being drawn backward. You can manipulate the expression and prove to yourself it is correct. Using some specific values, we get

$$x\vec{i} - h\vec{j} - 6\vec{n}_1 - {}_G\vec{r}_O = 0$$

We do not worry about $_G\vec{r}_O$ because we plan to differentiate to obtain velocity. Since $_G\vec{r}_O$ is constant, it will disappear anyway.

If you like to think of "walking" the mechanism, start at G and walk to H which is right a variable $X$ ($X\vec{i}$) then down to A ($-h\vec{j}$, $h$ is variable and the negative $\vec{j}$ says down) then along the link to O ($-6\vec{n}_1$) and back to G. You would get the same values we have.

Substituting in what $\vec{n}_1$ is (in terms of angle $\theta$) gives:

$$x\vec{i} - h\vec{j} - 6\left(-\sin\theta\vec{i} + \cos\theta\vec{j}\right) - {}_G\vec{r}_O = 0$$

Differentiating, we get

$$\dot{x}\vec{i} - \dot{h}\vec{j} - 6\left(-\cos\theta\vec{i} - \sin\theta\vec{j}\right)\dot{\theta} = 0$$

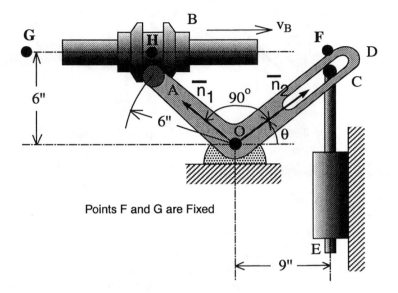

Figure 3.86: Points on a Bellcrank for Drawing Vector Loops

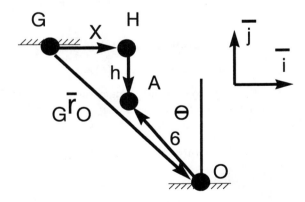

Figure 3.87: One Vector Loop for the Bellcrank

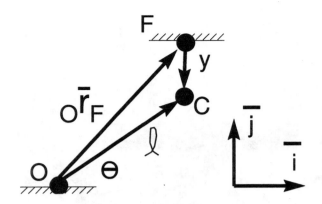

Figure 3.88: A Second Vector Loop for the Bellcrank

If we separate the horizontal and vertical components, we get

$$\dot{x} + 6\cos(\theta)\dot{\theta} = 0$$

$$-\dot{h} + 6\sin(\theta)\dot{\theta} = 0$$

These two equations contain two unknowns ($\dot{\theta}$, $\dot{h}$). We can solve for the two unknowns but unfortunately neither of these are the quantity we really want. We need more kinematic equations. Since we want to know the motion of C, we choose to write another loop using points O and C. Figure 3.88 shows this second loop. Mathematically, we can write this loop as follows:

$$_O\vec{r}_F + {}_F\vec{r}_C - {}_O\vec{r}_C = 0$$

which simplifies to

$$_O\vec{r}_F - y\vec{j} - l\left(\cos(\theta)\vec{i} + \sin(\theta)\vec{j}\right) = 0$$

Again, we do not worry about $_O\vec{r}_F$ because it is constant. The logic behind this loop is that the leg between O and C is there to bring in information about the angle $\theta$, which can be computed from the first loop. The leg between F and C is used to make it easy to specify the motion of C. The leg from O to F is simply a quick way to close the loop. Differentiating, we get

$$-\dot{y}\vec{j} - l\left(-\sin(\theta)\vec{i} + \cos(\theta)\vec{j}\right)\dot{\theta} - \dot{l}\left(\cos(\theta)\vec{i} + \sin(\theta)\vec{j}\right) = 0$$

Separating the components gives

$$l\sin(\theta)\dot{\theta} - \dot{l}\cos(\theta) = 0$$

$$-\dot{y} - l\cos(\theta)\dot{\theta} - \dot{l}\sin(\theta) = 0$$

Here $\dot{y}$ can be plugged into equation 3.32 to find the plunger velocity. Variable $\dot{l}$ is an unknown which we can determine using our equations should we need it in the future for some other calculation.

Differentiating the four velocity components, we get

$$\ddot{x} - 6\sin(\theta)\dot{\theta}^2 + 6\cos(\theta)\ddot{\theta} = 0$$

$$-\ddot{h} + 6\cos(\theta)\dot{\theta}^2 + 6\sin(\theta)\ddot{\theta} = 0$$

$$\dot{l}\sin(\theta)\dot{\theta} + l\cos(\theta)\dot{\theta}^2 + l\sin(\theta)\ddot{\theta} - \ddot{l}\cos(\theta) + \dot{l}\sin(\theta)\dot{\theta} = 0$$

$$-\ddot{y} - \dot{l}\cos(\theta)\dot{\theta} + l\sin(\theta)\dot{\theta}^2 - l\cos(\theta)\ddot{\theta} - \ddot{l}\sin(\theta) - \dot{l}\cos(\theta)\dot{\theta} = 0$$

**SOLUTION:**

Here $\ddot{x} = 0$ because the collar B moves with a constant velocity. We can solve the velocity components for $\dot{\theta}$, $\dot{l}$, $\dot{h}$ and $\dot{y}$ then use these values in the acceleration components to solve for $\ddot{\theta}$, $\ddot{l}$, $\ddot{h}$ and $\ddot{y}$. See Maple session 3.15.5.

Notice that we use the velocity components to determine the single dot terms and the acceleration components determine the double dot terms. Although these problems require a large amount of algebra, they are straightforward. You may find it helpful to use a computer tool to perform the algebraic manipulations.

**OTHER CONSIDERATIONS:**

Consider what would happen with other vector loops. In the above formulation, the problem required two vector loops. What would happen with other loops? Consider defining a loop which includes the known motion and the motion we want to find. For example a loop from G to H, H to A, A to O, O to C, C to F back to G. This one loop would include information about the motion of the collar (the given) and the plunger (what is required). Count the unknown variables in the loop.

| Leg | New Unknowns |
|---|---|
| G to H | none |
| H to A | $h$ |
| A to O | $\theta$ |
| O to C | $l$ |
| C to F | $y$ |
| F to G | none |

There are four unknowns and the loop provides only 2 scalar equations. We will need another loop. Now how can you know that there IS another loop? How do you know there isn't **suppose** to be four unknowns and two equations? What makes you think you are not wasting your time looking for a loop? Well, what is the [NIF]? The [NIF] is one which means there cannot be four unknown and unrelated flows such as $h$, $\theta$, $l$ and $y$, there can be only one flow. In this problem, since one flow has been given $(x)$, there must be a relationship somewhere that relates all the four variables together.

If you did write the one loop from G to H to A to O to C to F back to G. Then you know another loop is needed. So try G to H to A to O back to G, or from O to F to C back to O. Either one would work fine.

What about a loop from H to A to O to C to F back to H. The unknown variables here would be

| Leg | New Unknowns |
|---|---|
| H to A | $h$ |
| A to O | $\theta$ |
| O to C | $l$ |
| C to F | $y$ |
| F to H | none |

Still 4 unknown variables. You still need another loop. Is there anything wrong with any of these loops? No. If you do use the latter loop, just remember the derivative of the distance between F and H is $-v_B$. In other words, if B moves right, the distance between H and F is getting smaller.

**End 3.24** ▬▬▬▬▬▬▬▬▬▬▬▬▬▬▬▬▬▬▬

There is no difference between vector loops and geometry. A vector loop is simply a systematic method of applying the geometric constraints of a system to a problem. If you can easily see a geometric relationship when analyzing a system, use it. Vector loops are just a way of finding these geometric relationships when they are not easy to see. Most of the time, the only geometric relationships people are able to find involve triangles. Quite often students remember the law of sines, or the law of cosines and apply them to triangular shaped regions. There is nothing inherently wrong with this idea. Just be careful and make sure you know how to use loops for the problems which do not have easy geometry. We say be careful because there have been countless students who use what they thought were simple geometrical "tricks" to find answers quickly only to discover (after failing tests) that they have made incorrect assumptions about angles or lengths. "Gee, I could have sworn those angles were the same!"

**Example 3.25** ▬▬▬▬▬▬▬▬▬▬▬▬▬▬▬▬▬▬▬▬▬▬▬▬▬▬▬▬▬▬▬

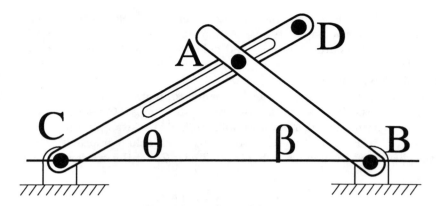

Figure 3.89: A Two Link Mechanism.

## Forces on a Mass.

Figure 3.89 shows one component, two links, of a machine. Suppose rod AB spins clockwise by an electric motor. Imagine that there is a massive blob stuck on the slotted member at point D. Maybe the blob is a tool which is performing a job, or maybe its a tennis ball being thrown, or maybe its just a big blob of gum. Anyway, suppose we want to know if it is going to stay on, or fly off. Suppose we know the blob will remain attached if the force between it and the slotted link is always less than F. How would we ever know what F is before it breaks? Well that is the subject of another course, but let's suppose we know F and the other parameters and determine if it breaks. What to do, what to do.

**Formulation:**

What we desire to do is determine the forces acting on the blob which means we must apply a conservation law. In fact since we must find the forces acting between the link and the blob, we should separate the blob from the link. In other words, we will choose the blob as a system. Why not take the slot as a system? Well, if we did, there would be more forces acting on the system than we care about. Let's suppose the blob is not touching anything but the link, and we'll ignore gravity. Why ignore gravity? Remember, in a machine, it doesn't take much speed before the forces due to acceleration exceed, by far, the weights.

Let's assume the motor rotates the rod AB at a constant rate, we'll talk about this a little later. What is the [NIF]? Its one. Since there is one [NIF] and one flow is given (rotation rate of rod AB), then all flows can be found without any conservation laws.

Draw the freebody diagram of the blob, the only thing on the freebody are the forces between the link and blob. Now apply the conservation of linear momentum. Its not difficult is it? What you will find is the forces are equal to the mass times acceleration of the blob. Write them down, pick any coordinate system, any direction you want, they are pretty simple equations. What is hard of course is trying to find the acceleration.

To find acceleration, simply write the position of point D relative to any fixed point then differentiate twice. We used maple for this. The program is given at the end of this example. We wrote the position relative to point C. Figure 3.90 shows some variable and parameter names. The position of point D is $d\cos(\theta)\vec{i} + d\sin\theta\vec{j}$. Now differentiate it twice and what unknowns will be present? The unknowns are two components of acceleration and second derivatives of $\theta$, three unknown flows. Too many.

Since the mechanism has one [NIF] with one flow specified, all three flows we mentioned (two components of acceleration and $\theta$) can be determined using kinematics. What we will do is construct a vector loop. The loop we used goes from C to A to B back to C. How many variables are in this loop? Three, one of which is known. The variables are $\theta$, $h$ and $\beta$. Angle $\beta$ is known, but how? Since $\dot{\beta}$ is known to be constant, integrate it with time to find $\beta$ as a function of time. What initial condition would you use for this integration? Since the vector loop is two dimensional, it can solve for the two remaining variables.

Differentiate the vector loop equation. Now what new variables are present? It will contain some

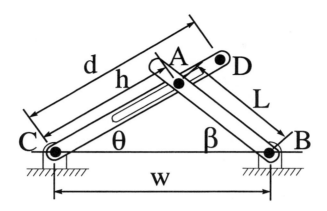

Figure 3.90: Some Coordinates for the Blob.

of the same old variables found in the vector loop and the new ones are their first derivatives ($\dot{\theta}$, $\dot{h}$ and $\dot{\beta}$). Of course $\dot{\beta}$ is known so the first derivative of the vector loop can be solved for $\dot{\theta}$ and $\dot{h}$. Differentiate the loop again bringing in $\ddot{\theta}$, $\ddot{h}$ and $\ddot{\beta}$. Knowing $\ddot{\beta}$ allows us to determine the other two.

Once we know $\theta$ and its derivatives we substitute into the expression for the acceleration of D and grind out the values of the forces. This is a typical solution process. Write the conservation equations. Write a position for whatever points we need the momentum and differentiate the positions to find velocities and accelerations. Usually there are too many unknown flows (the [NIF] tells us this) so we write loops or other kinematic equations and differentiate them. The process is relatively simple so where is the hard part? The hard part is trying not to make silly mistakes when differentiating and trying to solve the equations once we have them. Most of the time the difficulty in solving the equations arises when solving the position equations or the undifferentiated vector loops. See the program in 3.15.6 to see how all this comes together.

The plot output from the program appears in figure 3.91. Does the plot look like you expected? Do you believe it should be symmetric about 0 and 180 degrees? Why? Why does the magnitude of force peak as it does? Is this expected? Does it make sense? Why are there two peaks?

**End 3.25** ▄▄▄▄▄▄▄▄▄▄▄▄▄▄▄▄▄▄▄▄▄▄▄▄▄▄▄▄▄▄▄▄▄▄▄▄▄▄▄▄▄▄▄▄▄

The first step in relating flow variables is to express the length of ropes. If there are no ropes, try relating with a vector loop because they are quick and easy to use. However, vector loops do require a considerable amount of algebraic manipulation and therefore can be difficult to use for certain types of systems. We will discuss these systems and the alternative approaches for solving them when we encounter them later in the next section.

**Contact Relations**

**Objectives**
*When you complete this section, you should be able to*

1. *Define contact normals and tangent planes.*

2. *Define the types of contact between bodies.*

3. *Define the motion constraints present in various types of contact.*

4. *Use contact constraints to relate flow variables.*

In this text, we often deal with rigid bodies in contact with each other. Contact between rigid bodies occurs when points belonging to one body touch points belonging to another body. A point is a member of a body if it remains a fixed distance from all other points on the body. For example, figure 3.92 shows an aluminum plate with a hole drilled in its center. The molecules of aluminum

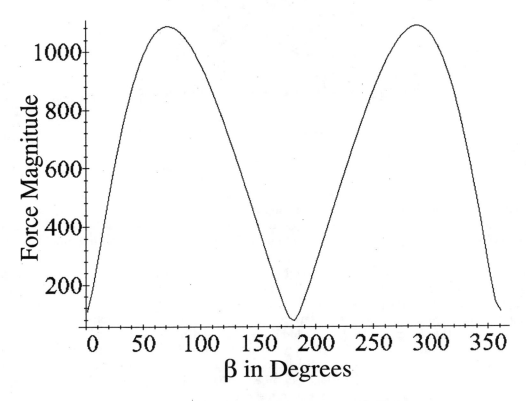

Figure 3.91: Magnitude of Force on the Blob.

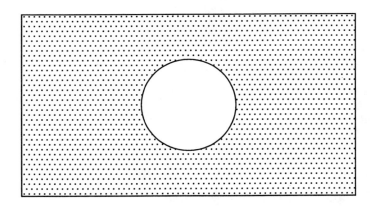

Figure 3.92: An Aluminum Plate with a Hole in its Center

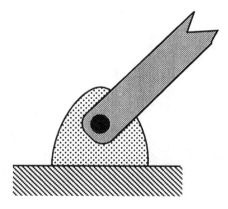

Figure 3.93: A Pin Connection Provides Permanent Contact Between Points on Two Bodies (The Normal Direction is Undefined and Unimportant)

are obviously members of the plate.[13] However, although not as obvious, the geometric center of the hole is also a member of the plate. A point does not have to be something that possesses mass. It can be a geometric point (*e.g.* the center of a circle).

If rigid bodies make contact, the contact can occur along a plane. For example a brick sitting on a flat table contacts over a plane. The plane between the brick and the table is called the contact plane. A line perpendicular to the contact plane is called the normal direction. Contact of this type presents no real new problems for us. For example, if the brick is stuck to the table, then the velocity and acceleration of the brick is the same as the table. If the brick slips on the table then we know the brick and table have the same velocity in the normal direction. The only thing we know about a sliding brick's acceleration is that it cannot accelerate into the table in the normal direction.

The more complex situation is when rigid bodies come together and contact along a line (like a cylinder on a flat table) or at a single point (like a sphere on a flat table). In either case, there will be an imaginary line passing through the contact points normal to the surfaces of both bodies. If a surface is curved, then there is a center of curvature and the normal line passes between the point of contact and the center of curvature. This normal line determines what we will call the normal direction. A plane tangent to the contact surface and perpendicular to the normal direction will be called the contact plane. A vector in the contact plane denotes the tangent direction.

When a point of one body touches a point on another body, the contact can be one of three types. Depending on the type of contact, we know something about the velocity and acceleration of the contacting points.

1. For two points in permanent contact (as figure 3.93 shows), the velocities and accelerations of the points are equal in all directions. As a proof, imagine two points at the geometric center of the pin right where the link and support meet. The point on the link side is a member of the link and the point on the support side is a member of the support. Since the two points are coincident, they are in contact (touching). The two contacting points will always be in contact (assuming the pin never breaks). They are in permanent contact and always have the same position. Likewise since their position is the same at all time (permanently), their velocities (changes in position) and accelerations (changes in velocity) are the same at all time. If their accelerations ever differed from one another, then the two points would soon be moving at differing speeds. If they begin to move at differing speeds, their positions would soon differ and hence the points would not be touching. But if the points ever stopped touching then they would not be permanently touching. So if the points ARE permanently touching THEN they will have identical velocities and accelerations in all directions.

2. If two points are not in permanent contact, then they are in temporary contact. One type of temporary contact is no slip. For two points in temporary contact without slip, the velocities are equal in all directions and the accelerations are equal in the tangential plane only. An

---

[13]On the macroscopic level, the molecules appear to be fixed distances from each other.

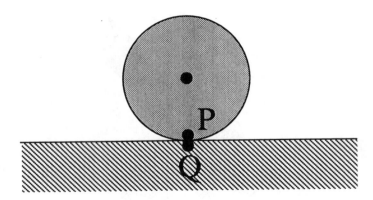

Figure 3.94: One Gear Rolling on Another Gear (Temporary Contact No Slip)Gear Teeth are not shown

example is a flat gear rolling on a rack (see figure 3.94). Point P is a dot painted on the gear and therefore is a point on the gear. Point Q is a dot painted on the rack and therefore is a point on the rack. At the instant shown, the two points happen to be touching; but as the gear moves, point P lifts up and is no longer in contact with Q. Thus, the points are in temporary contact. The gear teeth prevent slip between the gear and rack so the points are in temporary contact without slip. The normal direction is clearly vertical. The tangent plane is horizontal. Since the contacting bodies are rigid they cannot "ooze" into each other hence their velocities in the normal (vertical) directions are equal. By definition of slip, no slip also means that one point is not scooting passed the other, hence they have the same velocity in the tangent (horizontal) direction. It can be shown that the accelerations of the two points in the tangent direction are also equal as long as the points are in contact but their accelerations in the normal (vertical) direction are not necessarily equal. An argument offered as a simple "proof" follows (refer to figure 3.94). If the points P and Q had different accelerations in the horizontal direction then before long they would have a velocity in the horizontal direction and they would therefore be slipping which we assumed was not happening, therefore they have identical accelerations in the horizontal. If P and Q have the same accelerations vertically then a moment later they will continue to have identical velocities and hence identical positions so they are still in contact. If they are still in contact then nothing is new and we would apply the same argument to find that a moment later they are still in contact. Hence if the points do separate then their accelerations in the normal (vertical) MUST differ.

3. For two points in temporary contact with slip, (see figure 3.95), only the normal components of velocity are known to be equal. In this figure, we paint point P on the block and point Q on the floor. Because P and Q have the same position, they touch, however, because the block is moving to the right, P will not be touching Q for long.

Table 3.9 summarizes all the rigid body contact possibilities. Vector loops can be difficult to visualize when a body rotates, contains more than one special point, and has no permanent contact points. If you have difficulty using vector loops, try relating motions using information about rigid body contact. This approach involves the following steps

1. Construct position equations for each point in contact,

2. Differentiate each relation,

3. Determine the tangent and normal components of each velocity and acceleration, and finally

4. Equate the components that are known to be equal to each other.

The next example demonstrates this approach.

**Example 3.26** ————————————————————————

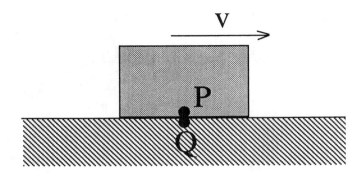

Figure 3.95: A Block Sliding on a Table

| Contact | Relationship Between the Motion of Contacting Points | | | |
| | Velocity | | Acceleration | |
| Type | Normal | Tangent | Normal | Tangent |
|---|---|---|---|---|
| Permanent | Identical | Identical | Identical | Identical |
| Temporary No Slip | Identical | Identical | Unknown | Identical |
| Temporary With Slip | Identical | Unknown | Unknown | Unknown |

Table 3.9: Relationship Between the Motion of Two Contacting Points

## A Gear Rolling on a Rack

**PROBLEM:**

The gear in figure 3.96 rolls on a stationary rack (represented by the shaded rectangle in the figure). Determine the velocity and acceleration of the center of the gear (point C in the figure). Let $R = 2$ [in] and $\dot{\theta} = -5$ [radians/s] (constant). It might help to look at figure 3.97 which shows the gear in a "different" position. Its key that you realize point P is painted on the circular gear and moves as the gear moves and point Q is painted on the rack, it sits still. At the moment shown in figure 3.96 points P and Q just happen to touch.

**FORMULATION:**

Since the system has one [NIF], we know it should be possible to relate the flow we know ($\dot{\theta} = -5$

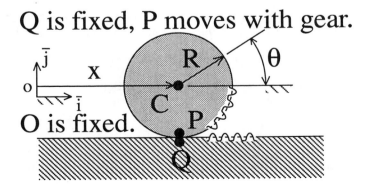

Figure 3.96: A Gear Rolling without Slip on a Stationary Rack

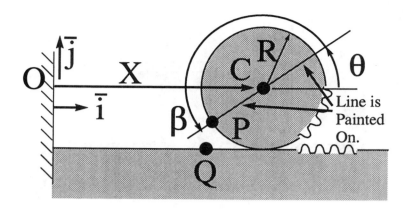

Figure 3.97: The Rolling Gear with Point P at an Arbitrary Location

constant) to the flow we want to know ($\dot{x}$ and $\ddot{x}$). The question is how to accomplish this. The shape and motion of the body indicate that vector loops may not be possible. In other words the gear has more than one special point (the center is known to move horizontally, and P touches the gear and the rack), the gear is rotating, and the gear does not have any permanent contact. We will demonstrate why vector loops do not work and how to use contact relationships to determine what we want.

First, we write a vector loop. There are several special points on the body ($o$, $C$, $P$ and $Q$). We will use these points to write a vector loop. One possible loop involving these points is

$$_o\vec{r}_Q + {}_Q\vec{r}_C + {}_C\vec{r}_o = 0$$

Now, substituting some values into this expression, we get that

$$_o\vec{r}_Q + x'\vec{i} + R\vec{j} - x\vec{i} = 0$$

Note that $x'$ is the horizontal distance from $Q$ to $C$. This distance is zero in the figure but it is not always zero. As the gear moves, point $C$ moves past $Q$ and $x'$ denotes this. Now, differentiating this expression (note that $_o\vec{r}_Q$ is constant since $Q$ is painted on the stationary rack)

$$0 + \dot{x}'\vec{i} + 0\vec{j} - \dot{x}\vec{i} = 0$$

which says $\dot{x}' = \dot{x}$. Big deal! We will have something just as useless if we differentiate again. Obviously, using a vector loop does not help us now.

To apply the contact relationships, we determine the velocities and accelerations of the contacting points, then use information from Table 3.9 to extract a relationship. For this problem, we will find the velocity and acceleration of points $P$ and $Q$. Since point $Q$ is painted on a fixed object, it is fixed so its velocity and acceleration are zero. The real problem then is to determine the motion of $P$.

Start by expressing the position of $P$ in some general location, then differentiating. Figure 3.97 shows $P$ is an arbitrary position (note that $\beta = \theta + 180^o$). The position of $P$ from fixed point $o$ is:

$$_o\vec{r}_P = x\vec{i} + R\cos(\beta)\vec{i} + R\sin(\beta)\vec{j} \tag{3.34}$$

Note that it is very important to draw $P$ in a **general** position because if you don't you may fool yourself into writing the position of $P$ as:

$$_o\vec{r}_P = x\vec{i} - R\vec{j}$$

which makes it look like the only variable on the right hand side is $x$. If you differentiate this faulty expression for position, you will get a bogus result. Now, return to the correct expression in equation 3.34 and differentiate to get

$$_o\vec{v}_P = \dot{x}\vec{i} - R\dot{\beta}\sin(\beta)\vec{i} + R\dot{\beta}\cos(\beta)\vec{j}$$

Differentiating again gives

$$_o\vec{a}_P = \ddot{x}\vec{i} - R\ddot{\beta}\sin(\beta)\vec{i} + R\ddot{\beta}\cos(\beta)\vec{j} - R\dot{\beta}^2\cos(\beta)\vec{i} - R\dot{\beta}^2\sin(\beta)\vec{j}$$

Now that differentiation is over, we can put $P$ in the special location where it is in contact with $Q$. This means $\theta = 90°$ and $\beta = 270°$. We have:

$$_o\vec{v}_P = \dot{x}\vec{i} + R\dot{\beta}\vec{i} + 0\vec{j} = \left(\dot{x} + R\dot{\beta}\right)\vec{i}$$

$$_o\vec{a}_P = \ddot{x}\vec{i} + R\ddot{\beta}\vec{i} + 0\vec{j} - 0\vec{i} + R\dot{\beta}^2\vec{j}$$

Now, since $P$ is in temporary contact no slip with $Q$, Table 3.9 tells us that their velocities must be equal.

$$_o\vec{v}_P = \left(\dot{x} + R\dot{\beta}\right)\vec{i} = 0 = {}_o\vec{v}_Q$$

Therefore, $R\dot{\beta} = -\dot{x}$. Now, since $\beta = \theta + 180°$, $\dot{\beta} = \dot{\theta}$ and $R\dot{\theta} = -\dot{x}$. According the Table 3.9, the acceleration of $P$ equals the acceleration of $Q$ in the direction tangent to contact $(\vec{i})$ therefore,

$$_o\vec{a}_P \cdot \vec{i} = \ddot{x} + R\ddot{\beta} = 0 = {}_o\vec{a}_Q \cdot \vec{i}$$

Hence, $\ddot{\beta} = \ddot{\theta} = -\frac{\ddot{x}}{R}$. Also, notice that the vertical acceleration of point $P$ is not zero. It is this acceleration that causes $P$ to break contact with $Q$.

**End 3.26** ────────────────────────────────

The results of the previous example may look so familiar to you that you see little reason for the added complexity used in the example. The problem is the handy dandy formula you may have memorized have many assumptions built into them. **If you can keep all the assumptions straight and apply the formula only on systems where they are applicable, fine, but this text chooses to derive the expressions one by one rather than memorizing minutia.** The following example will perhaps catch your attention to this point.

**Example 3.27** ────────────────────────────────

## A Gear Rolling on a Moving Rack

**PROBLEM:**

In example 3.26, we determined the velocity and acceleration of the gear in figure 3.96 on a stationary rack. In this example, the rack is no longer stationary. Determine the velocity and acceleration of the center of the gear (point C in the figure) if the rack moves to the left at 3 [in/s] and accelerates to the right at 2 [in/s$^2$]. Let $R = 2$ [in], $\dot{\theta} = -5$ [radians/s] (constant), $v_{\text{rack}} = -3\vec{i}$ [in/s], and $a_{\text{rack}} = 2\vec{i}$ [in/s$^2$].

**FORMULATION:**

The system is now two [NIF]. The rack can move, and if we stop the rack, the gear can still move. Because it is two [NIF], we have to know at least two motions to solve the problem. We are given two motions ($\dot{\theta}$ and $v_{\text{rack}}$), therefore, we can solve the problem.

Because its fixed to the rack, point Q has the same velocity and acceleration as the rack $\vec{v}_Q = -3\vec{i}$ and $\vec{a}_Q = 2\vec{i}$. The relations for the velocity and acceleration of point $P$ are identical to what they were before so we will simply repeat them.

$$_o\vec{v}_P = \left(\dot{x} + R\dot{\theta}\right)\vec{i}$$

$$_o\vec{a}_P = \ddot{x}\vec{i} + R\ddot{\theta}\vec{i} + R\dot{\theta}^2\vec{j}$$

Now, during the contact, the velocity of $P$ equals the velocity of $Q$.

$$(\dot{x} + R(-5))\vec{i} = -3\vec{i}$$

This tells us that $\dot{x} = -3 + 5R = 7$ [in/s]. The acceleration of $P$ equals the acceleration of $Q$ in the direction tangent to contact. Thus,

$$\ddot{x}\vec{i} + R(0)\vec{i} = 2\vec{i}$$

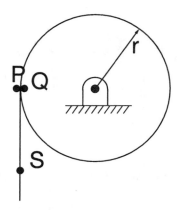

Figure 3.98: A Rope Contacts a Pulley. Points P and S are Painted on the Rope. Point Q is Painted on the Pulley.

This tells us that $\ddot{x} = 2$ [in/s²].

Notice that the gear center is not moving as fast as it was before. This is because we are pulling the rack out from under it. Also notice that C accelerates exactly the same as the rack, is this what you expected?

**End 3.27** ──────────────────────────────────────────────

Another application of the contact relations involves ropes and pulleys. Figure 3.98 shows a rope or belt coming off of a pulley. If you need to find a relationship between the angular motion of the pulley and the belt speed, you can use contact relationships. For example, provided the belt is not slipping on the pulley, point $P$ in figure 3.98 is in temporary contact without slip with the point $Q$. If the rope does not go limp or stretch then the motion of $S$ equals the motion of $P$.

## 3.9  Angular Velocity

**Objectives**
*When you complete this section, you should be able to*

1. *Define what a frame of reference means.*

2. *Recognize when bodies have angular velocity.*

3. *Classify the angular velocity of a body as simple or non-simple.*

4. *Calculate the magnitude of simple angular velocity.*

5. *Use angular velocity to calculate the derivative of a vector.*

───────────────────────────────────────────────────────────

This section provides a formal definition of angular velocity. It introduces the concept of relative angular velocities and how to add angular velocities. It also discusses the connection between angular velocity and the differentiation of vectors. Angular velocity is a measure of how fast a body is spinning. The spin a body appears to have is dependent on where we are when we try to measure it. For example, look around you for indications that the Earth is spinning. If we look only at objects on the Earth, we have no idea that the Earth is spinning. In fact, if we were not taught that the Earth rotates, we probably would not be aware of it. We can detect the Earth's rotation only if we look at objects beyond the Earth (such as the Sun).

Even then, it is impossible to determine which is moving – the Earth or the Sun.[14] For example, from the Earth's perspective (*i.e.* standing on the Earth), it appears the Sun spins about us and we do not spin at all. Conversely, observing from the Sun (use your imagination), the opposite is

───────────────────────────────

[14] Assuming Newton's laws are valid, a detailed study of the Sun's apparent trajectory will suggest the Earth is the one rotating.

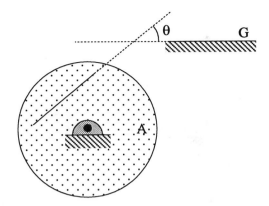

Figure 3.99: A Spinning Disk

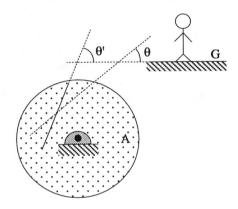

Figure 3.100: Angular Velocity of A Relative to G

true (it appears that the Earth spins but the Sun does not). A frame of reference means where you stand when you observe something.

Consider the two dimensional bodies of figure 3.99. Body A is pinned down but is allowed to spin. Suppose we want to determine the rate that disk A spins relative to (or as seen from) the ground G. Mathematically, we represent this as $^{G}\vec{\omega}^{A}$. The notation means stand on the body denoted by the left superscript and observe the rotation of the body denoted by the right superscript.[15] To formally determine the angular velocity of A relative to G, we paint two lines. One on body A and another on the ground G. Because the paint moves with the body it is on, it is a member of that body. If the line on A intersects the line on G, the lines will form an angle. This is $\theta$ in figure 3.99. If the lines do not intersect, $\theta$ is zero. The angle itself is of no interest to us, we care only about its change. If either body A or G moves such that the angle $\theta$ changes, there is an angular velocity. We determine the angular velocity direction (as a matter of habit) using the right hand rule. Curling the fingers of the right hand in the direction of increasing $\theta$ (counter clockwise), the right thumb points in the direction of the angular velocity vector (out of the page).

Assume that G (the ground) in figure 3.99 is motionless and that disk A spins at 1 [revolution/min] counterclockwise. As body A rotates, the angle $\theta$ changes into $\theta'$ as figure 3.100 shows. If we observe the change in angle by standing on G, as indicated by the stick person, the angle would appear to get bigger counterclockwise. Curving our right hand fingers counterclockwise (the direction of increasing $\theta$), our right thumb points out of the page. Thus, we say the angular velocity of A relative to (while standing on) G is

$$^{G}\vec{\omega}^{A} = 1 \text{ [rpm] } \vec{k}$$

where the unit vector $\vec{k}$ is out of the page. Now let's determine the change in $\theta$ relative to body A

---

[15] Notice that superscripts denote bodies whereas subscripts denote points.

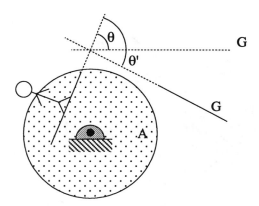

Figure 3.101: Angular Velocity of G Relative to A

as figure 3.101 shows. To do this, imagine standing on A and observing the change in angle between the two painted lines. We see the same rate of change (1 [rpm]), but this time the angle $\theta$ appears to get bigger in the clockwise direction. (Read this again and use your imagination!) Curving our right hand fingers in the direction of increasing $\theta$ (clockwise), our right thumb points into the page. This time we would write the angular velocity of G relative to (while standing on) A as

$$^A\vec{\omega}^G = -1 \text{ [rpm] } \vec{k}$$

Notice that $^G\vec{\omega}^A = -^A\vec{\omega}^G$. This equation means that the spin observed when standing on G and observing A is opposite that observed when standing on A and observing G. This identity applies to all bodies not just G and A and not just for two dimensions. As further proof, find a merry-go-round in a park. Spin the merry-go-round clockwise as viewed from the ground. Jump on the merry-go-round and observe the ground. You will notice the ground appears to rotate counterclockwise.

Notice that by definition, to determine the angular velocity of something we must paint a line on it. Therefore, points do not have angular velocity for the simple reason that we cannot paint a line on them (because a line needs at least two points to define it). Bodies have angular velocity. Points have velocity. Asking, "what is the angular velocity of **point** A?" does not make sense. If the left superscript of an angular velocity is missing, it implies that the angular velocity is relative to some stationary body.

We define the angular velocity of three dimensional objects in exactly the same way (*i.e.* by painting lines). However, expressing the angle between the lines can be difficult and angular velocity cannot be written as a simple change in angle. In fact, the only time that we can write an angular velocity as a rate of change of an angle is when the two bodies under consideration (those denoted by the superscripts) have a vector whose direction is fixed in both bodies. When there is a vector fixed in both bodies, the angular velocity is simple and simple angular velocities are a rate of change of an angle and point parallel to a fixed vector. When there isn't a vector fixed in both bodies, the angular velocity is non-simple. Two dimensional bodies, moving in a plane always have simple angular velocity or they have no angular velocity. If we roll the surface of a cone on a flat table, the motion has non-simple angular velocity. We will not discuss nonsimple angular velocity in this text.

Angular acceleration is defined as the rate of change of angular velocity. If a body has simple angular velocity, then angular acceleration is the second time rate of change of an angle. If the body has non-simple angular velocity, the angular acceleration may be very complex.

### 3.9.1  Relative Angular Velocities

Consider the two link device in figure 3.102. Let $\dot{\theta}$= 5 [radians/s] and $\dot{\alpha}$=5 [radians/s]. The angular velocity of body A relative to (observed from) body G is $^G\vec{\omega}^A$=5$\vec{k}$ and the angular velocity of B relative to (observed from) A is $^A\vec{\omega}^B = -5\vec{k}$. The angular velocity of B seen from the ground is $^G\vec{\omega}^A + ^A\vec{\omega}^B = ^G\vec{\omega}^B = 0$. In other words, if we stand on the ground and observe body B, it does not appear to rotate (use your imagination to convince yourself of this). Do not confuse angular velocity

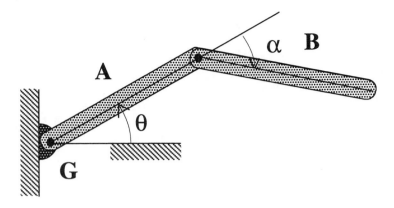

Figure 3.102: A Two Link Device

with motion on a curve, they are not equivalent. For example, the points of body B in figure 3.102 move on a curve, but B has no angular velocity.

In general, we can add angular velocities, as we did here, only if the left superscript of one equals the right superscript of another. Basically this addition formula (the superscripts) follow the same rules used for the subscripts of vector loops. In other words, the sum of the angular velocity of B relative to A and the angular velocity of A relative to G is the angular velocity of B relative to G. We can use the following memory trick to remember the addition formula for angular velocities. When we add angular velocities identical right and left superscripts cancel each other. The result is an angular velocity with a left superscript equal to the uncanceled left superscript and a right superscript equal to the uncanceled right superscript of the summed angular velocities. The following summations demonstrate this mnemonic:

$$^{A}\vec{\omega}^{B} + {}^{B}\vec{\omega}^{C} + {}^{C}\vec{\omega}^{D} + {}^{D}\vec{\omega}^{E} = {}^{A}\vec{\omega}^{E}$$

We can make a negative angular velocity positive by switching the superscripts.

$$^{B}\vec{\omega}^{C} = -{}^{C}\vec{\omega}^{B}$$

$$^{A}\vec{\omega}^{B} - {}^{C}\vec{\omega}^{B} = {}^{A}\vec{\omega}^{C}$$

These summations are incorrect

$$^{A}\vec{\omega}^{B} + {}^{B}\vec{\omega}^{C} + {}^{D}\vec{\omega}^{E} + {}^{C}\vec{\omega}^{D} \neq {}^{B}\vec{\omega}^{E}$$

$$^{A}\vec{\omega}^{B} + {}^{C}\vec{\omega}^{B} \neq {}^{A}\vec{\omega}^{C}$$

If a body has simple angular velocity, a similar addition formula for angular acceleration exists.

### 3.9.2  Differentiation of Unit Vectors

We can use angular velocity to reduce the algebra involved in calculating vector derivatives. Body B of figure 3.103 has an angular velocity relative to the ground (G) of $^{G}\vec{\omega}^{B} = \dot{\theta}\vec{n}_{k}$ (unit vector $\vec{n}_{k}$ points out of the page). $\vec{n}_{1}$ and $\vec{n}_{2}$ are unit vectors painted on body B. Expressing vector $\vec{n}_{1}$ and $\vec{n}_{2}$ mathematically, we get

$$\vec{n}_{1} = \cos(\theta)\vec{i} + \sin(\theta)\vec{j}$$

and

$$\vec{n}_{2} = -\sin(\theta)\vec{i} + \cos(\theta)\vec{j}$$

Differentiating these expressions gives

$$\dot{\vec{n}}_{1} = -\sin(\theta)\dot{\theta}\vec{i} + \cos(\theta)\dot{\theta}\vec{j} = \dot{\theta}\vec{n}_{2}$$

and

$$\dot{\vec{n}}_{2} = -\cos(\theta)\dot{\theta}\vec{i} - \sin(\theta)\dot{\theta}\vec{j} = -\dot{\theta}\vec{n}_{1}$$

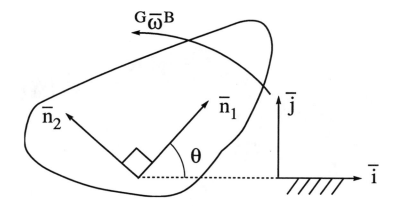

Figure 3.103: Two Unit Vectors Painted on Body B

After careful inspection, we find that these expressions are identical to

$$\dot{\vec{n}}_1 = {}^G\vec{\omega}^B \times \vec{n}_1$$

and

$$\dot{\vec{n}}_2 = {}^G\vec{\omega}^B \times \vec{n}_2$$

To convince yourself of this, consider the following:

1. The right hand rule gives the direction of $\omega$.

2. The definition of cross product states that $\vec{a} \times \vec{b} = |\vec{a}||\vec{b}|sin(\alpha)\vec{n}$ where $\alpha$ is the angle between $\vec{a}$ and $\vec{b}$ and $\vec{n}$ is the vector perpendicular to both $\vec{a}$ and $\vec{b}$ that points positively in the direction determined by the right hand rule.

Thus, $\dot{\vec{n}}_1 = {}^G\vec{\omega}^B \times \vec{n}_1$ equals $\dot{\vec{n}}_1 = |{}^G\vec{\omega}^B||\vec{n}_1| \sin 90^\circ \vec{n}_2$. Since the magnitude of ${}^G\vec{\omega}^B$ is $\dot{\theta}$ and the magnitude of a unit vector is one, this simplifies to $\dot{\vec{n}}_1 = (\dot{\theta})(1)(1)\vec{n}_2$.

Likewise, $\dot{\vec{n}}_2 = {}^G\vec{\omega}^B \times \vec{n}_2$ equals $\dot{\vec{n}}_2 = -|{}^G\vec{\omega}^B||\vec{n}_2| \sin 90^\circ \vec{n}_1$ Since the magnitude of ${}^G\vec{\omega}^B$ is $\dot{\theta}$ and the magnitude of a unit vector is one, this simplifies to $\dot{\vec{n}}_1 = -(\dot{\theta})(1)(1)\vec{n}_1$

The general method for differentiating unit vectors is

1. Find a body (B) in which the unit vector ($\vec{u}$) is fixed

2. Determine the angular velocity of B relative to the ground (${}^G\vec{\omega}^B$)

3. Calculate the derivative as the cross product of angular velocity and the vector.

$$\dot{\vec{u}} = {}^G\vec{\omega}^B \times \vec{u} \tag{3.35}$$

This method is applicable to both two and three dimensional systems.

When working some of the kinematics problems prior to this section, you probably noticed the incredible amount of algebra involved in computing velocity and acceleration. If you use rotating vectors and angular velocity, it can reduce the algebra. The next example demonstrates this.

## Example 3.28

### Velocity of a Peg in a Slot - Revisited

**PROBLEM:**

Determine the velocity and acceleration of the peg in figure 3.104 in terms of the angle $\theta$ and distance $r$.

**FORMULATION:**

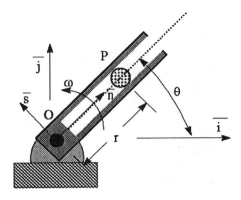

Figure 3.104: A Peg Moving in a Slot

The point P has two [NIF], that is why its velocity must be expressed in terms of two variables. We find the velocity by taking the derivative of the position $\vec{r}$ with respect to time. We can kinematically write the position of the peg at all time as

$$\vec{r} = r\vec{n}$$

with $\vec{n}$ fixed to the slotted member. The derivative of this vector is

$$\frac{d\vec{r}}{dt} = \dot{r}\vec{n} + r\dot{\vec{n}}$$

Note that we used the product formula to find this expression. Also note that one difference here from before is that the vector $\vec{n}$ is not constant like $\vec{i}$ and $\vec{j}$ are. That is why there is a time derivative of it. To evaluate the derivative of $\vec{n}$, recognize that the vector is fixed in the slot and the angular velocity of the slot is $\dot{\theta}\vec{k} = \omega\vec{k}$. Hence the time derivative is:

$$\frac{d\vec{r}}{dt} = \dot{r}\vec{n} + r\omega\vec{k} \times \vec{n}$$

Use the right hand rule to determine the cross product finding:

$$\frac{d\vec{r}}{dt} = \vec{v} = \dot{r}\vec{n} + r\omega\vec{s} \qquad (3.36)$$

Compare this to what we had when we were using $\vec{i}$ and $\vec{j}$. This is much simpler.

Now for the acceleration. Taking another derivative means:

$$\frac{d\vec{r}}{dt} = \ddot{r}\vec{n} + \dot{r}\dot{\vec{n}} + \dot{r}\omega\vec{s} + r\dot{\omega}\vec{s} + r\omega\dot{\vec{s}}$$

Using angular velocity gives:

$$\frac{d\vec{r}}{dt} = \ddot{r}\vec{n} + \dot{r}\omega\vec{k} \times \vec{n} + \dot{r}\omega\vec{s} + r\dot{\omega}\vec{s} + r\omega\omega\vec{k} \times \vec{s}$$

Taking the cross products gives:

$$\frac{d\vec{r}}{dt} = \ddot{r}\vec{n} + \dot{r}\omega\vec{s} + \dot{r}\omega\vec{s} + r\dot{\omega}\vec{s} - r\omega\omega\vec{n}$$

Finally gathering some terms:

$$\frac{d\vec{r}}{dt} = \left(\ddot{r} - r\omega^2\right)\vec{n} + \left(r\dot{\omega} + 2\dot{r}\omega\right)\vec{s} \qquad (3.37)$$

If you want to you can substitute $\vec{i}$ and $\vec{j}$ in instead of $\vec{n}$ and $\vec{s}$ to show that these are the same as equation 3.15.

**End 3.28**

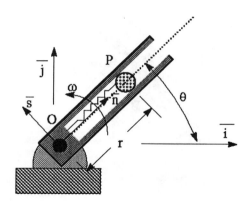

Figure 3.105: A Peg Moving in a Slot Connected With a Spring.

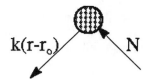

Figure 3.106: A Freebody Diagram of the Peg in the Slot.

# 3.10   SOLVING EQUATIONS OF MOTION

**Objectives**

*When you complete this section, you should be able to*

  *1.  Recognize the importance of coordinate systems in solving equations of motion.*

Up to this point, the differential equations we have solved have been relatively simple. Occasionally, the equations of motion end up as an ugly set of coupled differential equations. It is possible to solve many of these equations numerically however if you find a "good" set of coordinates, the equations sometimes simplify making them easier to solve. The following example shows this.

**Example 3.29** ──────────────────────────────────

## Motion of a Peg in a Slot

**Problem:**

Figure 3.105 may look familiar. Let's suppose there is a spring between points O and P and determine the motion of peg P. Let's say the slot rotates counterclockwise at a constant rate $\omega$. Just to keep this one simple, let's also ignore the weight of the peg and friction between the peg and slot. After we solve the problem, we'll see how reasonable it is to ignore the weight. We will also assume that sufficient initial conditions are given.

**Formulation:**

This is a two [NIF] machine but only one flow is specified therefore we must write a conservation equation. By the way, how many initial conditions do you think we will need? The peg has 3 [NIA] so we predict 3 initial conditions, two velocity components of the peg and the initial stretch of the spring. Figure 3.106 shows a freebody diagram of the peg.

To apply the conservation of linear momentum for the peg, we can look back at the previous examples of the peg in slot problem (examples 3.28 and 3.18) and use the change in momentum we found there. First we use the formulation using the $\vec{n}$ and $\vec{s}$ directions, see equation 3.37:

$$-k\left(r - r_o\right)\vec{n} + N\vec{s} = m\left(\ddot{r} - r\omega^2\right)\vec{n} + m\left(r\dot{\omega} + 2\dot{r}\omega\right)\vec{s}$$

Pulling off the components in the $\vec{n}$ direction gives:

$$-k\left(r - r_o\right) = m\left(\ddot{r} - r\omega^2\right) \tag{3.38}$$

Pulling off the components in the $\vec{s}$ direction gives:

$$N = m\left(r\dot{\omega} + 2\dot{r}\omega\right) \tag{3.39}$$

Notice that the first equation is second order the second is first order totaling three which matches the [NIA].

Notice that in equation 3.38 the only unknowns are $r$ and its derivatives. In fact it is a relatively easy differential equation to solve since it is a linear constant coefficient second order differential equation. What initial conditions do you need to solve it? You need two, say the initial value of $r$ (the initial stretch in the spring) and $\dot{r}$ (one of the initial velocity components). Just as we predicted. Let's let the initial values be $r = r_i$ and $\dot{r} = 0$. Maple (see program 3.15.9) says the solution is:

$$r = \frac{kr_o}{k - m\omega^2} - \frac{1}{2}\frac{\left[k\left(r_o - r_i\right) + mr_i\omega^2\right]}{k - m\omega^2}\left[\exp\sqrt{m\omega^2 - k}t + \exp-\sqrt{m\omega^2 - k}t\right] \tag{3.40}$$

Now think about solving equation 3.39. Since we know $\omega$ is constant, we know $\dot{\omega} = 0$ hence it reduces to:

$$N = 2m\dot{r}\omega$$

We have $\dot{r}$ by differentiating equation 3.40 so basically this tells us $N$. What happened to the third initial condition? Well it has been specified already. You should be expecting another "velocity" type initial condition, right? Well, we specified that when we said $\dot{\theta} = \omega$ for all time. This would have been obvious if the problem allowed $\omega$ to change we would have solved the second differential equation for $\omega$ and it would require us to have an initial condition for $\omega$. By the way, if $\omega$ had not been given as constant, then $N$ and $\dot{\omega}$ would have been unknown. What could we do? Is there another body to use? Is there something else needed? Is the problem ill defined? Well it isn't ill defined, but something would have to be given and yes there is another body. We will do this problem in example 3.31.

Before we look at the solution in detail, let's look at what happens if we use the $\vec{i}$ and $\vec{j}$ directions for our problem. Using equation 3.15 from example 3.18 we have:

$$k\left(r - r_o\right)\left(-\cos\theta\vec{i} - \sin\theta\vec{j}\right) + N\left(-\cos\theta\vec{i} + \sin\theta\vec{j}\right) =$$

$$m\left(-r\cos(\theta)\dot{\theta}^2 - r\sin(\theta)\ddot{\theta} - 2\dot{r}\sin(\theta)\dot{\theta} + \ddot{r}\cos(\theta)\right)\vec{i} +$$

$$m\left(-r\sin(\theta)\dot{\theta}^2 + r\cos(\theta)\ddot{\theta} + 2\dot{r}\cos(\theta)\dot{\theta} + \ddot{r}\sin(\theta)\right)\vec{j}$$

Pulling off the $\vec{i}$ components gives:

$$-k\left(r - r_o\right)\cos\theta - N\cos\theta = m\left(-r\cos(\theta)\dot{\theta}^2 - r\sin(\theta)\ddot{\theta} - 2\dot{r}\sin(\theta)\dot{\theta} + \ddot{r}\cos(\theta)\right)$$

Pulling off the $\vec{j}$ components gives:

$$-k\left(r - r_o\right)\sin\theta + N\sin\theta = m\left(-r\sin(\theta)\dot{\theta}^2 + r\cos(\theta)\ddot{\theta} + 2\dot{r}\cos(\theta)\dot{\theta} + \ddot{r}\sin(\theta)\right)$$

Unfortunately neither one is very simple. If we solve the first for $N$ and substitute it into the second (see Maple program 3.15.9) we get equation 3.38. Actually however if you don't want to hassle with the substitution, it is possible to find a numerical solution to these last couple of ugly equations. The only real problem with this approach is that you lose some insight to how the system behaves as we'll discuss next.

Let's look a little closer at the supposed solution given in equation 3.40. Is this solution really valid. It says $r$ does not depend explicitly on $\theta$ do you believe this? Would it depend on $\theta$ if we had kept gravity in the problem? What if we had included friction? We say yes, yes, and no. Also notice that there is something funny going on with the quantity $m\omega^2 - k$. Notice that for small values of $\omega$ less than $\sqrt{k/m}$, the exponent terms are complex numbers. If you recall, a complex (imaginary) exponential power is equivalent to a sine and cosine, hence for small $\omega$ the solution is oscillatory. This is perhaps what you would expect from a spring and mass. Also note that at low speeds (small $\omega$) if the initial stretch in the spring is just right, the value of $r$ never changes. Explain this and determine what initial condition is the special value.

If the value of $\omega$ increases beyond the $\sqrt{k/m}$ then one of the exponential terms has a positive power the other a negative. The one with a positive power blows up toward infinity. What does this mean? Could it be correct? What it means is the spring force is insufficient to hold the mass inward and make it accelerate in a tight radius hence the mass will move outward until it breaks the spring and flies out of the slot.

What happens to the solution when $\omega = \sqrt{k/m}$? Try taking the limit of equation 3.40 as $\omega \to \sqrt{k/m}$. The solution becomes:

$$r = \frac{1}{2}\frac{kr_o t^2}{m} + r_i$$

What this says is that if the spring freelength is zero then the mass's radius does not change but if the spring freelength is other than zero, the radius grows without bound. Does this really happen? Yes. In fact this is a pretty good approximation to many large rotating masses like turbines, compressors, pumps and the like. The shaft these machines rotate on is like a spring which is undeformed when straight, hence freelength of zero. Ugly things begin to happen when you spin the machine at or above the "natural" frequency of $\sqrt{k/m}$.

Now consider some of the assumptions we made. Of course the assumptions really depend on what is happening and we really cannot make general statements about them. What we will do is demonstrate some of the logic we use to make a decision about them. Let's say that the device is some machine running at "machine speeds". By the way the idle of a automobile engine is in the neighborhood of 1000 rpm so this would be a reasonable machine speed. Let's also say that the machine is not supposed to fly apart so $\omega < \sqrt{k/m}$. In other words it makes sense to say the spring must be at least as stiff as $m\omega^2$. Now suppose the spring is the minimum reasonable stiffness $(m\omega^2)$, how far does it have to deform to generate 10 times the weight? It must deform $10g/\omega^2$. Put some numbers in and you will find this is only 0.35 inches. Not very far. What this means is that for numbers you might find in a slow moving machine, it is not unreasonable to expect the spring force to be 10 times the weight. If this is the case, we would be willing to ignore the weight what about you? What do you think happens if the machine is supposed to be faster? Yes, the spring would have to be stiffer, hence any slight deflection is going to create large spring forces so we definitely would ignore weight. Do you think including weight would change the characteristic speed $\omega^2 = k/m$? No, because it is only a forcing term. For more details, we suggest you consult a text on differential equations [23]. What about the friction? If you have some values, test the magnitude of friction versus the magnitude of the spring.

**End 3.29** ————————————————————————————————————————

What the previous example demonstrated are the following:

1. If you can find the "right" set of coordinates, it is possible to reduce some ugly equations to a simpler form. Unfortunately it is not always possible to find the "right" set of coordinates. There are several standard sets which experts routinely use but there is no guarantee that they will work. If you use an algebra tool like Maple, you can try various coordinates easily. It is not uncommon for experts to spend considerable time searching for the right coordinates so don't get frustrated if you don't find the right set on the first try.

2. Although it is often possible to use any old coordinate and use numerical solutions to the differential equations, there is considerable information available from an nonnumerical solution. Imagine how many different numerical solutions you would have to perform before you uncovered the idea that $\omega^2 = k/m$ was a "special" speed.

# 3.11 ANGULAR MOMENTUM

**Objectives**
*When you complete this section, you should be able to*

1. *Calculate the angular momentum of a set of particles.*

2. *Calculate moment of inertia for bodies with elementary shapes.*

3. *Calculate moment of inertia for a set of bodies.*

4. *Calculate angular momentum of a body about a point in the body.*

5. *Calculate angular momentum of a body about a point which is not in the body.*

6. *Relate radius of gyration to moment of inertia.*

---

Angular momentum is a difficult concept for many students. By definition, angular momentum is equal to the moment of momentum. If you know how to calculate the moment of a force vector, you can calculate angular momentum. Simply replace the force vector with mass times velocity. Calculate the moment, and you have angular momentum (*i.e.* angular momentum$=\vec{r}\times m\vec{v}$). To use angular momentum effectively, you must be able to do two basic things: (1) calculate the angular momentum of a system, and (2) relate angular momentum to moment via a conservation equation. This section will discuss these two subjects separately.

Formulas for computing angular momentum are summarized in Table 3.10. [16] The equations given in the table for rigid bodies only apply to two dimensional bodies with simple angular velocity (rotation) about the third dimension. To effectively use the formula in table 3.10 find a line that matches the situation in your problem.

## 3.11.1 Finding Moment of Inertia

The formula in Table 3.10 include moment of inertia $I$. For a rigid body undergoing general three dimensional rotation, there are three "moments" of inertia and three "products" of inertia. These inertias depend on the body's mass distribution and are generally tabulated. If you draw an origin at the body's center of mass (cm), then the moments of inertia of the body at the center of mass are defined as (in Cartesian coordinates)

$$\bar{I}_x = \int r_x^2 dm = \int (y^2 + z^2)dm \tag{3.41}$$

$$\bar{I}_y = \int r_y^2 dm = \int (z^2 + x^2)dm \tag{3.42}$$

$$\bar{I}_z = \int r_z^2 dm = \int (x^2 + y^2)dm \tag{3.43}$$

the products of inertia of the body at the center of mass are defined as (in Cartesian coordinates)

$$\bar{I}_{xy} = \bar{I}_{yx} = \int xy\,dm \tag{3.44}$$

$$\bar{I}_{xz} = \bar{I}_{zx} = \int xz\,dm \tag{3.45}$$

$$\bar{I}_{yz} = \bar{I}_{zy} = \int yz\,dm \tag{3.46}$$

where $\int(\cdot)dm$ is an integral over the mass volume of the body. The coordinates appear in figure 3.107 along with $\vec{r} = r_x\vec{i} + r_y\vec{j} + r_z\vec{k}$. Most likely, you used these formula in a Calculus class when you

---

[16] The terms given in table 3.10 can be derived easiest by starting from equation 5.101 in Ginsberg [10] and assuming point A (used in Ginsberg) is either the "system" center of mass or a fixed point.

| System | Angular Momentum Point | Angular Momentum (H) | Comments |
|---|---|---|---|
| Particle $P$ | Any point $O$ | $\vec{H}_O = {}_O\vec{r}_P \times (m_P\vec{v}_P)$ | $m_P$ is mass, $\vec{v}_P$ is the velocity of $P$. |
| Particle $P$ | Any point $O$ | $|\vec{H}_O| = m_P L v_P$ | $L$ is perpendicular distance from $O$ to $P$. Use the right hand rule to determine the direction of H. |
| 2D Rigid Body | The center of mass | $H_{cm} = I_{cm}\omega$ | Two dimensional only. Direction is clockwise or counter clockwise depending on $\omega$. $I_{cm}$ is the moment of inertia of the body about its center of mass. |
| 2D Rigid Body | Any fixed point $O$ also fixed in the body | $H_O = I_O\omega$ | $I_O$ is the moment of inertia of the body about point $O$. Any point fixed in a body maintains constant distance from all other points in the body. |
| 2D Rigid Body | Any point $O$ | $\vec{H}_O = I_{cm}\omega\vec{k} +$ ${}_O\vec{r}_{cm} \times (m\vec{v}_{cm})$ | Two dimensional only. Vector $\vec{k}$ is parallel to the angular velocity. Point $cm$ is the body's center of mass. |
| Set of Objects | Any point $O$ | Sum of $\vec{H}_O$ for each object | It is possible to treat holes in a body as an object with a negative mass. |

Table 3.10: Formulas for Computing Angular Momentum.

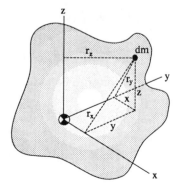

Figure 3.107: Coordinates Used for Mass Moments of Inertia

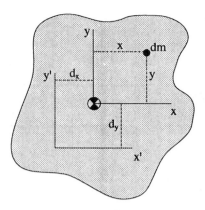

Figure 3.108: The Distances That We Use to Calculate the Products of Inertia About Alternative Axes

were learning about definite integrals. The skills you learned there apply here, you may want to review that material.

If the body happens to be two dimensional and experiences only two dimensional motion (such that the $z$ ($\vec{k}$) axis is normal to the body and all rotation occurs around $z$), then all products of inertia are zero and the only inertia that is important is $I_z$. This is given in equation 3.43. The value of $I_z$ is what is used in the formulas given in Table 3.10.

Rather than integrating to find the inertial properties, it is much easier to decompose the body of interest into several standard shapes, then look up the properties in a table. The basic idea here is that the moment of inertia of a set of bodies about some point $O$ is equal to the sum of the inertias of each body about $O$. For example, if an ugly shaped rigid body can be thought of as made up of a rectangle attached to a circle then the moment of inertia of the ugly body about $O$ equals the moment of inertia of the rectangle about $O$ plus that of the circle about $O$. Tables 3.11, 3.12 and 3.13 provide some useful inertial properties of bodies about axes that pass through the bodies).

If you notice, the axes shown in Tables 3.11 through 3.13 do not all point out of the page in the $z$ direction. The reason is because the bodies might be rotated about several axes. For example, consider a pipe. If the pipe is pointing out of the page, it would look like a ring. If it rotates about an axis out of the page and you need the moment of inertia, you would use the value $I_z = mr^2$ from the Circular Cylindrical Shell in Table 3.11. If however, the pipe is lengthwise on the paper so it looks like a stick and it rotates about an axis out of the page, you want the moment of inertia $I_x = \frac{1}{2}mr^2 + \frac{1}{12}mL^2$ from the same diagram in the same table.

Sometimes it is necessary to express the inertial properties of a body about a point other than the center of mass. This happens frequently when trying to find the moment of inertia of a composite body. To find the moment of inertia about a point other than the center of mass, use the parallel axis theorem. If you know the moment and/or product of inertia of a body at a center of mass axis, then you can shift the properties to any other point using the following formula (see figure 3.108):

$$I_x = \bar{I}_x + md_x^2 \tag{3.47}$$

$$I_y = \bar{I}_y + md_y^2 \tag{3.48}$$

$$I_z = \bar{I}_z + md_z^2 = \bar{I}_z + m\left(d_x^2 + d_y^2\right) \tag{3.49}$$

$$I_{yx} = I_{xy} = \bar{I}_{xy} + md_x d_y \tag{3.50}$$

$$I_{zx} = I_{xz} = \bar{I}_{xz} + md_x d_z \tag{3.51}$$

$$I_{yz} = I_{yz} = \bar{I}_{yz} + md_y d_z \tag{3.52}$$

where $\bar{I}$ is the inertia property at the center of mass and $m$ is the mass of the body. The distances $d$ appear in figure 3.108. Figure 3.108 shows only the $x - y$ plane, but it is a general result.

| Body of Mass ($m$) Center of Mass ($x', y', z'$) | | Mass Moments of Inertia & Location of Center of Mass |
|---|---|---|
| | Circular Cylindrical Shell Radius (r) | $I_x = I_y = \dfrac{1}{2}mr^2 + \dfrac{1}{12}mL^2$ $I_{x_1} = \dfrac{1}{2}mr^2 + \dfrac{1}{3}mL^2$ $I_z = mr^2$ $x' = y' = z' = 0$ |
| | Half Cylindrical Shell Radius (r) | $I_x = I_y = \dfrac{1}{2}mr^2 + \dfrac{1}{12}mL^2$ $I_{x_1} = I_{y_1} = \dfrac{1}{2}mr^2 + \dfrac{1}{3}mL^2$ $I_z = mr^2$ $I_{cm_z} = \left(1 - \dfrac{4}{\pi^2}\right)mr^2$ $x' = \dfrac{2r}{\pi}, y' = z' = 0$ |
| | Circular Cylinder Radius (r) | $I_x = I_y = \dfrac{1}{4}mr^2 + \dfrac{1}{12}mL^2$ $I_{x_1} = I_{y_1} = \dfrac{1}{4}mr^2 + \dfrac{1}{3}mL^2$ $I_z = \dfrac{1}{2}mr^2$ $x' = y' = z' = 0$ |
| | Half Circular Cylinder Radius (r) | $I_x = I_y = \dfrac{1}{4}mr^2 + \dfrac{1}{12}mL^2$ $I_{x_1} = I_{y_1} = \dfrac{1}{4}mr^2 + \dfrac{1}{3}mL^2$ $I_z = \dfrac{1}{2}mr^2$ $I_{cm_z} = \left(\dfrac{1}{2} - \dfrac{16}{9\pi^2}\right)mr^2$ $x' = \dfrac{4r}{3\pi}, y' = z' = 0$ |

Table 3.11: Inertial Properties of Various Geometries

| Body of Mass (m) Center of Mass $(x',y',z')$ | Mass Moments of Inertia & Location of Center of Mass |
|---|---|
| Rectangular Parallelepiped | $I_x = \frac{1}{12}m\left(a^2 + L^2\right)$ $I_y = \frac{1}{12}m\left(b^2 + L^2\right)$ $I_z = \frac{1}{12}m\left(a^2 + b^2\right)$ $I_{y_1} = \frac{1}{12}mb^2 + \frac{1}{3}mL^2$ $x' = y' = z' = 0$ |
| Uniform Slender Rod | $I_x = I_y = \frac{1}{12}mL^2$ $I_{y_1} = \frac{1}{3}mL^2$ $I_z = 0$ $x' = y' = z' = 0$ |
| Quarter-Circular Rod | $I_x = I_y = \frac{1}{2}mr^2$ $I_z = mr^2$ $x' = y' = \frac{2r}{\pi}, z' = 0$ |
| Half Torus | $I_x = I_y = \frac{1}{2}mR^2 + \frac{5}{8}ma^2$ $I_z = mR^2 + \frac{3}{4}ma^2$ $x' = \frac{a^2 + 4R^2}{2\pi R}, y' = z' = 0$ |

Table 3.12: Inertial Properties of Various Geometries

| Body of Mass $(m)$<br>Center of Mass $(x', y', z')$ | Mass Moments of Inertia &<br>Location of Center of Mass |
|---|---|
| Spherical Shell | $I_z = \dfrac{2}{3}mr^2$<br><br>$x' = y' = z' = 0$ |
| Sphere | $I_z = \dfrac{2}{5}mr^2$<br><br>$x' = y' = z' = 0$ |
| Conical Shell | $I_y = \dfrac{1}{4}mr^2 + \dfrac{1}{2}mh^2$<br><br>$I_{y_1} = \dfrac{1}{4}mr^2 + \dfrac{1}{6}mh^2$<br><br>$I_z = \dfrac{1}{2}mr^2$<br><br>$x' = y' = 0,\ z' = \dfrac{2h}{3}$ |
| Right Circular Cone | $I_y = \dfrac{3}{20}mr^2 + \dfrac{3}{5}mh^2$<br><br>$I_{y_1} = \dfrac{3}{20}mr^2 + \dfrac{1}{10}mh^2$<br><br>$I_z = \dfrac{3}{10}mr^2$<br><br>$x' = y' = 0,\ z' = \dfrac{3h}{4}$ |

Table 3.13: Inertial Properties of Various Geometries

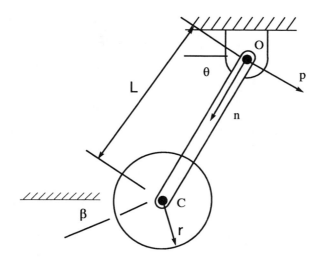

Figure 3.109: A Composite Body Made of a Rod of Length L and a Disk of Radius R.

Occasionally, inertial properties are given in terms of the radius of gyration. The radius of gyration measures the mass distribution of a body about an axis. The radius of gyration ($k$) for a body about a given axis is

$$k = \sqrt{\frac{I}{m}}$$

or

$$I = k^2 m.$$

where $I$ is the moment of inertia of a body about the given axis. The primary benefit of the radius of gyration is for tabulating data for structural shapes. For example, we may need to know $I$ for structural beams. If the beam is made of aluminum, it will have a different $I$ than a steel beam because the aluminum and steel have different masses. The radius of gyration is only a function of shape not mass, therefore if the aluminum and steel beams are the same shape and size, they will have the same $k$. Therefore, we can tabulate the radius of gyration of standard shapes without being concerned with the material in the body.

## Example 3.30

### A Multiple Component Body

**QUESTION:**
Determine the angular momentum of the body in figure 3.109 about point O. We will do it two ways, first assume the body is made of a disk pinned to a rod, second, assume the disk is welded to the rod ($\beta = \theta$).

**ANSWER:**
From table 3.10, use the last line to determine the angular momentum by summing the angular momentum of each piece. Therefore,

$$\vec{H}_{body,O} = \vec{H}_{rod,O} + \vec{H}_{disk,O}$$

For the rod, the fixed point $O$ is also fixed in the rod therefore its value can be found in the fourth row of the table:

$$\vec{H}_{rod,O} = I_{rod,O}\omega_{rod}\vec{k}$$

From table 3.12, the inertia of a slender rod about axis $y_1$ through the end of the rod is:

$$I_{rod,O} = \frac{1}{3}m_{rod}L^2$$

so:
$$\vec{H}_{rod,O} = \frac{1}{3} m_{rod} L^2 \dot{\theta} \vec{k}$$

For the disk, the fixed point $O$ is NOT fixed in the disk so we CANNOT use the same principle to find the angular momentum. Therefore, use the fifth row

$$\vec{H}_{disk,O} = I_{disk,cm} \omega_{disk} \vec{k} + {}_O \vec{r}_C \times m_{disk} \vec{v}_{disk,cm}$$

Matching the type of rotation of the disk to the cylinder rotating about the $z$ axis shown in figure 3.11, we find the moment of inertia to be:

$$I_{disk,cm} = \frac{1}{2} m_{disk} R^2$$

Now find the velocity of the disk's center as (we use angular velocity to take the derivative):

$$\vec{v}_C = \frac{d}{dt} \left( {}_O \vec{r}_C \right) = \frac{d}{dt} \left( L\vec{n} \right) = \dot{\theta} \vec{k} \times L\vec{n} = L\dot{\theta}\vec{p}$$

Now for the cross product:

$$_O \vec{r}_C \times m_{disk} \vec{v}_{disk,cm} = L\vec{n} \times m_{disk} L\dot{\theta}\vec{p} = m_{disk} L^2 \dot{\theta} \vec{k}$$

Combining all the terms for the disk gives its angular momentum as:

$$\vec{H}_{disk,O} = \left( \frac{1}{2} m_{disk} R^2 \dot{\beta} + m_{disk} L^2 \dot{\theta} \right) \vec{k}$$

If the disk is welded to the rod (part b), then $\dot{\beta} = \dot{\theta}$ so the result for the disk is:

$$\vec{H}_{disk,O} = \left( \frac{1}{2} m_{disk} R^2 \dot{\theta} + m_{disk} L^2 \dot{\theta} \right) \vec{k}$$

For the complete system then:

$$\vec{H}_{body,O} = \left( \frac{1}{3} m_{rod} L^2 \dot{\theta} + \frac{1}{2} m_{disk} R^2 \dot{\beta} + m_{disk} L^2 \dot{\theta} \right) \vec{k}$$

For part b, the complete system is:

$$\vec{H}_{body,O} = \left( \frac{1}{3} m_{rod} L^2 \dot{\theta} + \frac{1}{2} m_{disk} R^2 \dot{\theta} + m_{disk} L^2 \dot{\theta} \right) \vec{k}$$

Wow! What a mess. Another way to do part (b) is to realize the disk and rod are a single body, we could determine the moment of inertia of the composite body and multiply by $\dot{\theta}$ to arrive at the same answer. For example:

$$I_{disk,C} = \frac{1}{2} m_{disk} R^2$$

using the parallel axis theorem:

$$I_{disk,O} = \frac{1}{2} m_{disk} R^2 + m_{disk} L^2$$

For the rod:

$$I_{rod,O} = \frac{1}{3} m_{rod} L^2$$

Add them up for the composite body:

$$I_{body,O} = \frac{1}{2} m_{disk} R^2 + m_{disk} L^2 + \frac{1}{3} m_{rod} L^2$$

and from the table for angular momentum of a rigid body:

$$\vec{H}_{body,O} = \left( \frac{1}{2} m_{disk} R^2 + m_{disk} L^2 + \frac{1}{3} m_{rod} L^2 \right) \dot{\theta}$$

which is identical to the previous.

**End 3.30**

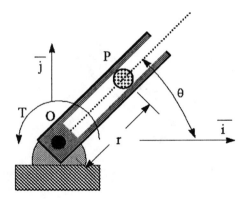

Figure 3.110: A Peg Moving in a Slot With an Applied Torque.

## 3.11.2 Conservation of Angular Momentum

**Objectives**
*When you complete this section, you should be able to*

1. *Express the conservation of angular momentum for a system.*

2. *Identify wise points to choose when expressing conservation of angular momentum.*

Although it is possible to express the conservation of angular momentum of any body about any point, there are some points that are wise to choose. For example, we can express conservation of angular momentum of a system about a point $O$ easiest if $O$ is:

1. A fixed point. It does not have to be on a body only fixed.

2. The center of mass of the complete system (everyting in the freebody diagram). It does not have to be fixed, just the center of mass.

3. A point with no acceleration.

4. A point with an acceleration provided the line of action of the acceleration passes through the mass center.

All of the time in this class, we will apply conservation of angular momentum using either a fixed point or the mass center. Up to now, the only time we have applied angular momentum (when we have summed moments) we applied it about any convenient point. This was possible because the systems in which moment was computed were all stationary. This means all points were fixed. If an object moves, be sure to apply conservation of angular momentum only about the center of mass or a fixed point.

**Example 3.31**

## Motion of a Peg in a Slot - Again!

**Problem:**
Figure 3.110 should look familiar. It is similar to example 3.29 except this time let's suppose there is a constant torque applied to the slot and determine the motion of peg P. As before there is also a spring attached to the peg. Just to keep this one simple, let's also ignore the weight of the peg and friction between the peg and slot. Let the mass properties of the slot be known (we know its size and shape).

**Formulation:**
This is a two [NIF] machine and no flows are specified therefore we must write a conservation equation. There is also 3 [NIA]. There are two bodies so we expect to need two freebodies before we

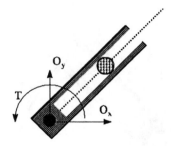

Figure 3.111: A Freebody Diagram of the Peg and Slot.

are finished. By the way, why didn't we use two freebodies in example 3.29? Well, in that problem the angle $\theta$ was specified which means there had to be a torque on the slotted shaft to make that motion happen. Had we used a second system, we would have written equations which would allow us to determine the torque and that's it. There would be nothing wrong with using the second system as long as we realized there has to be an unknown torque since there was a known flow. We avoided drawing the second freebody because we had not discussed angular momentum yet. Now we are ready so let's do it. Since there are potentially two systems to choose, let's spend a little time to see if we can wisely choose a system. If we choose the slot and peg together, the only external forces/torques are the pin reactions at O and the known applied torque $T$. Figure 3.111 shows a freebody diagram of the peg and slot.

If we apply linear momentum to the freebody in figure 3.111 all we would get is equations with unknown flow of the peg and unknown efforts $O_x$ and $O_y$ since we don't care about the forces, let's not waste time finding them. Consider the angular momentum of the system. Since O is fixed, let's sum moments about $O$ to find:

$$T\vec{k} = \frac{d}{dt}\left(I_O\omega\right)\vec{k} + \frac{d}{dt}\left(\vec{r}_P \times m\frac{d\vec{r}_P}{dt}\right)$$

Note that $I_O$ is the moment of inertia of the slot, $m$ is the mass of the peg, $\vec{r}_P$ is a vector from $O$ to $P$ and $\frac{d\vec{r}_P}{dt}$ is the velocity of the peg. Borrowing from example 3.28 we find the velocity of the peg from equation 3.36 (we could use the $\vec{i}$ and $\vec{j}$ equations too, its a little harder to find the cross product but not a big deal):

$$T\vec{k} = I_O\dot{\omega}\vec{k} + m\frac{d}{dt}\left(r\vec{n} \times (\dot{r}\vec{n} + r\omega\vec{s})\right)$$

Finally the cross product reduces to:

$$T\vec{k} = I_O\dot{\omega}\vec{k} + m\frac{d}{dt}\left(r^2\omega\right)\vec{k}$$

Well this is a first order equation (we need third order) plus it has two unknowns $r$ and $\omega$ so this one equation is not enough. We expected as much. Choose another system. Take the peg alone.

We have applied conservation of linear momentum for the peg already. Look back at example 3.29 and in particular equations 3.38 and 3.39. All equation 3.39 does if you recall is allow us to solve for $N$ (the force between the peg and slot). Who cares? Not us, so look only at equation 3.38 which is repeated here:

$$-k\left(r - r_o\right) = m\left(\ddot{r} - r\omega^2\right)$$

If you used the $\vec{i}$ and $\vec{j}$ formulation, you would need both equations and use some algebra to eliminate $N$ from the equations. Anyway this last equation is second order and contains the same two unknowns ($r$ and $\omega$). We are finished. Unfortunately these equations are nonlinear so there isn't much hope for solving them unless we do it numerically.

**End 3.31**

## 3.12 ENERGY METHODS

To apply conservation of energy or power to a system do the following.

1. Count the [NIF]. If there is more than 1 [NIF], chances are you will not find energy worthwhile because you only get one equation.

2. List all the forces that do work. If you cannot find the work or power done by these forces you cannot use energy.

3. Choose an initial and final time.

4. Apply the energy equation.

### Example 3.32

### Energy Accumulation in a Linear Spring

**PROBLEM:**

Determine the energy accumulated in a linear spring. Of course you probably have it memorized but let's derive it anyway.

**FORMULATION:**

We can determine the energy accumulated in a linear spring by expressing the conservation of energy for the spring in figure 3.16. Writing the conservation equation as a rate (power) gives:

**Conservation of Energy**

system[**spring fig.3.16**]    time period [ **differential**]

*input − output = accumulation*

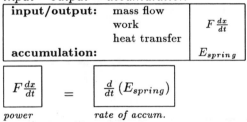

| **input/output:** | mass flow | |
| | work | $F\frac{dx}{dt}$ |
| | heat transfer | |
| **accumulation:** | | $E_{spring}$ |

$$\boxed{F\frac{dx}{dt}} \quad = \quad \boxed{\frac{d}{dt}\left(E_{spring}\right)}$$

*power*          *rate of accum.*

Note that the term $F\frac{dx}{dt}$ is power entering the spring, see figure 3.16. Since positive $F$ is in the same direction as the positive motion $x$, we have work going in. Since the spring is linear, the force in the spring is proportional to the deformation ($F = kx$). Also note that $F > 0$ when $x > 0$, see figure 3.16. Thus, we can simplify the conservation result to

$$kx\frac{dx}{dt} = \frac{d}{dt}\left(E_{spring}\right)$$

**SOLUTION:**

We can find the energy accumulated in the spring (from positions $x_0$ to $x_t$), by integrating for the time interval from 0 and $t$.

$$\int_{\tau=0}^{\tau=t}\left(kx\frac{dx}{d\tau}\right)d\tau = \Delta E_{spring} \overset{reduces}{\rightsquigarrow} \int_{x=x_o}^{x=x_t}(kx)dx = \Delta E_{spring} \overset{reduces}{\rightsquigarrow} \frac{1}{2}k\left(x_t^2 - x_o^2\right) = \Delta E_{spring}$$

If the initial condition of the spring was freelength then $x_o = 0$ and the initial energy (ignoring internal) is 0 therefore the energy after deforming to $x$ is $\frac{1}{2}kx^2$.

**End 3.32**

### Example 3.33

### Energy Dissipation in a Linear Damper

**PROBLEM:**

Unlike a spring, an ideal damper does not accumulate energy because an ideal damper is massless. This assumes that the fluid in the damper does not heat up and store the thermal energy. If we apply a force to a damper (see figure 3.20), we cause energy to enter the system through the work done by the force. The mechanical energy that enters a damper escapes as a less usable form of energy such as heat. Formulate a mathematical model of the amount of mechanical energy that a damper converts to other less usable forms.

**FORMULATION:**

We can determine the amount of mechanical energy a damper converts to other forms by accounting the mechanical energy of the damper in figure 3.20.

---

**Mechanical Energy  Accounting**
system[**damper fig.3.20**]      time period [ **differential**]

*input − output + generation − consumption = accumulation*

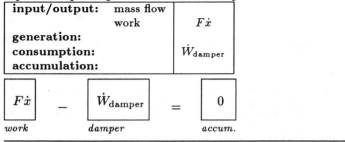

Since the force applied to a damper is proportional to the stretch (compression) velocity ($F = C\dot{x}$), we can simplify this conservation result to

$$C\dot{x}\dot{x} - \dot{W}_{\text{damper}} = 0$$

$$\dot{W}_{damper} = \dot{M}E_{\text{converted by damper}} = C\dot{x}^2$$

**End 3.33** ────────────────────────────────────────────────

One application of a spring is for storing energy in a mass. How could we store energy in the mass in figure 3.32? When the force is applied, the mass begins to move. As the mass moves, the force does work on the mass, since the force is in the same direction as the mass's displacement. The mass moves faster and faster as the force is applied because it stores the energy provided by the force. Unfortunately, this is not a beneficial energy storage system because the mass moves forever. Therefore, it would be difficult to harness the energy in the system. A better storage system would be allow the mass to move back and forth. In this case, the energy storage would always be conveniently located. To accomplish this back and forth motion, we allow $F$ to reverse direction. We apply $F$ to the right for a few seconds, then we apply it to the left for a few seconds. We repeat this action in an oscillatory manner. Will this work? Not really. What really happens is that $F$ is aligned with the mass's displacement for a while and adds energy to the mass during this time period, then $F$ is applied opposite the displacement as it brings the mass to a stop. During this time, $F$ removes energy.

What we really need is a means to continuously add energy to the mass without having the mass move continuously away. We need something to stop the mass for us. What if we use a spring? Figure 3.112 shows a reasonable system. We apply $F$ to the right until the spring stops the motion. When the motion stops, we reverse $F$ and pull left until the spring stops the mass again. We repeat this action in an oscillatory manner. In this case, $F$ is always in the same direction as the mass's displacement. Therefore, it always adds energy to the system. The mass will continue to gain energy for as long as we force it back and forth.

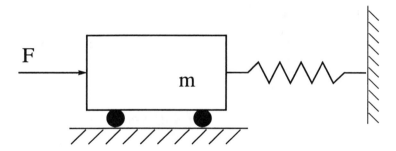

Figure 3.112: A Mass, Spring, and Force for Storing Energy

## 3.13 SUMMARY

The following list reviews the basic steps in applying conservation of momentum to a system.

1. Always count the [NIF]. The [NIA]is not as valuable. At this point you should be thinking about what assumptions you need to make and what simplifications you plan to use. In a problem where you do not know for certain what the [NIF]is, make a table of what the [NIF]is for various assumptions you make. Start the problem with the minimum [NIF]possibility.

2. Knowing the [NIF], what is given and what is asked for, determine whether or not you need conservation of momentum. If not skip to step 5.

3. Choose a system. If there are N bodies it may be necessary to choose N systems and apply conservation methods to each one. However it is sometimes possible to get all the information you need by cleverly choosing a system comprised of multiple bodies.

4. Isolate the system and draw a freebody diagram. When isolating a system, try to cut bodies loose such that a minimum number of unknowns are introduced. If a body has ideal connections, the cut points that introduce the fewest unknowns are almost always the idealized connections.

5. Choose a coordinate system that makes expressing motion easy. If there is no change in momentum, you may choose any direction you want as coordinate directions.

6. If you are writing conservation of momentum equations for the system, write them in the coordinate directions you chose in step 5.

7. Write the time rate of change of momentum in terms of the derivatives of position. Write angular momentum in terms of derivatives of angles.

8. Count the number of unknowns. If the number of unknown flow variables is greater than [NIF]minus the given flows, then you need to relate flows. To relate flows you should use the "length of rope" technique, vector loops, or contact relations.

9. If there are too many unknowns after relating motions, you must find more equations. Do this by applying the following steps.

    (a) Choose another system and express conservation equations for it.

    (b) We look for information that we may have overlooked in the problem statement.

    (c) Look for an additional conservation equation for the system.

## 3.14 Problems and Questions

An * after a problem number indicates a solution is available.

**Force and moment.**

1. * Is gravity an effort or a flow?

2. * Two people tug on a rope with 10 [lbf] in opposite directions. Determine the tension in the rope. The same rope is tied to a tree and one person tugs on the free end with 10 [lbf], what is the tension in the rope now.

3. * If

$$\vec{F}_B = 3t^2\vec{i} + 4t\vec{j} + 6\vec{k}$$

and the vector from A to B is

$$_A\vec{r}_B = 8t\vec{i} + 7\cos(2t)\vec{j} + 7\sin(2t)\vec{k},$$

then determine the torque of force $\vec{F}_B$ about point A.

**Friction**

4. * Jerk is the rate of change of acceleration. When stopping an automobile, if you apply a steady pressure on the brake pedal, a huge jerk occurs (which can be felt if you pay attention) just as the auto stops. Why?

5. * A man rides a bike up a steep hill. When he reaches the halfway point, he becomes too tired to continue peddling and begins to roll back down the hill. To stop this motion, he applies the brakes. Unfortunately, the bike falls over and the man slides down the hill.

   (a) Determine the direction of the friction force on the bike's tires as the man rides the bike up the hill.

   (b) Determine the direction of the friction force on the bike's tires when the bike begins to roll down the hill.

   (c) Determine the direction of the friction force from the hill to the tire when the brakes are applied.

   (d) Determine the direction of the friction force on the man as he slides down the hill.

6. * A military boxcar carries a crate of high explosives. The boxcar is part of a train carrying supplies to a military base. At a railroad crossing, the driver of an auto foolishly ignores the lights and tries to beat the train. The train's engineer quickly applies the train's brakes. This causes the crate of explosives to slide toward the front of the boxcar.

   (a) Determine the direction of the friction force on the crate.

   (b) Explain how we can determine what magnitude of deceleration will cause the crate in the boxcar to slip.

7. * The train described in question 6 begins to climb a hill. Determine the direction of the friction force on the crate.

8. * In machinery, a common operation is to slide one object over another. Imagine a precision machine performing a sliding operation. The objective is to slide one object over another a very small amount. One typical method for accomplishing this is to push it. It doesn't move so the machine pushes harder. If it still doesn't move the machine pushes a little harder. The basic idea is to increase the applied force very slowly so that the object being slid is "ever so slightly" out of static equilibrium to prevent it from accelerating out of control. Regardless of how slowly the force is increased however, eventually the object starts to move and it accelerates much more than desired and slides much further than intended. Why does this happen?

9. * Explain why during a panic stop "anti-lock" brakes on an automobile can stop the auto faster than regular brakes.

**FBD**

10. * Draw a freebody diagram that we can use to determine the reaction at $P$ for the crane in figure 3.113.

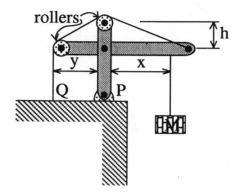

Figure 3.113:

11. For each of the objects in figure 3.114, complete the freebody diagram. Neglect the mass of the member unless otherwise specified.

**Linear momentum**

12. * Explain why we have the sensation of being thrown to the right when we turn to the left in an automobile.

13. * In figure 3.115, what would happen to the forces at point A if $\theta$ approached zero? Is this what we should expect? Explain.

14. * Determine the motion of the two blocks in figure 3.116. Let $m_1 = 12$ [kg], $m_2 = 4$ [kg], $\theta = 25°$, $\mu_{s1} = 0.25$, $\mu_{s2} = 0.5$.

15. * For the blocks shown in figure 3.117 determine the value of $P$ for impending motion. Let the friction coefficients between 1 and 2 and between 2 and the floor be different.

16. Often when we attempt to insert a peg into a tight fitting hole, the peg becomes jammed such that it cannot enter the hole. Regardless of how hard we push, the peg will not enter the hole. A schematic diagram of a peg entering a hole appears in figure 3.118. The figure shows a loosely fitting peg just so the angle $\theta$ is exaggerated enough to be obvious. For a tight fitting peg, $d < D$ but they are nearly equal causing $\theta$ to be nearly zero. The force $F$ is attempting to "push" the peg into the hole.

   (a) Show that for small values of $\theta$ a specific value of $l$ and at low speeds, the force $F$ factors out of the linear momentum equations.

   (b) What does it mean when the $F$ factors out?

   (c) What is the significance of the value of $l$ when $F$ factors out?

   (d) What does this analysis tell you about putting pegs into holes? What capability does it require?

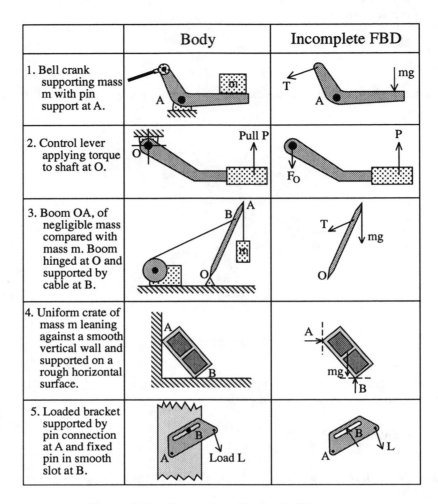

| | Body | Incomplete FBD |
|---|---|---|
| 1. Bell crank supporting mass m with pin support at A. | | |
| 2. Control lever applying torque to shaft at O. | | |
| 3. Boom OA, of negligible mass compared with mass m. Boom hinged at O and supported by cable at B. | | |
| 4. Uniform crate of mass m leaning against a smooth vertical wall and supported on a rough horizontal surface. | | |
| 5. Loaded bracket supported by pin connection at A and fixed pin in smooth slot at B. | | |

Figure 3.114: Incomplete Freebody Diagrams

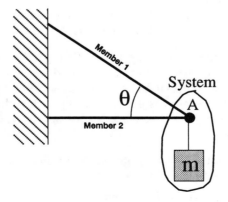

Figure 3.115:

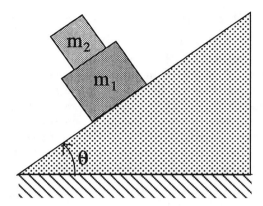

Figure 3.116:

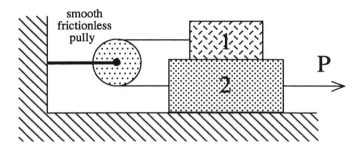

Figure 3.117:

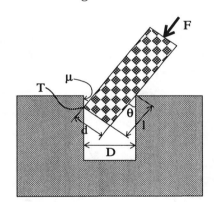

Figure 3.118:

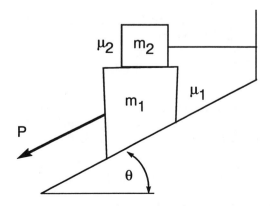

Figure 3.119:

17. Under what conditions, will the string connected to the top mass in figure 3.119 break? Let the string support $T$ pounds.

18. * You are to construct a "cruise control" for an automobile. The objective is to make the auto travel at a constant speed. The force pushing the car forward is under control of the engine's throttle position and you can control the throttle. As a result, you can apply a force to the car which is proportional to the car's speed. In other words the driving force $F$ is $kv$, where $v$ is speed and $k$ can be chosen by you. Wind gusts, air drag and the like push back on the car and are not predictable. Determine the value of $k$ in terms of parameters which makes the time constant of the automobile's speed control equal to some given value of $\tau$. HINT: If you put a first order differential equation in the form: $\dot{y} + \frac{1}{\tau}y = x$, $\tau$ is the time constant.

19. * Determine the force $P$ required to raise the 500-lb block (see figure 3.120). The coefficient of friction for all surfaces is 0.4. What is the minimum value of $\mu_s$ to allow the wedge to be "self locking".

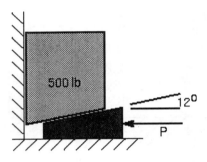

Figure 3.120:

20. * A man (see figure 3.121) pedals his bicycle up a grade on a slippery road at a steady speed. The man and bike have a combined mass of 82 kg with center at G. If the rear wheel is in impending slip, determine $\mu_s$ between the tire and road.

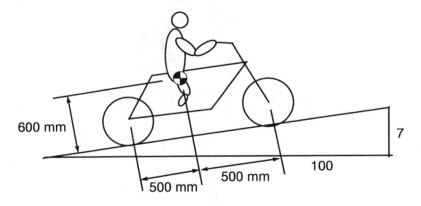

Figure 3.121:

21. * For the block shown in figure 3.122, determine the magnitude and direction of the friction force acting on the 90 kg block if (a) P=500N, (b) P=100N. Let $\mu_s = .2$ and $\mu_k = .17$. The force is applied with the block initially at rest. Also, if the block moves find its acceleration.

22. * Calculate the force P required to initiate motion of the 60-lb block up the 10 degree incline. The coefficient of static friction for each surface is 0.3. See figure 3.123.

23. * Figure 3.124 shows a support for a load with 4 different methods for anchoring the load. First decide quickly which of the four cases will produce the most moment at the support A. Next

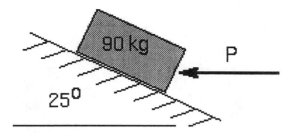

Figure 3.122:

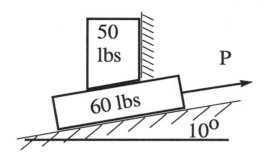

Figure 3.123:

for the case you have chosen, compute the moment at A and the force in the pin connection at point D. The radius of the pulley is 70 centimeters.

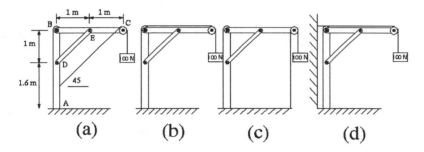

Figure 3.124:

**springs dampers**

24. Determine the motion of the masses in figure 3.125. Let $M_1$=25 [lbm], $M_2$=25 [lbm], $k_1$=2.5 [lbf/ft], $k_2$=2.5 [lbf/ft], $c_1$=0.5 [lbf s/ft], $c_2$=0.5 [lbf s/ft], F= 45 [lbf]. Also let everything start at rest with the springs undeformed.

25. * Optics-R-Us, your employer, has been asked to design a suspension system for a one ton reflective mirror for the new ICU telescope. The mirror must be isolated from building vibration induced by heavy trucks moving on the freeway nearby. Your supervisor plans to float the mirror on a fluid. The fluid can provide a drag force on the mirror proportional to the speed of motion between the building and the mirror. Your job is to determine the constant of proportionality for the drag force. The structural department estimates that the building velocity will be horizontal only, sinusoidal with a magnitude of "b" inches/second and frequency ranging from 0 to w radians/second. What constant of proportionality will keep the mirror velocity magnitude below "a" inches/second?

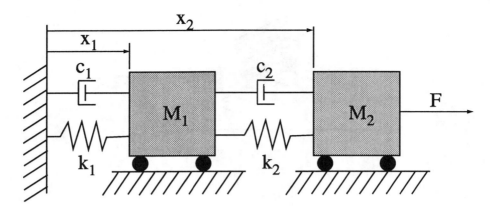

Figure 3.125:

26. For the system shown in figure 3.126, develop the equations of motion. F(t) is a given force. Discuss the [NIF] and [NIA], indicate appropriate flow variables, and discuss how the resulting equations relate to [NIF] and [NIA].

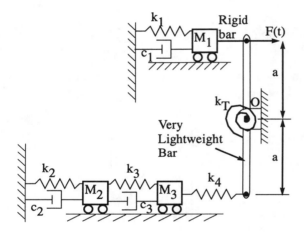

Figure 3.126:

27. Derive the equation that describes the motion of the device shown in figure 3.127.

**choosing systems**

28. If we assume that both members in figure 3.115 are truss elements, we can ignore any transverse (non-axial) forces in the members. How can we determine if this assumption is valid? Under what conditions could member 2 carry a transverse force?

29. *  If the supported load in figure 3.128 increases significantly, the frame will break. What is the most probable location for the break?

30. *  The truss in figure 3.129 supports a 25 [kg] mass. If the wall supports can only sustain 250 [N] of force, determine if the wall can support this mass. Explain what effect changing $\theta$ has on the distribution of forces in the truss. For what value of $\theta$ can the truss support this mass? Let $\theta=45°$.

31. *  Determine the reaction at $P$ and the tension in the tie line (at point $Q$) of the system in figure 3.130.

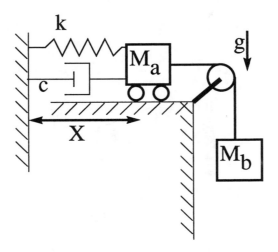

Figure 3.127:

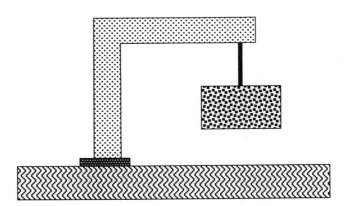

Figure 3.128:

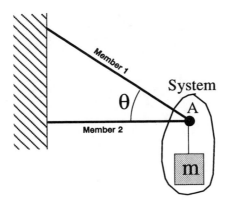

Figure 3.129:

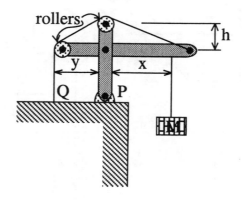

Figure 3.130:

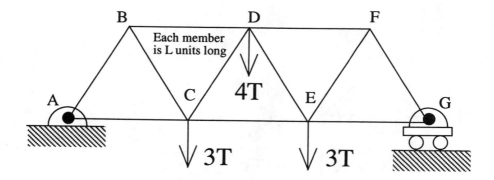

Figure 3.131:

32. * For the truss in figure 3.131, determine the load in member CE by writing the fewest equations possible.

33. For the inter-coastal canal lock gate in figure 3.132,

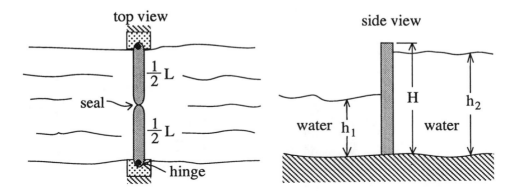

Figure 3.132:

(a) draw a freebody diagram

(b) determine the torque that the motors at the hinges must produce to keep the gate closed?

Hint: Water pressure increases with depth. For stationary liquids, the pressure is equal to $kd$ where $k$ is the density times the gravity constant $g$ and $d$ is the depth from the surface.

34. * Figure 3.133 shows a crane modeled as a truss. Determine if members $DE$, $DG$, and $HG$ are strong enough to handle the 2 [ton] tractor. Each member can handle a load of 5 [tons] tension and 4.5 [tons] compression. Note: Contrary to this problem, many members withstand more compression than tension.

35. * Figure 3.133 shows a crane modeled as a truss. If all members can support a load of 5 [tons] tension and 4.5 [tons] compression. What is the maximum tractor weight which can be supported?

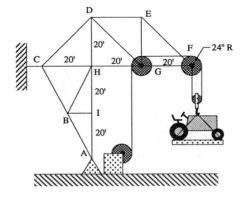

Figure 3.133:

36. * Calculate the components of all forces acting on each member of the loaded frame in figure 3.134.

37. * Calculate the moment $M$ that we have to apply at $A$ to keep the mechanism in figure 3.135 in static equilibrium at the position in the figure. Then, determine the reaction forces at pin $A$.

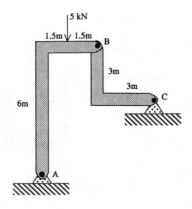

Figure 3.134:

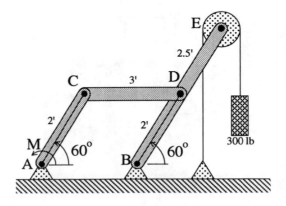

Figure 3.135:

38. * Calculate the $x$ and $y$ components of the force at $A$ and $C$ of the loaded frame in figure 3.136.

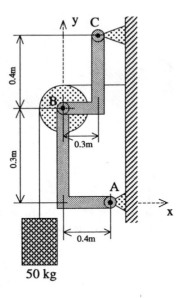

Figure 3.136:

39. * We tighten the turnbuckle (A) in figure 3.137 until it produces a tension of 1500 [lbs] in its cable. We fasten the two pulleys at (F) together and wrap the cables securely onto the pulleys. Calculate the $x$ and $y$ components of the force exerted by member BD on member CF.

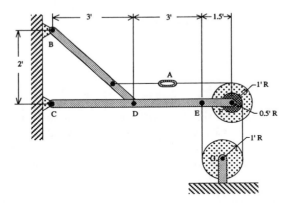

Figure 3.137:

40. If you were designing the tripod in figure 3.138 how much force would you make sure member A could support? The tripod has a height of $H$ and a base length of $L$. Member A is $h$ [units] above the ground. Use a safety factor of 2.

41. * Some of the indigenous people of a South American rain forest need to climb trees quickly. Due to the thick canopy however many of the vertical trees do not have limbs below 100 feet. To assist them in climbing a vertical, slightly less than shoulder diameter tree, the climber ties a strong cord around his ankles. The cord holds the ankles slightly wider than the tree diameter. Using a partially squatting stance (knees apart slightly) with one foot on each side of the tree, the climber can avoid sliding without exerting effort in his arms. Explain where the forces come from.

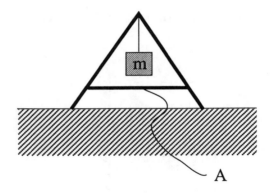

Figure 3.138:

42. * The coefficient of friction at both ends of the uniform slender bar shown in figure 3.139 is 0.5. Find the maximum forward acceleration that the truck may have without the bar moving relative to the truck.

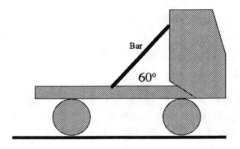

Figure 3.139:

**Kinematics**

43. * Collar A slides along a horizontal rod that is rotating as shown with an angular speed of 5 rad/s and an angular acceleration of 5 rad/s squared. At a distance of 7ft from the axis of rotation of the rod, the collar has a velocity of 10ft/s and an acceleration of 4 ft/s squared relative to the rod. Determine the velocity and acceleration of the collar at the instant shown.

44. * The system of masses in figure 3.141 starts from rest at the position in the figure. Determine the acceleration of mass C that causes the right side of mass A to slide to the right side of mass B in 4 [s]. Assume that the acceleration of C is constant.

45. * Bodies A, B, and C in figure 3.142 start from rest at a common elevation. Body A has a constant acceleration of 50[mm/s$^2$] downward. Body B has zero acceleration for the first 3 [s], after which its acceleration becomes 75[mm/s$^2$] upward. Determine the velocity of C after it has traveled a distance of 300 [mm].

46. For figure 3.143, if the bodies start at rest, determine the motion given the following. $m_A = 2$ [kg], $m_B = 3$ [kg], all $\mu_s = .8$, all $\mu_k = .75$, $m_C = 0$, $P = 75$ [N].

47. * For figure 3.143, determine the following:

   (a) is there a maximum value of $P$ for which there is no motion and if so find it in terms of the parameters.

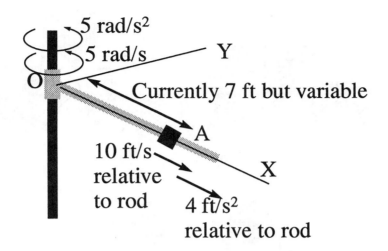

Figure 3.140:

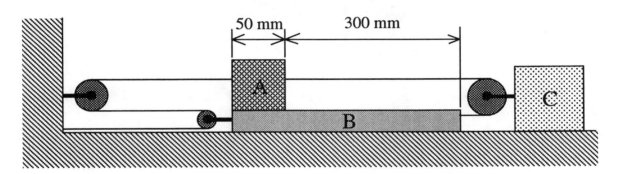

Figure 3.141:

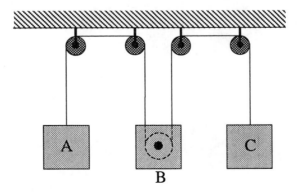

Figure 3.142:

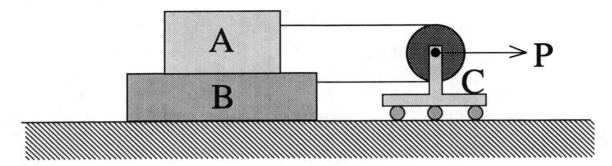

Figure 3.143:

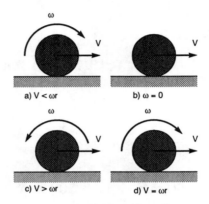

Figure 3.144:

(b) is there a minimum value of $P$ for which $A$ slides on $B$ and if so find it in terms of the parameters.

48. Determine the friction forces acting on the cylinders in figure 3.144.

49. * The elevator E shown in figure 3.145 is moving upward at a speed of 2 m/s and its speed is decreasing at the rate of 0.2 m/s squared. Determine the velocity and acceleration of the counterweight C.

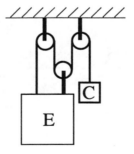

Figure 3.145:

50. * Figure 3.146 shows a wedge shaped cart accelerating to the right at a=1 m/sec squared. On the cart are two blocks connected by a rope. Block 1 has mass of 1 kg, block 2 is 2 kg. The coefficients of friction everywhere are 0.09 and 0.1, you decide which is kinetic and which is static. The angle of the wedge is 45 degrees. Assume that block 2 is sliding down the incline initially. Determine the acceleration of block 1. Assume the pulley is frictionless and the rope is inextensible.

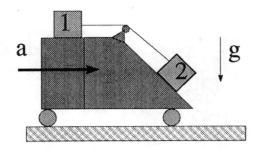

Figure 3.146:

51. * Two bodies A and B shown in figure 3.147 have masses of 40 kg and 30 kg, respectively. The coefficient of static friction for A is 0.3; the kinetic coefficient is 0.25. If the system is released from rest, determine (a) the acceleration of A; (b) the tension in the cable; (c) the velocity of B after 5 s of motion.

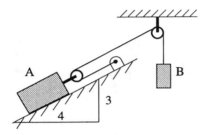

Figure 3.147:

52. * The 10-kg cart shown in figure 3.148 is being pulled to the left by a constant force $P = 10$ N applied to the end of the cord. If the cart starts from rest when $x = 8$ m, calculate and plot (a) the speed of the cart as a function of its position x between $-3 < x < 8$; HINT: If you use the chain rule, it is possible to show that the second time derivative of $x$ can be converted to a derivative with respect to $x$ as follows:

$$\frac{d^2x}{dt^2} = \frac{d\frac{dx}{dt}}{dt} = \frac{dV}{dt}\frac{dx}{dx} = \frac{dV}{dx}\frac{dx}{dt} = \frac{dV}{dx}V$$

(b) the position of the cart as a function of time between $0 < t < 5$.

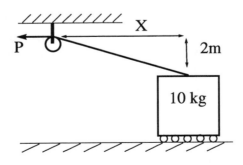

Figure 3.148:

53. * For a certain interval of motion, the slotted arm in figure 3.149, forces the pin (A) to move into the fixed parabolic slot. If we elevate the arm in the $y -$ direction at a constant rate of 30 [mm/s], determine the velocity $v$ and acceleration $a$ of the pin when $x = 60$ [mm].

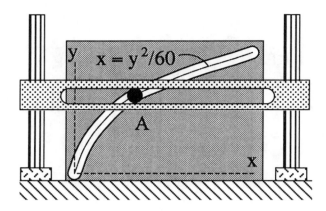

Figure 3.149:

54. For a certain interval of motion, the slotted arm in figure 3.149, forces the pin (A) to move into the fixed parabolic slot. If we elevate the arm in the $y-$ direction at a constant rate of 25 [in/s], plot the speed and the magnitude of acceleration of the pin versus $y$ from $y=0$ to $y=10$ in.

55. * The flanges of the collar (B) in figure 3.150 slide with a constant velocity $v_B$ to the right along the fixed shaft. The flanges guide the pin A in the bell-crank (AOD) of the figure. The radial slot in the bell-crank (AOD) positions the plugger's upper end. Determine the acceleration of point C on the plunger (CE). Let $v_B = 1.5$ [in/sec] and,

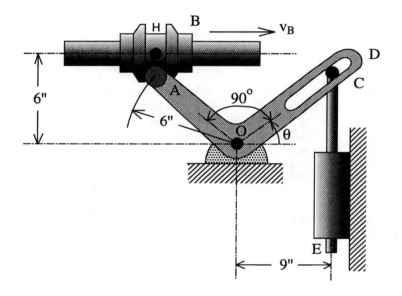

Figure 3.150:

(a) $\theta = 45^o$.
(b) $\theta = 35^o$,
(c) $\theta = 25^o$,
(d) $\theta = 15^o$,
(e) $\theta = 5^o$,
(f) $\theta = 0$,
(g) $\theta = -5^o$.

56. * The horizontal plunger (A) in figure 3.151 pushes the 70° bell crank (BOC). If the plunger has a velocity of 75 [mm/s] to the right and accelerates at a rate of 100 [mm/s²] when $\theta = 30°$, determine the bell crank's angular acceleration ($\ddot{\theta}$) at this instant.

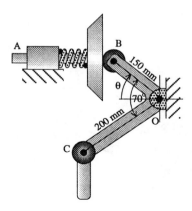

Figure 3.151:

57. The jet plane in figure 3.152 is flying at a constant altitude of $h$. The constant speed of the plane is unknown. If we are tracking the plane by radar from a point O, we can measure $\theta$ and all its derivatives. Determine $\ddot{r}$ and $|\vec{v}|$.

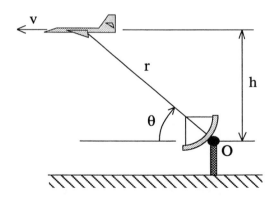

Figure 3.152:

58. * The slotted arm (OA) in figure 3.153 forces the small pin to move in the fixed spiral guide. We can define the distance from the pivot to the pin as

$$r = K\theta.$$

Arm (OA) starts from rest at $\theta = \frac{\pi}{4}$ and has a constant counterclockwise angular acceleration of $\ddot{\theta} = \alpha$. Determine the magnitude of the acceleration of the pin when $\theta = \frac{\pi}{2}$.

59. Figure 3.154 shows an $x - y$ plotter. The motor that drives the $x -$ axis of the plotter is malfunctioning. It will only move the pen in the $\pm x$ direction with a constant velocity of $\pm \dot{x}_o$. We want to get as much life out of the plotter as possible before we have to repair it. If all the figures that we have to draw are of the form $y = f(x)$, determine the required speed of the motor that drives the $y$ direction. We want to obtain smooth plots is this practical with the plotter malfunctioning in this way. Explain.

60. For a properly functioning plotter, like the one in figure 3.154, determine the relationship between the speed of the $x -$ axis and the speed of the $y -$ axis, if we describe the figures to be drawn in terms of polar coordinates (i.e., $r(t), \theta(t)$).

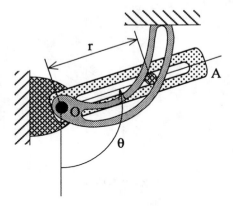

Figure 3.153:

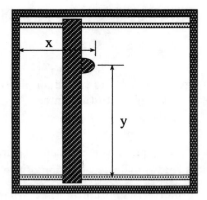

Figure 3.154: Pen plotter driven by a computer

61. A car rounds the turn of constant curvature between A and B with in figure 3.155 with a steady speed of 55 [mph]. If we mount an accelerometer in the car, what magnitude of acceleration would it record between A and B? The distance along the curve between A and B is 100 [ft].

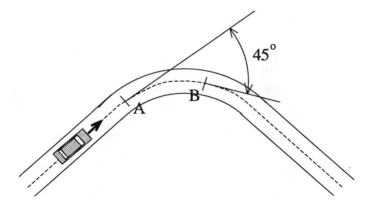

Figure 3.155:

62. A car rounds the turn of constant curvature between A and B with in figure 3.156 with a steady speed of 72 [km/hr] . If we mount an accelerometer in the car, what magnitude of acceleration would it record between A and B? The distance along the curve between A and B is 100 [m].

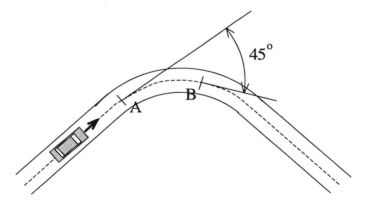

Figure 3.156:

63. * The pulleys in figure 3.157 transfer magnetic tape (which does not stretch) from reel A to reel B. At a certain instant point $P_1$ on the tape is in contact with pulley C and point $P_2$ is in contact with pulley D. If the normal component of acceleration of $P_1$ is 40 [m/s$^2$] and the tangential component of acceleration of $P_2$ is 30 [m/s$^2$] at this instant, determine the following:

   (a) the speed of the tape

   (b) the magnitude of the total acceleration of $P_1$

   (c) the magnitude of the total acceleration of $P_2$.

64. The satellite in figure 3.158 moves in an elliptical orbit around Earth. How fast is the satellite moving at point A? What is the radius of curvature of the satellite's orbit at point A? Is the radius of curvature equal to $r$? What is the rate of change of the speed at point A? Let $g_{\text{sea level}}$=32.17 [ft/s$^2$] and $r_{\text{Earth}}$=3958 [mi].

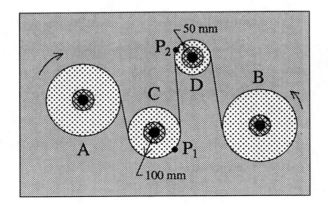

Figure 3.157:

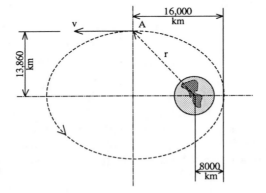

Figure 3.158:

65. Using figure 3.159, derive the vector relationships for the velocity and acceleration of point P in spherical coordinates $(r, \theta, \phi)$. The point is free to move in three dimensional space. The results for the spherical coordinates should be

$$\vec{v} = v_r \vec{r} + v_\theta \vec{\theta} + v_\phi \vec{\phi}$$

and

$$\vec{a} = a_r \vec{r} + a_\theta \vec{\theta} + a_\phi \vec{\phi}$$

where

$$v_r = \dot{r}$$

$$v_\theta = r\dot{\theta}\cos\phi$$

$$v_\phi = r\dot{\phi}$$

$$a_r = \ddot{r} - r\dot{\phi}^2 - r\dot{\theta}^2\cos^2(\phi)$$

$$a_\theta = \frac{\cos\phi}{r}\frac{d}{dt}(r^2\dot{\theta}) - 2r\dot{\theta}\dot{\phi}\sin\phi$$

and

$$a_\phi = \frac{2}{r}\frac{d}{dt}(r^2\dot{\phi}) + r\dot{\theta}^2\sin\phi\cos\phi.$$

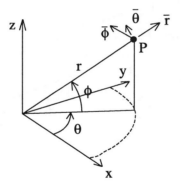

Figure 3.159:

66. * For the slider-crank mechanism in figure 3.160, AB = 2in, BP = 6 in, AB is rotating at a constant 1000 rpm counterclockwise. Plot the velocity $(\vec{V}_p)$ and acceleration $(\vec{a}_p)$ of the piston (point P) versus $\theta$ for $\theta = 0$ to 360 degrees. Does the solution appear correct? How does it compare with your intuition?

67. * For the linkage shown in figure 3.161, given a history of $\theta$, find $\beta$, $\dot{\beta}$ and $\ddot{\beta}$, $L$, $\dot{L}$ and $\ddot{L}$. Let the distance between A and B be 2 inches and between A and C be 4 inches.

68. * Figure 3.160 shows a slider crank. The crank (the left hand body with length $L_1$) is driven at a constant speed of 2 rad/sec by an electric motor (not shown). What is the maximum force on the pin connecting the piston to the $L_2$ rod? Assume all weights are negligible except the piston which has a mass of $M$. Let $L_1 = 1$ and $L_2 = 2$ feet.

69. * Two disks A and B with radii of 4ft and 3ft (shown in figure 3.162 are in contact with each other without slipping. Disk A starts from rest and rotates with a constant angular acceleration of 5 rad/sec squared as shown. Determine the angular velocity and angular acceleration of disk B just after disk A turns 5 revolutions.

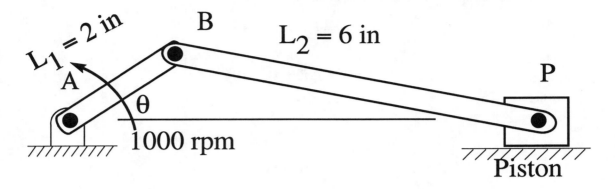

Figure 3.160:

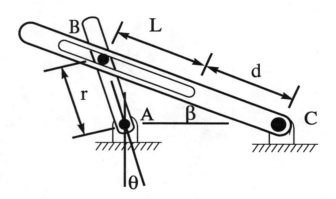

Figure 3.161:

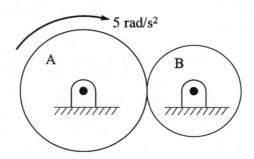

Figure 3.162:

70. * A double pulley shown in figure 3.163 serves as a hoisting mechanism. The rotating motion of the outer pulley is controlled by a cable connected to the inner pulley as shown. A point A along the cable has a constant acceleration of 1 ft/s squared and an initial velocity of 1 ft/s, in the direction indicated. Determine (a) the number of revolutions executed by the double pulley in 5 s; (b) the vertical displacement of the load after 5 s; (c) acceleration of point B at 5s. Note that $\theta$ is zero at time = 0 so B is located to the right of the pulley support at initial time.

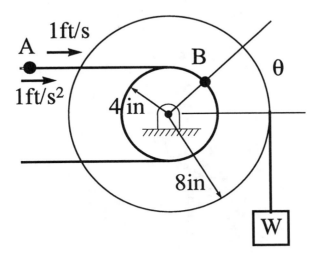

Figure 3.163:

71. * A wheel radius = 0.5m rolls without slipping on a belt that moves to the right at a constant speed of 0.2m/s. At the instant shown, the center O of the wheel has a velocity of 0.3 m/s to the left and a deceleration of 1 m/s squared to the right. Determine the acceleration of point A.

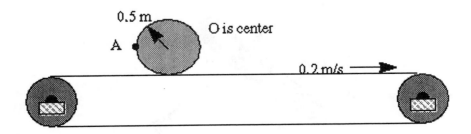

Figure 3.164:

72. * A 600 mm diameter wheel rolls without slipping on a horizontal surface. The 1-m long rod AB is attached to the wheel at a point 250 mm from its center, and end A slides freely along the surface. If the center of the wheel has a constant speed of 1.2 m/s to the right and $\theta = 0$ when $t = 0$, calculate and plot (a) the velocity of end A of the rod as a function of time $(0 < t < 3)$; (b) the angular acceleration of the rod as a function of time $(0 < t < 3)$.

73. * The mechanism shown in figure 3.166 is supposed to convert the rotary motion of the arm AB into translational motion of the plunger CD. At the instant shown, the small peg holding the two slots together is located 0.2 m from A and $\dot{\theta}$ is constant at 12 rad/s with $\theta$ 60 degrees. Is it possible, with the information given, to determine the velocity and acceleration of the plunger (point D)? If so, do it.

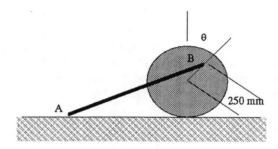

Figure 3.165:

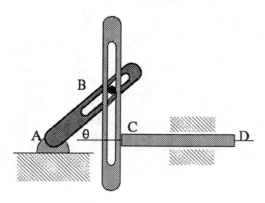

Figure 3.166:

74. Link AB (see figure 3.167)is 150 mm long and rotates through a limited range of the angle $\beta$, and its end A causes the slotted link AC to rotate also. For the instant represented where $\beta$ is 60 deg and $\dot{\beta} = 0.6$ rad/s clockwise and constant, determine the corresponding values of $\dot{r}$, $\ddot{r}$, $\dot{\theta}$ and $\ddot{\theta}$.

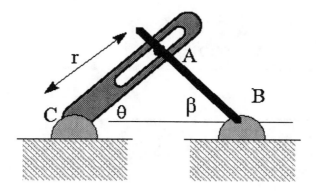

Figure 3.167:

**Angular velocity**

75. Determine the angular velocity (relative to the ground) of the disk in figure 3.168, if

  (a) the disk slips at P and rotates at a rate of $\omega$, and the vertical bar rotates at a rate of $\Omega$.

  (b) the disk rolls without slipping at P, and the bar rotates at a rate of $\Omega$.

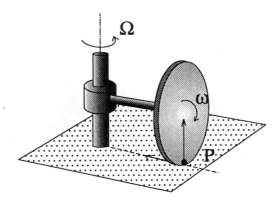

Figure 3.168:

76. Figure 3.169 shows a carnival ride.

    (a) Determine the angular velocity of compartment C relative to the operator on the ground. The angles increases in the direction that the figure shows. Be sure to label all the coordinate systems that are necessary.

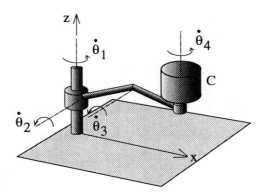

Figure 3.169:

    (b) If a rider moves into compartment C with a constant velocity $v$ relative to C, what is the time rate of change of their velocity as observed by the operator on the ground.

77. The circular disk in figure 3.170 rotates about the $x-$ axis with an angular velocity of $p$. At the same time the supporting arm and yoke rotate about the z axis with a constant angular velocity of $\Omega$. Determine the angular velocities of all the bodies.

78. Figure 3.171 shows a 1 kg particle P sliding on a very smooth rod. The rod is rotating at 5 radians/second constant. A string runs down the rod and is attached to the particle. The string is fed down the rod at a constant rate of 3 meters/second. The string can withstand 25 Newtons before it snaps. At time $= 0$, the particle is located $r_o$ down the tube. Determine the value of $r$ when the string snaps. Ignore gravity.

**Angular momentum**

79. For the truck and package of figure 3.172, if the truck stops on a downhill stretch of road, determine the following:

    (a) the height to width ratio of the package if it just begins to tip as the truck stops

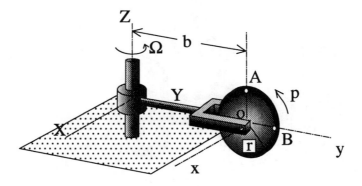

Figure 3.170:

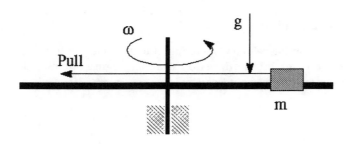

Figure 3.171:

(b) the maximum tolerable deceleration under this condition.

Express the answer in terms of the road's slope.  What happens if the truck is on an uphill road?

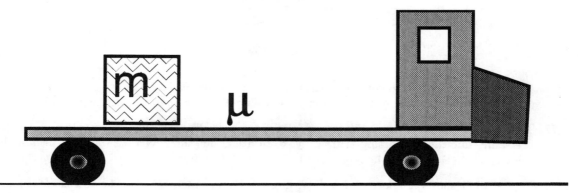

Figure 3.172:

80.  * The motor and antenna shown in figure 3.173 are floating in space and are symmetrical into the plane of the paper (after all this is just a class).  The rotating disk is known as an inertia wheel or reaction wheel.  The antenna has a mass of 25 kg and a mass moment of inertia about its center of mass (point B) of 50 kg $-$ m$^2$.  The little rod connecting the inertia wheel to the antenna is very lightweight and is 6 meters long (not very little is it).  There is a motor (not shown) which connects the inertia wheel to the little connecting rod.  There are two main parts of the motor.  One part is rigidly connected to the inertia wheel, when the wheel turns that part of the motor turns too.  The motor part attached to the inertia wheel and the inertia

wheel have a combined mass of 15 kg and a combined radius of gyration of 0.5 meters about its combined center of mass at A. The second part of the motor is attached to the connecting rod. It has a mass of 5 kg with essentially no dimension (its a concentrated mass) located at point A. When the motor is off, all masses have the same angular velocity. When the motor runs it causes the inertia wheel to have an angular velocity relative to the antenna equal to the running speed of the motor. At initial time, the motor is turned off and a meteor hits the antenna causing all of the masses to suddenly have an absolute angular velocity of 10 rpm counterclockwise. How fast, and in what direction, should the motor turn to cause the antenna to have 0 absolute angular velocity. What effect does the rotation caused by orbital motion have on the result? Why would you include or exclude it? Discuss the advantages and disadvantages of using inertia wheels to stabilize space structures.

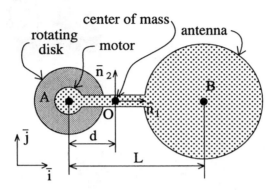

Figure 3.173:

81. The motor M in figure 3.174 drives disk A. Disk A turns disk B. There is no slip between A and B. Disk A, its attached shaft, and the motor armature weigh 36 [lbs] and have a combined radius of gyration of 3.4 [in]. Disk B weighs 48 [lbs] and has a radius of gyration about axis $O - O$ of 18 [in]. Before the motor starts, the entire assembly is rotating as a unit about axis $O - O$ with a speed of $\omega^o = 30$ [rev/min] in the direction the figure shows. The motor $M$ has an operating speed of 1720 [rev/min] with respect to C in the direction the figure shows. Determine the new rotational speed of arm C when the motor starts.

82. The disk in figure 3.175 has a radius of $r$. It moves inside the large ring of radius of $R$. The coefficient of friction between the two bodies is $\mu$. There is a torque applied to the ring that is very large compared to the weight of the ring. If the ring's angular velocity is $\Omega$ and the disk's $\omega$ for the position in the figure, determine if the disk slips. If it is slipping, explain what else we would need to know about the present state of the bodies in order to solve for the motion of the ring.

83. * Figure 3.176 shows a ladder that is $L$ [ft] long. The ladder rests against a wall and the floor. If the ladder begins to slip down the wall, determine an equation for the angle $\theta$ as a function of time.

84. Under some conditions it is possible to bounce a rubber ball under a table and have it return to the approximate location where it began, see figure 3.177. Explain how this can happen.

85. * Figure 3.178 shows a schematic of a gear box. The figure does not show the gear teeth. As the pinion gear D rotates, the rack R translates side to side. Gear D is driven by an electric motor (not shown). It is possible to write an equation for the motion of the pinion gear D and put it in the form $M\ddot{\theta} = T$ where $M$ consists of parameters (masses and lengths), $\ddot{\theta}$ is the angular acceleration of the pinion and $T$ is the torque provided by the motor. Determine $M$.

86. * Figure 3.179 shows an air compressor. There is air inside the cylinder that creates a force like a spring attached as the figure shows. The spring is unstretched at $\theta = \pi$. The spring is always

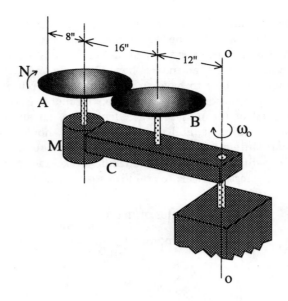

Figure 3.174:

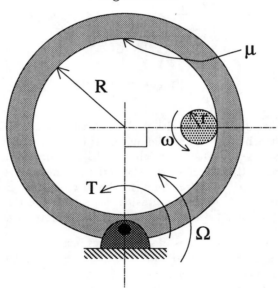

Figure 3.175:

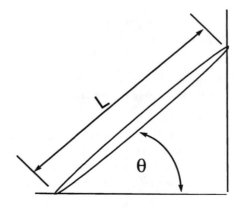

Figure 3.176:

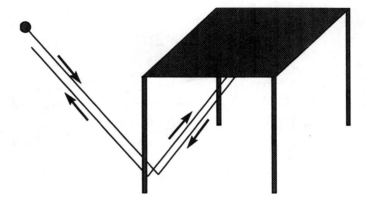

Figure 3.177:

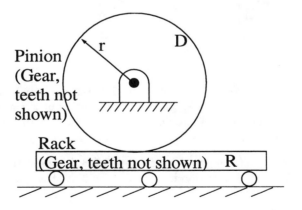

Figure 3.178:

connected to the piston. There is an electric motor (not shown) that applies a constant torque T=3 ft-lbs to the crank (the crank is the rod that is $L_1 = 1$ ft long). Let all the components be massless except for the crank. You may also ignore gravity and assume everything starts from rest at $\theta = 0$, let $L_2 = 2$ feet and the spring constant is 3 lbs/ft. How massive does the crank have to be if the difference between the minimum and maximum rotational speed of the crank during the second revolution is to be no more than 1 radian/second?

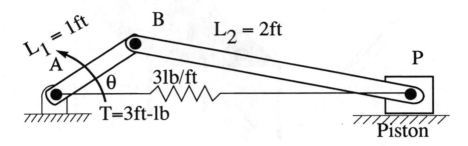

Figure 3.179:

87. * A spool is shown in figure 3.180 with a massless cord wrapped around the inner drum. End B of the cord is fixed as shown. The spool rests on an inclined plane. The angle $\theta$ is slowly increased (plane pivoted). For each case (a) and (b), find the angle $\theta$ at which motion commences, and find the time required for the mass center to travel 3 ft down the incline. Let the weight of the spool = 0.25 lb, R = 1 in, r = 0.5 in, centroidal moment of inertia = $2.2 \times 10^{-5}$ lb-ft-sec squared, $\mu_s = 0.3$ and $\mu_k = 0.25$.

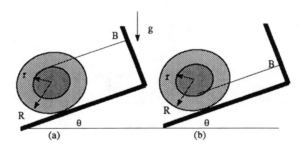

Figure 3.180:

88. * Two weights $W_1$ and $W_2$ are connected by an inextensible cord that passes over a pulley shown in figure 3.181. The pulley weighs W and its radius of gyration is $K$. Determine the acceleration of the left mass if they are released from rest and the cord does not slip on the pulley. Also find the tensions on both sides of the pulley.

89. * A uniform slender rod, 10 ft long and weighting 90 LB, is supported by wires attached to its ends, see figure 3.182. Find the tension in the left wire just after the right wire is cut. Assume the wires to be inextensible.

90. * An industrial fan (see figure 3.183) weighing 250 LB is mounted to the end of a beam which is part of the structure of a building. The fan blades have been balanced to minimize vibration transmitted to the structure when the fan is operating. One day the fan is operating at 600 rpm when one of the blade tips breaks off (a 2oz piece), causing the fan to be unbalanced. The fan is immediately turned off and allowed to coast to a stop. We need to know the vertical amplitude of vibration produced by the imbalance at the end of the beam for several rotational speeds during coastdown. The beam is modeled as a linear spring. The mathematical model is shown in figure 3.184. The spring stiffness is 1200 LB/in; the damper constant is 0.6 LB

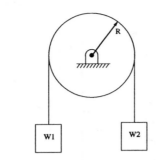

Figure 3.181:

Figure 3.182:

sec/in. Determine the amplitude of the steady state vibration at 600, 500, 400, all the way to 0 rpm. Plot this amplitude versus speed in rpm. What observations can be made about the vibration behavior?

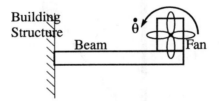

Figure 3.183:

91. * A light rod of length $l$ with a concentrated end mass M, is welded to a vertical shaft turning at constant speed $\omega$ as figure 3.185 shows. Find the force and moment exerted by the vertical rod onto the angled shaft. Include the effect of gravity.

92. * The cylindrical object shown in figure 3.186 is originally resting on a stationary conveyor belt. The cylinder has a centroidal radius of gyration of $k_o$. The large radius of the cylinder is R, the small radius is r. The cylinder has a rope wound around its middle like a yo-yo. At time 0, the conveyor begins accelerating to the left at a known constant amount a. At the same instant a person pulls on the cord with a known force T. The person pulling the cord moves with the yo-yo so the cord is always vertical. The belt has a coefficient of friction of $\mu_s$ and $\mu_k$. All masses, distances and radii of gyrations are known. Determine the angular acceleration of the yo-yo. You may assume the yo-yo remains in contact with the conveyor. If you make any other assumptions, you must test them.

93. * Figure 3.187 shows a slider crank. Both members are uniform rods and have mass. Gravity points toward the bottom of the page. The crank (body 1) rotates at a known constant speed. Determine the forces present in both pin joints at the instant $\theta$ becomes zero. Express your answer using only the following quantities, masses and lengths of the rods, gravity constant and speeds.

94. * Figure 3.188 shows two cylindrical objects. They both have known mass, radii and centroidal radii of gyration. The coefficient of friction (static and kinetic) between the disks is also known.

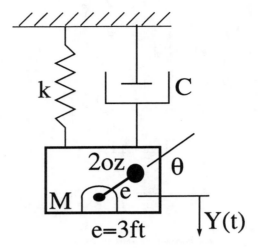

Figure 3.184:

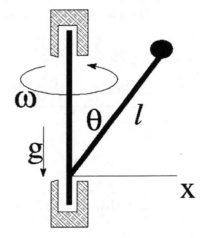

Figure 3.185:

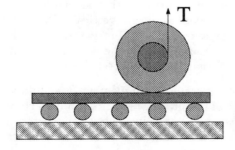

Figure 3.186:

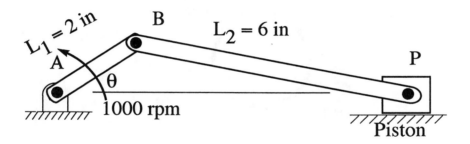

Figure 3.187:

At first, the left body (body 1) is rotating clockwise with a known speed and the right object (body 2) is stationary. Disk 1 is then slowly moved toward disk 2 until they make contact. Once they are in contact, neither center moves. Disk 1 slows down and disk 2 speeds up until eventually slipping stops and the disks continue at constant spin. Determine the angular velocity of each disk after slipping stops.

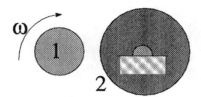

Figure 3.188:

95. * Figure 3.189 shows a uniform slender rod with length of L, initially resting on the floor with $\theta = 45$ degrees. The static coefficient of friction between the rod and floor is 0.1. The rod has mass 1 kg. If the string is suddenly cut, determine whether the rod begins to slip immediately after the cut.

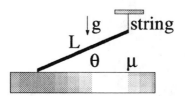

Figure 3.189:

96. * Figure 3.190 shows a 2 kg hoop rolling on the ground. A force of F=10 N is applied to the top of the hoop as shown. Determine the minimum coefficient of static friction required to keep the hoop from sliding. The centroidal radius of gyration of the hoop is r.

97. For figure 3.191 the angle $\theta$ is equal to $\pi/4$ radians. Determine the acceleration of the box when $\theta = 45$ degrees (that's $\pi/4$ radians). The mass is 1 kg, $\mu_s$ and $\mu_k = 0.1$. At time = 0, the mass is located 1 meter from the pivot.

**Energy**

98. Body A in figure 3.192 has a mass of 10 [kg]. The spring constant of the spring acting on body A is 150 [N/m]. At t=0, the system is in equilibrium, and the tension in the rope is zero.

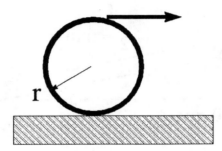

Figure 3.190:

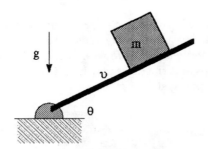

Figure 3.191:

Then, at t=0$^+$ the drum starts to rotate and produces a constant tension in the rope of 220 [N]. Determine the velocity of $A$ after it has moved 1.5 [m] up the smooth slot.

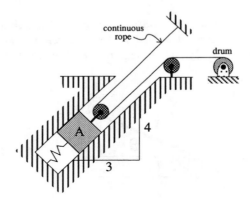

Figure 3.192:

99. * Body A in figure 3.193 has a mass of 20 [lbm]. The spring constant of the spring acting on body A is 100 [lbf/ft]. At t=0, the system is in equilibrium, and the tension in the rope is zero. Then, at t=0$^+$ the drum starts to rotate and produces a constant tension in the rope of 50 [N]. Determine the velocity of $A$ after it has moved 5 [ft] up the smooth slot.

100. * Bodies A and B in figure 3.194 have a mass of 15 [kg] and 7 [kg] respectively. The spring constant of the spring acting on body A is 8.0 [N/m]. The plane is frictionless. In the position shown, body $A$ is at rest, and the tension in the spring is 60 [N]. Determine the following

   (a) the velocity of A as it passes under the small peg at C

   (b) the tension in the cord as A passes under C.

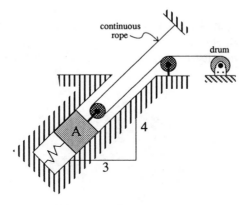

Figure 3.193:

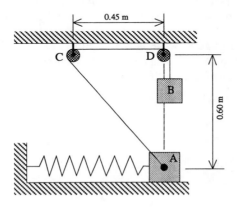

Figure 3.194:

101. Bodies A and B in figure 3.194 have a mass of 20 [lbm] and 16 [lbm] respectively. The spring constant of the spring acting on body A is 12 [lbf/ft]. The plane is frictionless. In the position shown, body A is at rest, and the tension in the spring is 20 [lbf]. Determine the following

    (a) the velocity of A as it passes under the small peg at C

    (b) the tension in the cord as A passes under C.

102. The solid square block in figure 3.195 is initially balanced as shown. It gets bumped ever so slightly and rotates clockwise from its rest position until it hits its supporting surface. Determine the angular velocity of the block as its edge $OC$ becomes horizontal (just before

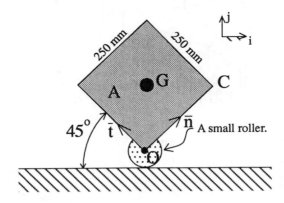

Figure 3.195:

the block strikes the supporting surface). Let $m = 20$ [kg], $g = 9.81$ [m/s], $|OC|=0.50$ [m], $\frac{|OC|}{2}=0.25$ [m], $I_G = \frac{2}{6}m|OC|^2$ [kg m$^2$].

103. We throw a ball horizontally against the floor in figure 3.196. If the floor is frictionless, determine the maximum possible rebound height. What conditions would cause the ball to rise less than the maximum height that we calculated?

104. * How fast are the rods (see figure 3.197) rotating just before it strikes the floor. It begins from a VERY SMALL clockwise rotation from a vertical position.

105. * The 10-kg cart shown in figure 3.198 is being pulled to the left by a constant force $P = 10$ N applied to the end of the cord. If the cart starts from rest when $x = 8$ m, calculate and plot the speed of the cart as a function of its position x between $-3 < x < 8$.

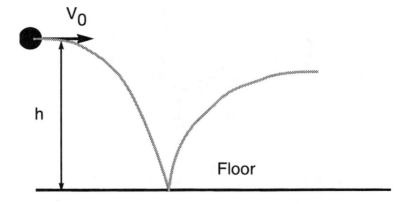

Figure 3.196:

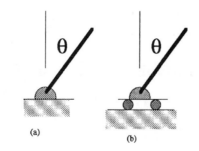

Figure 3.197:

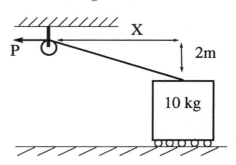

Figure 3.198:

## 3.15 Maple Programs

This section includes the Maple files referenced in the chapter.

### 3.15.1 File: stick.ms - Maple program for Example 3.6.

Here are the parameters.

```
>  restart;
>  Degrees := Pi / 180:
>  g:=9.81:m1:=5:m2:=2:theta:=15 * Degrees:mus1:=0.1:mus2:=0.3:
```

Here are the linear momentum equations.

```
>  e1 := -f2 - m2*g*sin(theta) = m2*diff(diff(x2(t),t),t);
```

$$e1 := -f2 - 4.905000000\sqrt{6}\,(1 - \frac{1}{3}\sqrt{3}) = 2\,(\frac{\partial^2}{\partial t^2}\,\text{x2}(t))$$

```
>  e2 := - m2*g*cos(theta)+N2 = 0;
```

$$e2 := -4.905000000\sqrt{6}\,(1 + \frac{1}{3}\sqrt{3}) + N2 = 0$$

```
>  e3 := -m2*g*sin(theta)-m1*g*sin(theta) + f1 = (m1+m2)*diff(diff(x1(t),t),t);
```

$$e3 := -17.16750000\sqrt{6}\,(1 - \frac{1}{3}\sqrt{3}) + f1 = 7\,(\frac{\partial^2}{\partial t^2}\,\text{x1}(t))$$

```
>  e4 := -m2*g*cos(theta)-m1*g*cos(theta)+N1 = 0;
```

$$e4 := -17.16750000\sqrt{6}\,(1 + \frac{1}{3}\sqrt{3}) + N1 = 0$$

Here are the maximum friction forces.

```
>  f2max := abs(mus2*N2);f1max := abs(mus1*N1);
```

$$f2max := .3\,|N2|$$

$$f1max := .1\,|N1|$$

Now set positions to constants and solve for N1,N2, f1,f2.

```
>  ans := simplify(solve(subs({x1(t)=0,x2(t)=0},{e1,e2,e3,e4}),{N1,N2,f1,f2 }));
```

$$ans := \{N2 = 18.95146471,\ N1 = 66.33012649,\ f2 = -5.078029665,\ f1 = 17.77310383\}$$

```
> check := simplify(subs(ans,[f2max,f1max]));
```
$$check := [5.685439413, 6.633012649]$$

## 3.15.2    File: botslide.ms - Maple Program for Example 3.6.

Here are the parameters.
```
> Degrees := Pi / 180:
> g:=9.81:m1:=5:m2:=2:theta:=15 *Degrees:
> mus1:=0.1:mus2:=0.3:muk1:=0.9*mus1:muk2:=0.9* mus2:
```
Here are the linear momentum equations.
The equations are the same as for the sticking case.
```
> e1 := -f2 - m2*g*sin(theta) = m2*diff(diff(x2(t),t),t);
```
$$e1 := -f2 - 4.905000000\ \sqrt{6}\ (1 - \frac{1}{3}\ \sqrt{3}) = 2\,(\frac{\partial^2}{\partial t^2}\ x2(t))$$
```
> e2 := - m2*g*cos(theta)+N2 = 0;
```
$$e2 := -4.905000000\ \sqrt{6}\ (1 + \frac{1}{3}\ \sqrt{3}) + N2 = 0$$
```
> e3 := -m2*g*sin(theta)-m1*g*sin(theta) + f1 = (m1+m2)*diff(diff(x1(t),t),t);
```
$$e3 := -17.16750000\ \sqrt{6}\ (1 - \frac{1}{3}\ \sqrt{3}) + f1 = 7\,(\frac{\partial^2}{\partial t^2}\ x1(t))$$
```
> e4 := -m2*g*cos(theta)-m1*g*cos(theta)+N1 = 0;
```
$$e4 := -17.16750000\ \sqrt{6}\ (1 + \frac{1}{3}\ \sqrt{3}) + N1 = 0$$
Here are the maximum friction forces.
```
> f2max := abs(mus2*N2);f1max := abs(mus1*N1);
```
$$f2max := .3\ |N2|$$
$$f1max := .1\ |N1|$$
Here is the difference from before. Set positions to be equal, set f1 to kinetic and solve for N1,N2,f2 and acceleration.
```
> ans := simplify(solve(subs({x1(t)=x2(t),f1=muk1*N1},{e1,e2,e3,e4}),
> {N1,N2,f2,diff(diff(x2(t),t),t)}));
```

$$ans :=$$
$$\{N2 = 18.95146471,\ f2 = -1.705631824,\ \frac{\partial^2}{\partial t^2}\ x2(t) = -1.686198920,\ N1 = 66.33012649\}$$
```
> check := simplify(subs(ans,f2max));
```
$$check := 5.685439413$$

## 3.15.3    File: springd.ms - Maple Program for Example 3.7.

Here is the equation for motion of the mass.
```
> restart;
> eq:= m*diff(diff(y(t),t),t)+ c*diff(y(t),t)+k*y(t)=m*g+k*Lo;
```
$$eq := m\,(\frac{\partial^2}{\partial t^2}\ y(t)) + c\,(\frac{\partial}{\partial t}\ y(t)) + k\,y(t) = m\,g + k\,Lo$$
Solve it assuming it is dropped from the ceiling from rest.
```
> ans := simplify(dsolve({eq,y(0)=0,D(y)(0)=0},y(t)));
```

$$ans := y(t) = -\frac{1}{2}(-2\ \sqrt{c^2 - 4\,m\,k}\ m\,g - 2\ \sqrt{c^2 - 4\,m\,k}\,k\,Lo + \%2\ \sqrt{c^2 - 4\,m\,k}\ m\,g$$
$$+ \%2\ \sqrt{c^2 - 4\,m\,k}\,k\,Lo + \%2\,c\,m\,g + \%2\,c\,k\,Lo - \%1\,c\,m\,g - \%1\,c\,k\,Lo$$
$$+ \%1\ \sqrt{c^2 - 4\,m\,k}\,m\,g + \%1\ \sqrt{c^2 - 4\,m\,k}\,k\,Lo)/(\sqrt{c^2 - 4\,m\,k}\,k)$$
$$\%1 := e^{(-1/2\ \frac{(c + \sqrt{c^2 - 4\,m\,k})\,t}{m})}$$
$$\%2 := e^{(-1/2\ \frac{(c - \sqrt{c^2 - 4\,m\,k})\,t}{m})}$$

Pick some parameters and plot the response. Notice that without damping (c=0), the mass oscillate:

```
>   params := {m=1,g=981/100,c=0,k=1,Lo=2}:
>   pans := simplify(subs(params,ans));
>   plot(25 - rhs(pans),t=0..4*Pi);
```

$$pans := y(t) = -\frac{1181}{100}\cos(t) + \frac{1181}{100}$$

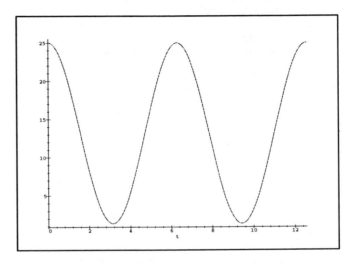

Notice that the mass never stops, there is nothing to remove energy from the mass so it goes forever. Of course this is impractical, friction or something will convert motion into thermal energy if nothing else and it will eventually stop.

Add a little damping (small c) and the system settles where the spring force balances the weight. Notice that the system still oscillates. This is called underdamped. (c is small)

```
>   params := {m=1,g=981/100,c=1/2,k=1,Lo=2}:
>   pans := simplify(evalf(subs(params,ans)));
```

$$pans := y(t) = (-5.904999999 + 1.524664444\,I)(e^{((-.2500000000 + .9682458365\,I)\,t)}$$
$$+ .4841229185\,I\,e^{((-.2500000000 - .9682458365\,I)\,t)} + .8750000003\,e^{((-.2500000000 - .9682458365\,I)\,t)}$$
$$- 1.875000001 - .4841229185\,I)$$

Note that maple uses complex numbers to find the solution. We take the real part of the solution and ignore the imaginary part.

```
>   plot(25 - Re(rhs(pans)),t=0..4*Pi);
```

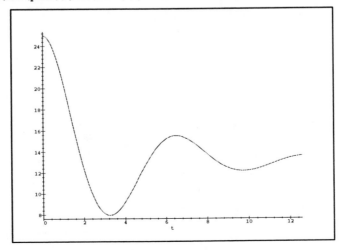

   When the radical becomes zero, there are repeated roots in the transient solution. This condition is called critical damping. Note that the system no longer oscillates. This occurs at a special value of c.

> ```
  params := {m=1,g=981/100,c=2,k=1,Lo=2}:
  ```

> ```
  pans := simplify(dsolve({subs(params,eq),y(0)=0,D(y)(0)=0},y(t)));
  ```

$$pans := y(t) = -\frac{1181}{100}\left(-e^t + 1 + t\right)e^{(-t)}$$

> ```
  plot(25 - Re(rhs(pans)),t=0..4*Pi);
  ```

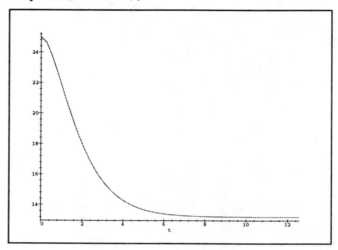

   When the radical becomes positive, there are two distinct roots. This condition is called overdamped. The system begins to look like a first order system.

> ```
  params := {m=1,g=981/100,c=3,k=1,Lo=2}:
  ```

> ```
  pans := simplify(subs(params,ans));
  ```

> ```
  plot(25 - Re(rhs(pans)),t=0..4*Pi);
  ```

$$pans := y(t) = -\frac{1181}{1000}$$
$$\frac{\left(-2\sqrt{5} + e^{(1/2\,(-3+\sqrt{5})\,t)}\,\sqrt{5} + 3\,e^{(1/2\,(-3+\sqrt{5})\,t)} - 3\,e^{(-1/2\,(3+\sqrt{5})\,t)} + e^{(-1/2\,(3+\sqrt{5})\,t)}\,\sqrt{5}\right)}{\sqrt{5}}$$

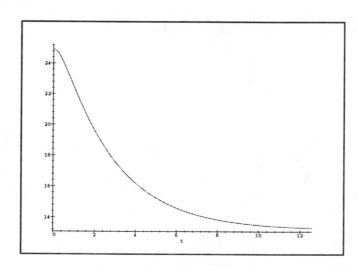

### 3.15.4   File: padball.ms - The Maple Program for Example 3.8.

```
> restart:with(plots):
```

Here is the equation for motion of the mass.

```
> z(t) := sin(omega*t):  eq:= m*diff(diff(y(t),t),t)+k*y(t)=k*z(t) - m*g - k*Lo;
```

$$eq := m\left(\frac{\partial^2}{\partial t^2}\,y(t)\right) + k\,y(t) = k\sin(\omega\,t) - m\,g - k\,Lo$$

Solve it assuming it is initially at rest in static equilibrium.

```
> ans := simplify(dsolve({eq,y(0)=-Lo-m/k*g,D(y)(0)=0},y(t)));
```

$$ans := y(t) = \frac{1}{2}(2\,k^2\sin(\omega\,t)\,\sqrt{-m\,k}\,e^{(\frac{\sqrt{-m\,k}\,t}{m})} - k^2\,\omega\,m\,e^{(2\,\frac{\sqrt{-m\,k}\,t}{m})}$$
$$- 2\,k^2\,Lo\,\sqrt{-m\,k}\,e^{(\frac{\sqrt{-m\,k}\,t}{m})} + k^2\,\omega\,m - 2\,m\,g\,k\,\sqrt{-m\,k}\,e^{(\frac{\sqrt{-m\,k}\,t}{m})}$$
$$+ 2\,m\,k\,Lo\,\omega^2\,\sqrt{-m\,k}\,e^{(\frac{\sqrt{-m\,k}\,t}{m})} + 2\,m^2\,g\,\omega^2\,\sqrt{-m\,k}\,e^{(\frac{\sqrt{-m\,k}\,t}{m})})e^{(-\frac{\sqrt{-m\,k}\,t}{m})}\Bigg/(k\,\sqrt{-m\,k}$$
$$(k - \omega^2\,m))$$

Pick some parameters.

```
> params := {m=1,g=981/100,k=1,Lo=2}:  pans := simplify(subs(params,ans));
```

$$pans := y(t) = -\frac{1}{100}\frac{-100\,\omega\sin(t) + 100\sin(\omega\,t) - 1181 + 1181\,\omega^2}{-1 + \omega^2}$$

Notice from the plot the only significant motion occurs near the frequency omega = 1.

```
> plot3d(rhs(pans),t=0..8*Pi,omega=0..2);
```

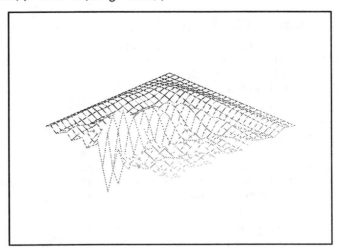

This is what it looks like when forcing it at the natural frequency.

```
> natural:=limit(rhs(pans),omega=1); out:=plot(natural,t=0..10*Pi):input:=plot(sin(t), t=0..10*Pi
display({out,input});
```

$$natural := \frac{1}{2}\sin(t) - \frac{1181}{100} - \frac{1}{2}\cos(t)\,t$$

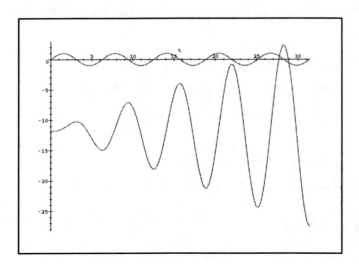

Now notice what happens at low frequency. The sine terms are vastly different frequencies and they are both small amplitude. There simply is little energy in the system.

> subs(omega=1/100,pans);

$$y(t) = -\frac{100}{9999}\sin(t) + \frac{10000}{9999}\sin(\frac{1}{100}t) - \frac{1181}{100}$$

Here is what happens at high frequency. Notice that as omega increases, the sine magnitudes diminish.

> subs(omega=2,pans);subs(omega=3,pans);subs(om ega=10,pans);

$$y(t) = \frac{2}{3}\sin(t) - \frac{1}{3}\sin(2\,t) - \frac{1181}{100}$$

$$y(t) = \frac{3}{8}\sin(t) - \frac{1}{8}\sin(3\,t) - \frac{1181}{100}$$

$$y(t) = \frac{10}{99}\sin(t) - \frac{1}{99}\sin(10\,t) - \frac{1181}{100}$$

### 3.15.5   File: bellcms.ms - The Maple Program for Example 3.24.

The following uses some linear algebra functions so we can define
a vector as [x,y]. Really the only function we need is the evalm
function that allows us to add [x,y] + [z,w] = [x+z y+w].
The polar definition allows us to easily specify vectors.

> with(linalg):  polar:=(r,theta)->[r*cos(theta),r*sin(theta)]:

The following allows easier typing.

> x:=X(t):h:=H(t):the:=theta(t):y:=Y(t):l:=L(t)  :

The following allows us to refer to the derivatives as xdot xddot etc.

> pd:={diff(x,t)=xdot, diff(h,t)=hdot, diff(the,t)=thetadot,
> diff(y,t)=ydot, diff(l,t)=ldot, diff(x,t,t)=xddot, diff(h,t,t)=hddot,
> diff(the,t,t)=thetaddot, diff(y,t,t)=yddot, diff(l,t,t)=lddot}:

Define a few vectors, the last two are constant.

> n1:=polar(1,Pi/2+the);rgo:=[rgox,rgoy];rof:=[ rofx,rofy];

$$n1 := [-\sin(\theta(t)),\ \cos(\theta(t))]$$

$$rgo := [rgox,\ rgoy]$$

$$rof := [rofx,\ rofy]$$

This is the first vector loop since Maple defaults to setting an
equation to 0, we need not include it.

> e1:=[x,0]-[0,h]-6*n1-rgo;v1:=diff(e1,t);a1:=d iff(v1,t);

$$e1 := [-rgox + 6\sin(\theta(t)) + X(t),\ -rgoy - 6\cos(\theta(t)) - H(t)]$$

$$v1 := [6\cos(\theta(t))\,(\frac{\partial}{\partial t}\theta(t)) + (\frac{\partial}{\partial t}X(t)),\ 6\sin(\theta(t))\,(\frac{\partial}{\partial t}\theta(t)) - (\frac{\partial}{\partial t}H(t))]$$

$$a1 := [-6\sin(\theta(t))\,(\frac{\partial}{\partial t}\theta(t))^2 + 6\cos(\theta(t))\,(\frac{\partial^2}{\partial t^2}\theta(t)) + (\frac{\partial^2}{\partial t^2}X(t)),$$

$$6\cos(\theta(t))\,(\frac{\partial}{\partial t}\theta(t))^2 + 6\sin(\theta(t))\,(\frac{\partial^2}{\partial t^2}\theta(t)) - (\frac{\partial^2}{\partial t^2}H(t))]$$

This is the second vector loop.

```
> e2:=rof-[0,y]-polar(l,the); v2:=diff(e2,t); a2:=diff(v2,t);
```

$$e2 := [-L(t)\cos(\theta(t)) + rofx, \; -L(t)\sin(\theta(t)) - Y(t) + rofy]$$

$$v2 := [-(\frac{\partial}{\partial t}L(t))\cos(\theta(t)) + L(t)\sin(\theta(t))\,(\frac{\partial}{\partial t}\theta(t)),$$

$$-(\frac{\partial}{\partial t}L(t))\sin(\theta(t)) - L(t)\cos(\theta(t))\,(\frac{\partial}{\partial t}\theta(t)) - (\frac{\partial}{\partial t}Y(t))]$$

$$a2 := [-(\frac{\partial^2}{\partial t^2}L(t))\cos(\theta(t)) + 2\,(\frac{\partial}{\partial t}L(t))\sin(\theta(t))\,(\frac{\partial}{\partial t}\theta(t)) + L(t)\cos(\theta(t))\,(\frac{\partial}{\partial t}\theta(t))^2$$

$$+ L(t)\sin(\theta(t))\,(\frac{\partial^2}{\partial t^2}\theta(t)), -(\frac{\partial^2}{\partial t^2}L(t))\sin(\theta(t)) - 2\,(\frac{\partial}{\partial t}L(t))\cos(\theta(t))\,(\frac{\partial}{\partial t}\theta(t))$$

$$+ L(t)\sin(\theta(t))\,(\frac{\partial}{\partial t}\theta(t))^2 - L(t)\cos(\theta(t))\,(\frac{\partial^2}{\partial t^2}\theta(t)) - (\frac{\partial^2}{\partial t^2}Y(t))]$$

This substitutes in our dot terminology just to make it look
better. It also add the vectors together using the evalm function.

```
> sv1:=subs(pd,evalm(v1));sa1:=subs(pd,evalm(a1 ));
```

$$sv1 := [6\cos(\theta(t))\,thetadot + xdot, \; 6\sin(\theta(t))\,thetadot - hdot]$$

$$sa1 := [-6\sin(\theta(t))\,thetadot^2 + 6\cos(\theta(t))\,thetaddot + xddot,$$

$$6\cos(\theta(t))\,thetadot^2 + 6\sin(\theta(t))\,thetaddot - hddot]$$

```
> sv2:=subs(pd,evalm(v2));sa2:=subs(pd,evalm(a2 ));
```

$$sv2 :=$$
$$[-ldot\cos(\theta(t)) + L(t)\sin(\theta(t))\,thetadot, \; -ldot\sin(\theta(t)) - L(t)\cos(\theta(t))\,thetadot - ydot]$$

$$sa2 := [-lddot\cos(\theta(t)) + 2\,ldot\sin(\theta(t))\,thetadot + L(t)\cos(\theta(t))\,thetadot^2$$
$$+ L(t)\sin(\theta(t))\,thetaddot, \; -lddot\sin(\theta(t)) - 2\,ldot\cos(\theta(t))\,thetadot$$
$$+ L(t)\sin(\theta(t))\,thetadot^2 - L(t)\cos(\theta(t))\,thetaddot - yddot]$$

Now solve the four scalar velocity equations for the
four unknown dot terms.

```
> vans:=simplify(solve({sv1[1],sv1[2],sv2[1],s v2[2]},
> {thetadot,hdot,ldot,ydot}));
```

$$vans := \{ldot = -\frac{1}{6}\frac{L(t)\sin(\theta(t))\,xdot}{\cos(\theta(t))^2}, \; thetadot = -\frac{1}{6}\frac{xdot}{\cos(\theta(t))}, \; ydot = \frac{1}{6}\frac{L(t)\,xdot}{\cos(\theta(t))^2},$$

$$hdot = -\frac{\sin(\theta(t))\,xdot}{\cos(\theta(t))}\}$$

Now solve the four scalar acceleration equations for the four
unknown ddot terms.

```
> aans:=subs(xddot=0,simplify(
> solve({sa1[1],sa1[2],sa2[1],sa2[2]},
> {thetaddot,hddot,lddot,yddot})));
```

$$aans := \{thetaddot = \frac{\sin(\theta(t))\,thetadot^2}{\cos(\theta(t))},$$

$$lddot = \frac{1}{6}\frac{6\,L(t)\,thetadot^2 + 12\,ldot\sin(\theta(t))\,thetadot\cos(\theta(t))}{\cos(\theta(t))^2}, \; hddot = 6\frac{thetadot^2}{\cos(\theta(t))},$$

$$yddot = -\frac{1}{6}\frac{12\,ldot\cos(\theta(t))\,thetadot + 6\,L(t)\sin(\theta(t))\,thetadot^2}{\cos(\theta(t))^2}\}$$

### 3.15.6   File: linkslot.ms - The Maple Program for Example 3.25.

This is file LinkSlot.mws

```
> restart;with(linalg):
> i := vector([1,0,0]): j:=vector([0,1,0]): k:= vector([0,0,1]):
> dot := (VectorOne,VectorTwo) -> dotprod(VectorOne,VectorTwo,'orthogonal'):
> cross := (a,b) -> crossprod(a,b):
> VectorEqn := (VectorEquation,VectorDirection) -> dot(lhs(VectorEquation),VectorDirection)
= dot(rhs(VectorEquation),VectorDirection):
```

Warning, new definition for norm

Warning, new definition for trace

The following allows me to type th instead of theta(t) all the time. Similarly for the other variables which are functions of time. This program was written after the problem was formulated so we know what variables are important.

```
> th:=theta(t):  H := h(t):  b:= beta(t):
```

The following are short cuts for typing cos and sin.

```
> c := (a) -> cos(a):  s := (a) -> sin(a):
```

The following are definitions for unit vectors which make typing a little easier. The names should be obvious, alongslot points from C to A along the slot, etc. The cross product is used in the definition of some to avoid stupid mistakes like an incorrect negative sign or sine cosine mixup. You could type them in manually if you want to. Not all of these unit vectors are used but what the heck. One of them is printed just so you can see what they look like.

```
> alongslot := c(th)*i + s(th)*j:  perpslot := cross(k,alongslot):
> alongrod := -c(b)*i + s(b)*j:  perprod := cross(k,alongrod);
```
$$perprod := [-\sin(\beta(t)), -\cos(\beta(t)), 0]$$

The following are positions and accelerations of D. I don't print the acceleration because it is ugly. Notice the [1,0,0] is the i unit vector, the [0,1,0] is the j. Since I formulated the problem before writing the program, I know I need the acceleration of D and the vector loop already.

```
> posd := d*alongslot; acceld := diff(posd,t,t):
```
$$posd := d\left(\cos(\theta(t))\,i + \sin(\theta(t))\,j\right)$$

Here is the vector loop. Notice I use the unit vectors defined above to avoid typing mistakes.

```
> loop := H*alongslot - L*alongrod - w*i = 0;
```
$$loop := \mathrm{h}(t)\left(\cos(\theta(t))\,i + \sin(\theta(t))\,j\right) - L\left(-\cos(\beta(t))\,i + \sin(\beta(t))\,j\right) - w\,i = 0$$

Here is what I know about the angle beta.

```
> param := {diff(b,t)=omega, diff(b,t,t)=0}:
```

Note I take derivatives of the loop and stick in the values of Beta derivatives. I print the first derivative for your reference but the second derivative is real ugly.

```
> vloop := subs(param,diff(loop,t));
```

$$vloop := (\frac{\partial}{\partial t}\,\mathrm{h}(t))\left(\cos(\theta(t))\,i + \sin(\theta(t))\,j\right)$$
$$+ \mathrm{h}(t)\left(-\sin(\theta(t))\,(\frac{\partial}{\partial t}\,\theta(t))\,i + \cos(\theta(t))\,(\frac{\partial}{\partial t}\,\theta(t))\,j\right) - L\left(\sin(\beta(t))\,\omega\,i + \cos(\beta(t))\,\omega\,j\right) = 0$$

```
> aloop := subs(param,diff(loop,t,t)):
```

When you try to solve the equations by hand you should look for a set which is simple. With maple this is not really necessary but I'll do it here just for reference. You can play with these if you want but you will find the i and j components are about as nice as you'll find. Here I compare i and j to one component along the slot. You can see how much nicer the i and j equations look. Remember the function VectorEqn(vloop,i) sets the i component on the left side of equation vloop equal to the i component on the right side of vloop.

```
> VectorEqn(vloop,i);
```

$$(\frac{\partial}{\partial t}\,\mathrm{h}(t))\cos(\theta(t)) - \mathrm{h}(t)\sin(\theta(t))\,(\frac{\partial}{\partial t}\,\theta(t)) - L\sin(\beta(t))\,\omega = 0$$

```
>   VectorEqn(vloop,j);
```

$$(\frac{\partial}{\partial t}\,\mathrm{h}(t))\sin(\theta(t)) + \mathrm{h}(t)\cos(\theta(t))\,(\frac{\partial}{\partial t}\,\theta(t)) - L\cos(\beta(t))\,\omega = 0$$

```
>   VectorEqn(vloop,alongslot);
```

$$((\frac{\partial}{\partial t}\,\mathrm{h}(t))\cos(\theta(t)) - \mathrm{h}(t)\sin(\theta(t))\,(\frac{\partial}{\partial t}\,\theta(t)) - L\sin(\beta(t))\,\omega)\cos(\theta(t))$$

$$+ ((\frac{\partial}{\partial t}\,\mathrm{h}(t))\sin(\theta(t)) + \mathrm{h}(t)\cos(\theta(t))\,(\frac{\partial}{\partial t}\,\theta(t)) - L\cos(\beta(t))\,\omega)\sin(\theta(t)) = 0$$

Now solve the i and j components of the first derivative of the loop for the first derivative terms.

```
>   vsol := solve({VectorEqn(vloop,i),VectorEqn(vloop,j)},
>   {diff(th,t), diff(H,t)});
```

$$vsol := \{\frac{\partial}{\partial t}\,\mathrm{h}(t) = L\,\omega\,(\cos(\beta(t))\sin(\theta(t)) + \cos(\theta(t))\sin(\beta(t))),$$

$$\frac{\partial}{\partial t}\,\theta(t) = -\frac{L\,\omega\,(-\cos(\theta(t))\cos(\beta(t)) + \sin(\beta(t))\sin(\theta(t)))}{\mathrm{h}(t)}\}$$

Aha! It looks like these can be simplified a little.

```
>   vsol := simplify(vsol,trig);
```

$$vsol := \{\frac{\partial}{\partial t}\,\mathrm{h}(t) = L\,\omega\cos(\beta(t))\sin(\theta(t)) + L\,\omega\cos(\theta(t))\sin(\beta(t)),$$

$$\frac{\partial}{\partial t}\,\theta(t) = -\frac{L\,\omega\,(-\cos(\theta(t))\cos(\beta(t)) + \sin(\beta(t))\sin(\theta(t)))}{\mathrm{h}(t)}\}$$

Solve the same components of the second derivative of the loop for the unknown second derivative terms.

```
>   asol := solve({VectorEqn(aloop,i),VectorEqn(aloop,j)},
>   {diff(th,t,t), diff(H,t,t)});
```

$$asol := \left\{ \frac{\partial^2}{\partial t^2}\,\mathrm{h}(t) = \mathrm{h}(t)\,(\frac{\partial}{\partial t}\,\theta(t))^2 - L\sin(\beta(t))\,\omega^2\sin(\theta(t)) + \cos(\theta(t))\,L\cos(\beta(t))\,\omega^2, \right.$$

$$\left. \frac{\partial^2}{\partial t^2}\,\theta(t) = -\frac{2\,(\frac{\partial}{\partial t}\,\mathrm{h}(t))\,(\frac{\partial}{\partial t}\,\theta(t)) + \cos(\theta(t))\,L\sin(\beta(t))\,\omega^2 + L\cos(\beta(t))\,\omega^2\sin(\theta(t))}{\mathrm{h}(t)} \right\}$$

Again simplify a little.

```
>   asol := simplify(asol,trig);
```

$$asol := \left\{ \frac{\partial^2}{\partial t^2}\,\mathrm{h}(t) = \mathrm{h}(t)\,(\frac{\partial}{\partial t}\,\theta(t))^2 - L\sin(\beta(t))\,\omega^2\sin(\theta(t)) + \cos(\theta(t))\,L\cos(\beta(t))\,\omega^2, \right.$$

$$\left. \frac{\partial^2}{\partial t^2}\,\theta(t) = -\frac{2\,(\frac{\partial}{\partial t}\,\mathrm{h}(t))\,(\frac{\partial}{\partial t}\,\theta(t)) + \cos(\theta(t))\,L\sin(\beta(t))\,\omega^2 + L\cos(\beta(t))\,\omega^2\sin(\theta(t))}{\mathrm{h}(t)} \right\}$$

The real difficulty usually comes from solving the positions. First try to find the cleanest looking ones. i and j are the best I could find.

```
>   VectorEqn(loop,i); VectorEqn(loop,j);
```

$$\mathrm{h}(t)\cos(\theta(t)) + L\cos(\beta(t)) - w = 0$$

$$\mathrm{h}(t)\sin(\theta(t)) - L\sin(\beta(t)) = 0$$

Unfortunately these are nonlinear so they can be hard to solve. Let's just try something simple. This part can be very tough in some problems. This time its not too bad.

```
>   possol := solve({VectorEqn(loop,i), VectorEqn(loop,j)},{H,th});
```

$$possol := \{\theta(t) = \arctan(\%1\sin(\beta(t))\,L,\, -\%1\,(L\cos(\beta(t)) - w)),\; \mathrm{h}(t) = \frac{1}{\%1}\}$$

$$\%1 := \mathrm{RootOf}((L^2 - 2\,L\cos(\beta(t))\,w + w^2)\,\_Z^2 - 1)$$

Notice the "general" solution maple finds for h(t). Note the denominator. If it is ever 0 I will have difficulty using this equation. When will it be zero? When beta is zero and 180 degrees. Note that this happens when the triangle shaped vector loop collapses into a single line. This is called a singularity and they can be very nasty, especially when you don't know they exist and are not expecting them. How did I find it? Well I asked maple for the solution at beta = 0 (not shown, I was fiddling with the equations as I wrote this program) and maple complained about dividing by zero. When this happened I looked at the equations for a little while and found this little difficulty. Should I have expected it? Yes, but though I didn't, I easily discovered it when I computed a couple of numbers. Notice that when the singularity does occur, it isn't difficult to find the value of theta and h its just that the equations I derived won't do it. We have to treat it as a special case or else avoid it.

Now to continue, shove all of these solutions (theta, h and their derivatives) into the expression for mass times acceleration of point D. I should put in the possol too but if I did it would look really ugly, so I leave it out for right now.

```
> ans := mass * simplify(subs(vsol,subs(asol,acceld)));
```

$$
\begin{aligned}
ans := &-mass\, d\, L\, \omega^2 (6\cos(\theta(t))^3\, L\, i \cos(\beta(t))^2 - L\, j \sin(\theta(t)) \cos(\beta(t))^2 \\
&- 3\cos(\theta(t))^2\, j\, L \sin(\theta(t)) - 4\, L\, j \cos(\theta(t)) \cos(\beta(t)) \sin(\beta(t)) - i \cos(\beta(t))\, h(t) \\
&+ i \cos(\beta(t))\, h(t) \cos(\theta(t))^2 + 2\, L\, i \cos(\beta(t)) \sin(\beta(t)) \sin(\theta(t)) + 3\cos(\theta(t))\, L\, i \\
&- 3\, L \cos(\theta(t))^3\, i - 5\, L \cos(\theta(t))\, i \cos(\beta(t))^2 - 6\cos(\theta(t))^2\, L\, i \cos(\beta(t)) \sin(\beta(t)) \sin(\theta(t)) \\
&- \sin(\theta(t))\, i \cos(\theta(t)) \sin(\beta(t))\, h(t) + 6 \sin(\theta(t))\, L\, j \cos(\theta(t))^2 \cos(\beta(t))^2 \\
&+ 6 \cos(\theta(t))^3\, j\, L \sin(\beta(t)) \cos(\beta(t)) + \cos(\theta(t))^2\, j \sin(\beta(t))\, h(t) \\
&+ \cos(\theta(t))\, j \cos(\beta(t)) \sin(\theta(t))\, h(t) + \sin(\theta(t))\, L\, j) \Big/ h(t)^2
\end{aligned}
$$

You can determine whatever component of force you want here by dotting the ans with any unit vector direction you want. Just for kicks, I find the ones parallel and perp. to the slot.

```
> force1 := dot(ans,alongslot);
```

$$
\begin{aligned}
force1 := &-mass\, d\, L\, \omega^2 (6\cos(\theta(t))^3\, L \cos(\beta(t))^2 - \cos(\beta(t))\, h(t) + \cos(\beta(t))\, h(t) \cos(\theta(t))^2 \\
&+ 2\, L \cos(\beta(t)) \sin(\beta(t)) \sin(\theta(t)) + 3\cos(\theta(t))\, L - 3\, L\cos(\theta(t))^3 - 5\, L \cos(\theta(t)) \cos(\beta(t))^2 \\
&- 6\cos(\theta(t))^2\, L \cos(\beta(t)) \sin(\beta(t)) \sin(\theta(t)) - \sin(\theta(t)) \cos(\theta(t)) \sin(\beta(t))\, h(t))\cos(\theta(t)) \\
&\Big/ h(t)^2 - mass\, d\, L\, \omega^2 (-L \sin(\theta(t)) \cos(\beta(t))^2 - 3\cos(\theta(t))^2\, L \sin(\theta(t)) \\
&- 4\, L \cos(\theta(t)) \cos(\beta(t)) \sin(\beta(t)) + 6 \sin(\theta(t))\, L \cos(\theta(t))^2 \cos(\beta(t))^2 \\
&+ 6\cos(\theta(t))^3\, L \sin(\beta(t)) \cos(\beta(t)) + \cos(\theta(t))^2 \sin(\beta(t))\, h(t) \\
&+ \cos(\theta(t)) \cos(\beta(t)) \sin(\theta(t))\, h(t) + \sin(\theta(t))\, L)\sin(\theta(t)) \Big/ h(t)^2
\end{aligned}
$$

Let's see if it will simplify.

```
> force1 := simplify(force1,trig);
```

$$
\begin{aligned}
force1 := mass\, d\, L^2\, \omega^2 \big( &2\cos(\theta(t)) \cos(\beta(t)) \sin(\beta(t)) \sin(\theta(t)) + \cos(\theta(t))^2 - 2\cos(\theta(t))^2 \cos(\beta(t))^2 + \cos(\beta(t))^2 - 1 \\
&\big) \Big/ h(t)^2
\end{aligned}
$$

```
> force2 := mass*dot(ans,perpslot);
```

$$
\begin{aligned}
force2 := mass( &mass\, d\, L\, \omega^2 (6\cos(\theta(t))^3\, L \cos(\beta(t))^2 - \cos(\beta(t))\, h(t) \\
&+ \cos(\beta(t))\, h(t) \cos(\theta(t))^2 + 2\, L \cos(\beta(t)) \sin(\beta(t)) \sin(\theta(t)) + 3\cos(\theta(t))\, L - 3\, L\cos(\theta(t))^3 \\
&- 5\, L \cos(\theta(t)) \cos(\beta(t))^2 - 6\cos(\theta(t))^2\, L \cos(\beta(t)) \sin(\beta(t)) \sin(\theta(t)) \\
&- \sin(\theta(t)) \cos(\theta(t)) \sin(\beta(t))\, h(t))\sin(\theta(t)) \Big/ h(t)^2 - mass\, d\, L\, \omega^2 (-L \sin(\theta(t)) \cos(\beta(t))^2 \\
&- 3\cos(\theta(t))^2\, L \sin(\theta(t)) - 4\, L \cos(\theta(t)) \cos(\beta(t)) \sin(\beta(t)) \\
&+ 6 \sin(\theta(t))\, L \cos(\theta(t))^2 \cos(\beta(t))^2 + 6\cos(\theta(t))^3\, L \sin(\beta(t)) \cos(\beta(t)) \\
&+ \cos(\theta(t))^2 \sin(\beta(t))\, h(t) + \cos(\theta(t)) \cos(\beta(t)) \sin(\theta(t))\, h(t) + \sin(\theta(t))\, L)\cos(\theta(t)) \Big/ \\
&h(t)^2)
\end{aligned}
$$

```
> force2 := simplify(force2,trig);
```

$$\begin{aligned}force2 := -mass^2\, d\, L\, \omega^2 \big(\sin(\theta(t))\cos(\beta(t))\,\mathrm{h}(t) - 2\,L\cos(\beta(t))\sin(\beta(t)) \\ + 4\,L\cos(\theta(t))^2\cos(\beta(t))\sin(\beta(t)) - 2\sin(\theta(t))\cos(\theta(t))\,L \\ + 4\sin(\theta(t))\,L\cos(\theta(t))\cos(\beta(t))^2 + \cos(\theta(t))\sin(\beta(t))\,\mathrm{h}(t)\big) \Big/ \mathrm{h}(t)^2\end{aligned}$$

Try some parameters so I can plot the force magnitude.

```
> values := {mass = 1, L=.25, d = 3.5, w = 3, omega = 60}:
```

Substitute the values and the position solution into the two force components.

```
> f1 := evalf(subs(values,subs(possol,force1))):
> f2 := evalf(subs(values,subs(possol,force2))):
```

Compute the magnitude of the force as a function. This makes is easier to build a table of values.

```
> mags := (val) -> evalf(sqrt(evalf(subs(beta(t)=val,f1^2+f2^2)))):
> ffun1 := (val) -> evalf(subs(beta(t)=val,f1)):
> ffun2 := (val) -> evalf(subs(beta(t)=val,f2)):
```

The number of data points I want in the plot.

```
> num := 80:
```

Form a table of angles from 1 to 361 degrees. Note I avoid 0 and 180 because of the singularity. Try plugging either value into the possol equation, you'll see it divides by zero.

```
> angles := [seq(360/num*i+1,i=0..(num+1))]:
> anglesinradians := evalm(Pi/180*angles):
```

Determine the force magnitude for each angle in the table.

```
> Mags := map(mags,anglesinradians):
> F1 := map(ffun1,anglesinradians):
> F2 := map(ffun2,anglesinradians):
```

Make a table like [ [angle, force], [angle, force] ]

```
> tab := [seq([angles[i],Mags[i]],i=1..(num+1))]:
> tab1 := [seq([angles[i],F1[i]],i=1..(num+1))]:
> tab2 := [seq([angles[i],F2[i]],i=1..(num+1))]:
```

Plot the table of data.

```
> plot(tab,title='Force Magnitude',
> font=['TIMES','ROMAN',12],color='BLACK');
```

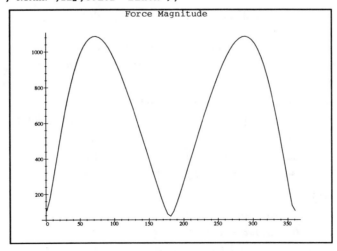

```
> plot(tab1,title='Force Along Slot');
```

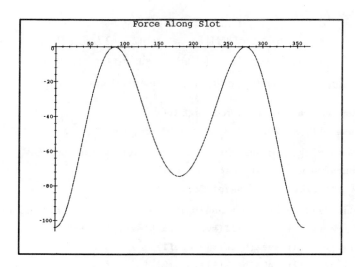

```
>  plot(tab2,title='Force Perpendicular to Slot');
```

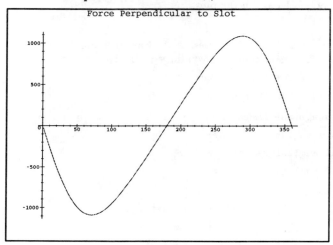

### 3.15.7    File: gizmo.ms - The Maple Program for Example 3.10.

```
>  restart;
```

Linear momentum of mass 1.  We do not plan to solve the differential equations so we will not waste time writing the variables as functions of time.

```
>  e1 := k1*(H-y1-h-r*theta-L1)-k3*(y1-L3)-m1*g = m1*y1ddot;
```
$$e1 := k1\,(H - y1 - h - r\,\theta - L1) - k3\,(y1 - L3) - m1\,g = m1\,y1ddot$$

Linear momentum of mass 2.

```
>  e2 := k2*(H-y2-h+r*theta-L2) - m2*g = m2*y2ddot;
```
$$e2 := k2\,(H - y2 - h + r\,\theta - L2) - m2\,g = m2\,y2ddot$$

Angular momentum of pulley.

```
>  e3 := k1*(H-y1-h-r*theta-L1)*r - k2*(H-y2-h+r*theta-L2)*r = 0;
```
$$e3 := k1\,(H - y1 - h - r\,\theta - L1)\,r - k2\,(H - y2 - h + r\,\theta - L2)\,r = 0$$

```
>  eqns := {e1,e2,e3}:
```

At static equil. all derivatives are zero.

```
>  static := {y1ddot = 0, y2ddot = 0}:
```

Substitute zero for derivatives and solve for the variables.

```
>  staticans := solve(subs(static,eqns),{y1,y2,theta});
```

$staticans := \{$

$$\theta = -\frac{k1\ m2\ g + k3\ m2\ g - k3\ k1\ H + k3\ k1\ h + k3\ k1\ L1 + k3\ L3\ k1 - m1\ g\ k1}{k3\ k1\ r},$$

$$y1 = \frac{m2\ g + k3\ L3 - m1\ g}{k3}, y2 = -(-2\ k2\ H\ k3\ k1 + 2\ k2\ h\ k3\ k1 + k2\ k1\ m2\ g$$

$$+ k2\ k3\ m2\ g + k2\ k3\ k1\ L1 + k2\ k3\ L3\ k1 - k2\ m1\ g\ k1 + k2\ L2\ k3\ k1 + m2\ g\ k3\ k1)/(k2$$

$$k3\ k1)\}$$

Define the static equilibrium values of the variables.

```
>  y1s := subs(staticans,y1):
>  y2s := subs(staticans,y2):
>  thetas := subs(staticans,theta):
```

Define NEW variables as New one = Old one - Static value or Old one = New + static.

```
>  newvars := {y1 = Y1 + y1s, y2 = Y2 + y2s, theta = Theta + thetas, y1ddot = Y1ddot, y2ddot
= Y2ddot}:
>  final := simplify(subs(newvars,eqns));
```

$$final := \{-k2\ (Y2 - \Theta\ r) = m2\ Y2ddot, -Y1\ k1 - \Theta\ k1\ r - Y1\ k3 = m1\ Y1ddot,$$
$$r\ (-Y1\ k1 - \Theta\ k1\ r + Y2\ k2 - k2\ \Theta\ r) = 0\}$$

Ta Da!

### 3.15.8  File: polar.ms - The Maple Program for Example 3.18.

```
>  restart;
```

Define position. R and Theta are functions of time. Define linear momentum and its derivative
This is a silly way to define vectors but it can be done. Other maple programs use more
sophistication.

```
>  pos := r(t)*(cos(theta(t))*i + sin(theta(t))*j); linmom := m*diff(pos,t); ans := diff(linmom,t)
```

$$pos := r(t)\,(\cos(\theta(t))\,i + \sin(\theta(t))\,j)$$

$linmom :=$
$$m\,((\frac{\partial}{\partial t}\,r(t))\,(\cos(\theta(t))\,i + \sin(\theta(t))\,j) + r(t)\,(-\sin(\theta(t))\,(\frac{\partial}{\partial t}\,\theta(t))\,i + \cos(\theta(t))\,(\frac{\partial}{\partial t}\,\theta(t))\,j))$$

$ans := m((\frac{\partial^2}{\partial t^2}\,r(t))\,(\cos(\theta(t))\,i + \sin(\theta(t))\,j)$
$$+ 2\,(\frac{\partial}{\partial t}\,r(t))\,(-\sin(\theta(t))\,(\frac{\partial}{\partial t}\,\theta(t))\,i + \cos(\theta(t))\,(\frac{\partial}{\partial t}\,\theta(t))\,j) + r(t)$$
$$(-\cos(\theta(t))\,(\frac{\partial}{\partial t}\,\theta(t))^2\,i - \sin(\theta(t))\,(\frac{\partial^2}{\partial t^2}\,\theta(t))\,i - \sin(\theta(t))\,(\frac{\partial}{\partial t}\,\theta(t))^2\,j + \cos(\theta(t))\,(\frac{\partial^2}{\partial t^2}\,\theta(t))\,j))$$

Here is the I component.

```
>  simplify(subs({j=0,i=1},ans));
```

$$m\,(\frac{\partial^2}{\partial t^2}\,r(t))\,\cos(\theta(t)) - 2\,m\,(\frac{\partial}{\partial t}\,r(t))\,\sin(\theta(t))\,(\frac{\partial}{\partial t}\,\theta(t)) - m\,r(t)\,\cos(\theta(t))\,(\frac{\partial}{\partial t}\,\theta(t))^2$$
$$- m\,r(t)\,\sin(\theta(t))\,(\frac{\partial^2}{\partial t^2}\,\theta(t))$$

Here is the j component.

```
>  simplify(subs({j=1,i=0},ans));
```

$$m\,(\frac{\partial^2}{\partial t^2}\,r(t))\,\sin(\theta(t)) + 2\,m\,(\frac{\partial}{\partial t}\,r(t))\,\cos(\theta(t))\,(\frac{\partial}{\partial t}\,\theta(t)) - m\,r(t)\,\sin(\theta(t))\,(\frac{\partial}{\partial t}\,\theta(t))^2$$
$$+ m\,r(t)\,\cos(\theta(t))\,(\frac{\partial^2}{\partial t^2}\,\theta(t))$$

## 3.15.9   File: rotate.ms - The Maple Program for Example 3.29.

```
> restart;
```
Use the following statements to enable using dot products.
```
> with(linalg):
```

Warning, new definition for norm

Warning, new definition for trace

I don't like typing dotprod all the time so I define dot.
```
> dot := (VectorOne,VectorTwo) -> dotprod(VectorOne,VectorTwo,'orthogonal'):
```
The following looks strange but all it does is take a
vector equation like x=y and some desired direction n and returns x dot n = y dot n.  Essentially
it takes the n direction on the left = n direction on the right.
```
> VectorEqn := (VectorEquation,VectorDirection) -> dot(lhs(VectorEquation), VectorDirection)
= dot(rhs(VectorEquation),VectorDirection):
```
Define i j k vectors.
```
> i := [1,0,0]:  j:=[0,1,0]:  k:= [0,0,1]:
```
Define the n and s directions.
```
> n := cos(theta(t))*i + sin(theta(t))*j; s := -sin(theta(t))*i+cos(theta(t))*j;
```
$$n := \cos(\theta(t))\,[1,\,0,\,0] + \sin(\theta(t))\,[0,\,1,\,0]$$
$$s := -\sin(\theta(t))\,[1,\,0,\,0] + \cos(\theta(t))\,[0,\,1,\,0]$$

The forces acting on the peg.
```
> fspring := -K*(r(t)-ro)*n;
```
$$fspring := -K\,(\mathrm{r}(t) - ro)\,(\cos(\theta(t))\,[1,\,0,\,0] + \sin(\theta(t))\,[0,\,1,\,0])$$
```
> fnorm := N * s;
```
$$fnorm := N\,(-\sin(\theta(t))\,[1,\,0,\,0] + \cos(\theta(t))\,[0,\,1,\,0])$$

Find linear momentum by differentiating position.
```
> pos := r(t)*n;
```
$$pos := \mathrm{r}(t)\,(\cos(\theta(t))\,[1,\,0,\,0] + \sin(\theta(t))\,[0,\,1,\,0])$$
```
> lm := m*diff(pos,t);
```

$$lm := m((\frac{\partial}{\partial t}\mathrm{r}(t))\,(\cos(\theta(t))\,[1,\,0,\,0] + \sin(\theta(t))\,[0,\,1,\,0]) + \mathrm{r}(t)(-\sin(\theta(t))\,(\frac{\partial}{\partial t}\theta(t))\,[1,\,0,\,0]$$
$$+ \cos(\theta(t))\,[0,\,0,\,0] + \cos(\theta(t))\,(\frac{\partial}{\partial t}\theta(t))\,[0,\,1,\,0] + \sin(\theta(t))\,[0,\,0,\,0]))$$

The following is pretty ugly so we don't print it.
All it is is Force = ddt of Linear mom.
```
> eom := simplify(fspring+fnorm = diff(lm,t)):
```
The following works well.  Notice each equation has only one double derivative.
From above eom is a vector equation, if you printed it you would find several [1,0,0] and
[0,1,0] (i and j vectors) all over the place.  What we do now is take the forces in the n direction
set them equal to the ddt of l.m.  in the n direction, then do the same for the s direction.
```
> en := simplify(VectorEqn(eom,n));
```
$$en := -K\,\mathrm{r}(t) + K\,ro = m\,(\frac{\partial^2}{\partial t^2}\mathrm{r}(t)) - m\,\mathrm{r}(t)\,(\frac{\partial}{\partial t}\theta(t))^2$$
```
> es := simplify(VectorEqn(eom,s));
```
$$es := N = 2\,m\,(\frac{\partial}{\partial t}\mathrm{r}(t))\,(\frac{\partial}{\partial t}\theta(t)) + m\,\mathrm{r}(t)\,(\frac{\partial^2}{\partial t^2}\theta(t))$$
sub in theta double dot is 0 and theta dot is omega
```
> param := {diff(theta(t),t,t)=0,diff(theta(t),t)=omega};
```
$$param := \{\frac{\partial}{\partial t}\theta(t) = \omega,\; \frac{\partial^2}{\partial t^2}\theta(t) = 0\}$$

```
>  eq:=subs(param,en);
```

$$eq := -K\,\mathrm{r}(t) + K\,ro = m\,(\frac{\partial^2}{\partial t^2}\,\mathrm{r}(t)) - m\,\mathrm{r}(t)\,\omega^2$$

```
>  ans := dsolve({eq,r(0)=ri,D(r)(0)=0},r(t));
```

$ans :=$

$$\mathrm{r}(t) = \frac{-K\,ro - \dfrac{1}{2}\dfrac{\%2\,e^{\%1}\,K}{-K+\omega^2\,m} + \dfrac{1}{2}\dfrac{\%2\,e^{\%1}\,\omega^2\,m}{-K+\omega^2\,m} - \dfrac{1}{2}\dfrac{\%2\,e^{(-\%1)}\,K}{-K+\omega^2\,m} + \dfrac{1}{2}\dfrac{\%2\,e^{(-\%1)}\,\omega^2\,m}{-K+\omega^2\,m}}{-K+\omega^2\,m}$$

$$\%1 := \frac{\sqrt{m\,(-K+\omega^2\,m)}\,t}{m}$$

$$\%2 := K\,ro - ri\,K + ri\,\omega^2\,m$$

Look what happens as omega approaches sqrt of K/m, as it approaches the "natural frequency"
The natural frequency is the frequency the system wants to vibrate at naturally.

```
>  limit(rhs(ans),omega=sqrt(K/m));
```

$$\frac{1}{2}\frac{K\,ro\,t^2 + 2\,ri\,m}{m}$$

The solution gets bigger with time. This is what we call unstable. What is happening is it is rotating fast enough that the spring cannot hold the mass back. There is too much acceleration for the force. The following show plots with low omega and high omega.

```
>  plot(subs({K=1,m=1,omega=1/2,ro=1,ri=3/4},r hs(ans)), t=0..20);
>  plot(subs({K=1,m=1,omega=2,ro=1,ri=3/4},rhs (ans)), t=0..2);
```

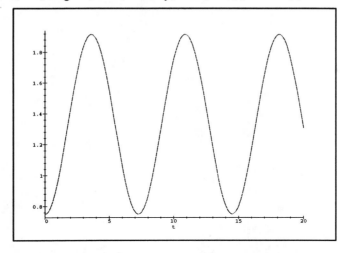

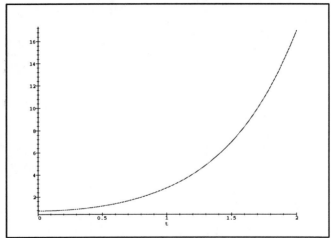

Ok we are done but will play with the equations for a little bit.  If you simply blast into the vector equation an take the i and j components.  You've got a mess.  Look...

```
>  a := simplify(VectorEqn(subs(param,eom),i));
```

$$a := -K\, \mathrm{r}(t) \cos(\theta(t)) + K\, ro \cos(\theta(t)) - N \sin(\theta(t)) =$$
$$m\,(\frac{\partial^2}{\partial t^2}\, \mathrm{r}(t)) \cos(\theta(t)) - 2\, m\,(\frac{\partial}{\partial t}\, \mathrm{r}(t)) \sin(\theta(t))\, \omega - m\, \mathrm{r}(t) \cos(\theta(t))\, \omega^2$$

```
>  b := simplify(VectorEqn(subs(param,eom),j));
```

$$b := -K\, \mathrm{r}(t) \sin(\theta(t)) + K\, ro \sin(\theta(t)) + N \cos(\theta(t)) =$$
$$m\,(\frac{\partial^2}{\partial t^2}\, \mathrm{r}(t)) \sin(\theta(t)) + 2\, m\,(\frac{\partial}{\partial t}\, \mathrm{r}(t)) \cos(\theta(t))\, \omega - m\, \mathrm{r}(t) \sin(\theta(t))\, \omega^2$$

Notice that the n and s equations look better.  Either set of equations can be solved, however the n and s directions are easiest.

# Chapter 4

# MULTIDISCIPLINE EXAMPLES

## 4.1  Electro-Mechanical Machines

**Objectives**

*When you complete this section, you should be able to*

1. *Explain what an electro-mechanical transducer is.*

2. *Model certain classes of DC motors and generators.*

3. *List various ways to increase torque in DC motors.*

4. *Explain why motors are "rated" at a specific voltage.*

5. *Describe the effect of driving a DC motor at larger than its rated voltage.*

---

This section discusses machines that convert electrical energy into mechanical energy. A transducer is a device that performs this conversion. An electric motor (generator) converts electric energy to mechanical energy and *vice versa*. The schematic in figure 4.1 is an engineering representation of a typical electric motor/generator. We can readily derive the equations that govern this class of transducer with the conservation of charge and conservation of angular momentum. What follows is a brief discussion of these relationships. For more details on electric machines see Smith [33]. Typically, DC motors operate when a stationary magnet attracts a rotating magnet. The stationary parts of the motor are often called the stator. The rotating components make up the rotor. The stator magnet can be generated in a number of ways. The simplest field generator is a permanent magnet. The $R$ and $L$ result from winding a wire around a rotating shaft. Figure 4.2 shows a simplified diagram of a DC Permanent Magnet Motor. The stator consists of a permanent magnet with the north and south poles shown. The rotor is a magnetic material with a wire

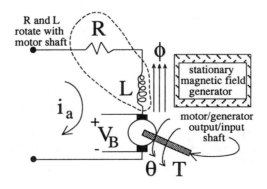

Figure 4.1: A Typical DC Electric Motor/Generator

417

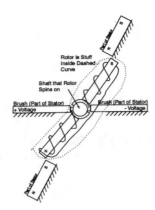

Figure 4.2: A Functional Diagram of a DC Permanent Magnet Motor.

wound around it.  Its an electromagnet.  The electromagnet and its associated wire rotate.  The two ends of the rotor wire make a connection to the electrical power supply by connecting them to two semicircular metal pieces that rotate with the rotor and in turn rub on two stationary brushes.  In the configuration shown, the electrical polarity and electromagnet windings are such that an electromagnetic north pole is in the vicinity of the stator south pole (similarly for the other poles) which causes the rotor to be attracted to the stator.  The semicircular rotor pieces are constructed such that just as the rotor and stator poles "align", the electrical polarity reverses causing what used to be the electromagnetic north to become south (vice versa for the other electromagnetic pole) causing the rotor to be repelled from the stator which causes it to continue rotation.

One of the problems with this type of motor is that the rubbing of the brushes causes friction and the brushes wear out.  Although electrical motors are some of the cleanest type of motion producing devices, this type of motor is one of the dirtiest due to the brush wear.  What we have described is a two pole motor. It is called this because there is a single electromagnet.  What do you think are some characteristics of this type motor?  What do you think is the advantage of building a motor with more than two poles?  Many industrial quality motors have more than two poles.  Motors you find in many children's toys are two pole. Why do you think is the case.

What do you think impacts the torque that a motor produces?  Since torque is the moment of a force, there are two ways to increase torque.  You can increase the force (the magnetic attraction between the rotor and stator magnets) or you can increase the "moment arm" or the location that the forces are applied.  If you look at high torque electric motors, you will find that they have a large diameter.  Surprised?  You better not be!  Now consider the second alternative, make the magnetic attraction force larger.  To do this you can increase the strength (the B field) of the stator.  This is what is done in what is called rare earth motors. The concept behind superconducting motors is to produce a very high B constant electromagnet by coiling a superconducting wire around a magnetic material.

Once you select the motor diameter and the type of stator magnet, how else might you change the motor torque?  The only other major adjustment is in the strength of the rotor magnet.  It turns out that the torque relationship for a rotating electro-mechanical transducer is:

$$T(t) = K_m \phi(t) i_a(t) \tag{4.1}$$

where $i_a$ is the current flowing in the armature (another name for the rotor) circuit, $\phi(t)$ is the magnetic flux[1] of the external field (the stator's field), and $K_m$ is a constant.[2]  $K_m$ accounts for the type of material, shape, size and other factors in the motor.  We have discussed things like motor diameter (which is part of $K_m$) and the strength of the stator field ($\phi(t)$ characterizes that).  Why is the rotor magnetic strength proportional to current?  Look back at the discussion on inductors.

We can generate the magnetic flux $\phi(t)$ using a permanent magnet or an electro-magnet.  For the case of an electromagnet, we can represent the current in a coil of the stator field circuit (the magnetic flux generator in figure 4.1) as follows:

$$\phi(t) = K_f i_f(t) \tag{4.2}$$

where $i_f(t)$ is the current flowing in the stator field circuit, $K_f$ is another constant (another approximation). The constant $K_f$ accounts for the geometry and orientation of the field windings and other lumped factors.

---

[1] The integral over area of flux is the magnetic field ($\vec{B}$).

[2] Actually saying $K_m$ is a constant is an approximation.  The actual value can vary due to several factors which include rotor position and temperature.

Combining equations 4.1 and 4.2 gives

$$T(t) = K_m K_f i_f(t) i_a(t) \tag{4.3}$$

The current in the field windings is sometimes a constant, separately excited field, which produces a constant flux $\phi$. If the stator field is constant, we can write equation 4.3 as

$$T(t) = K_t i_a(t) \tag{4.4}$$

where $K_t = K_m K_f i_f(t)$ is called the torque constant.

The coil labeled $L$ rotates with the motor shaft and is in the presence of the stator magnetic field generated by the stator field circuit. The inductor (the coil forming $L$) is a conductor. As the motor turns, the magnetic field on the wire forming $L$ changes (due to motion of the rotor). Why? Look at one point on the wire that right now is near the north pole of the stator. Once the rotor turns 180 degrees, it is near the south pole. If that is not a change in magnetic flux, what is? This changing flux induces a voltage in $L$. The voltage the machine generates is

$$V_B = K_B \dot\theta \tag{4.5}$$

where $\dot\theta$ is the angular speed of the machine. $K_B$ is the voltage constant which depends on configuration and lumps several factors together. Why should the generated voltage be proportional to the speed of the machine? Its because the rate of change of the flux is proportional to the machine's motion. Note that we can use the machine to generate electricity.

What is confusing to many students is the fact that torque and voltage generation occur at the same time. When the machine is acting like a motor, we apply voltage to the brushes causing current to flow. If the rotor is not held still, the motor turns. As the motor turns, it generates a voltage (sometimes called a backemf). The backemf is specified by equation 4.5 and if the applied voltage is larger than the backemf the current flows from the positive to the negative terminal producing a driving torque which attempts to drive the motor faster. If we rotate the motor's rotor by hand, we force the coil $L$ to turn and the relation given in equation 4.5 is the voltage output from the machine. If we connect the brushes forming a circuit, current will flow. The current will cause a torque given by equation 4.3. We must overcome this amount of torque to keep the rotor moving. When we operate the machine as a motor, equation 4.3 is the output and equation 4.5 is the voltage that must be overcome by the electrical driving source.

To determine positive/negative directions for the backemf and torque do this. Pick one brush call it the positive brush and apply a positive voltage to it while grounding the other. Call the direction the motor turns positive. When the motor turns in the positive direction, a positive backemf is generated on the positive terminal.

Applying the conservation of energy to the transducer element (we do not include the inductor or the resistor in the system), we can determine a relationship between the motor constants. To do this, we assume that the machine has a constant stator field (constant stator magnet)

**Conservation of Energy**
system[**motor tranducer fig.4.1**]      time period [**differential**]

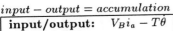

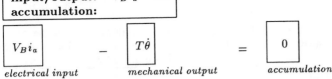

| $V_B i_a$ | $-$ | $T\dot\theta$ | $=$ | $0$ |
|---|---|---|---|---|
| *electrical input* | | *mechanical output* | | *accumulation* |

From the conservation law, we have

$$V_B i_a = T\dot\theta \tag{4.6}$$

or after using equations 4.5 and 4.4

$$K_B \theta i_a = K_t i_a \dot\theta. \tag{4.7}$$

This implies that $K_B$ [volt sec/rad]$= K_t$ [N m/A]. Notice that in metric units, the constants are identical. If we express mechanical power in a non-metric unit, a correction factor will be present due to the unit conversion.

What happens to the electrical energy applied to the brushes of the motor? Some of it (hopefully a lot of it) is converted into mechanical energy (torque and motion) but some of it is converted into thermal energy via the resistance from the wire forming the rotor's electromagnet. A motor is rated at a particular voltage because that is a value of voltage that can be applied continuously and not cause the wire to melt. If you wanted to apply a voltage above the rated, what do you need to pay attention to? You need to keep

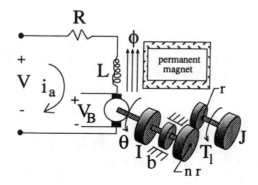

Figure 4.3: A Typical Electric Motor and Load

the wire cool. How? Let the motor rest and cool or maybe blow air on the motor. What benefit is there to applying a larger than rated voltage? You get more current, which means more torque. This does of course have a limit, the magnetic field strength eventually saturates and does not increase forever. There are other (weird) problems with high current such as demagnetizing the stator magnet but we'll not discuss these.

## Example 4.1

## An Electric Motor Driving a Mechanical Load

**PROBLEM:**

The motor and load in figure 4.3 is a typical representation of an electric motor driving a mechanical load. Formulate a mathematical model of this system. Assume that the motor is a permanent magnet DC motor.

**FORMULATION:**

The first step in solving this problem is to determine the [NIA] and [NIF] of the system. The independent accumulation terms in this system are

1. the kinetic energy of the rotor and load

2. the electrical energy accumulation in the inductor.

We ignore the thermal buildup in the motor. Thus, the [NIA] of the system is two. If we cut the circuit, then the load can still move. Therefore, the [NIF] of the system is two. We can formulate a mathematical model of this system using

1. the constraint relationships 4.4 and 4.5

2. the conservation of charge for the circuit

3. the conservation of angular momentum for the gear train.

Assuming the motor drives its load through a simple gear box, we will start with an analysis of the gears. Figure 4.4 shows the freebody diagrams of the gears in the gear train. Applying the conservation of angular momentum for the gear on the left in figure 4.4 (the one connected to the motor), we find that

**Conservation of Angular Momentum (Vector) [Gear Center**
system[**motor gear fig.4.4**]        time period [**differential**]

*input − output = accumulation*

| **input/output:** | mass flow | |
| | external moment | $T_{\text{gen}} + Fnr - b\dot{\theta}$ |
| | gravity moment | |
| **accumulation:** | | $\frac{d}{dt}\left(I\dot{\theta}\right)$ |

$$z: \quad \boxed{T_{\text{gen}}}_{motor} \quad + \quad \boxed{Fnr}_{external} \quad - \quad \boxed{b\dot{\theta}}_{friction} \quad = \quad \boxed{\frac{d}{dt}\left(I\dot{\theta}\right)}_{accumulation}$$

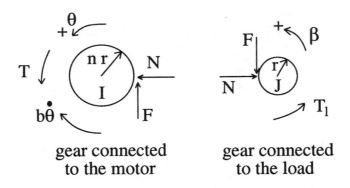

Figure 4.4: A Freebody Diagram of the Gears

$$T_{\text{gen}} + Fnr - b\dot{\theta} = \frac{d}{dt}\left(I\dot{\theta}\right) \tag{4.8}$$

where $T_{\text{gen}}$ is the torque generated by the motor, $b$ is the viscous friction coefficient (representing friction inside the motor itself), and $I$ is the mass moment of inertia of the armature, shaft, and gear. $F$ is the interaction force between the gears and $nr$ is the radius of the driving gear. Similarly, the result for the right gear of figure 4.4 is

**Conservation of Angular Momentum (Vector) [Gear Center]**
system[**gear load f.ch3motorfbd**]　　　　time period [**differential**]

*input − output = accumulation*

| **input/output:** | mass flow | |
| | external moment | $T_l + Fr\,\vec{k}$ |
| | gravity moment | |
| **accumulation:** | | $\frac{d}{dt}\left(J\dot{\beta}\right)\,\vec{k}$ |

$$z: \quad \boxed{T_l} \;+\; \boxed{Fr} \;=\; \boxed{\frac{d}{dt}\left(J\dot{\beta}\right)}$$
$$\quad\; \textit{load} \quad\;\; \textit{external} \qquad \textit{accumulation}$$

$$Fr + T_l = \frac{d}{dt}\left(J\dot{\beta}\right) \tag{4.9}$$

where $\dot{\beta}$ is the angular velocity of the load, $T_l$ is the external load torque caused by (for example) something touching the load, and $J$ is the mass moment of inertia of the output shaft and gear. Multiplying equation 4.9 by $n$ and subtracting the result from equation 4.8 eliminates the gear interaction forces and gives

$$T_{\text{gen}} - nT_l - b\dot{\theta} = \frac{d}{dt}\left(I\dot{\theta} - nJ\dot{\beta}\right) \tag{4.10}$$

Why might you be interested in the interaction force $F$?

The obvious kinematic constraint between the gears is[3]

$$nr\dot{\theta} = -r\dot{\beta} \tag{4.11}$$

Combining equations 4.10 and 4.11 and taking the indicated derivative gives

$$T_{\text{gen}} - nT_l - b\dot{\theta} = (I + Jn^2)\ddot{\theta} \tag{4.12}$$

which is the equation for the mechanical system. Now can we solve this equation? No because we do not know what $T_{\text{gen}}$ is. How can we find it? It is given by equation 4.1, but that requires us to know current. What we need is to turn attention to the electrical part of the problem.

---

[3]The system's [NIF] tells us that we need to use kinematics because we have too many flow variables.

Accounting for electrical energy for the circuit in figure 4.3 is[4] (we could use KVL which is a little simpler)

**Electrical Energy Accounting**
system[**circuit fig.4.3**]        time period [**differential**]

$input - output + generation - consumption = accumulation$

| | |
|---|---|
| **input/output:** | $Vi_a - V_B i_a$ |
| **generation:** | |
| **consumption:** | $i_a^2 R$ |
| **accumulation:** | $\frac{d}{dt}\left(\frac{1}{2}Li_a^2\right)$ |

$$\boxed{Vi_a - V_B i_a} \quad - \quad \boxed{i_a^2 R} \quad = \quad \boxed{\frac{d}{dt}\left(\frac{1}{2}Li_a^2\right)}$$

$\quad$ input/output $\qquad\qquad$ consumption $\qquad\qquad$ accumulation

$$Vi_a - i_a^2 R - V_B i_a = \frac{d}{dt}\left(\frac{1}{2}Li_a^2\right)$$

This reduces to

$$V - i_a R - V_B = L\frac{d}{dt}(i_a) \qquad\qquad (4.13)$$

Count the unknowns, there are too many (don't count the constants since we would look them up in the motor catalog). We are looking for more equations but the [NIA] is satisfied. The [NIF] is also satisfied. What to do? We need some constraints. We need something to relate the mechanics to the electrical. We need the constraint relationships 4.4 and 4.5, relating the electrical and mechanical variables. Plugging into equations 4.12 and 4.13 gives

$$V - K_B\dot{\theta} = Ri + L\frac{di}{dt}$$

and

$$K_t i - nT_l - b\dot{\theta} = (I + Jn^2)\ddot{\theta}.$$

These are a set of coupled differential equations of second order. The two state variables are $\dot{\theta}$ and $i$. The reader might try to derive equation 4.12 by taking the gear train as a system. Look at the effect of the gear box. If the speed of the motor is reduced (this is often the case because DC motors spin fast) then $n > 1$ which means $J$ (the inertia of the gears) is "amplified" by the gears. In the case of a robot, $n$ can be greater than 100 which means the value of $I$ is essentially negligible compared to the gear.
**End 4.1**

## Example 4.2

# An Electric Hoist

**PROBLEM:**

We want to lift the mass $m_2$ using the pulleys and motor in figure 4.5. The motor generates a tension in the cable proportional to the current flowing in the electrical circuit. The fuse is rated at $i_f$. This means that the conductor melts when the current $i$ exceeds $i_f$. The mechanical motion of the motor also generates a *back emf* – a voltage ($V_b$) — proportional to the rotational speed of the motor. The polarities of the motor are such that positive current $i$ produces a tension ($F$). Furthermore, the motion of $m_1$ to the left produces a positive $V_b$. Determine the initial acceleration of block 2 (assuming that it starts from rest). Also, determine the maximum speed that $m_2$ can achieve if the applied voltage $V$ is instantaneous and constant.

**FORMULATION:**

To accomplish our objective, we obviously need to predict the motion of $m_2$. To do this, we will construct a model of the system. Since the acceleration of $m_2$ interests us, we will want to write our conservation equations in terms of rates. Before we begin writing our equations, we will determine the [NIA] of the system. This helps us think about the problem before we jump into the middle of it. The masses can accumulate linear momentum and energy. We can quantify these using $x$ and $\dot{x}$. Therefore, the [NIA] thus far is two. The electrical circuit has no accumulation terms. Therefore, we get no increase in [NIA]. In summary, the system is two [NIA].

---
[4]The system's [NIA] tells us that we need another differential equation.

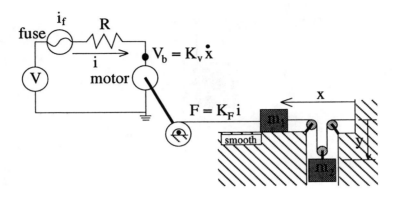

Figure 4.5: An Electric Hoist

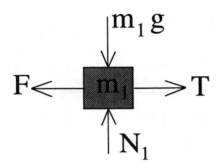

Figure 4.6: Mass One Freebody Diagram

We have several constraint equations in this system. One constraint is $V_b = K_v \dot{x}$. A second constraint is $F = K_F i$. The third important constraint is the cord connecting the two masses. The cord constraint can be expressed as

$$c_1 + 2y + x - c_2 = L$$

where $c_1$ and $c_2$ are some constants and $L$ is the constant cord length. Differentiating, this becomes $2\dot{y} + \dot{x} = 0$. Taking a system comprised of $m_1$ only yields the freebody diagram in figure 4.6. Expressing linear momentum in the horizontal direction for $m_1$ (taking left as positive) gives

**Conservation of Linear Momentum (Vector)**
system[**mass one fig.4.6**]  time period [**differential**]

*input − output = accumulation*

| input/output: | mass flow external forces gravity force | $F - T\,\vec{i}$ |
|---|---|---|
| accumulation: | | $\frac{d}{dt}(m_1 \dot{x})\,\vec{i}$ |

| | | | |
|---|---|---|---|
| $x$: | $\boxed{F - T}$ | $=$ | $\boxed{\frac{d}{dt}(m_1 \dot{x})}$ |
| | *external* | | *accumulation* |

$$F - T = \frac{d}{dt}(m_1 \dot{x})$$

Now consider the electrical components. If you are wondering why we have ignored $m_2$, do not worry we will get back to it. If you are *not* wondering why we ignored $m_2$, good! You are being set up for something. Now we have a slight problem. If the current $i$ exceeds $i_f$, then the fuse will blow and the current will fall

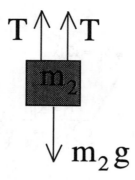

Figure 4.7: Freebody Diagram of Mass Two

to zero. If $i$ does not exceed $i_f$, then the fuse will not blow and current may flow. Since we do not know which will happen, we must assume one or the other, determine the response, and check our assumption. If we assume the fuse will not blow, we can check our assumption by guaranteeing that $i$ is below $i_f$ which will be pretty easy to do. Assuming that the fuse does not blow, we can account electrical energy equation for the resistor as follows:

---

**Electrical Energy  Accounting**
system[ **resistor fig.4.5** ]        time period [**differential**]

*input − output + generation − consumption = accumulation*

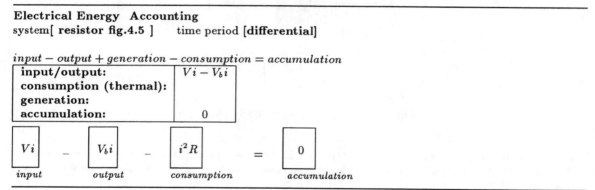

| input/output: | $Vi - V_b i$ |
| :--- | :--- |
| **consumption (thermal):** | |
| **generation:** | |
| **accumulation:** | 0 |

$$Vi \quad - \quad V_b i \quad - \quad i^2 R \quad = \quad 0$$

input          output          consumption          accumulation

---

$$Vi - V_b i - i^2 R = 0$$

Notice this is an algebraic equation which is what we expected since there are no storage elements in the electrical domain. Now, we have two equations with too many variables. To help the situation, we apply the constraint equations. Using the constraints, we can write our model as follows:

$$K_F i - T = \frac{d}{dt}\left(m_1(-2\dot{y})\right)$$

$$V - K_v \dot{x} - iR = 0 = V - K_v(-2\dot{y}) - iR \tag{4.14}$$

Now this is two equations with three unknowns $i$, $\dot{y}$, and $T$. We obviously need another equation. Often, students panic at a point like this and *assume* that something must be zero. What we should do is determine what else we can write equations for. In this case, we forgot to include $m_2$ in our analysis. Taking $m_2$ as our system, we can draw the freebody diagram that appears in figure 4.7. Applying linear momentum in the vertical direction for $m_2$ gives (we will take positive downward as figure 4.5 shows)

**Conservation of Linear Momentum (Vector)**
system[ $m_2$ **fig.4.7** ]     time period [**differential**]

*input − output = accumulation*

| input/output: | mass flow | |
|---|---|---|
| | external forces | $-2T$ |
| | gravity force | $+m_2 g$ |
| accumulation: | | $\frac{d}{dt}(m_2 \dot{y})$ |

$$ y: \quad \boxed{-2T} \quad + \quad \boxed{m_2 g} \quad = \quad \boxed{\frac{d}{dt}(m_2 \dot{y})} $$

$$ \underset{external}{} \qquad\qquad \underset{gravity}{} \qquad\qquad \underset{accumulation}{} $$

$$ -2T + m_2 g = \frac{d}{dt}(m_2 \dot{y}) $$

Finally, we now have three equations with three unknowns. If we had applied enough forethought, we would have realized that we can eliminate the cable tension $T$ from our analysis by expressing the energy equation for a system comprised of the two masses together. Just for practice we will demonstrate how to do this.

**Conservation of Energy**
system[**both masses fig.4.5**]     time period [**differential**]

*input − output = accumulation*

| input/output: | mass flow | |
|---|---|---|
| | work | $F\dot{x} - T\dot{x} - 2T\dot{y}$ |
| | heat transfer | |
| accumulation: | | $\frac{d}{dt}\left(\frac{1}{2}m_1\dot{x}^2 + \frac{1}{2}m_2\dot{y}^2 - m_2 gy\right)$ |

$$ \boxed{F\dot{x} - T\dot{x} - 2T\dot{y}} \quad = \quad \boxed{\frac{d}{dt}\left(\frac{1}{2}m_1\dot{x}^2 + \frac{1}{2}m_2\dot{y}^2 - m_2 gy\right)} $$

$$ \underset{work}{} \qquad\qquad \underset{accumulation}{} $$

$$ F\dot{x} - T\dot{x} - 2T\dot{y} = \frac{d}{dt}\left(\frac{1}{2}m_1\dot{x}^2 + \frac{1}{2}m_2\dot{y}^2 - m_2 gy\right) $$

We can simplify this result using the cord constraint.

$$ F(-2\dot{y}) - T(-2\dot{y}) - 2T\dot{y} = \frac{d}{dt}\left(\frac{1}{2}m_1(-2\dot{y})^2 + \frac{1}{2}m_2\dot{y}^2 - m_2 gy\right) $$

which becomes

$$ -2F\dot{y} = \frac{d}{dt}\left(\frac{1}{2}\left(m_2 + 4m_1\right)\dot{y}^2 - m_2 gy\right) $$

and finally

$$ -2F = \left(m_2 + 4m_1\right)\ddot{y} - m_2 g \tag{4.15} $$

This is exactly what we get when we eliminate the tension from the two momentum equations. The point here is that although forethought can help us choose a system that eliminates one or more of the variables, it is not necessary to invest significant amounts of time in selecting the proper system. Another point is that if we inadvertently overlook one part of the system (as we did by ignoring $m_2$), a count of unknowns and equations will quickly let us know that something is wrong. Continuing with our primary objective, we use equation 4.14 to determine $i$ as

$$ i = \frac{V + 2K_v\dot{y}}{R} $$

Using this in equation 4.15 gives (after a little algebra)

$$ m_2 g - \frac{2K_F V}{R} = (4m_1 + m_2)\ddot{y} + \frac{4K_F K_v}{R}\dot{y} \tag{4.16} $$

We can us this latter equation to answer our questions. If the system starts from rest then the initial acceleration of mass two is

$$ \ddot{y} = \frac{m_2 g - \frac{2K_F V}{R}}{4m_1 + m_2} $$

If $2K_FV > Rm_2g$, then the acceleration is negative. This means that the mass rises. Therefore, if we apply enough voltage, then the mass will rise. If we do not apply enough voltage, then the mass will fall. We have merely determined the value of this voltage.

If we desire to know the maximum speed of $m_2$, we must find the maximum $\dot{y}$. Now the maximum (or minimum) will occur when $\dot{y}$ quits changing. Mathematically this is when we hit a *critical* point defined as $\frac{d}{dt}(\dot{y}) = 0$ which means the same thing. So the maximum speed will occur when $\ddot{y} = 0$ or (using equation 4.16)

$$\dot{y}_{\max} = \frac{m_2gR - 2K_FV}{4K_FK_v}$$

Notice that if the voltage is sufficient to raise the mass, the maximum velocity will be upward as well.

It is interesting to note that even if the voltage is insufficient to raise the mass, there will be a terminal velocity. For example, suppose $m_2gR - 2K_FV$ is positive so that there will be an initial positive acceleration (the mass falls, remember positive is downward). As the mass falls, $\dot{x}$ becomes negative. The negative $V_b$, creates a larger voltage drop across $R$ which results in more current flow. The increase in current causes an increase in force $F$, which tends to decrease the mass's acceleration. This will continue until the acceleration becomes zero (maximum speed) or the fuse blows. We can see this in equation 4.16 as well. Since $m2gR - 2K_FV$ is positive, $\ddot{y}$ is positive causing $\dot{y}$ to become positive. After motion starts downward the acceleration is

$$\ddot{y} = \frac{m_2g - \frac{2K_FV}{R} - \frac{4K_FK_v}{R}\dot{y}}{4m_1 + m_2}$$

So the effect of the positive (downward) motion ($\dot{y}$) is to decrease the acceleration, as speed increases, acceleration decreases until it becomes zero and speed stops changing. We can apply a similar analysis to the upward motion. The effect of motion upward is to increase (make more downward) acceleration so the initially negative acceleration (upward) becomes more positive (downward) until it hits zero. The final item to consider is to check our assumption that the fuse does not blow. From the equation for current we have at the initial time $i = V/R$. Provided this is less than $i_f$, the fuse will not immediately blow. Now suppose the motion begins. If motion occurs downward, the fuse will blow if

$$i_f < \frac{V + 2K_v\dot{y}}{R}$$

or

$$\dot{y} > \frac{i_fR - V}{2K_v}$$

Based on our prediction of the maximum speed attainable, we have

$$\dot{y} < \frac{m_2gR}{4K_FK_v} - \frac{V}{2K_v}$$

Clearly this means the fuse will blow if

$$\frac{i_fR - V}{2K_v} < \dot{y} < \frac{m_2gR}{4K_FK_v} - \frac{V}{2K_v}$$

which we can reduce to

$$i_f < \frac{m_2g}{2K_F}$$

On the other hand, if the motion occurs upward, then $\dot{y}$ is negative and the fuse will blow if ever

$$i_f < -\frac{V + 2K_v\dot{y}}{R}$$

or

$$-\dot{y} > \frac{i_fR + V}{2K_v}$$

which we can write as

$$-\frac{i_fR}{2K_v} - \frac{V}{2K_v} > \dot{y}$$

Now based on the maximum speeds that we determined above, we know that if the mass is traveling upward, then the speed is bounded by the following:

$$\dot{y} > \frac{m_2gR}{4K_FK_v} - \frac{V}{2K_v}$$

This implies the fuse will blow if it is possible that

$$-\frac{i_fR}{2K_v} - \frac{V}{2K_v} > \dot{y} > \frac{m_2gR}{4K_FK_v} - \frac{V}{2K_v}$$

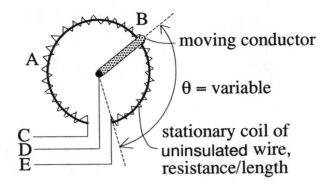

Figure 4.8: An Angular Potentiometer, When $\theta$ Changes, the Resistance Between D and E Changes.

which reduces to

$$-i_f > \frac{m_2 g}{2 K_F}$$

Clearly, since the quantity on the left is always negative and the quantity on the right is always positive, the inequality can never be satisfied. Therefore, the upward speed can never cause the fuse to blow. What happens for upward motion is the current decreases from the initial amount until it hits a minimum. This minimum current generates just enough force to overcome the gravity load pulling on $m_2$.

**End 4.2**

## 4.2 Sensors

**Objectives**
*When you complete this section, you should be able to*

1. *Explain what sensors are and why linear sensors are desirable.*

2. *Calculate the response of sensor components.*

This section discusses sensors. A sensor is a multidiscipline device that converts one quantity into another. For example, an accelerometer converts acceleration into an electrical voltage, a strain gauge converts strain into voltage. A thermocouple converts temperature into a voltage.

Not all sensors convert into the electrical domain. For example, some pressure sensors convert pressure into a mechanical motion and some temperature sensors convert temperature into motion. Chances are, if its old enough, the thermostat controlling the air conditioner in your home converts temperature into motion.

**Example 4.3**

### Measuring Position with a Potentiometer (POT)

**PROBLEM:**
Develop the mathematical formulation necessary to measure position using the potentiometer (POT) in figure 4.8.

A potentiometer (POT) is a variable resistor used for measuring positions). The POT in the figure consists of a coil of wire wound around a fixed circular member (A) and a wiper (B). The wire has a resistance per unit length of $\rho$. Typically, we apply an ideal voltage source ($V$) to terminal C, and we ground terminal E.[5]

When the wiper contacts the end near E, the voltage on terminal D is near ground. When the wiper contacts the end near C, the voltage on terminal D is near $V$. At intermediate positions, the voltage on terminal D is related to the position $\theta$. The resistance between terminals C and E is constant ($R_p$). The

---

[5]We can assume that the ground terminal is an ideal effort source where the specified effort is zero. Problems can arise when the ground connection is not actually ideal.

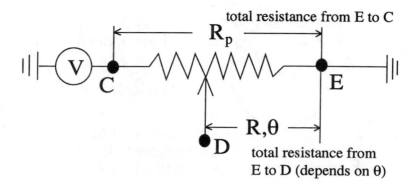

Figure 4.9: A Simplified Schematic of the Potentiometer (POT).

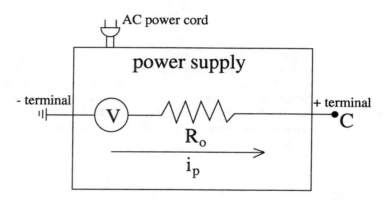

Figure 4.10: A Realistic Model of the Source (the Power Supply), $R_o$ is Usually Small.

resistance between E and D varies with $\theta$. We can express it as $R = \rho(\frac{D}{2}\theta)$ where $D$ is the diameter of the circular member.[6]

## FORMULATION:
## The Simplified Approach (For Ideal Sources and Loads)

We can model the behavior of the POT using the simplified schematic in figure 4.9. This circuit assumes that we have an ideal volt meter connected at terminal D, and an ideal source ($V$) applied to terminal C. In actual practice, the source (the power supply) may have a non-zero output impedance (see figure 4.10) $R_o$ that prevents it from supplying voltage $V$ to terminal C unless $i_p$ is zero. We can model the actual output of the source using the circuit in figure 4.10. Figure 4.11 shows the output impedance of this non-ideal circuit in terms of the output voltage versus the current that flows through the source.

To measure the voltage on terminal D, we have to attach a real device between terminals D and E. A real device has a finite input impedance and therefore causes a finite current to flow through the volt meter. We can model the actual input to the volt meter using the circuit in figure 4.12. Figure 4.13 shows the input impedance of this non-ideal meter in terms of the input voltage versus the current that flows through the meter.

If we include the non-ideal power supply and meter in our previous schematic representation of the POT (figure 4.9), we get the circuit in figure 4.14. Since we are assuming non-ideal sources and meters, we have to include them in our analysis. Note that eventually, we will have to resort to assuming that something is ideal.

---

[6]We use $\frac{D}{2}\theta$ because it is the formula for the length of the circular arc (this formula is easy to remember because it simplifies to the formula for the circumference of a circle for $\theta = 2\pi$). We can verify this formula by checking to see that the units of the expression are consistent. Thus, $\rho\frac{D}{2}\theta$ has to have the same units as $R$ and it does.

$$\frac{[resistance]}{[length]}([length]) = [resistance]$$

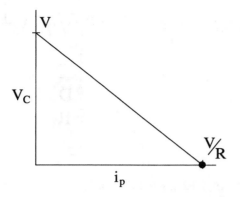

Figure 4.11: The Output Impedance Curve of the Source, the Slope is Usually Small.

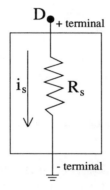

Figure 4.12: A Realistic Model of the Volt Meter.

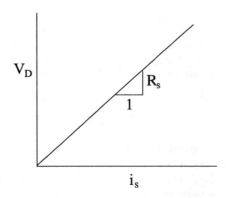

Figure 4.13: The Input Impedance Curve of the Volt Meter.

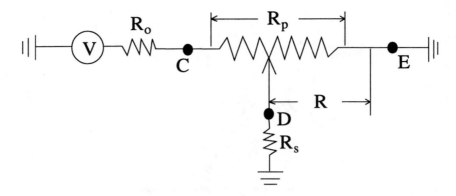

Figure 4.14: A Realistic Schematic of the Potentiometer.

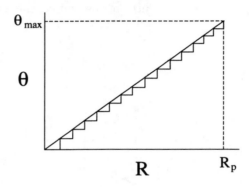

Figure 4.15: The Resistance Characteristics of the Potentiometer.

We now have a practical schematic representation of the POT. The next step is to develop relations using this model between what we know and what we want to determine. Since we want the POT to find $\theta$ by sensing the voltage at D, we need to find a relation between these two quantities. If we relate the motion of the wiper to the resistance of the wire $R$, we get the following expressions:

$$R = 0 \text{ at } \theta = 0$$

$$R = R_p \text{ at } \theta = \theta_{\max}$$

where $\theta_{\max}$ is the maximum angle of motion. The assumption involved with these expressions is that the wiper can reach both ends of the POT. Using these expressions and assuming that $R$ is proportional to $\theta$ (which means the uninsulated wire is wound uniformly along the POT), we arrive at the following expression for the position:

$$\theta = \frac{\theta_{\max}}{R_p} R \tag{4.17}$$

The plot of equation 4.17 is the solid straight line in figure 4.15.

Since the wiper only contacts one side of the windings, the actual relation between the motion and the resistance looks more like the jagged curve in figure 4.15 (the height of each little step is very small). However, our assumption that the resistance is straight is not a bad approximation of this stair step reality.

At this point, we do not have enough equations to determine position by measuring the voltage at terminal D. To achieve this result, we have to find a relation between the voltage at D and resistance $R$. Applying the conservation of charge at nodes C, D, and E of the circuit in figure 4.9, gives us the equation we need.

**Conservation of Charge**

system[ **node E fig.4.9** ]      time period [**differential**]

$input - output = accumulation$

| input/output: node E | $\frac{V_C - V_D}{R_p - R} - \frac{V_D - V_E}{R}$ |
|---|---|
| accumulation: | 0 |

| $\frac{V_C - V_D}{R_p - R} - \frac{V_D - V_E}{R}$ | $=$ | 0 |
|---|---|---|
| *in/out* | | *accumulation* |

$$\frac{V_C - V_D}{R_p - R} - \frac{V_D - V_E}{R} = 0$$

If we assume that the source is ideal ($V_C = V$ and $V_E = 0$) and solve for $V_D$, this equation simplifies to

$$V_D = \frac{R}{R_p} V \tag{4.18}$$

This gives us a simple linear relationship between the desired quantity ($R$) and the measured quantity ($V_D$). We can use this result and equation 4.17 to solve for position as follows:

$$\theta = \frac{\theta_{\max} V_D}{V}.$$

We like simple linear relationships between what we know and what we want to know because we can make one measurement, do one multiplication and have what we want. Remember, this linear relationship results from the fact that we ignored the output impedance of the source and the input impedance of the volt meter.

# The Realistic Approach (For non-Ideal Sources and Loads)

If we let the power supply (the source) and sensor (the load) be more realistic, the system becomes more complicated. Figure 4.14 shows the circuit for this realistic system. The conservation of charge for this circuit tells us that

**Conservation of Charge**

system[ **node C fig.4.14** ]      time period [**differential**]

$input - output = accumulation$

| input/output: node C | $\frac{V - V_C}{R_o} - \frac{V_C - V_D}{R_p - R}$ |
|---|---|
| accumulation: | 0 |

| $\frac{V - V_C}{R_o} - \frac{V_C - V_D}{R_p - R}$ | $=$ | 0 |
|---|---|---|
| *in/out* | | *accumulation* |

$$\frac{V - V_C}{R_o} - \frac{V_C - V_D}{R_p - R} = 0$$

**Conservation of Charge**

system[ **node D fig.4.14** ]      time period [**differential**]

$input - output = accumulation$

| input/output: node D | $\frac{V_C - V_D}{R_p - R} - \frac{V_D - V_E}{R} - \frac{V_D}{R_s}$ |
|---|---|
| accumulation: | 0 |

| $\frac{V_C - V_D}{R_p - R} - \frac{V_D - V_E}{R} - \frac{V_D}{R_s}$ | $=$ | 0 |
|---|---|---|
| *in/out* | | *accumulation* |

$$\frac{V_C - V_D}{R_p - R} - \frac{V_D - V_E}{R} - \frac{V_D}{R_s} = 0$$

Solving these equations for $V_D$ gives the following:

$$V_D = \frac{R_s R V}{R_p R_s + R_o R_s + R R_p + R R_o - R^2} \tag{4.19}$$

Likewise, solving for $R$ gives

$$R = \left[\left\{[(4R_p + 4R_o)R_s + R_p^2 + 2R_oR_p + R_o^2]V_D^2 + (-2R_p - 2R_o)R_sVV_D\right.\right.$$

$$\left.\left. + R_s^2V^2\right\}^{\frac{1}{2}} + (R_p + R_o)V_D - R_sV\right]\frac{1}{2V_D}$$

We can relate this last equation to position as follows (all the change is at the end of the expression):

$$\theta = \left[\left\{[(4R_p + 4R_o)R_s + R_p^2 + 2R_oR_p + R_o^2]V_D^2 + (-2R_p - 2R_o)R_sVV_D\right.\right.$$

$$\left.\left. + R_s^2V^2\right\}^{\frac{1}{2}} + (R_p + R_o)V_D - R_sV\right]\frac{R_p}{2V_D\theta_{\max}}$$

Notice that the simple relationship between position, $\theta$, and $V_D$ that we developed in the first approach becomes much more complicated for the non-ideal POT. Fortunately, most of the terms in this equation are constants. Therefore, we can reduce it to

$$\theta = \frac{\sqrt{AV_D^2 + BV_D + C} + DV_D - E}{2V_D}\left(\frac{R_p}{\theta_{\max}}\right)$$

where $A$, $B$, $C$, $D$ and $E$ are constants. The expression looks better with the constants but it is definitely not linear. It is a nonlinear relation between the voltage we read and the angle we want. It takes much more time to compute what we want than it does with a linear relationship.

**VERIFICATION:**

We can check this result by verifying that it simplifies to equation 4.18 for an ideal source and load. As the source (the power supply) approaches ideal, its output impedance approaches zero ($R_o \to 0$ and $V_C \to V$). Likewise, if the sensor approaches ideal, its input impedance approaches $\infty$ ($R_s \to \infty$). If we take the limit as the source (the power supply) and the sensor approach ideal, equation 4.19 becomes

$$R = \frac{R_pV_D}{V},$$

which is identical to equation 4.18.

**End 4.3** ───────────────────────────────────────────

Everything in the Universe interacts. Nothing is totally removed or isolated; however, depending on the circumstances, the coupling may not be significant. This example shows that if the output impedances of driving devices (the power supply) are small and the input impedances of driven devices (the voltage meter) are large, a system can be satisfactorily modeled as being isolated as shown for the POT in figure 4.9.

The realistic POT (with a non-ideal power supply [the source] and voltage meter [the load]) has a complex relationship between the measured quantity (terminal D voltage) and the desired quantity ($\theta$). Because the relationship between $V_D$ and $\theta$ for a realistic POT is not the equation of a line, it is said to be non-linear. To use the POT effectively, we want to use a linear approximation because it is much easier to use. If we approximate $\theta$ with equation 4.18, we will have a non-linear error. The following example shows the analysis of this non-linear error.

**Example 4.4** ───────────────────────────────────────────

## Measurement Errors In a POT

When a POT is used as a sensor, some object moves the wiper causing $R$ to change. The objective is to measure the wiper voltage ($V_D$) and use it to compute $R$. If this can be done, then we can determine where the wiper is by measuring voltage. The realistic POT (with a non- ideal power supply and non-ideal voltage meter) has a complex relationship between the $V_D$ and $R$ (see equation 4.19). Because the relationship between $V_D$ and $R$ for a realistic POT is not the equation of a line, it is said to be a non-linear sensor. We want to use a linear approximation for relating $R$ to $V_D$ (like in the ideal pot, see equation 4.18) because it is much easier to use. If we approximate a real system using equation 4.18, we will have an error.

**PROBLEM:**

We desire for equation 4.18 to estimate the value of $V_D$ determined by equation 4.19 to within a 5% error. To do this, we will determine constraints on $R_s$ and $R_o$.

**SOLUTION:**

We will express error as the difference between equation 4.18 and 4.19. Thus, the percentage error is

$$E = \left(\frac{V_{D,linear} - V_{D,non-linear}}{V_{D,linear}}\right)100 \qquad (4.20)$$

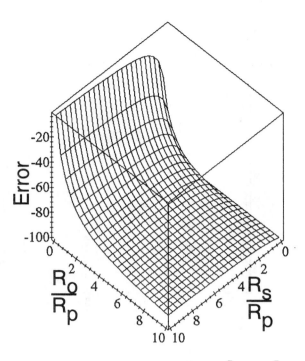

Figure 4.16: Maximum Error Versus $\frac{R_s}{R_p}$ and $\frac{R_o}{R_p}$.

The parameters that we have to determine in the error equation are the resistances $R_o$, $R_s$ and $R$. Since $R$ changes as we use the POT, we have to vary it to determine the maximum error.

We can maximize the error with respect to $R$ by finding the maximum of the error curve. The maximum of a curve can only occur at its end points or the points where the derivative of the function is zero or undefined. If you check the error at the endpoints (at $R = 0$ and $R = R_p$) you will find zero error. For points in the middle we look for the values of $R$ that have a first derivative of zero and a negative second derivative (concave down).[7] We used Maple to compute the derivative of the error (the derivative of equation 4.20). The program is given in 4.4.1. The program shows that the error has a zero derivative when:

$$R = \frac{1}{2}\left(R_p + R_o\right)$$

If we substitute this value of $R$ into the error equation we find the worst case error to occur at:

$$E_{\max} = -100\frac{4\,R_o\,R_s + R_p{}^2 + 2\,R_p\,R_o + R_o{}^2}{4\,R_p\,R_s + 4\,R_o\,R_s + R_p{}^2 + 2\,R_p\,R_o + R_o{}^2}$$

This expression still contains three variables. To simplify it further, define $r_o = \frac{R_o}{R_p}$ and $r_s = \frac{R_s}{R_p}$. After substituting these in, $R_p$ factors out leaving:

$$E_{\max} = -\frac{400\,r_o\,r_s + 100 + 200\,r_o + 100\,r_o{}^2}{4\,r_s + 4\,r_o\,r_s + 1 + 2\,r_o + r_o{}^2} \tag{4.21}$$

Now this last step is important to discuss. The operation we performed eliminated one variable. How did this happen? We wrote the equation in nondimensional form. This is very common practice and you will see it often. Note however that although the variable $R_p$ is gone, when and if we solve for $r_o$ and $r_s$ we will only know the ratio of $R_o$ and $R_s$ to $R_p$. You see, the variable $R_p$ is not totally gone.

Since equation 4.21 contains only two independent quantities, we can plot it to get some idea of what it looks like. Figure 4.16 shows what the expression looks like. Notice that as $R_s$ gets larger and as $R_o$ shrinks, the error gets better. This should be expected because large $R_s$ and small $R_o$ are close to ideal. Note also that $R_o$ has the greatest effect and that $R_s$ has little effect when it is greater than $4R_p$.

---

[7]We check the second derivative because a zero derivative can also be a minimum if the second derivative is positive (concave up). In this example we will not bother with computing the second derivative since there will be only one value of $R$ giving a zero derivative.

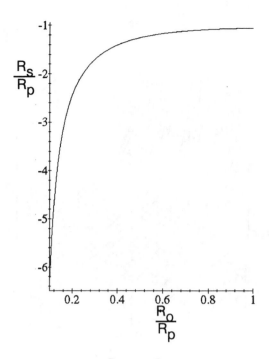

Figure 4.17: Value of $\frac{R_s}{R_p}$ and $\frac{R_o}{R_p}$ That Provide 5% Error.

To get a little more detailed, we can look for the values of $r_s$ and $r_o$ that satisfy the condition that error is -5% (error is never positive, see figure 4.16). The maple program tells us the expression for 5% error is:

$$r_s = -\frac{19 + 38\,r_o + 19\,{r_o}^2}{76\,r_o - 4} \qquad (4.22)$$

Notice that when $r_o = \frac{1}{19}$ there is not a value of $r_s$ that keeps error beneath 5%. Figure 4.17 shows what equation 4.22 looks like for values above $R_o = \frac{1}{10}$.

In conclusion, if the system is designed properly, it is possible to achieve small error.

**End 4.4** ───────────────────────────────────────────────

# Example 4.5 ───────────────────────────────────

## A Position Sensor

**PROBLEM:**

Figure 4.18 shows a sensor capable of locating the center of a light spot. Essentially the device is a large photodiode. Figure 4.19 shows a simplified schematic diagram of the device. When light rays strike the top resistive layer, they produce a photocharge at the point they strike. This photocharge in turn causes a current to flow in the direction indicated. Since the top surface is resistive, the resistance from the location of the photocharge to terminals A and B varies with the light location as indicated. Determine the response of this device when we place it in a circuit.

**FORMULATION:**

Although the device is designed as a large photodiode, it is not perfect. As a result, the device has a behavior which is not exactly like a perfect diode. We ignored these difficulties before, but now we will consider them just for practice. Figure 4.20 shows an equivalent (approximate) representation of the photosensor in a simple battery and resistor circuit. We approximate the device with a current source $i_S$, whose magnitude depends on the light intensity. Because the device is fairly large and because there will be charges on the top and bottom layer, we assume it has a capacitance $C_S$.[8] The resistance $R_S$ is the shunt resistance and is typically large. The diode represents an ideal diode, but we have chosen to ignore it in our

───────────────────────

[8]This is typically called the junction capacitance. All semiconductors have some value of junction capacitance which is one factor influencing their response speed. In the position sensor, the capacitance is actually distributed across the entire face of the device but we choose to lump it all at one location.

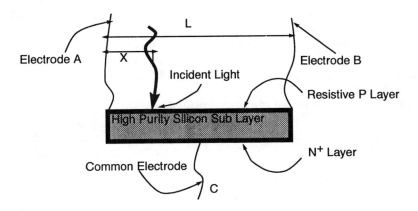

Figure 4.18: A Light Sensitive Sensor

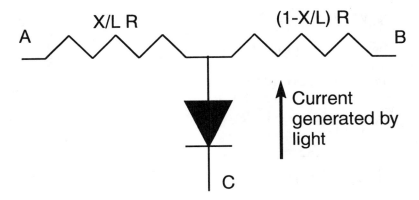

Figure 4.19: Schematic Diagram of a Light Sensitive Sensor

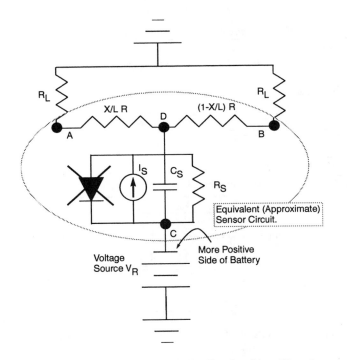

Figure 4.20: A Light Sensitive Device in a Circuit

analysis so we cross it out. We anticipate that the voltage at D will never exceed the voltage at C ($V_R$), hence the diode will do nothing.

Note that the circuit is one [NIA], with four [NIF]. To analyze the circuit, we will apply conservation of charge for three systems – nodes A, B, and D.

Applying the conservation of charge for node D, we find that

**Conservation of Charge**
system[**node D fig.4.14** ]      time period [**differential**]

*input − output = accumulation*

| input/output: node D | $i_S + C_S\left(\dot{V}_C - \dot{V}_D\right) + \frac{(V_C-V_D)}{R_S} - \frac{V_D}{\left[(1-\frac{X}{L})R+R_L\right]} - \frac{V_D}{\left(\frac{X}{L}R+R_L\right)}$ |
| accumulation: | 0 |

$$\boxed{i_S + C_S\left(\dot{V}_C - \dot{V}_D\right) + \frac{(V_C-V_D)}{R_S}} - \boxed{\frac{V_D}{\left[(1-\frac{X}{L})R+R_L\right]} - \frac{V_D}{\left(\frac{X}{L}R+R_L\right)}} = \boxed{0}$$

*input*          *output*          *accumulation*

$$i_S + C_S\left(\dot{V}_C - \dot{V}_D\right) + \frac{(V_C - V_D)}{R_S} - \frac{V_D}{\left[\left(1-\frac{X}{L}\right)R + R_L\right]} - \frac{V_D}{\left(\frac{X}{L}R + R_L\right)} = 0 \qquad (4.23)$$

Applying the conservation of charge for node B, we find that

**Conservation of Charge**
system[**node B fig.4.14** ]      time period [**differential**]

*input − output = accumulation*

| input/output: node B | $\frac{V_D-V_B}{R\left(1-\frac{X}{L}\right)} - \frac{V_B}{R_L}$ |
| accumulation: | 0 |

$$\boxed{\frac{V_D-V_B}{R\left(1-\frac{X}{L}\right)}} - \boxed{\frac{V_B}{R_L}} = \boxed{0}$$

*output*          *input*          *accumulation*

$$\frac{V_D - V_B}{R\left(1-\frac{X}{L}\right)} - \frac{V_B}{R_L} = 0$$

Applying the conservation of charge for node A, we find that

**Conservation of Charge**
system[**node A fig.4.14** ]      time period [**differential**]

*input − output = accumulation*

| input/output: node A | $\frac{L(V_D-V_A)}{RX} - \frac{V_A}{R_L}$ |
| accumulation: | 0 |

$$\boxed{\frac{L(V_D-V_A)}{RX}} - \boxed{\frac{V_A}{R_L}} = \boxed{0}$$

*input*          *output*          *accumulation*

$$\frac{L(V_D - V_A)}{RX} - \frac{V_A}{R_L} = 0$$

We can solve these last two equations to determine the voltages at nodes A and B.

$$V_A = \frac{L R_L V_D}{L R_L + RX}$$

and

$$V_B = \frac{L R_L V_D}{L R + L R_L - RX}$$

We can use these equations to eliminate $V_D$ from equation 4.23 and find the solution of $X$ as (using a lot of algebra)

$$\frac{X}{L} = \frac{-\left(\frac{R_L}{R}V_A\right) + V_B + \frac{R_L}{R}V_B}{V_A + V_B}$$

As a matter of convenience, we can substitute $u = \frac{X}{L} - 1/2$ ($u$ conveniently ranges from -1/2 to +1/2) to obtain

$$u = \frac{\left(1/2 + \frac{R_L}{R}\right)(-V_A + V_B)}{(V_A + V_B)}$$

From this last equation, we can easily determine the position of the light spot on the sensor by calculating the difference $V_A - V_B$ and dividing by $V_A + V_B$.

The circuit between nodes A and B is a voltage divider. Voltage dividers are commonly found in circuits that require an adjustable voltage and in sensors.

**End 4.5** ━━━━━━━━━━━━━━━━━━━━━━━━━━━━━━━━━━━━━━━━━━━━━━

## 4.2.1 Thermal-Mechanical Machines

## Example 4.6 ━━━━━━━━━━━━━━━━━━━━━━━━━━━━━━━━━━━━━━━━━━━━━━

### Analyzing a Water Rocket

**PROBLEM STATEMENT:**

We need to predict the motion of a toy air/water powered rocket. The toy will operate by placing some tap water into a cylindrical plastic container and pressurizing it. The weight of the toy is 2 ounces. The total length of the toy (which tells you something about the amount of air and water it can hold) is 6 inches. The toy is a cylinder (to keep it simple) with cross sectional area of 4 square inches. The maximum air pressure which can be safely put into the toy is 30 psi gage. The mass flow from a hole at the bottom of the toy occurs with 100% efficiency (no energy loss). The opening area at the bottom of the toy is 0.25 square inches. The air pumped into the toy at the beginning is 70° F. There is aerodynamic drag on the rocket which is proportional to the area of the toy times the speed squared. The drag constant of proportionality is 1 ounce per (10 miles per hour per inch) squared. Initially the rocket is half full of water and the air is at maximum pressure. Determine the speed of the toy when all the water is expelled.

**FORMULATION:**

The rocket's vertical motion is one flow. Stop the rocket and the water can still flow out of the toy. Prevent the water flow and heat might still flow. Think about it though. The motion will probably occur very quickly, how much thermal energy do you think would flow? We will ignore the heat transfer. So what else can flow? Nothing, there are 2 [NIF]. We will use $y$ measured from the ground to the center of mass of the plastic rocket (up positive) to represent the flow of the rocket. The value $h$ (measured from the bottom of the plastic) will represent the level of water in the plastic. $h$ is zero when no water is in the rocket and is positive up.

What can accumulate and what are their state variables? Potential energy in the rocket's mass given by $y$. Another is kinetic energy given by $\frac{d}{dt}(y)$. Internal energy in the air inside the rocket specified by temperature $T$. Mass of water in the rocket specified by $h$. Kinetic energy of the water, given by the derivatives of $y$ and $h$. What else can you think of? We bet whatever it is, it can be quantified by the five variables already given. The [NIA] is 5.

Let's start by expressing conservation of what we know is accumulating. Start with the water in the rocket.

$$-\dot{m}_x = \rho A_{\text{toy}} \frac{d}{dt}(h) \tag{4.24}$$

where $\rho$ is the water density, $\dot{m}_x$ is the rate that mass exits the rocket and $A_{\text{toy}}$ is the toy's cross sectional area. This has two variables. Next is the energy of the air (assume it's an ideal gas and no heat transfer):

$$P A_{\text{toy}} \frac{d}{dt}(h) = \frac{d}{dt}(m_{\text{air}} C_v T) \tag{4.25}$$

The term on the left is the work done by pressure $P$ as it pushes against the moving water surface. The term $C_v$ is the constant volume specific heat for air. Now for the momentum of the rocket (energy would be fine too):

$$-f_{\text{drag}} - m_{\text{toy}} g - m_{\text{water}} g - \dot{m}_x v_{\text{water}} = \frac{d}{dt}\left(m_{\text{toy}} \dot{y} + m_{\text{water}} \frac{d}{dt}(y - L/2 + h/2)\right) \tag{4.26}$$

Note that we are ignoring the mass of the air since it is small compared to everything else. Here $f_{\text{drag}}$ is the drag force, $m_{\text{toy}}$ is the mass of the plastic, $m_{\text{water}}$ is the mass of water inside the toy, and $v_{\text{water}}$ is the velocity of the water jet exiting the plastic (positive up so we expect to compute it to be a negative number). $L$ is the length of the plastic. Finally for the energy of the water inside the rocket:

$$-(P - P_{\text{atm}}) A_{\text{toy}} \frac{d}{dt}(h) - \dot{m}_x g y - \frac{1}{2}\dot{m}_x v_{\text{water}}^2 =$$

$$\frac{d}{dt}\left(m_{\text{water}}\,g\,(y - L/2 + h/2) + \frac{1}{2}m_{\text{water}}\frac{d}{dt}\left((y - L/2 + h/2)\right)^2\right) \qquad (4.27)$$

Note that we are ignoring internal energy and density changes in the water which is why we can account for the enthalphy of the exit stream by subtracting atmospheric pressure from the air pressure.

Now count the unknowns. They are: pressure $P$, $h$ and its derivatives, $\dot{m}_x$, $y$ and its derivatives, $v_{\text{water}}$, $m_{\text{water}}$, $f_{\text{drag}}$, $m_{\text{air}}$, and $T$. We have 9 total unknowns but only 4 equations. There are 4 flow variables in the set but we should have only 2 so we need some flow constraints.

One easy flow constraint is the following:

$$\dot{m}_x = -\rho A_{\text{toy}}\dot{h}$$

Unfortunately, this is the same as the conservation of mass equation 4.24 so we cannot count it. Note that the negative is required since $\dot{m}_x > 0$ means flow is leaving the rocket which can only happen if $\dot{h} < 0$. Another flow constraint is:

$$\rho A_{\text{hole}}\,v_{\text{water}} = \rho A_{\text{toy}}\dot{h} \qquad (4.28)$$

This last equation comes from a relation between volume flow and speed of an incompressible fluid flowing through an area. Are the + and - signs correct in this equation? Yes. The other flow constraint involves the velocity of the water stream. To find the velocity of the water stream we could define a position and differentiate or do the following. Let $d$ be the distance from the ground to a water molecule that came out of the nozzle. The volume of what came out since the molecule popped out (call the volume $q$) is equal to the nozzle area times the distance from the bottom of the rocket to the water molecule. Hence:

$$q = A_{\text{hole}}\,(y - L/2 - d)$$

The volume flow rate is related to the mass exiting as:

$$\rho\frac{d}{dt}\,(q) = \dot{m}_x$$

Hence:

$$\dot{m}_x = \rho\frac{d}{dt}\,(A_{\text{hole}}\,[y - L/2 - d]) = \rho A_{\text{hole}}\,(\dot{y} - \dot{d}) = \rho A_{\text{hole}}\,(\dot{y} - v_{\text{water}}) \qquad (4.29)$$

The time derivative of $d$ is the velocity of the water because $d$ is the position of the water and therefore its derivative is velocity. Does equation 4.29 make sense? If $v_{\text{water}} = \dot{y}$ (if the water is moving at the same speed as the rocket) how much water comes out? Zero. Is that what the equation says? What if $v_{\text{water}} = 0$, does it say that water comes out only if the rocket moves up? Yes. Does it also say that if the rocket is stationary ($\dot{y} = 0$) that the water comes out only if $v_{\text{water}} < 0$? Yes. We have confidence in the equation.

So far we are up to 6 equations with 9 unknowns and we have all the flow constraints we expect. We need 3 more equations. Are they conservation? Probably not because our number of differential equations matches the [NIA]. Well there has got to be some constraints what are they? Well according to the problem statement, the drag force is related to the motion:

$$f_{\text{drag}} = K A_{\text{toy}}\dot{y}^2 \qquad (4.30)$$

Two left. What about the mass of water:

$$m_{\text{water}} = \rho A_{\text{toy}}h \qquad (4.31)$$

What's left? Well think about the air. If we know the temperature $T$ and the value of $h$ we would know the temperature and volume of the air. These are two "properties" of the air, shouldn't we know all other properties like pressure? Yes, write the ideal gas law:

$$PV = m_{\text{air}}RT \qquad (4.32)$$

Here $V$ is the air volume (a new variable) and $R$ is the gas constant. We need an equation for air volume:

$$V = A_{\text{toy}}L - A_{\text{toy}}h \qquad (4.33)$$

Now we have a balance in the equations and the unknowns and will attempt a solution.

**SOLUTION:**

We solved this using the maple program 4.4.2. We did make some simplifications to help maple. For example, we tossed out all second and higher derivatives of $h$. Maple had a hard time manipulating the differential equations properly when these derivatives remained. Perhaps we could have manually manipulated the equations but we thought we would try the assumption first. What we will look at is a plot of $h$ versus time and hope that it is nearly straight which means the higher derivatives of $h$ are small. We'll keep our

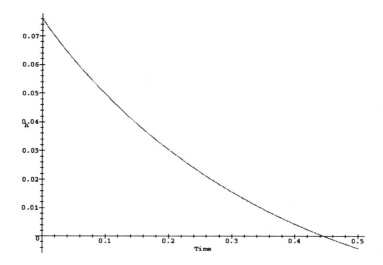

Figure 4.21: Water Height $h$ Versus Time. When its Zero, the Rocket is Empty.

fingers crossed. Second, we decided to drop all the energy storage terms from equation 4.27, after all we think the water will be gone in such a short period of time, they will not amount to much. Again we should verify this assumption. Third, maple had some difficulty with the square velocity term in equation 4.27 and refused to eliminate some of the common terms so we simplified it by hand. Finally since we ignored the kinetic energy of the water inside the rocket, the [NIA] should drop by one and we observed that the conservation of mass equation is eliminated in the process of eliminating variables. We solved our simplified equations and our results can be found in figures 4.21 through 4.24.

Notice the curvature of the plot of $h$ versus time (figure 4.21). There is a slight curvature indicating that the second derivative of $h$ is not zero. We determine the time when the rocket is empty by looking for the zero crossing. We figure it to be approximately 0.44 seconds. Figure 4.22 shows the vertical speed of the rocket as a function of time up to when it is empty. Notice that we predict the rocket is already slowing by the time the water is completely expelled. This means there is insufficient thrust at the end to keep the rocket accelerating upward. How can we change this? What do you think would happen if we started with a higher pressure or less water? Figure 4.23 shows the temperature in Celsius. Note that the temperature gets below freezing. Do you think this is realistic? Do you think the water will freeze in the rocket and not come out? How long is the water in the rocket? We don't think it has time to freeze. Figure 4.24 shows that the pressure never hits a vacuum, which is a good thing. A negative gage pressure would mean the water is sucked into the rocket rather than being expelled.

The rocket does not perform well. What would you do to make it perform better? How would you test your ideas? Do you think we made any mistakes in this problem formulation? How would you know? How could you test the effect of some of our simplifications?

**End 4.6**

## 4.3  Questions and Problems

An * after a problem number indicates a solution is available.

1. Modify equations 4.1-4.5 to make them suitable for the dc motor in figure 4.25. This is a series motor. The field circuit is in series with the armature circuit.

2. Verify analytically the maximum error, found in example 4.4.

3. Suppose we change the load on a position sensitive device to that of the figure. Determine equations which define the voltage response at nodes A and B. What is the effect of the additional capacitors? What happens if the light striking the sensor blinks on and off?

4. In the circuit of figure 4.20 explain what would happen at nodes A and B as $R_L \to \infty$.

5. Explain what the effect of $R_S$ and $C_S$ is in the circuit of figure 4.20.

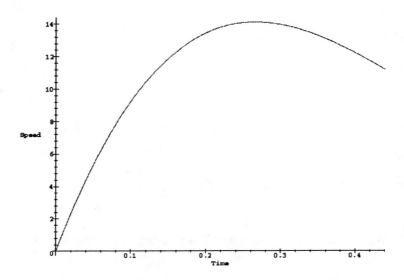

Figure 4.22: Rocket Speed Versus Time.

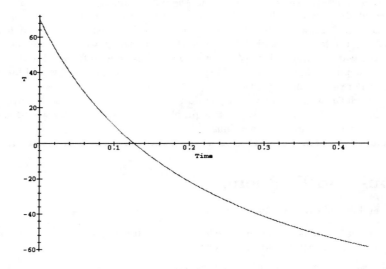

Figure 4.23: Temperature Versus Time.

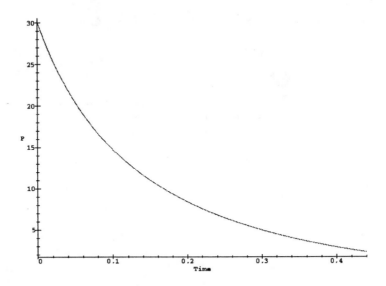

Figure 4.24: Pressure (Gage) Versus Time.

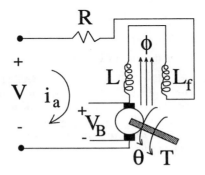

Figure 4.25:

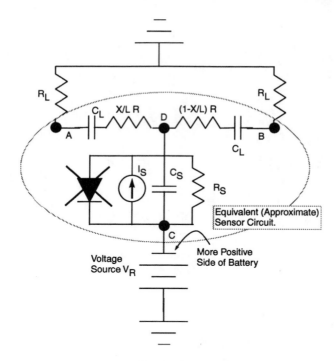

Figure 4.26:

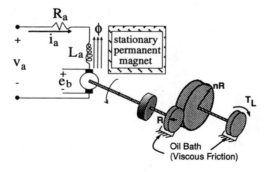

Figure 4.27:

6. Explain what the effect of $V_R$ is in the circuit of figure 4.20.

7. We work as a supervisor for GREAP JUICER INCORPORATED. A summer co-op student has just finished the work required in example 1.17. The president of the company feels that there might be an error in the solution. We know that the co-op student used third party software to arrive at the solution. We decide to verify the solution by deriving it analytically. Further analysis of equations 1.11 and 1.12 tells us that the problem is simply a maximum minimum problem. Using these facts, determine if the co-op student's solution is correct.

8. Which of the following would cause more thermal energy generation in a DC motor, a motor with its rated voltage applied and being held still or one with its rated voltage running at a high speed?

9. Consider the permanent magnet DC motor driving a load (see figure 4.27). Determine the relationship between the input voltage and load speed. What is the maximum speed of the system? The motor's inertia is small and the load inertia is large. Discuss the effects of the gear reduction in the system.

10. Calculate what $V_x$ will be when the mass is in static equilibrium. The resistance per unit length of the coil is $\rho$. Refer to figure 4.28. The resistor of length $L$ is a POT.

11. Determine the maximum altitude of the rocket in example 4.6.

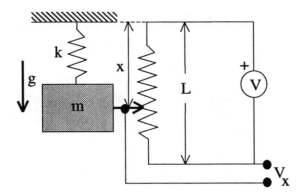

Figure 4.28:

12. Determine the optimal amount of water in the rocket of example 4.6. Define optimal as one that achieves maximum altitude.

13. Determine the optimal orifice diameter in the rocket of example 4.6. Optimal is one that achieves maximum altitude.

14. Determine the entropy change in the air in the rocket of example 4.6. Discuss your findings.

15. Suppose you have a POT with $R_p = 1000$ Ohms. Suppose further that you have an ideal voltage source, a volt meter with input impedance of 100 Ohms, an ideal OpAmp and several assorted resistors. Configure the POT so that it has less than 5% nonlinear error.

## 4.4  Maple Programs

This section includes the Maple files referenced in the chapter. Plots generated by Maple are displayed in the text, not in this section.

### 4.4.1   File: pot.ms - Maple program for Example 4.4.

```
>  restart:
```
First the linear solution.
```
>  vdlinear := R/Rp*V:
```
Now the nonlinear one.
```
>  vdnonlinear := Rs*R*V/(Rp*Rs+Ro*Rs+R*Rp+R*Ro-R^2):
```
definition of error
```
>  err := simplify((vdnonlinear - vdlinear)/vdlinear)*100;
```

$$err := -100\,\frac{Ro\,Rs + R\,Rp + R\,Ro - R^2}{Rp\,Rs + Ro\,Rs + R\,Rp + R\,Ro - R^2}$$

take derivative with respect to R (finding the R that makes error maximum)
```
>  Rbad := solve(diff(err,R)=0,R);
```

$$Rbad := \frac{1}{2}\,Rp + \frac{1}{2}\,Ro$$

sub in the worse value of R into error
```
>  baderr := simplify(subs(R=Rbad,err));
```

$$baderr := -100\,\frac{4\,Ro\,Rs + Rp^2 + 2\,Rp\,Ro + Ro^2}{4\,Rp\,Rs + 4\,Ro\,Rs + Rp^2 + 2\,Rp\,Ro + Ro^2}$$

Simplify by replacing ro with ro/rp and rs with rs/rp
```
>  error:=simplify(subs(Ro=ro*Rp,subs(Rs=rs*Rp,b aderr)));
>  plot3d(error,rs=0..10,ro=0..10);
```

$$error := -100\,\frac{4\,ro\,rs + 1 + 2\,ro + ro^2}{4\,rs + 4\,ro\,rs + 1 + 2\,ro + ro^2}$$

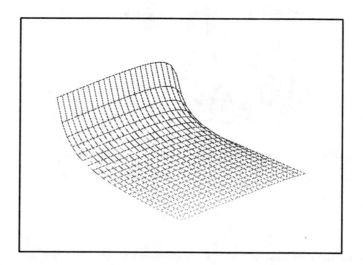

find the value of rs/rp for 5% error

```
> roeqn := simplify(solve(error=-5,rs));
```

```
> plot(roeqn,ro=1/10..1);
```

$$roeqn := -\frac{19}{4}\frac{1 + 2\,ro + ro^2}{19\,ro - 1}$$

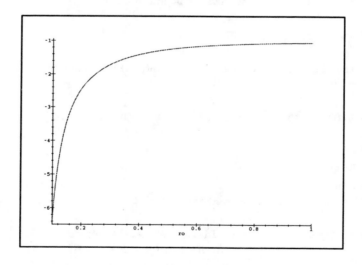

## 4.4.2   File: rocket.ms - Maple Program for the Water Rocket.

This is file rocket.mws.

```
> restart;with(plots):read('c:\\math\\units .msh'):
```
$$Units\ Version\ Ia$$

There are 2dof. I will use y[toy] for the position of the toy and h as the level of water inside the rocket. h = 0 when all the water is gone.

Here is the conservation of mass. rho is h2o density, a[toy] is toy area.

```
> mass := -RateOfMassOut = rho*a[toy]*diff(h(t),t);
```

$$mass := -RateOfMassOut = \rho\,a_{toy}\,(\frac{\partial}{\partial t}\,\mathrm{h}(t))$$

Here is the orifice equation given in the footnote. P is pressure, q is volume flow rate, a[noz] is nozzle area.

```
> orifice := P(t)*q = 1/2*rho*q*(q/a[noz])^2:
> orifice := sqrt(2*P(t)/rho)*a[noz] = q;
```

$$orifice := \sqrt{2}\sqrt{\frac{P(t)}{\rho}}\,a_{noz} = q$$

Here is the linear momentum equation. fdrag is the drag force, m are masses, v[h2o] is the velocity of the water leaving, L is the length of the toy.

```
> mom := -fdrag-m[toy]*g -m[h2o](t)*g - RateOfMassOut*v[h2o] = diff(m[toy]*diff(y[toy](t),t)
+m[h2o](t)*diff(y[toy](t)-L/2+h(t)/2,t),t);
```

$$mom := -fdrag - m_{toy}\,g - m_{h2o}(t)\,g - RateOfMassOut\,v_{h2o} =$$

$$m_{toy}\left(\frac{\partial^2}{\partial t^2}y_{toy}(t)\right) + \left(\frac{\partial}{\partial t}m_{h2o}(t)\right)\left(\left(\frac{\partial}{\partial t}y_{toy}(t)\right)+\frac{1}{2}\left(\frac{\partial}{\partial t}h(t)\right)\right) + m_{h2o}(t)\left(\left(\frac{\partial^2}{\partial t^2}y_{toy}(t)\right)+\frac{1}{2}\left(\frac{\partial^2}{\partial t^2}h(t)\right)\right)$$

This is the conservation of energy for the air. cv is the constant volume heat capacity for air, T is temperature.

```
> enerair := (P(t)+Pa) *a[toy]*diff(h(t),t)=m[air]*cv*diff(T(t),t);
```

$$enerair := (P(t) + Pa)\,a_{toy}\left(\frac{\partial}{\partial t}h(t)\right) = m_{air}\,cv\left(\frac{\partial}{\partial t}T(t)\right)$$

Enter the values of some constants. a is area. h is the distance from the bottom of the rocket to the water level in the rocket. Initially the rocket is half full of water. Rair is the constant in the ideal gas law (PV=mRT), c[drag] is the drag coefficient, g is gravity constant.

```
> a[noz]:=(.25*inch)^2:hinit := L/2:Pinit := (30)*pound/inch^2: Pa := 14.7*pound/inch^2:
Tinit := (70+460)*rankine:
> rho:=62.38*poundmass/foot^3:a[toy]:=4*inch^2: Rair:=.37*pound*foot^3/(poundmass*inch^2*rankine
u/(poundmass*rankine):c[drag]:=1/16*pound/(10*mile/hour*inch)^2:m[toy] :=2/16*poundmass:g:=981/100,
```

y[toy] is the vertical position of the toy. There are 2 NIF, the rocket and the water so there should be 2 flow variables. I use y and h.

```
> v[toy] := diff(y[toy](t),t);
```

$$v_{toy} := \frac{\partial}{\partial t}y_{toy}(t)$$

This is the drag force.

```
> fdrag := c[drag]*a[toy]*v[toy]*v[toy];
```

$$fdrag := \frac{12665625}{227611648}\frac{Newton\,Second^2\left(\frac{\partial}{\partial t}y_{toy}(t)\right)^2}{Meter^2}$$

Next determine the q in terms of mass flow rate out. Notice that q is a flow but I relate it to y and h since there is only 2 dof and I have decided to use h and y as my variables. q is the volume flow rate of water. I use Q in the equation to avoid problems with circular definitions.

```
> q := solve(rho*Q = RateOfMassOut,Q);
```

$$q := .001000768846\frac{RateOfMassOut\,Meter^4}{Newton\,Second^2}$$

Now determine the velocity of the water which leaves the nozzle. I would like to express the position of the water and differentiate it, but it is not clear what the position is in general terms so I will go about it in a round about way.

Let d be the distance from the ground to a water molecule that came out of the nozzle.The volume of what comes out is equal to the nozzle area times the distance from the bottom of the rocket to d. Since q is the rate of volume, take the rate of volume and set it to q.

```
> VolumeLeavingRocket := a[noz]*(y[toy](t)-L/2-d(t));
```

$$VolumeLeavingRocket := .00004032250000\,Meter^2\left(y_{toy}(t) - \frac{381}{5000}Meter - d(t)\right)$$

```
> v[h2o] := evalf(solve(diff(VolumeLeavingRocket,t)=q,diff(d(t),t)));
```

$$v_{h2o} := -.4960009920\,10^{-7}($$

$$-.2016125\,10^8\,Newton\,Second^2\left(\frac{\partial}{\partial t}y_{toy}(t)\right) + .500384423\,10^9\,RateOfMassOut\,Meter^2)/($$

$$Newton\,Second^2)$$

Another way to do it is to say the velocity of the water exiting relative to the rocket is q/a[nozzle] downward so the absolute velocity of the water is the velocity of the rocket minus q/a. Notice it is the same as above.

```
>   v[h2o] := v[toy] - q/a[noz];
```

$$v_{h2o} := (\frac{\partial}{\partial t} y_{toy}(t)) - 24.81911702 \frac{RateOfMassOut\ Meter^2}{Newton\ Second^2}$$

Now I find the masses.

```
>   m[h2o] := (t) -> rho*a[toy]*h(t);
>   m[air] := (Pinit+Pa)*(L-hinit)*a[toy]/(Rair*Tinit);Howmany(m[air],poundmass);
```

$$m_{h2o} := t \to \rho\, a_{toy}\, h(t)$$

$$m_{air} := .0007180144390 \frac{Newton\ Second^2}{Meter}$$

$$.001582950875$$

Here is the ideal gas law to determine the pressure in terms of volume and temperature.

```
>   P(t) := m[air]*Rair * T(t)/(a[toy]*(L-h(t)))-Pa;
```

$$P(t) := 79.75843047 \frac{Newton\ T(t)}{Meter\ Kelvin\ (\frac{381}{2500}\ Meter - h(t))} - 101352.5789 \frac{Newton}{Meter^2}$$

```
>   eqns := [mass,orifice,mom,enerair]:  evalf(eqns);
```

$$\left[ -1.\, RateOfMassOut = 2.578657410 \frac{Newton\ Second^2\ (\frac{\partial}{\partial t} h(t))}{Meter^2}, .1803970108\ 10^{-5}( \right.$$

$$(79.75843047 \frac{Newton\ T(t)}{Meter\ Kelvin\ (.1524000000\ Meter - 1.h(t))} - 101352.5789 \frac{Newton}{Meter^2})$$

$$Meter^4/(Newton\ Second^2))^{1/2} Meter^2 = .001000768846 \frac{RateOfMassOut\ Meter^4}{Newton\ Second^2},$$

$$-.05564576818 \frac{Newton\ Second^2\ \%1^2}{Meter^2} - .5562176437\ Newton$$

$$- 25.29662919 \frac{Newton\ h(t)}{Meter}$$

$$- 1.\, RateOfMassOut\ (\%1 - 24.81911702 \frac{RateOfMassOut\ Meter^2}{Newton\ Second^2}) =$$

$$.05669904625 \frac{Newton\ Second^2\ (\frac{\partial^2}{\partial t^2} y_{toy}(t))}{Meter}$$

$$+ 2.578657410 \frac{Newton\ Second^2\ (\frac{\partial}{\partial t} h(t))\ (\%1 + .5000000000\ (\frac{\partial}{\partial t} h(t)))}{Meter^2}$$

$$+ 2.578657410 \frac{Newton\ Second^2\ h(t)\ ((\frac{\partial^2}{\partial t^2} y_{toy}(t)) + .5000000000\ (\frac{\partial^2}{\partial t^2} h(t)))}{Meter^2},$$

$$.2058277960 \frac{Newton\ Meter\ T(t)\ (\frac{\partial}{\partial t} h(t))}{Kelvin\ (.1524000000\ Meter - 1.h(t))} =$$

$$\left. .5110434063 \frac{Newton\ Meter\ (\frac{\partial}{\partial t} T(t))}{Kelvin} \right]$$

$$\%1 := \frac{\partial}{\partial t} y_{toy}(t)$$

```
>   RateOfMassOut := solve(mass,RateOfMassOut);
```

$$RateOfMassOut := -2.578657410 \frac{Newton\ Second^2\ (\frac{\partial}{\partial t} h(t))}{Meter^2}$$

```
> EQ := subs(diff(h(t),t,t)=0,{orifice,enerair,mom});
```

$$EQ := \left\{ .2058277960 \frac{Newton\ Meter\ \mathrm{T}(t)\left(\frac{\partial}{\partial t}\,\mathrm{h}(t)\right)}{Kelvin\left(\frac{381}{2500}\,Meter - \mathrm{h}(t)\right)} = .5110434063 \frac{Newton\ Meter\left(\frac{\partial}{\partial t}\,\mathrm{T}(t)\right)}{Kelvin}, \right.$$

$$.1275599497\,10^{-5}\sqrt{2}$$

$$\frac{\sqrt{\dfrac{\left(79.75843047\,\dfrac{Newton\ \mathrm{T}(t)}{Meter\ Kelvin\left(\frac{381}{2500}\,Meter - \mathrm{h}(t)\right)} - 101352.5789\,\dfrac{Newton}{Meter^2}\right)Meter^4}{Newton\ Second^2}}}{}$$

$$Meter^2 = -.002580640000\,Meter^2\left(\frac{\partial}{\partial t}\,\mathrm{h}(t)\right), -\frac{12665625}{227611648}\,\frac{Newton\ Second^2\ \%1^2}{Meter^2}$$

$$-\frac{44497411497}{80000000000}\,Newton - 25.29662919\,\frac{Newton\ \mathrm{h}(t)}{Meter}$$

$$+\,2.578657410\,\frac{Newton\ Second^2\left(\frac{\partial}{\partial t}\,\mathrm{h}(t)\right)\left(\%1 + 64.00000001\left(\frac{\partial}{\partial t}\,\mathrm{h}(t)\right)\right)}{Meter^2} =$$

$$\frac{45359237}{800000000}\,\frac{Newton\ Second^2\left(\frac{\partial^2}{\partial t^2}\,y_{toy}(t)\right)}{Meter}$$

$$+\,2.578657410\,\frac{Newton\ Second^2\left(\frac{\partial}{\partial t}\,\mathrm{h}(t)\right)\left(\%1 + \frac{1}{2}\left(\frac{\partial}{\partial t}\,\mathrm{h}(t)\right)\right)}{Meter^2}$$

$$\left.+\,2.578657410\,\frac{Newton\ Second^2\ \mathrm{h}(t)\left(\frac{\partial^2}{\partial t^2}\,y_{toy}(t)\right)}{Meter^2}\right\}$$

$$\%1 := \frac{\partial}{\partial t}\,y_{toy}(t)$$

After you check units, execute the following.

```
> Newton:=1:Meter:=1:Second:=1:Kelvin:=1:
> ans := dsolve(EQ union {h(0)=hinit,y[toy](0)=0,D(y[toy])(0)=0,T(0)=Tinit},{T(t),h(t),y[toy
]( t)},type=numeric);
```

$$ans := \mathbf{proc}(rkf45\_x)\ \dots\ \mathbf{end}$$

```
> odeplot(ans,[t,h(t)],0..0.5,labels=['Time','h ']);
```

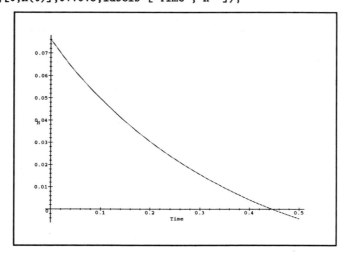

```
> ans(.44);tend:= .44;
```

$$[t = .44, \; h(t) = .0005118955708453946, \; T(t) = 223.0226979825033,$$
$$y_{toy}(t) = 2.174070473903799, \; \frac{\partial}{\partial t} y_{toy}(t) = 4.999205147040635]$$

$$tend := .44$$

```
> PosToy := (time) -> (subs(ans(time),y[toy](t)))/foot;
```
$$PosToy := time \rightarrow \frac{\mathrm{subs}(\mathrm{ans}(time), \; y_{toy}(t))}{foot}$$

```
> VelToy := (time) -> (subs(ans(time),diff(y[toy](t),t)))/(mile/hour);
```
$$VelToy := time \rightarrow \frac{\mathrm{subs}(\mathrm{ans}(time), \; \frac{\partial}{\partial t} y_{toy}(t)) \; hour}{mile}$$

```
> Temp := (time) -> (subs(ans(time),T(t)))/rankine -460;
```
$$Temp := time \rightarrow \frac{\mathrm{subs}(\mathrm{ans}(time), \; T(t))}{rankine} - 460$$

```
> Press := (time) -> (subs(ans(time),P(t))/(pound/inch^2));
```
$$Press := time \rightarrow \frac{\mathrm{subs}(\mathrm{ans}(time), \; P(t)) \; inch^2}{pound}$$

```
> plot(PosToy,0..tend,labels=['Time','Position' ]);plot(VelToy,0..tend,labels=['Time','Speed']);p
ls=['Time','T']);plot(Press,0..tend,labels=['Time','P']);
```

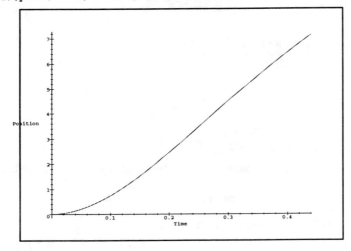

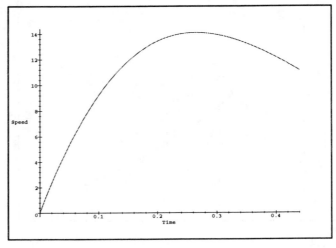

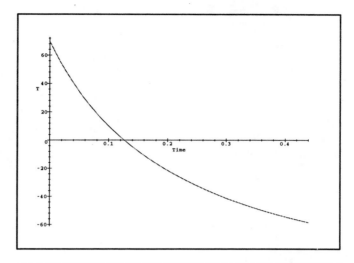

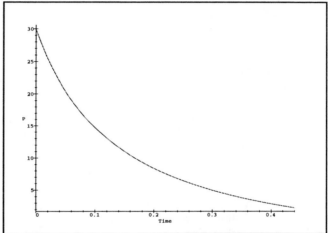

# Appendix A

# Miscellaneous Data

**The International System of Units (SI)**

| Dimension | Abbreviation | Unit |
|---|---|---|
| **Fundamental Dimensions** | | |
| absolute temperature | [K] | kelvin |
| amount of substance | [mol] | mole |
| electric current | [A] | Ampere (amp) |
| electric potential | [V] | Volt |
| length | [m] | meter |
| luminous intensity | [cd] | candela |
| mass | [kg] | kilogram |
| plane angle | [rad] | radian |
| solid angle | [sr] | steradian |
| time | [s] | second |
| | | |
| **Derived Dimensions** | | |
| electric capacitance | $[F] \equiv [A\ s/V]$ | Farad |
| electric charge | $[C] \equiv [A\ s]$ | Coulomb |
| electric conductance | $[S] \equiv [A/V]$ | Siemens |
| electric inductance | $[H] \equiv [V\ s/A]$ | Henry |
| electric resistance | $[\Omega] \equiv [V/A]$ | Ohm |
| energy, heat, work | $[J] \equiv [N\ m] \equiv [V\ A\ s]$ | Joule |
| force | $[N] \equiv [kg\ m/s^2]$ | Newton |
| frequency | $[Hz] \equiv [cycles/s]$ | Hertz |
| luminous flux | $[l] \equiv [cd\ sr]$ | lumen |
| magnetic flux | $[Wb] \equiv [V\ s]$ | Weber |
| power | $[W] \equiv [J/s] \equiv [V\ A] \equiv [N\ m/s]$ | Watt |
| pressure | $[Pa] \equiv [N/m^2]$ | Pascal |
| | $[MPa] \equiv 10^6 [Pa]$ | Mega Pascal |
| | $[Bar] \equiv \frac{1}{10}[MPa]$ | Bar |
| | $[atm] \equiv 0.10135\ [MPa]$ | Atmospheres |
| temperature | $[C] \equiv K\text{-}273.15$ | Celsius |

## The American Engineering System of Units (AES)

### Only Those Differing From SI Are Listed

| Dimension | Abbreviation | Unit |
|---|---|---|
| **Fundamental Dimensions** | | |
| absolute temperature | [R] | Rankine |
| force | $[lb_f]$ | pound force |
| length | [ft] | foot |
| | | |
| **Derived Dimensions** | | |
| area | $[acre] \equiv 43560 \ [ft^2]$ | acre |
| energy, heat, work | $[BTU] \equiv 778.16 \ [lb_f \ ft]$ | British Thermal Unit |
| length | $[in] \equiv \frac{1}{12}[ft]$ | Inch |
| | $[yd] \equiv 3 \ [ft]$ | Yard |
| | $[mi] \equiv 5280 \ [ft]$ | Mile |
| mass | $[oz] \equiv \frac{1}{16} \ [lb_m]$ | ounce |
| | $[slug] \equiv 32.174 \ [lb_m]$ | slug |
| power | $[hp] \equiv 550 \ [lb_f \ ft/s]$ | Horse Power |
| | $[hp] \equiv 0.7068 \ [BTU/s]$ | |
| pressure | $[psi] \equiv [lb_f/in^2]$ | PSI |
| | $[atm] \equiv 14.7 \ [psi]$ | Atmospheres |
| temperature | $[°F] \equiv °R-459.67$ | Fahrenheit |
| volume | $[gal] \equiv 0.133681 \ [ft^3]$ | gallon |

## Conversions Between The SI and AES System

| Dimension | Conversion | | |
|---|---|---|---|
| absolute temperature | [R] | = | $\frac{5}{9}$[K] |
| force | [lb$_f$] | = | 4.448 [N] |
| length | [ft] | = | 0.3048 [m] |
| | [in] | = | 2.54 [cm] |
| | [mi] | = | 1.609 [km] |
| energy, heat, work | [BTU] | = | 1055.04 [J] |
| | [lb$_f$ ft] | = | 1.356 [J] |
| mass | [slug] | = | 14.59 [kg] |
| power | [hp] | = | 745.7 [W] |
| pressure | [psi] | = | 6895 [Pa] |
| temperature | $\frac{5}{9}$([° F]−32) | = | [°C] |

| Prefix | Name | Factor of Base Unit |
|---|---|---|
| E | Exa | $10^{18}$ |
| P | Peta | $10^{15}$ |
| T | Tera | $10^{12}$ |
| G | Giga | $10^{9}$ |
| M | Mega | $10^{6}$ |
| my | myria | $10^{4}$ |
| k | kilo | $10^{3}$ |
| h | hecto | $10^{2}$ |
| da | deka | $10^{1}$ |
| | | 1 |
| d | deci | $10^{-1}$ |
| c | centi | $10^{-2}$ |
| m | milli | $10^{-3}$ |
| $\mu$ | micro | $10^{-6}$ |
| n | nano | $10^{-9}$ |
| p | pico | $10^{-12}$ |
| f | femto | $10^{-15}$ |
| a | atto | $10^{-18}$ |

# Bibliography

[1] Beachley, N.H. and H.L. Harrison, *Introduction to Dynamic System Analysis*. Harper and Row, 1978.

[2] Covey, Stephen R., *The 7 Habits of Highly Effective People; Restoring the Character Ethic, Powerful Lessons in Personal Change*. Simon & Schuster, 1989.

[3] *Cryodata (*and a computer code for refrigerants*)*, GASPAK.
Web Site : http://www.csn.net/partners (Software Available, $595), June 1996.

[4] Daubert, T. E., R. P. Danner, H. M. Sibul, and C. C. Stebbins, *Physical and Thermodynamic Properties of Pure Chemicals: Data Compilation*. Taylor & Francis, Bristol, PA (1,405 chemicals, extant 1995).

[5] DIPPR *Data Compilation of Pure Compound Properties*. ASCII Files, National Institute of Science and Technology, Standard Reference Data, Gaithersburg, MD (1,458 chemicals, extant 1995); also see the Student DIPPR Database, PC-DOS version (100 common chemicals for teaching purposes) from the same source.

[6] Douglas, James M., *Conceptual Design of Chemical Processes*. McGraw-Hill, 1988.

[7] *Electronics Workbench EDA*,
Web Site : http://www.interactiv.com, June 1998.

[8] Felder, Richard M., and Ronald W. Rousseau, *Elementary Principles of Chemical Processes*, Second Edition, John Wiley & Sons, 1986.

[9] Fogler, H. Scott and Steven E. LeBlanc, *Strategies for Creative Problem Solving*. Prentice-Hall, 1995.

[10] J. H. Ginsberg, *Advanced Engineering Dynamics*. Harper & Row, 1988.

[11] Glover, Charles, and Harry L. Jones, *Conservation Principles for Continuous Media*, Fourth Edition. McGraw-Hill, 1997.

[12] Glover, Charles, Kevin M. Lunsford, and John A. Fleming, *Conservation Principles and the Structure of Engineering*, Fifth Edition. McGraw-Hill, 1996.

[13] Hawking, Stephen W., *A brief history of time : from the big bang to black holes*, Bantam Books, 1988.

[14] Kane, T.R., P.W. Litkins, and D.A. Levinson, *Spacecraft Dynamics*. McGraw-Hill, 1983.

[15] Kays, W. M. and A. L. London, *Compact Heat Exchangers*, Third Edition. McGraw-Hill, 1984.

[16] Koen, Billy Vaughn, *Definition of the Engineering Method*. American Society for Engineering Education, Washington D.C., 1987, c1985.

[17] Krieth, Frank and William Z. Black, *Basic Heat Transfer*. Harper & Row, 1980.

[18] McNeill, Barry W. and Lynn Bellamy, *Engineering Core Workbook for Active Learning, Assessment and Team Training*, Fourth Edition. Custom Academic Publishing Company, Oklahoma City, Oklahoma 73102, Spring 1996 Web Site : http://ece.eas.asu, June 1996.

[19] *Macsyma*. Symbolics, Inc., 8 New England Executive Park East, Burlington, MA. 08103, 1989.

[20] *Marks' Standard Handbook for Mechanical Engineers*, Eighth Edition. McGraw-Hill, 1978.

[21] *Materials Safety Data Shet*, [MSDS]
Web Site : gopher://gopher.chem.utah.edu:70/11/MSDS, June 1996.

[22] Matthews, C. S. and C. O. Hurd, *Transactions of the AICHE*. Vol. 42, No. 55 1946.

[23] Nagle, Norman, *Fundamentals Of Differential Equations And Boundary Value Problems*, Second Edition, Addison Wesley, 1996.

[24] *NIOSH Pocket Guide to Chemical Hazards*. DHHS (NIOSH) Publication No. 90-117 (a public domain document; copies available from: NIOSH Publications Dissemination, 4676 Columbia Parkway, Cincinnati, OH 45226.

[25] *Perry's Chemical Engineers' Handbook*, Sixth Edition. McGraw-Hill, 1984.

[26] Peters, M. S. and K. D. Timmerhaus, *Plant Design and Economics for Chemical Engineers*, Fourth Edition. McGraw-Hill, 1991.

[27] Pollock, T.C., *Properties of Matter*. McGraw-Hill, 1991.

[28] Reid, Robert C., John M. Prausnitz, and Bruce E. Poling, *The Properties of Gases & Liquids*, Fourth Edition. McGraw-Hill, 1987.

[29] Reynolds, William C. and Henry C. Perkins, *Engineering Thermodynamics*, Second Edition, McGraw-Hill, 1977.

[30] Rizzoni, Giorgio, *Principles and Applications of Electrical Engineering*, Second Edition, Irwin, 1996.

[31] Rosenberg, R.C. and D.C. Karnopp, *Introduction to Physical System Dynamics*. McGraw-Hill, 1983.

[32] Smith, J. M. , H. C. Van Ness, and M. M. Abbott, *Introduction to Chemical Engineering Thermodynamics*, Fifth Edition. McGraw-Hill, 1996.

[33] Smith, R.J., Circuits, *Devices, and Systems*, Third Edition. John Wiley & Sons, 1976.

[34] Starfield, Anthony M., Karl A. Smith, and Andrew L. Blelock, *How to Model It: Problem Solving for the Computer Age*. Burgess Publishing, 1994.

[35] U.S. Department of Health and Human Services, National Institute for Occupational Safety and Health, *NIOSH Manual of Analytic Methods*, Fourth Edition. NIOSH, 1994.

[36] Valkenburg, M.E.V., *Network Analysis,* Third Edition. Prentice Hall, 1974.

[37] Wylen, G.J.V. and R.E. Sonntag, *Fundamentals of Classical Thermodynamics*. John Wiley and Sons, 1973.

# Index